Birkhäuser

Trends in Mathematics

Research Perspectives

Research Perspectives collects core ideas and developments discussed at conferences and workshops in mathematics, as well as their increasingly important applications to other fields. This subseries' rapid publication of extended abstracts, open problems and results of discussions ensures that readers are at the forefront of current research developments.

Marcelo R. Ebert • Uwe Kähler • Irene Sabadini •
Joachim Toft
Editors

New Tools in Mathematical Analysis and Applications

Proceedings of the 14th ISAAC Congress 2023, Ribeirão Preto, Brazil

Birkhäuser

Editors
Marcelo R. Ebert
Department of Computer Science and Mathematics
University of São Paulo
Ribeirão Preto, Brazil

Uwe Kähler
University of Aveiro
Aveiro, Portugal

Irene Sabadini
Department of Mathematics
Polytechnic University of Milan
Milan, Italy

Joachim Toft
Department of Mathematics
Linnaeus University
Växjö, Sweden

ISSN 2297-0215 ISSN 2297-024X (electronic)
Trends in Mathematics
ISSN 2509-7407 ISSN 2509-7415 (electronic)
Research Perspectives
ISBN 978-3-031-77049-4 ISBN 978-3-031-77050-0 (eBook)
https://doi.org/10.1007/978-3-031-77050-0

This book is published under the imprint Birkhäuser, www.birkhauser-science.com by the registered company Springer Nature Switzerland AG
The registered company address is: Gewerbestrasse 11, 6330 Cham, Switzerland

Preface

This volume contains selected contributions from participants of the 14th International ISAAC Congress, which was organized at Department of Computer Science and Mathematics at University of São Paulo, Campus Ribeirão Preto, from July 17 to July 21, 2023. The ISAAC congress series is the main bi-annual event of the International Society for Analysis, its Applications and Computation. This edition continued the successful series of meetings previously held in: Delaware (USA, 1997), Fukuoka (Japan, 1999), Berlin (Germany, 2001), Toronto (Canada, 2003), Catania (Italy, 2005), Ankara (Turkey, 2007), London (UK, 2009), Moscow (Russia, 2011), Krakow (Poland 2013), Macau (China, 2015), Växjö (Sweden, 2017), Aveiro (Portugal, 2019), and Ghent (Belgium, 2021). The 14th ISAAC Congress was an important scientific event that promoted communication of advances in mathematical analysis, its applications, and its interactions with computation, encouraging further research progress. Mathematicians from different parts of the world had the opportunity to present their results and new ideas. It was also a great possibility for young mathematicians to contact experts in a variety of fields. In total, there were 275 participants from all continents who registered to take part in the conference. There were 215 talks, consisting of 9 plenary lectures and 206 contributed talks, contributing to the 16 thematic sessions, some of which were organized by the special interest groups of the society. Following a well-established tradition within society, an award is presented to one or various outstanding young mathematicians. The ISAAC award of the 14th ISAAC Congress was shared between Hubert Lacoin (IMPA, Brazil) and Danylo Radchenko (University of Lille, France). The following sessions contributed to the present volume.

- Complex Geometry, organized by Alexander Schmitt.
- Complex Variables and Potential Theory, organized by Tahir Aliyev Azeroglu, Massimo Lanza de Cristoforis, Anatoly Golberg, and Sergiy Plaksa.
- Function Spaces and their Applications to Nonlinear Evolutional Equations, organized by Mitsuru Sugimoto and Baoxiang Wang.
- Harmonic Analysis and Related topics, organized by Lucas Oliveira and Tiago Picon.

- Partial differential equations on curved spacetimes, organized by Anahit Galstyan, Makoto Nakamura, and Karen Yagdjian.
- PDEs in Fluid Mechanics, organized by Lucas Catão de Freitas Ferreira, Nikolai Vasilievich Chemetov, Gabriela Planas, and Anna Laura Mazzucato.
- Recent Progress in Evolution Equations, organized by Marcello D'Abbicco, Alessandro Palmieri, and Michael Reissig.
- Stochastic Processes, organized by Benedetta Ferrario, Fernanda Cipriano, Fernando Pigeard de Almeida Prado, and Rafael Andres Rosales Mitrowsky.
- Wavelet theory and its Related Topics, organized by Keiko Fujita and Akira Morimoto.

We would like to thank the organizers of all sessions of the congress for their invaluable work and efforts. They very much supported the congress organization by inviting participants, planning their sessions, and selecting speakers. During the congress itself, they did an excellent job organizing the chairing of their meetings. The session organizers were also responsible for the refereeing process of the contributions to this proceedings volume.

Ribeirão Preto, Brazil — Marcelo R. Ebert
Aveiro, Portugal — Uwe Kähler
Milan, Italy — Irene Sabadini
Växjö, Sweden — Joachim Toft
July 2024

Acknowledgements

We would like to thank the local organizers of the 14th ISAAC Congress for their invaluable work and efforts to the success of the conference and to the ISAAC Board members for the Scientific advices to the Conference.

We also would like to thank the sponsors of the 14th ISAAC Congress: Coordenação de Aperfeiçoamento de Pessoal de Nível Superior (CAPES), Conselho Nacional de Desenvolvimento Cientíco e Tecnológico (CNPq), Fundação de Amparo à Pesquisa do Estado de São Paulo (FAPESP), Instituto Nacional de Ciência e Tecnologia de Matemática (INCTMat), International Society for Analysis its Applications and Computation (ISAAC), Pró—Reitoria de Pesquisa e Inovação (PRPI-USP), and Università degli Studi di Padova.

Finally, we thank the staff from Springer/Birkhäuser publishing house for all the support we have received and for the fruitful co-operation in preparing the final version of the Proceedings.

Contents

Part I
Complex Geometry

On the Jacobian Variety of the Accola-Maclachlan Curve of Genus Four

Robert Auffarth, Sebastián Reyes-Carocca, and Anita M. Rojas

Abstract In this short note, we study the Jacobian variety of the Accola-Maclachlan curve of genus four and obtain explicitly its Poincaré isogeny decomposition. More precisely, we show that its Jacobian variety is isomorphic to the product of two abelian surfaces that are simple, and provide explicitly a Riemann matrix for each one of the involved Abelian surfaces.

1 Introduction and Statement of the Result

A celebrated theorem due to Hurwitz, which goes back to the late nineteenth century [10], states that the automorphism group of a compact Riemann surface X of genus $g \geqslant 2$ is finite, and that this order satisfies

$$|\mathrm{Aut}(X)| \leqslant 84(g-1).$$

The smallest genus for which this upper bound is attained is three: the Klein's quartic

$$x^3y + y^3z + z^3x = 0$$

represents a compact Riemann surface of genus three with 168 automorphisms. This famous curve was found by Klein himself a few years earlier than Hurwitz' Theorem was proved. It was not until seven decades later that MacBeath proved in [13] that there are infinitely many values of g for which the Hurwitz' bound is attained.

Let N_g denote the maximum among the orders of the automorphism groups of compact Riemann surfaces of a fixed genus $g \geqslant 2$. Since the Hurwitz Theorem provides a uniform upper bound for N_g, the problem of determining a lower bound

R. Auffarth · S. Reyes-Carocca · A. M. Rojas (✉)
Universidad de Chile, Santiago, Chile
e-mail: rfauffar@uchile.cl; sebastianreyes.c@uchile.cl; anirojas@uchile.cl

M. R. Ebert et al. (eds.), *New Tools in Mathematical Analysis and Applications*, Research Perspectives, https://doi.org/10.1007/978-3-031-77050-0_1

for N_g arise naturally. This problem was successfully solved independently by Accola in [1] and Maclachlan in [14]. More precisely, they showed that for each $g \geqslant 2$, the compact Riemann surface X_g of genus g represented by the algebraic curve

$$y^2 = x^{2g+2} - 1$$

is endowed with a group of automorphisms of order $8(g+1)$; hence, they proved that

$$N_g \geqslant 8(g+1).$$

Nowadays, the curve X_g is known as the *Accola-Maclachlan curve* of genus g.

The role of X_g became even more important because of the fact that for infinitely many values of g the inequality above turns into an equality (see [1] and [14]), showing that $8(g+1)$ is the best lower uniform bound for N_g. In the early nineties, the problem of the uniqueness of X_g was addressed by Kulkarni in [12]. Indeed, he proved that if $g \not\equiv 1 \bmod 4$ and g is sufficiently large, then X_g is the unique compact Riemann surface of genus g endowed a group of automorphisms of order $8(g+1)$.

We denote by J_X the Jacobian variety of a compact Riemann surface X. The rule

$$X \mapsto J_X \quad \text{defines a map} \quad t : \mathcal{M}_g \to \mathcal{A}_g$$

from the moduli space of compact Riemann surfaces of genus g into the moduli space of principally polarised abelian varieties of dimension g. Torelli's theorem states that t is injective, namely, each isomorphism between Jacobian varieties (as principally polarised abelian varieties) comes from an isomorphism between the corresponding Riemann surfaces. In other words, a Riemann surface is determined by its Jacobian variety:

$$J_X \cong J_{X'} \iff X \cong X'.$$

Let $\mathcal{H}_g$ denote the Siegel upper space of $g \times g$ matrices; that is, the space of $g \times g$ complex symmetric matrices with positive definite imaginary part. The fact that

$$\mathcal{A}_g \text{ is isomorphic to } \mathcal{H}_g/\mathrm{Sp}(2g, \mathbb{Z}),$$

where $\mathrm{Sp}(2g, \mathbb{Z})$ denotes the symplectic group, shows that each principally polarised abelian variety A can be described in terms of (the equivalence class of) a matrix

$$Z_A \in \mathcal{H}_g.$$

Such a matrix is called a *Riemann matrix* for A. However, the problem of determining explicitly a Riemann matrix for a given abelian variety is very difficult, and has been successfully solved only for Jacobians of a few famous Riemann surfaces.

By using the theory of Fuchsian groups, a general –but implicit– form of Riemann matrices for the Jacobian variety of the Accola-Maclachlan curves was determined by Bujalance, Costa, Gamboa and Riera in [7]. They also used their results in order to provide an explicit Riemann matrix for J_{X_2}. See also [15, §.3]. Recently, an explicit Riemann matrix for J_{X_4} was provided in [17] by considering a symplectic representation of the automorphism group of X_4 and supported by computational routines developed in [4].

We recall that an abelian variety is termed *simple* if it does not contain proper non-trivial abelian subvarieties. Poincaré's irreducibility theorem states that if A is an abelian variety then there are positive integers n_j and simple abelian subvarieties A_j of A such that

$$A \sim A_1^{n_1} \times \cdots \times A_r^{n_r}$$

where $\sim$ stands for isogeny. The decomposition above is called the *Poincaré isogeny decomposition* of A. It is worthwhile recalling here that both the problem of deciding whether or not a given abelian variety is simple, and the determination of its Poincaré isogeny decomposition are difficult tasks.

As we shall discuss later, the Jacobian variety J_{X_g} is known to be non-simple (this follows after noticing that X_g has a non-central involution and therefore covers a positive genus Riemann surface) and to have complex multiplication (this follows, for instance, from [21, Theorem 4 (3)]). In addition, for the case of genus $g = 2$ one has that

$$J_{X_2} \sim E_{-\omega} \times E_{\omega^2}$$

where ω is a third root of unity. Observe that the Riemann matrices $(-\omega)$ and (ω^2) for the involved elliptic curves are defined over $\bar{\mathbb{Q}}$ as being elliptic curves with complex multiplication. This fact, together with a result due to Beauville [3], implies that

$$J_{X_2} \text{ is isomorphic to a product } E \times E'$$

where $E \sim E'$ are elliptic curves. Such curves can be chosen as $E_{i\sqrt{3}/3}$ and $E_{i\sqrt{3}}$.

This short note is devoted to generalise the aforementioned facts, and to prove the following theorem.

Theorem 1 *Let J_{X_4} denote the Jacobian variety of the Accola-Maclachlan curve of genus four. There is an isomorphism of complex tori*

$$J_{X_4} \cong S_1 \times S_2$$

where S_1 and S_2 are isomorphic abelian subvarieties of J_{X_4} of dimension two that are simple. If ζ denotes a fifth root of unity, then the Riemann matrices for S_1 and S_2 are

$$Z_1 = \begin{pmatrix} -8\zeta^3 + 4\zeta^2 - 4\zeta - 2 & -4\zeta^3 - 4\zeta^2 - 6 \\ -4\zeta^3 - 4\zeta^2 - 6 & 4\zeta^3 + 4\zeta + 2 \end{pmatrix}$$

$$Z_2 = \begin{pmatrix} -\frac{4}{5}\zeta^3 + \frac{8}{5}\zeta^2 + \frac{4}{5}\zeta + \frac{2}{5} & -\frac{4}{5}\zeta^2 - \frac{4}{5}\zeta - \frac{2}{5} \\ -\frac{4}{5}\zeta^2 - \frac{4}{5}\zeta - \frac{2}{5} & \frac{4}{5}\zeta^3 + \frac{4}{5}\zeta^2 + \frac{8}{5}\zeta + \frac{4}{5} \end{pmatrix}$$

Observe that the theorem above provides explicitly the Poincaré isogeny decomposition of J_{X_4}. At the end of this paper we shall briefly discuss the situation for J_{X_6} as well as some open questions for the general case.

2 Preliminaries

2.1 The Accola-Maclachlan Curve of Genus g

Let $g \geqslant 2$ be an integer and let $T \subset \mathbb{H}$ be a hyperbolic triangle of angles $\frac{\pi}{2}, \frac{\pi}{4}, \frac{\pi}{2g+2}$. If R_1, R_2 and R_3 are the reflexions in each one of the sides of T then

$$\gamma_1 = R_3 \circ R_1, \ \ \gamma_2 = R_1 \circ R_2 \ \text{ and } \ \gamma_3 = R_2 \circ R_3$$

generate a discrete group Γ of automorphisms of $\mathbb{H}$. It follows that the quotient $\mathbb{H}/\Gamma$ has a structure of (compact) Riemann orbifold. The group Γ, usually called a *triangle Fuchsian group of signature* $(0; 2, 4, 2g+2)$, has a canonical presentation

$$\Gamma = \langle \gamma_1, \gamma_2, \gamma_3 : \gamma_1^2 = \gamma_2^4 = \gamma_3^{2g+2} = \gamma_1\gamma_2\gamma_3 = 1 \rangle.$$

In addition, the elements of Γ of finite order are the elements conjugate to powers of γ_1, γ_2 and γ_3. We now consider the finite group G of order $8(g+1)$ with presentation

$$G = \langle a, b, c : a^{2g+2} = b^2 = c^2 = [a, b] = [b, c] = cacab = 1 \rangle$$

(a semidirect product of $\mathbb{Z}_{2g+2} \times \mathbb{Z}_2$ and $\mathbb{Z}_2$). The correspondence

$$\gamma_1 \mapsto c, \ \ \gamma_2 \mapsto ca^{-1}, \ \ \gamma_3 \mapsto a$$

defines a epimorphism $\Gamma \to G$ whose kernel K does not have nontrivial elements of finite order. Such an epimorphism is usually called a *surface-kernel epimorphism*. It then follows from the uniformisation theorem that the compact Riemann surface

$$X := \mathbb{H}/K$$

is endowed with a group of automorphisms Γ/K isomorphic to G. Moreover, the associated branched regular covering map induced by the inclusion $K \leqslant \Gamma$

$$X = \mathbb{H}/K \to \mathbb{H}/\Gamma \cong X/G \cong \mathbb{P}^1$$

ramifies over exactly three values, and it is locally $2:1$, $4:1$ and $2g+2:1$ over the branch values respectively. For more details about signatures, covering maps and Fuchsian groups, see, for instance, [6, Section 2] and the references therein. It follows from the Riemann-Hurwitz formula that the genus of X is g. Note that X has a central involution, represented by $b \in G$, fixing exactly $2g+2$ points. Hence, X is hyperelliptic or, equivalently, X is isomorphic to the normalisation of an algebraic curve of the form

$$y^2 = (x - t_1)\cdots(x - t_{2g+2})$$

where $t_1, \ldots, t_{2g+2}$ are pairwise distinct complex numbers. In addition, since $G/\langle y \rangle$ is isomorphic to a dihedral group, after a suitable Möbius transformation, the complex numbers $t_1, \ldots, t_{2g+2}$ can be taken as the $(2g+2)$-th roots of unity. Consequently, X is isomorphic to the Accola-Maclachlan curve X_g. It follows from classical results due to Singerman [20] that the (full) automorphism group of X_g is isomorphic to G.

2.2 *Riemann Matrices and Abelian Varieties*

Let $T = V/\Lambda$ be a complex torus of dimension g. If we choose bases

$$B = \{v_1, \ldots, v_g\} \text{ and } L = \{\lambda_1, \ldots, \lambda_{2g}\}$$

of V and Λ as a complex vector space and as a $\mathbb{Z}$-module respectively, then the *period matrix* for T with respect to these bases is the $g \times 2g$ matrix defined as

$$\Pi = (\pi_{i,j}) \text{ where } \lambda_j = \pi_{1,1}v_1 + \cdots + \pi_{1,g}v_g \text{ for each } j.$$

Moreover, by choosing appropriate bases, the period matrix has the form

$$\Pi = (\mathrm{I}_g \; Z) \text{ where } \det(\mathrm{Im}(Z)) \neq 0,$$

with I_g denoting the identity $g \times g$ matrix.

Let $f : T \to T'$ be a homomorphism of complex tori of dimension g and g', and let Π and Π' be their period matrices with respect to given bases. One can associate to f two linear maps: the *analytic representation* and the *rational representation*

$$\rho_a(f) : V \to V' \text{ and } \rho_r(f) : \Lambda \to \Lambda',$$

represented by a $g' \times g$ complex matrix and by a $2g' \times 2g$ integral matrix respectively. The fundamental relation they satisfy is given by

$$\rho_a(f)\, \Pi = \Pi'\, \rho_r(f).$$

Observe that if f is an isomorphism then $g' = g$, the matrix $\rho_a(f)$ is nonsingular, and the matrix $\rho_r(f)$ is unimodular.

A complex *abelian variety* is a complex torus which is also a complex projective algebraic variety. Each abelian variety $T = V/\Lambda$ admits a *polarisation*, that is, a non-degenerate real alternating form Θ on V such that for all $v, w \in V$

$$\Theta(iv, iw) = \Theta(v, w) \quad \text{and} \quad \Theta(\Lambda \times \Lambda) \subset \mathbb{Z}.$$

If each elementary divisor of $\Theta|_{\Lambda\times\Lambda}$ equals 1 then Θ is called *principal* and T is called a *principally polarised abelian variety*; from now on, we write *ppav* for short. In this case, there exists a basis for Λ such that the matrix for $\Theta_{\Lambda\times\Lambda}$ is given by

$$J = \begin{pmatrix} 0 & \mathrm{I}_g \\ -\mathrm{I}_g & 0 \end{pmatrix} \quad \text{where} \quad g = \dim(X);$$

such a basis is called *symplectic*. In addition, there exist a basis for V with respect to which the period matrix for T is $\Pi = (\mathrm{I}_g\ Z)$ where

$$Z \in \mathcal{H}_g = \{Z \in \mathrm{M}(g, \mathbb{C}) : Z = Z^t,\ \mathrm{Im}(Z) > 0\},$$

with Z^t denoting the transpose matrix of Z. By an *isomorphism of ppavs* we mean an isomorphism of the underlying complex tori preserving the involved polarisations. Observe that if $(\mathrm{I}_g\ Z_i)$ is the period matrix for T_i then an isomorphism $T_1 \to T_2$ is given by invertible matrices

$$M \in \mathrm{GL}(g, \mathbb{C}) \quad \text{and} \quad R \in \mathrm{GL}(2g, \mathbb{Z}) \quad \text{such that} \quad M(\mathrm{I}_g\ Z_1) = (\mathrm{I}_g\ Z_2)R. \tag{1}$$

As R preserves the polarisation J, it must satisfy $R^t J R = J$ and hence it belongs to $\mathrm{Sp}(2g, \mathbb{Z})$. It follows from (1) that the orbits of the action $\mathrm{Sp}(2g, \mathbb{Z}) \times \mathcal{H}_g \to \mathcal{H}_g$ given by

$$\left(R = \begin{pmatrix} A & B \\ C & D \end{pmatrix}, Z\right) \mapsto R \cdot Z := (A + ZC)^{-1}(B + ZD)$$

are in bijective correspondence with the isomorphism classes of ppavs of dimension g. Hence

$$\mathcal{H}_g \to \mathcal{A}_g := \mathcal{H}_g/\mathrm{Sp}(2g, \mathbb{Z})$$

is the moduli space of isomorphism classes of ppavs of dimension g. See [5] and [16] for more details.

3 Proof of the Theorem

The Decomposition of J_{X_4} We recall that the automorphism group of the Accola-Maclachlan curve X_4 of genus four is isomorphic to

$$G = \langle a, b, c : a^{10} = b^2 = c^2 = [a, b] = [b, c] = cacab = 1 \rangle,$$

and that the action of G on X_4 is represented by the surface-kernel epimorphism

$$\theta = (c, ca^{-1}, a)$$

We write $J_{X_4} = V/\Lambda$ where V is the g-dimensional complex vector space of holomorphic 1-forms on X_4 and L is the first (integral) homology group of X_4.

We follow the routines developed in [4] to ensure that there is a symplectic basis $\gamma = \{\alpha_1, \ldots, \gamma_4, \beta_1, \ldots, \beta_4\}$ for L (and hence $\{\alpha_1, \ldots, \alpha_4\}$ is a basis for V) in such a way that the rational representation

$$\rho_r : G \to \mathrm{GL}(L \otimes \mathbb{Q})$$

induced by the action of G is given by

$$\rho_r(a) = \begin{pmatrix} 0 & 1 & 1 & 1 & -1 & 0 & 0 & 0 \\ 1 & 0 & 0 & 0 & 0 & -1 & 0 & 0 \\ 0 & 0 & 0 & 0 & 0 & 1 & -1 & 0 \\ 0 & 0 & 0 & 0 & 0 & 0 & 1 & -1 \\ 1 & 0 & 0 & 0 & 0 & 0 & 0 & 0 \\ 0 & 1 & 1 & 1 & 0 & 0 & 0 & 0 \\ 0 & 0 & 1 & 1 & 0 & 0 & 0 & 0 \\ 0 & 0 & 0 & 1 & 0 & 0 & 0 & 0 \end{pmatrix} \text{ and } \rho_r(c) = \begin{pmatrix} 0 & 0 & 0 & 0 & 0 & 0 & -1 & 1 \\ 0 & -1 & 0 & 0 & 0 & 0 & 0 & 0 \\ 0 & 0 & 0 & 0 & 1 & 0 & 0 & 0 \\ 0 & 1 & 1 & 1 & -1 & 0 & 0 & 0 \\ 0 & 0 & 1 & 0 & 0 & 0 & 0 & 0 \\ 0 & 0 & 0 & 0 & 0 & -1 & 0 & 1 \\ -1 & 0 & 0 & 0 & 0 & 0 & 0 & 1 \\ 0 & 0 & 0 & 0 & 0 & 0 & 0 & 1 \end{pmatrix}$$

(observe that $b = (ac)^2$).

Although an explicit Riemann matrix for J_{X_4} has been found in [17], we prefer to compute it again here by finding the fixed points in $\mathcal{H}_4$ by the action of

$$G_r = \langle \rho_r(a), \rho_s(c) \rangle \leqslant \mathrm{Sp}(8, \mathbb{Z}).$$

In other words, a period matrix for J_{X_4} is the (unique) solution Z of the equation

$$R \cdot Z = R \text{ for all } R \in G_r.$$

Explicitly, if ζ a fifth root of unity then

$$Z = \begin{pmatrix} -2\zeta^3+\zeta^2-\zeta-\frac{1}{2} & -\zeta^3-\zeta^2-2 & \zeta^3+\zeta^2+\frac{3}{2} & \zeta^3+\zeta^2+\frac{3}{2} \\ -\zeta^3-\zeta^2-2 & \frac{6}{5}\zeta^3+\frac{2}{5}\zeta^2+\frac{8}{5}\zeta+\frac{4}{5} & -\frac{3}{5}\zeta^3-\frac{1}{5}\zeta^2-\frac{4}{5}\zeta-\frac{2}{5} & -\frac{7}{5}\zeta^3+\frac{1}{5}\zeta^2-\frac{6}{5}\zeta-\frac{3}{5} \\ \zeta^3+\zeta^2+\frac{3}{2} & -\frac{3}{5}\zeta^3-\frac{1}{5}\zeta^2-\frac{4}{5}\zeta-\frac{2}{5} & \frac{4}{5}\zeta^3+\frac{3}{5}\zeta^2+\frac{7}{5}\zeta+\frac{7}{10} & \frac{6}{5}\zeta^3-\frac{3}{5}\zeta^2+\frac{3}{5}\zeta+\frac{3}{10} \\ \zeta^3+\zeta^2+\frac{3}{2} & -\frac{7}{5}\zeta^3+\frac{1}{5}\zeta^2-\frac{6}{5}\zeta-\frac{3}{5} & \frac{6}{5}\zeta^3-\frac{3}{5}\zeta^2+\frac{3}{5}\zeta+\frac{3}{10} & \frac{4}{5}\zeta^3+\frac{3}{5}\zeta^2+\frac{7}{5}\zeta+\frac{7}{10} \end{pmatrix}.$$

We point out that the period matrix $\Pi = (\mathrm{I}_4\, Z)$ we found for J_{X_4} allows us to recover each β_j in terms of a linear combination of $\alpha_1, \ldots, \alpha_4$.

We consider the subgroup $H_1 = \langle c \rangle$ of G. Following [17, Theorem 3], there is an isogeny

$$J_{X_4} \to J_{Y_1}^2$$

where Y_1 is the genus two quotient Riemann surface associated to the action of H_1 on X_4. In addition, if we consider the subgroup $H_2 := H_1^a = \langle a^2bc \rangle$ of G and accordingly we denote the corresponding quotient Riemann surface by Y_2, then J_{Y_1} and J_{Y_2} are isomorphic (as ppavs). Hence, there is an isogeny

$$\tilde{\varphi} : J_{Y_1} \times J_{Y_2} \to J_{X_4}.$$

We claim that $\tilde{\varphi}$ is an isomorphism of complex tori.

In order to prove the claim, we start by recalling that the regular covering map

$$\pi_j : X_4 \to Y_j \quad \text{for} \quad j = 1, 2$$

given by the action of H_j on X_4 ramifies over exactly two values (corresponding to the image of the two fixed points of H_j). It follows from [5, Proposition 11.4.3] that the pull-back $\pi_j^* : J_{Y_j} \to J_{X_4}$ is injective and consequently it induces an isomorphism onto its image, namely,

$$J_{Y_j} \cong S_j := \pi_j^*(J_{Y_j}) \subset J_{X_4} \text{ for } j = 1, 2.$$

In particular, as JY_1 and JY_2 are isomorphic ppavs, S_1 and S_2 are isomorphic as tori. The subvarieties S_1 and S_2 of JX_4 can be described in a very explicit way as follows. We denote by

$$\Phi : \mathbb{Q}[G] \to \mathrm{End}_{\mathbb{Q}}(J_{X_4})$$

the $\mathbb{Q}$-algebra homomorphism induced by the action of G on X_4. Then

$$S_j := \mathrm{Im}(\Phi(p_{H_j})) \text{ where } p_{H_j} = \tfrac{1}{2}\Sigma_{h \in H_j} h \in \mathbb{Q}[G]$$

(see [8] for more details). Further, we can describe $S_j = V_j/\Lambda_j$ in terms of coordinates as follows. The rational representation of p_{H_j} is

$$\rho_r(p_{H_j}) = \tfrac{1}{2}\Sigma_{h\in H_j}\rho_r(h) \in \mathrm{Sp}(8, \mathbb{Q})$$

and the coordinates of Λ_j with respect to the symplectic basis γ are given by

$$B_1 = \begin{pmatrix} 0&0&0&1&0&0&0&0 \\ 1&0&0&0&0&0&-1&0 \\ 1&0&0&0&0&1&1&2 \\ 0&0&1&0&1&0&0&0 \end{pmatrix} \quad \text{and} \quad B_2 = \begin{pmatrix} 0&0&0&1&-1&0&0&0 \\ 0&1&1&0&0&0&0&0 \\ 1&0&0&0&0&0&0&1 \\ 0&0&0&0&0&1&1&0 \end{pmatrix}$$

Once a period matrix Π for J_{X_4} is given, the procedures developed in [18] (assisted by the algorithms in [19]) allow us to identify the relations between the coordinates captured in the matrices B_j. We obtain that

1. the induced polarisation on the subvarieties S_j is twice a principal one (as expected by Birkenhake and Lange [5, Lemma 12.3.1]), and that
2. Riemann matrices for S_1 and S_2 are

$$Z_1 = \begin{pmatrix} -8\zeta^3 + 4\zeta^2 - 4\zeta - 2 & -4\zeta^3 - 4\zeta^2 - 6 \\ -4\zeta^3 - 4\zeta^2 - 6 & 4\zeta^3 + 4\zeta + 2 \end{pmatrix}$$

$$Z_2 = \begin{pmatrix} -\frac{4}{5}\zeta^3 + \frac{8}{5}\zeta^2 + \frac{4}{5}\zeta + \frac{2}{5} & -\frac{4}{5}\zeta^2 - \frac{4}{5}\zeta - \frac{2}{5} \\ -\frac{4}{5}\zeta^2 - \frac{4}{5}\zeta - \frac{2}{5} & \frac{4}{5}\zeta^3 + \frac{4}{5}\zeta^2 + \frac{8}{5}\zeta + \frac{4}{5} \end{pmatrix}$$

The fact that each S_j is an abelian subvariety of J_{X_4} implies that the sum induces an isogeny

$$\varphi : S_1 \times S_2 \to J_{X_4}.$$

Observe that $\varphi \circ (\pi_1^*, \pi_2^*) = \tilde{\varphi}$. The rational representations of φ is

$$\rho_r(\varphi) = \begin{pmatrix} 0&1&0&0&1&0&1&0 \\ 0&0&0&1&0&0&0&0 \\ 0&0&0&1&0&1&0&0 \\ 1&0&1&0&0&0&0&0 \\ 0&0&-1&0&0&1&0&0 \\ 0&0&0&0&1&0&0&1 \\ 0&-1&0&0&1&0&0&1 \\ 0&0&0&0&2&0&1&0 \end{pmatrix}.$$

Finally, since the determinant of the latter matrix equals one (and this number agrees with the order of $\ker(\varphi)$), we conclude that φ is an isomorphism.

The Simplicity of S_1 and S_2 Since S_1 and S_2 are isomorphic tori, it is enough to show that S_1 is simple. To accomplish this task, we employ the equivalence between the following statements, proved in [2, Proposition 4.4].

1. The abelian surface represented by the Riemann matrix

$$\begin{pmatrix} \tau_1 & \tau_2 \\ \tau_2 & \tau_3 \end{pmatrix}$$

admits a sub elliptic curve.

2. There exists a 6-uple $(a_{12}, a_{13}, a_{14}, a_{23}, a_{24}, a_{34})$ of rational numbers satisfying

$$-1 = a_{13} + a_{24} \tag{i}$$

$$(\tau_1\tau_3 - \tau_2^2)a_{12} - a_{14}\tau_1 + a_{13}\tau_2 - a_{24}\tau_2 + a_{23}\tau_3 + a_{34} = 0 \tag{ii}$$

$$a_{14}a_{23} - a_{13}a_{24} + a_{12}a_{34} = 0. \tag{iii}$$

We apply the criterion above with the Riemann matrix for S_1

$$\tilde{Z} = \tfrac{1}{2}Z_1 = \begin{pmatrix} -4\zeta^3 + 2\zeta^2 - 2\zeta - 1 & -2\zeta^3 - 2\zeta^2 - 3 \\ -2\zeta^3 - 2\zeta^2 - 3 & 2\zeta^3 + 2\zeta + 1 \end{pmatrix}$$

which has determinant $\det(\tilde{Z}) = 2\zeta^3 + 2\zeta^2 + 2$. By replacing the entries of $\tilde{Z}$ in equation (ii), we obtain a linear combination over the rationals of the form

$$A + B\zeta + C\zeta^2 + D\zeta^3 = 0.$$

Now, the fact that $\{1, \zeta, \zeta^2, \zeta^3\}$ is linearly independent over $\mathbb{Q}$ yields four equations, labeled as (2.b) to (2.e) below. The system turns into

$$a_{13} + a_{24} = -1 \tag{2a}$$

$$a_{14} + a_{23} = 0 \tag{2b}$$

$$a_{12} - a_{13} - a_{14} - a_{24} = 0 \tag{2c}$$

$$a_{12} - a_{13} + 2a_{14} + a_{23} + a_{24} = 0 \tag{2d}$$

$$2a_{12} - 3a_{13} + a_{14} + a_{23} + 3a_{24} + a_{34} = 0 \tag{2e}$$

$$a_{14}a_{23} - a_{13}a_{24} + a_{12}a_{34} = 0 \tag{2f}$$

Observe that equations (2.a) to (2.e) give rise to a linear system of equations in six variables, whose solutions are

$$(a_{12}, a_{13}, a_{14}, a_{23}, a_{24}, a_{34}) = (\mu, \tfrac{\mu-1}{2}, 0, 0, -\tfrac{1+\mu}{2}, \mu) \text{ where } \mu \in \mathbb{C}.$$

We replace this information in equation (2.f) to obtain that

$$\tfrac{\mu-1}{2}\cdot\tfrac{\mu+1}{2}+\mu^2=0 \text{ and therefore } 5\mu^2=1.$$

Hence, the system does not have any rational solution, showing that S_1 is simple. Since S_1 is isomorphic to S_2, we get that S_2 is also simple. Hence, we have the Poincaré decomposition of JX_4.

4 Final Comments

We end this article by mentioning some comments and remarks.

CM Type and Riemann Matrices It is known that J_{X_g} admits complex multiplication, for each $g \geqslant 2$. For instance, this property follows from [21, Theorem 4], after noticing that the regular covering map $X_g \to X_g/\langle a, b\rangle \cong \mathbb{P}^1$ ramifies over three values. If we further assume that $g+1$ is prime then, as proved in [17], there is an isogeny between J_{X_g} and J_Y^2, where Y is the quotient Riemann surface associated to the action of $\langle c\rangle$ on X_g. Thus, Y admits complex multiplication too. Moreover, by the results of [9, §4], Y is defined over $\bar{\mathbb{Q}}$. Now, since J_Y admits complex multiplication and is defined over $\bar{\mathbb{Q}}$, a result due to Wolfart [21] ensures that each Riemann matrix for J_Y is a $\bar{\mathbb{Q}}$-rational point of the Siegel upper half-space. For genus four we went further, and obtained that a Riemann matrix for J_Y is

$$\begin{pmatrix} -4\zeta^3+2\zeta^2-2\zeta-1 & -2\zeta^3-2\zeta^2-3 \\ -2\zeta^3-2\zeta^2-3 & 2\zeta^3+2\zeta+1 \end{pmatrix}$$

which is, as expected, a $\bar{\mathbb{Q}}$-rational point of $\mathcal{H}_2$.

The Case of JX_6 and Some Open Problems Assume that $g+1$ is a prime number. By arguing as done in the case of genus four, it can be seen that

$$J_{Y_1}\times J_{Y_2}\cong A_1\times A_2\sim J_{X_g}, \tag{3}$$

where A_1 and A_2 are abelian subvarieties of J_{X_g} of dimension $\frac{g}{2}$. These subvarieties are, in general, not complementary. In fact, they are associated to pairs of subgroups H_1 and H_2 of G conjugate to $\langle c\rangle$ and the corresponding idempotents p_{H_1} and p_{H_2} do not sum the identity. It would be interesting to understand the extent to which our methods can be generalised to the general case. For instance, in the case of genus six, by choosing appropriate subgroups H_1 and H_2 as before, it can be seen that the symplectic representation of the corresponding isogeny in (3) has determinant one. Hence, the isogeny is an isomorphism.

In fact, more can be said about the case $g=6$. By calculating a period matrix for A_1 (which is isogenous to A_2) and calculating its Néron-Severi group, it can be seen that its Picard number is $9 = 3^2$. By the main theorem of [11], we have that this implies that A_1 (and therefore A_2) is isogenous to the self-product of an elliptic curve (with complex multiplication). In particular, JX_6 itself is isogenous to the self-product of an elliptic curve (with complex multiplication), as happens in the case $g = 2$. This naturally leads us to the following amusing question that we would like to answer in the future:

Question How many primes p exist such that $p \equiv 3 \bmod 4$ and JX_{p-1} splits isogenously as the product of elliptic curves (with complex multiplication)?

Acknowledgments The authors would like to thank the referee for their thorough reading of the paper and for their helpful comments. This work was partially supported by ANID/CONICYT Fondecyt Regular Grants 1220099, 1230708, 1230034 and 1220997, and MATH-AmSud Project 22-MATH-03.

References

1. Accola, R.D.M.: On the number of automorphisms of a closed Riemann surface. Trans. Amer. Math. Soc. **131**, 398–408 (1968)
2. Auffarth, R., Lange, H., Rojas, A.M.: A criterion for an abelian variety to be non-simple. J. Pure Appl. Algebra **221**(8), 1906–1925 (2017)
3. Beauville, A.: Some surfaces with maximal Picard number. J. Éc. Polytech. Math. **1**, 101–116 (2014)
4. Behn, A., Rodríguez, R.E., Rojas, A.M.: Adapted hyperbolic polygons and symplectic representations for group actions on Riemann surfaces. J. Pure Appl. Algebra **217**(3), 409–426 (2013)
5. Birkenhake, C., Lange, H.: Complex Abelian Varieties, 2nd edn. Grundlehren der Mathematischen Wissenschaften, vol. 302, xii+635pp. Springer, Berlin (2004). ISBN: 3-540-20488-1
6. Broughton, S.A.: Classifying finite group actions on surfaces of low genus. J. Pure Appl. Algebra **69**(3), 233–270 (1991)
7. Bujalance, E., Costa, A.F., Gamboa, J.M., Riera, G.: Period matrices of Accola-Maclachlan and Kulkarni surfaces. Ann. Acad. Sci. Fenn. Math. **25**(1), 161–177 (2000)
8. Carocca, A., Rodríguez, R.E.: Jacobians with group actions and rational idempotents. J. Algebra **306**(2), 322–343 (2006)
9. González-Diez, G.: Variations on Belyi's theorem. Q. J. Math. **57**(3), 339–354 (2006)
10. Hurwitz, A.: Ueber algebraische Gebilde mit eindeutigen Transformationen in sich (in German). Math. Ann. **41**(3), 403–442 (1892)
11. Katsura, T.: On the structure of singular abelian varieties. Proc. Japan Acad. **51**(4), 224–228 (1975)
12. Kulkarni, R.S.: A note on Wiman and Accola-Maclachlan surfaces. Ann. Acad. Sci. Fenn. Ser. A I Math. **16**(1), 83–94 (1991)
13. Macbeath, A.M.: On a theorem of Hurwitz. Proc. Glasgow Math. Assoc. **5**, 90–96 (1961)
14. Maclachlan, C.: A bound for the number of automorphisms of a compact Riemann surface. J. Lond. Math. Soc. **44**, 265–272 (1969)
15. Marchisio, M.R.: Abelian surfaces and products of elliptic curves. Boll. Unione Mat. Ital. Sez. B Artic. Ric. Mat. **1**(2), 407–427 (1998)
16. Oort, F.: Singularities of coarse moduli schemes. Semin. d'Algebre, Paul Dubreil, Paris 1975–76 (29eme Annee), Lect. Notes Math. **586**, 61–76 (1977)

17. Reyes-Carocca, S., Rojas, A.M.: On large prime actions on Riemann surfaces. J. Group Theory **25**(5), 887–940 (2022)
18. Rodríguez, R.E., Rojas, A.M.: Period matrices for abelian varieties: algorithms and applications (preprint, 2024)
19. Rodríguez, R.E., Rojas, A.M.: Period matrices (Magma code). https://github.com/rojas-ani/Period-matrices
20. Singerman, D.: Finitely maximal Fuchsian groups. J. Lond. Math. Soc. **6**, 29–38 (1972)
21. Wolfart, J.: Triangle groups and Jacobians of CM type (2000). https://www.math.uni-frankfurt.de/~wolfart/Artikel/jac.pdf

T-Invariant Points on T-Varieties

Pedro L. del Angel R.

Abstract Let $T \cong \mathbb{G}_m^d$ be a d-dimensional torus, defined over $\mathbb{C}$ and let X be a non-singular projective scheme of finite type, defined over $\mathbb{C}$. Under some (presumably mild) assumptions, we give a description of the T–invariant subschemes of X.

MSC 2020 Subject Classification Primary 14L30, Secondary: 14M99

1 Introduction

In recent years, there has been great interest in the study of T–varieties. No doubt the study of toric varieties has concentrated most of the efforts, in part due to the lack of appropriate techniques to study T–varieties of higher complexity. Here we present some techniques with the hope that they will prove useful when studying T–varieties of higher complexity. In what follows $T \cong \mathbb{G}_m^d$ will be a d-dimensional torus and X will be a non-singular projective scheme of finite type. Further, we will assume that there is an effective action of T on X (i.e. the only subgroup of T acting trivially on X is the identity). We give a description of the T–invariant subschemes of X.

To avoid excess of technicalities, all varieties will be defined over $\mathbb{C}$ and we will use X (and T) to denote both, the scheme over $\mathbb{C}$ as well as the underlying algebraic variety. Variety (respectively subvariety) will always mean irreducible.

2 Invariant Partitions

Assume that there exist a finite partition $\mathfrak{P}$ of X by locally closed subschemes which are T-invariant.

Pedro L. del Angel R. (✉)
CIMAT, A.C., Guanajuato, Mexico
e-mail: luis@cimat.mx

M. R. Ebert et al. (eds.), *New Tools in Mathematical Analysis and Applications*, Research Perspectives, https://doi.org/10.1007/978-3-031-77050-0_2

Lemma 1 *Any irreducible closed subscheme $Y \subset X$ is of the form*

$$Y = \overline{Y \cap X_\beta}$$

for some $X_\beta \in \mathfrak{P}$.

Proof Observe that $X = \sqcup_{X_\beta \in \mathfrak{P}} X_\beta$, therefore

$$\begin{aligned} Y &= Y \cap \left(\sqcup_{X_\beta \in \mathfrak{P}} X_\beta\right) \\ &= Y \cap \overline{\left(\sqcup_{X_\beta \in \mathfrak{P}} X_\beta\right)} \\ &= \sqcup_{X_\beta \in \mathfrak{P}} \overline{(Y \cap X_\beta)} \end{aligned}$$

since $\mathfrak{P}$ is finite. But Y is irreducible, hence there exist one $X_\beta \in \mathfrak{P}$ such that $Y = \overline{(Y \cap X_\beta)}$. □

Lemma 2 *Assume that $X_\beta \in \mathfrak{P}$ is such that for any $x, y \in X_\beta$ one has*

$$\dim \mathrm{Stab}_T(x) = \dim \mathrm{Stab}_T(y). \tag{1}$$

If furthermore X is affine or normal, then the (universal) geometric quotient

$$\mathcal{X}_\beta = X_\beta / T$$

exist.

Proof By Sumihiro [7] if X_β is normal, for free if X_β is affine, any point $x \in X_\beta$ is pre-stable in the sense of Mumford [6] and the claims follows. □

Corollary 3 *Assume that X_β satisfies the conditions of Lemma 2. Further, assume that $Y \subset X$ is an irreducible, closed, T-invariant subscheme of X such that $Y \cap X_\beta$ is dense in Y. If X_β is affine or Y is normal, then the (universal) geometric quotient*

$$\mathcal{Y}_\beta = (Y \cap X_\beta)/T$$

exist and is irreducible.

Proof Since X_β is locally closed, then $Y \cap X_\beta$ is a non empty open subscheme of Y (and a closed subscheme of X_β) and therefore irreducible and T-invariant. On the other hand, if $y_1, y_2 \in Y \cap X_\beta$, then

$$\dim \mathrm{Stab}_T(y_1) = \dim \mathrm{Stab}_T(y_2).$$

by hypothesis, and the proof follows as in Lemma 2. □

Moreover, the converse is also true.

Corollary 4 *Let $X_\beta \in \mathfrak{P}$ and assume it satisfies the conditions of Lemma 2. Then for any irreducible closed subscheme $\mathcal{Y} \subset \mathcal{X}_\beta$, the scheme $\overline{\mathcal{Y} \times_{\mathcal{X}_\beta} X_\beta}$ is an irreducible, T–invariant, closed subscheme of X.*

Proof Let $\pi : X_\beta \to \mathcal{X}_\beta$ be the natural projection, $y \in \mathcal{Y}$ and $x \in (X_\beta)_y$. Then the following are cartesian diagramms

$$\begin{array}{ccccc} T \cdot x & \longrightarrow & \mathcal{Y} \times_{\mathcal{X}_\beta} X_\beta & \longrightarrow & X_\beta \\ \downarrow & & \downarrow & & \downarrow \\ y & \longmapsto & \mathcal{Y} & \longrightarrow & \mathcal{X}_\beta \end{array}$$

i.e., the fiber of $\mathcal{Y} \times_{\mathcal{X}_\beta} X_\beta \to \mathcal{Y}$ over y can be identified with $T \cdot x$, the orbit of x under the action of T. But $T \cdot x$ is irreducible since T is, therefore $\mathcal{Y} \times_{\mathcal{X}_\beta} X_\beta \to \mathcal{Y}$ is a surjective morphism over an irreducible scheme which has irreducible fibers, hence $\mathcal{Y} \times_{\mathcal{X}_\beta} X_\beta$ and also its closure $\overline{\mathcal{Y} \times_{\mathcal{X}_\beta} X_\beta}$ are irreducible. Since $\mathcal{Y} \times_{\mathcal{X}_\beta} X_\beta$ can be identified with the preimage of $\mathcal{Y}$ on X_β, then $\mathcal{Y} \times_{\mathcal{X}_\beta} X_\beta$ is T–invariant and so is its closure. □

Theorem 5 *Assume that every $X_\beta \in \mathfrak{P}$ satisfies the conditions of Lemma 2. Then any irreducible T–invariant, normal subscheme of X is of the form*

$$Y = \overline{\mathcal{Y} \times_{\mathcal{X}_\beta} X_\beta}$$

for some irreducible, closed subscheme $\mathcal{Y} \subset \mathcal{X}_\beta$, for some $X_\beta \in \mathfrak{P}$. If every X_β is affine we do not need to assume normality of Y.

Proof It follows at once from lemma 1 and corollary 3, since

$$Y \cap X_\beta = \mathcal{Y} \times_{\mathcal{X}_\beta} X_\beta$$

where $Y = \overline{Y \cap X_\beta}$ and $\mathcal{Y} = \left(Y \cap X_\beta\right)/T$. □

Corollary 6 *With the hypothesis of Theorem 5 and keeping its notation, any irreducible, closed (normal) T–invariant subvariety $Y \subset X$ is of the form*

$$Y = \overline{\overline{\{\eta\}} \times_{\mathcal{X}_\beta} X_\beta}$$

for some $X_\beta \in \mathfrak{P}$ and some $\eta \in \mathcal{X}_\beta$.

Proof Take η to be the generic point of $\mathcal{Y}$. □

This allows us to define, for any X_β satisfying the hypothesis of Lemma 2, a map

$$\mathcal{X}_\beta \xrightarrow{\mathbb{V}_\beta} X^T$$

$$\eta \longmapsto \zeta$$

where ζ is the generic point of $\overline{\overline{\{\eta\}} \times_{\mathcal{X}_\beta} X_\beta}$.

Corollary 7 *Assume that every $X_\beta \in \mathfrak{P}$ satisfies the conditions of Lemma 2. If X_β is affine for every β, then*

$$X^T = \sqcup_{X_\beta \in \mathfrak{P}} \mathbb{V}_\beta(\mathcal{X}_\beta).$$

Finally, we can consider the cycle class map $X^T \to H_*(X)^T$, thence he have

Corollary 8 *With the hypothesis of Theorem 5, there is a map*

$$\sqcup_{X_\beta \in \mathfrak{P}} \mathcal{X}_\beta \xrightarrow{\sqcup \mathbb{V}_\beta} H_*(X)^T$$

$$\eta \longmapsto [\overline{\{\zeta\}}]$$

3 First Examples

For any subgroup $H < T$, consider the set

$$X_H := \{x \in X \mid \mathrm{Stab}_T(x) = H\}.$$

Let $\mathcal{N}$ be the net of *significant* subgroups of T, i.e.,

$$\mathcal{N} = \{H < T \mid X_H \neq \emptyset\}.$$

Lemma 9 *For any $H \in \mathcal{N}$, the set X_H is T-invariant.*

Proof Let $H \in \mathcal{N}$, $x \in X_H$ and $t \in T$, then an easy computation shows that $\mathrm{Stab}_T(tx) = t \cdot \mathrm{Stab}_T(x) \cdot t^{-1}$, but $t \cdot \mathrm{Stab}_T(x) \cdot t^{-1} = \mathrm{Stab}_T(x)$ because T is commutative. □

Remark 1 Obviously we have $X = \sqcup_{H \in \mathcal{N}} X_H$.

If $\mathcal{N}$ is finite and X_H is locally closed for any $H \in \mathcal{N}$, then we can take $\mathfrak{P} = \mathcal{N}$.

Examples

1. If X is a toric variety, any X_H is just a T–orbit and the classes in the image of $\sqcup_H \mathbb{V}_H$ generate the homology of X.
2. If X is a Grassmannian variety, then the X_H's are nothing more than the tiny Schubert cells, see for instance [4].
3. If X is a flag variety of the form GL_n/P, then we can take $\mathfrak{P}$ to be the partition given by the (multi) tiny Schubert cells, which is finer than $\mathcal{N}$ in this case. See [5].

4 Mumford's Like Examples

Following Mumford, let us define, for every $i \in \mathbb{N} \cup \{0\}$, the set

$$X_{(i)} = \{x \in X \mid \dim(\mathrm{Stab}_T(x)) = i\}.$$

Lemma 10 *The sets $X_{(i)}$ are T–invariant for every integer i.*

Proof If $x \in X_{(i)}$ and $t \in T$, then

$$\mathrm{Stab}_T(tx) = t\,\mathrm{Stab}_T(x)t^{-1} = \mathrm{Stab}_T(x)$$

since T is abelian. □

Remark 2 Obviously we have $X = \sqcup_{i \in \mathbb{N}\cup\{0\}} X_{(i)}$ is a *finite* union.

One can take $\mathfrak{P}$ to be the collection of $X'_{(i)}s$, provided they are locally closed.

Examples

4. If X is the symplectic Grassmannian $SpG(2, 2n)$ of 2–dimensional isotropic subspaces in a $2n$ dimensional vector space with a symplectic form Q, we can replace the $X_{(i)}$ by the finer partition given by the tiny BC^n Schubert cells, see [3], which is nevertheless coarser than the partition $\mathcal{N}$ of the previous paragraph.
5. If X is a flag variety of the form SpG_{2n}/P, we can again work with the partition given by the (multy) tiny Schubert cells, see [2].

5 Final Comments

One can replace $\mathbb{C}$ by any algebraically closed field K and the results remain true (with exactly the same proofs), except for Corollary 8, where we get uniform geometric quotients instead of universal geometric quotients.

We do not really need X to be of finite type, as long as the set of its geometric points is dense.

For any affine scheme X over $\mathbb{C}$ with a T–action, a uniform categorical quotient X/T always exists, but could be very bad behaved (can be a very "pre", pre-scheme, in words of Mumford). Altmann and Hausen have shown, using polyhiedral divisors (see [1]), that for any affine, normal T-scheme X, there exist a normal algebraic variety $\widetilde{X}$ that admits a good quotient $\widetilde{X} \to \widetilde{\mathcal{X}}$ which is normal and semi-projective, and such that there is a rational, T-equivariant contraction $\widetilde{X} \to X$.

Acknowledgments I like to thank Prof. Alexander Schmitt for the kind invitation to speak at the Complex Geometry session as part of the 14th ISAAC Congress 2023 at USP in Ribeirão Preito (Brazil) as well as the local organising committee and the ISAAC board for the great hospitality. I also like to thank the Centro de Investigaci6n en Matemáticas for financial support, Luis Yair Meza for fruitful discussions and the referee for his useful comments and remarks, specially for calling my attention towards [1].

References

1. Altmann, K., Hausen, J.: Polyhiedral divisors and algebraic torus actions. Math. Ann. **334**, 557–607 (2006)
2. del Angel R.P.L, Meza-Perez L.Y.: T-invariant points of symplectic Flag varieties [in preparation]. CIMAT & UJAT (2024)
3. del Angel, P.L., Javier Elizondo, E., Garay, C., Zaldivar, F.: On T–invariant subvarieties of symplectic grassmannians and representability of rank 2 symplectic matroids over $\mathbb{C}$. J. Algebra **640**, 300–325 (2024)
4. Elizondo, J.E., Fink, A., Garay, C.: Matroids and the space of torus-invariant subvarieties of the Grassmannian with given homology class DOI: 10.48550/arXiv.2112.15334
5. Meza-Perez, L.Y.: (Co)homologia de algunas T-variedades [Unpublished doctoral dissertation]. Universidad juarez autonoma de tabasco (UJAT), Mexico (2024)
6. Mumford, D.: Geometric Invariant Theory. Ergebnisse der Mathematik und ihrer Grenzgebiete, vol. 34, 2nd edn. Springer, Berlin (1982)
7. Sumihiro, H.: Equivariant completion I and II. J. Math. Kyoto **14**, 1 (1974); **15**, 573 (1975)

Computations on the Hilbert-Mumford Criterion for Grassmanians

Cesare Goretti

Abstract Given a representation $\rho : \mathrm{GL}(s) \to \mathrm{GL}(n_2)$, there is an action of $\mathrm{GL}(n_1)\times\mathrm{GL}(s)$ on $X = \mathrm{P}(\mathrm{Hom}(\mathbb{C}^{n_1}, \mathbb{C}^{n_2}))$ via $(g_1, g_2)\cdot[f] = [\rho(g_2)\circ f\circ g_1^{-1}]$. We explicitly prove that there is a correspondence between semistable points $x \in X$ and semistable points in the Grassmanian of n_1-planes in $\mathbb{C}^{n_2}$. This is done by computing the Hilbert-Mumford weight in the two different scenarios.

1 Introduction

In mathematics, it is often needed to consider objects up to isomorphism. It is therefore useful to have a space that parameterizes isomorphism classes. However, in order to use practically the parameterizing space, it is also desirable that it has 'good enough' properties.

A good reference containing all the following theory is Chapter 1 in [3]. In the case where X is an affine variety and G is a reductive group acting on it, the categorical quotient $X/\!\!/G$ solves the problem. It is defined as the affine variety whose regular functions are the elements of $\mathbb{C}[X]^G$, the G-invariant regular functions on X. The categorical quotient parameterizes closed orbits in X.

On the other hand, if X is an algebraic variety, then to have a good quotient (compare Section 1.4 in [3]) one has to restrict the study to points with a property called semistability. This property is defined in terms of a line bundle L on X and a linearization $\overline{\sigma}$ on L of the action of G. A point $x \in X$ is said to be semistable if there exist a positive natural number m and a G-invariant section $s \in H^0(L^{\otimes m})$ such that $x \in X_s = \{x \in X \mid s(x) \neq 0\}$ and X_s is affine. Semistable points of X with respect to $\overline{\sigma}$ will be denoted by $X^{\mathrm{ss}}_{\overline{\sigma}}$. One can define the good quotient of $X^{\mathrm{ss}}_{\overline{\sigma}}$, denoted by $(X/\!\!/_{\overline{\sigma}}G, \pi_{\overline{\sigma}})$, by gluing local good quotients of the affine varieties X_s. The variety $X/\!\!/_{\overline{\sigma}}G$ is quasi-projective and, if X is projective and L is ample, then it

C. Goretti (✉)
Freie Universität, Berlin, Germany
e-mail: cesare.goretti@fu-berlin.de

M. R. Ebert et al. (eds.), *New Tools in Mathematical Analysis and Applications*, Research Perspectives, https://doi.org/10.1007/978-3-031-77050-0_3

is also projective. Moreover, if the line bundle L is ample, then the Hilbert-Mumford criterion gives a computational way to check if a point is semistable. We give a brief description of the Hilbert-Mumford criterion in Sect. 2.2.

If G is a reductive group acting on a vector space V via σ, then σ induces a natural action of G on $\mathbb{P}(V)$. Here, as it often happens in the literature, we have denoted the space of lines in the dual of V by $\mathbb{P}(V)$. On the other hand, we will denote by $\mathrm{P}(V)$ the space of lines in V. Moreover, there is a natural linearization of σ in the line bundle $\mathcal{O}_{\mathbb{P}(V)}(1)$ which we will still denote by σ.

Let's consider now the following situation. Let G and H be reductive groups, and $\sigma : G\times H \to GL(V)$ an action of $G\times H$ on a vector space V. We get an action σ_1 of G and an action σ_2 of H on V. It is well known that the good quotient of $\mathbb{P}(V)$, with respect to $(G \times H)$, can be constructed in two ways: either as $\mathbb{P}(V)/\!\!/_{\sigma}(G \times H)$ or as $Y/\!\!/_{\sigma_2}H$, where Y stands for $X/\!\!/_{\sigma_1}G$. In particular, $\mathbb{P}(V)^{\mathrm{ss}}_{\sigma} = \mathbb{P}(V)^{\mathrm{ss}}_{\sigma_1} \cap \pi_{\sigma_1}^{-1}(Y^{\mathrm{ss}}_{\sigma_2})$. This can be proven by looking at invariant sections, see Proposition 1.3.1 in [2] for the details.

We are interested in the application of the aforementioned theorem in the following setting. Let ρ be a fixed representation $\rho : \mathrm{GL}(s) \to \mathrm{GL}(n_2)$ homogeneous of degree a, i.e., $\rho(z\,\mathrm{Id}) = z^a\,\mathrm{Id}$. Then, there is a natural action $\tilde{\rho}$ of $\mathrm{GL}(n_1) \times \mathrm{GL}(s)$ on $X = \mathrm{P}(\mathrm{Hom}(\mathbb{C}^{n_1}, \mathbb{C}^{n_2}))$ via $(g_1, g_2) \cdot [f] = [\rho(g_2) \circ f \circ g_1^{-1}]$.

Define $Y = X/\!\!/\,\mathrm{GL}(n_1)$ as the Grassmanian of n_1 planes in $\mathbb{C}^{n_2}$. Then, we will explicitly prove that $X^{\mathrm{ss}}_{\tilde{\rho}} = X^{\mathrm{ss}} \cap \pi^{-1}(Y^{\mathrm{ss}})$, where X^{ss} denotes the semistable points of X with respect to the $\mathrm{GL}(n_1)$ action. This is done by checking the stability of a point via the Hilbert-Mumford criterion.

Explicit computations, via the Hilbert-Mumford criterion, for the stability of a point in $\mathbb{P}(V)$ with respect to an action of $G \times \mathbb{C}^*$, have been carried out by Okonek, Schmitt and Teleman in [2], Theorem 1.4.1. In all our statements we are comparing the two Hilbert-Mumford weights for a fixed one parameter subgroup λ_2 of $\mathrm{SL}(s)$. Hence, we are checking a $\mathrm{SL}(n_1) \times \mathbb{C}^*$-stability. However, there are two main differences between [2] and our work. First, our $\mathbb{C}^*$-action is induced by the $\mathrm{SL}(s)$-action. Therefore, we have already picked a line bundle and a linearization of the action, as explained at the beginning of Sect. 3. This means that we don't have anymore the freedom of multiplying it by a character of $\mathbb{C}^*$, as it happens in [2]. Secondly, as we state in Remark 4, in our proof we find estimates between the two Hilbert-Mumford weights which work also for negative weights, i.e., for objects that are unstable.

Our main motivation comes from two papers of A. Schmitt, [5] and [4]. There, he constructs a moduli space for general coherent systems. A general coherent system is a pair (E, Γ), where E is a rank s vector bundle on a curve and Γ is a vector subspace of $H^0(E_\rho)$, the vector bundle induced by E and ρ. The stability notion for general coherent systems is obtained by doing a linear combination, via a rational positive parameter δ, of the usual one for vector bundles and the Hilbert-Mumford weight of Γ. The two papers use different approaches which should correspond to see Γ either as a point in the Grassmanian or as a map $H^0(\mathbb{C}^{\oplus k}) \to H^0(E_\rho)$. In particular, for small values of δ we can find semistable general coherent systems (E, Γ) with Γ unstable. For this reason, we believe that our computations can be used to establish an explicit link between the two different notions of stability presented in the aforementioned papers.

2 Preliminaries

2.1 *One Parameter Subgroups of SL(n)*

Test objects for the Hilbert-Mumford criterion are one parameter subgroups of G. We will be mainly interested in the case in which G is $\mathrm{SL}(n)$. In such a case, one parameter subgroups are well understood.

Let $\lambda : \mathbb{C}^* \to \mathrm{SL}(n)$ be a one parameter subgroup. Then, there exists a basis $(w_1, \ldots, w_n)$ of $\mathbb{C}^n$ and integers $(\gamma_1, \ldots, \gamma_n)$, $\gamma_1 \leq \gamma_2 \leq \ldots \leq \gamma_n$, such that

$$\lambda(z)w_i = z^{\gamma_i} w_i, \ i = 1, \ldots, n,$$

and

$$\sum_{i=1}^{n} \gamma_i = 0.$$

Given a one parameter subgroup λ of $\mathrm{SL}(s)$, then $\rho(\lambda)$ is a one parameter subgroup of $\mathrm{SL}(n_2)$. This is clear from the fact that $\rho : \mathrm{SL}(s) \to \mathrm{SL}(n_2)$. Indeed, if $\rho(\mathrm{SL}(s)) \not\subseteq \mathrm{SL}(n_2)$, then $\det \circ \rho$ would give a non-trivial character of $\mathrm{SL}(s)$ but such character does not exist.

2.2 *The Hilbert-Mumford Criterion*

The Hilbert-Mumford criterion is a powerfull tool to check the (semi)stability of a point x in a projective variety, with respect to a linearized action in an ample line bundle.

Let X be a projective variety on which a reductive affine algebraic group G acts. Let L be an ample line bundle in which we have linearized the action. Given a point $x \in X$ and a one parameter subgroup λ of G, we define x_∞ as

$$\lim_{z\to\infty} \lambda(z) \cdot x =: x_\infty.$$

The point x_∞ is a fixed point for the $\mathbb{C}^*$-action induced by λ. This means that the linearized action fixes L_{x_∞}. Therefore, there exists a $\gamma \in \mathbb{Z}$ such that $\lambda(z)$ acts on L_{x_∞} via multiplication by z^γ.

The Hilbert-Mumford weight of x with respect to the one parameter subgroup λ is defined as $-\gamma$ and it is denoted by $\mu(x, \lambda)$.

The Hilbert-Mumford criterion relates the (semi)stability of a point with the Hilbert-Mumford weight:

Theorem 1 *Let X, L, G be as before. Then, a point $x \in X$ is (semi)stable if and only if for every nontrivial one parameter subgroup λ of G the following inequality holds:*

$$\mu(x, \lambda)(\geq)0.$$

Here, the symbol $(\geq)$ means that the weak inequality "$\geq$" is requested for semistable, while a strict inequality "$>$" is requested for stable.

Given a vector space V on which G acts, Exercise 1.5.1.1 in [3] gives an explicit computation for the Hilbert-Mumford weights of points in $\mathbb{P}(V)$ with respect to the linearized action on $\mathcal{O}_{\mathbb{P}(V)}(1)$.

Note that in our setting we have an action of G on V and we would like to study stability of points in $\mathrm{P}(V)$, the space of lines in V. If one wants to phrase it in the more familiar context of the literature, it is enough to note that there is an isomorphism between $\mathrm{GL}(V)$ and $\mathrm{GL}(V^\vee)$ as well as between $\mathrm{P}(V)$ and $\mathbb{P}(V^\vee)$.

Therefore, we can rephrase the results of Exercise 1.5.1.1 as follows. Given a one parameter subgroup λ of G, pick a basis $(v_1, \ldots, v_n)$ of V on which the action diagonalizes: $\lambda(z) \cdot v_i = z^{\gamma_i} v_i, \gamma_1 \leq \ldots \leq \gamma_n$. Given $[v] \in \mathrm{P}(V)$, let $v = \sum_{i=1}^n a_i v_i$ be any lift in V. Then, the Hilbert-Mumford weight of $[v]$ with respect to λ is expressed by the following:

$$\mu([v], \lambda) = \max\{\gamma_i \mid a_i \neq 0\}.$$

3 Computations

As stated in the last part of the introduction, we are interested in the case in which $X = \mathrm{P}(\mathrm{Hom}(\mathbb{C}^{n_1}, \mathbb{C}^{n_2}))$, ρ is a fixed homogeneous representation of degree a, $\mathrm{GL}(n_1) \times \mathrm{GL}(s)$ acts on X via $(g_1, g_2) \cdot [f] = [\rho(g_2) \circ f \circ g_1^{-1}]$ and $Y = X /\!\!/ \mathrm{GL}(n_1)$ is defined as the Grassmanian of n_1 planes in $\mathbb{C}^{n_2}$.

Since both the centers of $\mathrm{GL}(n_1)$ and of $\mathrm{GL}(s)$ act trivially on X, we have to carefully pick the linearized action. It has to be done in such a way that all the one parameter subgroups of these centers have weight zero, otherwise we will have no semistable points. Once we have done this, we can restrict our study to the subgroups $\mathrm{SL}(n_1)$ and $\mathrm{SL}(s)$.

Let σ be the standard linearized action on $\mathcal{O}_X(1)$. It induces a linearization σ^{k_1} on $\mathcal{O}_X(k_1)$ via the k_1th symmetric power. We modify this linearization using the character $k_2\det$ of $\mathrm{GL}(n_1)$ and the character $k_3(\det \circ \rho)$ of $\mathrm{GL}(s)$ for some good choice of parameters k_1, k_2, k_3. Here, a good choice means, as we mentioned earlier, that one parameter subgroups of the center must have weight zero. By checking this condition first for one parameter subgroups of the center of $\mathrm{GL}(n_1)$ and then of $\mathrm{GL}(s)$ one finds two relations:

$$\begin{aligned} -k_1 + n_1 k_2 &= 0, \\ a k_1 + a n_2 k_3 &= 0. \end{aligned} \tag{1}$$

In particular, we can choose $k_1 = n_1 n_2, k_2 = n_2, k_3 = -n_1$.

Similarly, to study semistability on Y we will use the Plücker embedding, linearize the action on $\mathcal{O}_Y(k_1')$ and multiply it by the character $k_3'(\det \circ \rho)$. By imposing again that the one parameter subgroup of the center of $\mathrm{GL}(s)$ have weight zero one finds:

$$n_1 a k_1' + n_2 a k_3' = 0. \tag{2}$$

As it should be, our previous choice of linearization induces a choice on the Grassmanians given by $k_1' = n_2$ and $k_3' = -n_1$.

With this choice of $k_1, k_2, k_3, k_1', k_3'$ we can restrict our attention only on the action of $\mathrm{SL}(n_1) \times \mathrm{SL}(s)$ and $\mathrm{SL}(s)$.

Note that the characters we have used to modify the standard linearization, once restricted to these groups, are trivial. For this reason, it is enough to study the Hilbert-Mumford weights on $\mathcal{O}_X(n_1 n_2)$ and $\mathcal{O}_Y(n_2)$ with respect to the standard linearization. However, these are nothing more than multiples of the usual Hilbert-Mumford weights on $\mathcal{O}_X(1)$ and $\mathcal{O}_Y(1)$. Hence, in what follows we will study the linearized $(\mathrm{SL}(n_1) \times \mathrm{SL}(s))$- and $\mathrm{SL}(s)$-actions on these last two bundles, as it will lead to the same semistable points.

Remark 1 Note that Y is an orbit space. This is because $\mathrm{SL}(n_1)$-stability and semistability coincide. A point $x \in X$ is $\mathrm{SL}(n_1)$-(semi)stable if and only if $x = [f]$ with f injective, compare Lemma 2.4.4 in [2]. It is easy to see that a one parameter subgroup which has a negative weight on the kernel of f and a positive one on the complementary destabilizes $[f]$. On the other hand, if f is injective, then for every nontrivial one parameter subgroup λ_1, f will be supported on all the eigenvectors. In particular, it will be supported on both a positive and a negative one, from which stability follows. For this reason, given a point $y \in Y$, we will say that $[f] \in X$ is in the orbit y if and only if f is injective and $[f]$ is mapped to y, i.e., the space y, as a point in the Grassmanian, is the image of the function f.

A first guess of a statement with the Hilbert-Mumford criterion that would explain Proposition 1.3.1 in [2] in our setting could be the following: "Let λ_2 be a one parameter subgroup of $\mathrm{SL}(s)$ and y a point in Y, then $\mu(y, \lambda_2) \geq 0$ if and only if $\mu([f], \lambda_1, \lambda_2) \geq 0$ for every $[f]$ in the orbit y and every λ_1 a one parameter subgroup of $\mathrm{SL}(n_1)$".

Unfortunately this statement is not true; see Remark 5. However, it is not that far from what really happens which is described by the following theorem and the next lemmas:

Theorem 2 *Given a point $y \in Y$ and a one parameter subgroup λ_2 of* SL(s)*, the inequality $\mu([f], (\lambda_1, n\lambda_2)) \geq 0$ holds for every $[f]$ in the orbit y, for every one parameter subgroup λ_1 of* SL(n_1) *and for every $n \in \mathbb{Z}_{>0}$ if and only if $\mu(y, n\lambda_2) \geq 0$ is satisfied for every $n \in \mathbb{Z}_{>0}$.*

Proof Apply Lemma 1 and Corollary 1. □

Remark 2 Remember that the Hilbert-Mumford weight satisfies $\mu(x, n\lambda) = n\mu(x, \lambda)$ for every $n \in \mathbb{Z}_{>0}$. Therefore, the factor n in Theorem 2, which comes from the factor n_1 in Corollary 1, can be regarded as a denominator for the one parameter subgroup λ_1. For this reason, one way to explain Theorem 2 is to say that, it is not enough to study $\mu([f], (\lambda_1, \lambda_2))$ for every one parameter subgroup λ_1 of SL(n_1), but we also have to consider rational one parameter subgroups.

Lemma 1 *Given a point $y \in Y$ and a one parameter subgroup λ_2 of* SL(s)*, if the inequality $\mu(y, \lambda_2) \geq 0$ holds, then $\mu([f], (\lambda_1, \lambda_2)) \geq 0$ for every $[f]$ in the orbit y and for every one parameter subgroup λ_1 of* SL(n_1).

Proof Let's start by recalling the value of $\mu(y, \lambda_2)$ computed in [1, Section 4]. It can be reformulated in the following way. Let $[f]$ be any element in the orbit y, pick any basis of $\mathbb{C}^{n_1}$ and a basis of eigenvectors $\{v_i\}_{i=1,\ldots,n_2}$ in $\mathbb{C}^{n_2}$ for the one parameter subgroup $\rho(\lambda_2)$. If A is the matrix associated with f in these bases, then

$$\mu(y, \lambda_2) = \max_{|I|=n_1} \left\{ \sum_{i \in I} \gamma_i | \det(A_I) \neq 0 \right\}.$$

In this last expression, A_I denotes the n_1-dimensional minor of the matrix A obtained by choosing the rows associated with the set of indices I. As before, γ_i is the weight associated with v_i. It can also be understood as follows: the columns of A can be used as a basis of the image of f and hence, by taking the exterior product, we obtain the point $y \in \bigwedge^{n_1} \mathbb{C}^{n_2}$. We can decompose each column vector using the basis $\{v_i\}_{i=1,\ldots,n_2}$ and obtain coordinates with respect to the basis of eigenvectors of $\bigwedge^{n_1} \mathbb{C}^{n_2}$ of the form $v_I = v_{i_1} \wedge \cdots \wedge v_{i_{n_1}}$. The weight associated with v_I is clearly $\sum_{i \in I} \gamma_i$ and the coordinate of y in the direction v_I is non zero if and only if $\det(A_I) \neq 0$. Note that $\det(A_I) \neq 0$ implies that there exists at least one way of pairing each row i_k in I with a column $l_k \in \{1, \ldots, n_1\}$ such that $a_{i_k, l_k} \neq 0$ and $l_{k_1} \neq l_{k_2}$ if $k_1 \neq k_2$.

For what concerns the SL$(n_1) \times$ SL(s) (semi)stability of points $x \in X$ there is not much to say. From a one parameter subgroup (λ_1, λ_2) of SL$(n_1) \times$ SL(s) we get bases of eigenvectors: $\{w_j\}_{j=1,\ldots,n_1}$ of $\mathbb{C}^{n_1}$ with weights δ_j and $\{v_i\}_{i=1,\ldots,n_2}$ of $\mathbb{C}^{n_2}$ with weights γ_i. Therefore, we obtain a basis of eigenvectors $v_i \otimes w_j^\vee$ with weight $\gamma_i - \delta_j$. If A is the matrix that expresses the map f with respect to this basis we obtain $\mu([f], (\lambda_1, \lambda_2)) = \max\{\gamma_i - \delta_j | a_{i,j} \neq 0\}$.

We will prove now that given $y \in Y$ and λ_2 a one parameter subgroup of $\mathrm{SL}(s)$ such that $\mu(y, \lambda_2) \geq 0$ then $\mu([f], (\lambda_1, \lambda_2)) \geq 0$ for every $[f]$ in the orbit y and every one parameter subgroup λ_1 of $\mathrm{SL}(n_1)$.

We fix λ_1 a one parameter subgroup of $\mathrm{SL}(n_1)$ and we will prove that, under the hypothesis, $\mu([f], (\lambda_1, \lambda_2)) \geq 0$ holds.

Let, as before, $\{w_j\}_{j=1,\dots,n_1}$ be a basis of eigenvectors for λ_1 and $\{v_i\}_{i=1,\dots,n_2}$ a basis of eigenvector for $\rho(\lambda_2)$. We denote with A the matrix that expresses f in these bases. From the inequality $\mu(y, \lambda_2) \geq 0$ we know that there is a set of indices I such that $\det(A_I) \neq 0$ and $\sum_{i\in I} \gamma_i \geq 0$. From what we have said before, we can associate every row $i_k \in I$ with a different column l_k such that $a_{i_k,l_k} \neq 0$. By hypothesis and the definition of $\mu(y, \lambda_2)$ we have

$$\mu(y, \lambda_2) = \sum_{i\in I} \gamma_i \geq 0. \tag{3}$$

Now, since λ_1 is a one parameter subgroup of $\mathrm{SL}(n_1)$, the fundamental identity $\sum_{k=1}^{n_1} \delta_{l_k} = 0$ holds. Therefore we can rewrite (3) as

$$\sum_{i\in I} \gamma_i = \sum_{k=1}^{n_1} \gamma_{i_k} - \sum_{k=1}^{n_1} \delta_{l_k} = \sum_{k=1}^{n_1} (\gamma_{i_k} - \delta_{l_k}) \geq 0. \tag{4}$$

In particular, there will be an integer k such that $\gamma_{i_k} - \delta_{l_k} \geq 0$ and $a_{i_k,l_k} \neq 0$. From that, we deduce the following sequence of inequalities:

$$\mu([f], (\lambda_1, \lambda_2)) \geq \gamma_{i_k} - \delta_{l_k} \geq 0,$$

and this yields the result. □

Remark 3 From Lemma 1 we can deduce two additional observations. First, if $\mu(y, \lambda_2)$ is strictly greater than 0, then so is $\mu([f], (\lambda_1, \lambda_2))$. This, together with Lemma 2, gives a result analogous to Theorem 2 for stability. Secondly, we remark that there exists a lower bound for $\mu([f], (\lambda_1, \lambda_2))$. Indeed, from the two inequalities:

$$\mu([f], (\lambda_1, \lambda_2)) \geq \max_k \{\gamma_{i_k} - \delta_{l_k}\}$$

and

$$\mu(y, \lambda_2) = \sum_{i\in I} \gamma_i = \sum_{k=1}^{n_1} (\gamma_{i_k} - \delta_{l_k}) \leq n_1 \max_k \{\gamma_{i_k} - \delta_{l_k}\},$$

we obtain

$$\mu([f], (\lambda_1, \lambda_2)) \geq \frac{\mu(y, \lambda_2)}{n_1}. \tag{5}$$

To deal with the other implication of Theorem 2, we will prove a stronger result expressed by the following lemma.

Lemma 2 *Let y be a point in Y, $[f] \in X$ an element in the orbit y and λ_2 a one parameter subgroup of* $\mathrm{SL}(s)$. *Then, there exists a one parameter subgroup λ_1 of* $\mathrm{SL}(n_1)$ *such that $\mu([f], (\lambda_1, n_1\lambda_2)) = \mu(y, \lambda_2)$.*

The following corollary is an immediate consequence.

Corollary 1 *Let y be a point in Y, $[f] \in X$ an element in the orbit y and λ_2 a one parameter subgroup of* $\mathrm{SL}(s)$. *If $\mu([f], (\lambda_1, n_1\lambda_2)) \geq 0$ for every one parameter subgroup λ_1 of* $\mathrm{SL}(n_1)$, *then $\mu(y, \lambda_2) \geq 0$.*

Remark 4 Note that Inequality (5) in Remark 3 and Lemma 2 hold even if $\mu(y, \lambda_2)$ is negative.

Proof of Lemma 2 To prove the lemma we will use that for every $g \in \mathrm{SL}(n_1)$ the following identity holds: $\mu(g \cdot [f], (\lambda_1, \lambda_2)) = \mu([f], (g^{-1}\lambda_1 g, \lambda_2))$.

We are assuming that f is injective, therefore $\mu(y, \lambda_2)$ is well defined. Let I be a set of indices that realizes $\mu(y, \lambda_2)$ and A the matrix that expresses f with respect to the standard basis of $\mathbb{C}^{n_1}$ and the basis $\{v_i\}_{i=1\dots,n_2}$ of eigenvectors of $\mathbb{C}^{n_2}$ for the one parameter subgroup $\rho(\lambda_2)$. Each vector v_i is associated with the weight γ_i. Up to an element $g \in \mathrm{SL}(n_1)$, we can pretend that the minor A_I is the identity matrix. In particular, we have $a_{i_k,k} \neq 0$ for every k and $a_{i_l,k} = 0$, for every l different from k.

We fix now the one parameter subgroup λ_1 of $\mathrm{SL}(n_1)$ which has the standard basis as eigenvectors and weights δ_j defined by

$$\begin{aligned} \delta_j &= n_1\gamma_{i_j} - \mu(y, \lambda_2) \qquad \forall j \neq n_1, \\ \delta_{n_1} &= -\, n_1 \sum_{k=1}^{n_1-1} \gamma_{i_k} + (n_1 - 1)\mu(y, \lambda_2). \end{aligned} \tag{6}$$

We prove that, for this choice of λ_1, the value $\mu([f], (\lambda_1, n_1\lambda_2))$ is indeed $\mu(y, \lambda_2) = \sum_{k=1}^{n_1} \gamma_{i_k}$.

To prove that, we will show that $\{n_1\gamma_i - \delta_j \mid a_{i,j} \neq 0\}$ is bounded by $\mu(y, \lambda_2)$.

Note that, $\mu(y, \lambda_2)$ is exactly the weight that acts in position (i_k, k) for every k from 1 to n_1. This, together with our assumption on A_I ensure that we only have to check the weights in position (i, j) with $i \notin I$.

The proof goes by contradiction: let's pretend there exists an index $l \notin I$ such that the matrix A has a non zero entry $a_{l,j}$ and $\mu([f], (\lambda_1, n_1\lambda_2)) = n_1\gamma_l - \delta_j > \mu(y, \lambda_2)$. Therefore, we get $n_1\gamma_l - \delta_j > n_1\gamma_{i_j} - \delta_j$ and we deduce $\gamma_l > \gamma_{i_j}$.

Moreover, the minor of the matrix A obtained by choosing the l-th row, instead of the i_j-th one, has non zero determinant. This is clear since the row i_j is in the span of this new set of rows. However, this contradicts the maximality of $\sum_{i \in I} \gamma_i$. □

It is interesting to note that $\mu([f], \lambda_1, \lambda_2) \geq 0$ for every one parameter subgroup λ_1 of $\mathrm{SL}(n_1)$ is not a sufficient condition for $\mu(y, \lambda_2) \geq 0$. The following remark gives a counterexample.

Remark 5 Consider the following situation of planes in $\mathbb{C}^3$:

- $n_1 = 2$;
- $n_2 = 3$;
- $s = 3$;
- $\rho = \mathrm{Id}$;
- $y = \mathrm{Span}\{e_1, e_2\}$;
- $f = \begin{pmatrix} 1 & 0 \\ 0 & 1 \\ 0 & 0 \end{pmatrix}$;
- $\lambda_2(z) = \begin{pmatrix} z^{-1} & 0 & 0 \\ 0 & 1 & 0 \\ 0 & 0 & z \end{pmatrix}$.

It is clear that $\mu(y, \lambda_2) = -1 + 0 = -1 < 0$.

We will prove now that $\mu([f], (\lambda_1, \lambda_2)) \geq 0$ for every one parameter subgroup λ_1 of SL(2). To do so we will restrict our study to one parameter subgroups in the maximal torus of diagonal matrices and we will do that for any $\tilde{f}$ in the orbit of f.

The map $\tilde{f}$ will have the shape

$$\tilde{f} = \begin{pmatrix} a_{1,1} & a_{1,2} \\ a_{2,1} & a_{2,2} \\ 0 & 0 \end{pmatrix}$$

with

$$a_{1,1}a_{2,2} - a_{2,1}a_{1,2} \neq 0. \tag{7}$$

Let λ_1 be a one parameter subgroup of SL(2) of the shape

$$\lambda_1(z) = \begin{pmatrix} z^{\gamma} & 0 \\ 0 & z^{-\gamma} \end{pmatrix}.$$

We can now compute $(\lambda_1, \lambda_2)(z) \cdot \tilde{f}$, it is equal to

$$\lambda_2(z)\tilde{f}\lambda_1^{-1}(z) = \begin{pmatrix} z^{-\gamma-1}a_{1,1} & z^{\gamma-1}a_{1,2} \\ z^{-\gamma}a_{2,1} & z^{\gamma}a_{2,2} \\ 0 & 0 \end{pmatrix}.$$

We will prove that there are no values of γ and $\tilde{f}$ such that $\mu([\tilde{f}], (\lambda_1, \lambda_2)) < 0$. Hence, we try to find a couple $(\gamma, \tilde{f})$ for which $\mu([\tilde{f}], (\lambda_1, \lambda_2)) < 0$ and we show that this leads to a contradiction. Since at least one between γ and $-\gamma$ is surely greater or equal than zero, we must have that the corresponding element of the matrix is zero.

Let's first consider the case $\gamma \leq 0$ and $a_{2,1} = 0$. From Eq. (7), we get that both $a_{1,1}$ and $a_{2,2}$ are not zero. In particular, this tells us that, to have $\mu([\tilde{f}], (\lambda_1, \lambda_2)) < 0$, we need

$$\begin{cases} -\gamma - 1 < 0 \\ \gamma < 0 \end{cases} .$$

However, this system doesn't have any solution in the integers. Similarly, if $\gamma > 0$ and $a_{2,2} = 0$, we get that the following system is the condition to have $\mu([\tilde{f}], (\lambda_1, \lambda_2)) < 0$. Once again, it does not have integer solutions:

$$\begin{cases} \gamma - 1 < 0 \\ \gamma > 0 \end{cases} . \tag{8}$$

Therefore, we have proved that $\mu([f], (\lambda_1, \lambda_2)) \geq 0$ for every one parameter subgroup λ_1 of $\mathrm{SL}(n_1)$. Note that, as it should be by Corollary 1, we can actually find a destabilizing one parameter subgroup λ_1 when we look at $\mu([f], (\lambda_1, 2\lambda_2))$.

In that case, the system of Eq. (8) will have the shape:

$$\begin{cases} \gamma - 2 < 0 \\ \gamma > 0 \end{cases} .$$

It is clear that this time the system has an integer solution. Hence, $\gamma = 1$ gives us a destabilizing one parameter subgroup for which $\mu([f], (\lambda_1, 2\lambda_2)) < 0$.

It is easy to generalize this example for higher values of n_1 and n_2.

In particular, this shows that $\mu([f], (\lambda_1, \lambda_2)) \geq 0$ for every one parameter subgroup λ_1 of $\mathrm{SL}(n_1)$ does not imply that $\mu(y, \lambda_2) \geq 0$.

Acknowledgments I would like to thank A. Schmitt for the invitation to participate as a speaker at this conference. I also thank an anonymous reviewer who suggested including Remark 2. Funded by the Deutsche Forschungsgemeinschaft (DFG, German Research Foundation) under Germany's Excellence Strategy—The Berlin Mathematics Research Center MATH+ (EXC-2046/1, project ID: 390685689). Founded by the Deutsche Forschungsgemeinschaft (DFG, German Research Foundation) project: New developments in the moduli theory of decorated sheaves (Project number: 471183228).

References

1. Mumford, D., Fogarty, J., Kirwan, F.: Geometric Invariant Theory. Ergebnisse der Mathematik und ihrer Grenzgebiete (2) [Results in Mathematics and Related Areas (2)], vol. 34, 3rd edn. Springer, Berlin (1994)
2. Okonek, C., Schmitt, A., Teleman, A.: Master spaces for stable pairs. Topology **38**(1), 117–139 (1999). https://doi.org/10.1016/S0040-9383(98)00006-8
3. Schmitt, A.H.W.: Geometric Invariant Theory and Decorated Principal Bundles. Zurich Lectures in Advanced Mathematics. European Mathematical Society (EMS), Zürich (2008). https://doi.org/10.4171/065
4. Schmitt, A.H.W.: Notes on coherent systems. Rev. Mat. Teor. Apl. **28**(1), 1–38 (2021). https://doi.org/10.15517/rmta.v28i1.42154
5. Schmitt, A.: A general notion of coherent systems. Rev. Mat. Iberoam. **38**(6), 1783–1826 (2022). https://doi.org/10.4171/RMI/1314

Rational Homotopy and Hodge Theory of Moduli Stacks of Principal G-Bundles

Pedro L. del Angel Rodriguez and Frank Neumann

Abstract For a semisimple complex algebraic group G we determine the rational cohomology and the Hodge-Tate structure of the moduli stack $\mathcal{B}un_{G,X}$ of principal G-bundles over a connected smooth complex projective variety X of special type using the homotopy theory of the underlying topological stack.

MSC 2020 Subject Classification: Primary 14D20, 14A20 Secondary 14C30, 14H60

1 Introduction

The moduli stack of principal G-bundles over a complex algebraic curve plays a fundamental role in algebraic geometry and mathematical physics. In their seminal paper [3] of 1983, Atiyah and Bott, applying methods from Morse theory to the Yang-Mills functional over a compact Riemann surface (or a projective smooth algebraic curve) C, compute the cohomology of the moduli space of principal $U(n)$-bundles over C and outlined the essential steps towards a similar computation for general semisimple Lie groups. Later, in 2006, Dhillon [8] computed the Betti numbers as well as the Hodge numbers of the moduli space of stable vector bundles of rank r and degree d over a smooth projective curve C and, in 2008, Arapura and Dhillon [2] went a step further to determine the relative André motive as well as the rational cohomology of the moduli stack of G-bundles over the universal curve.

Inspired by these previous works, we will show how, for a semisimple complex algebraic group G, we can determine the Hodge-Tate structure of the moduli stack

P. L. del Angel Rodriguez
Centro de Investigación en Matemáticas (CIMAT), Guanajuato, México
e-mail: luis@cimat.mx

F. Neumann (✉)
Dipartimento di Matematica 'Felice Casorati', Università di Pavia via Ferrata, Pavia, Italy
e-mail: frank.neumann@unipv.it

M. R. Ebert et al. (eds.), *New Tools in Mathematical Analysis and Applications*, Research Perspectives, https://doi.org/10.1007/978-3-031-77050-0_4

$\mathcal{B}un_{G,X}$ of principal G-bundles over a connected smooth complex projective variety X of a special type depending on the underlying topological CW-structure.

A key ingredient of our work will be the determination of the rational homotopy type of the moduli stack $\mathcal{B}un_{G,X}$ and its underlying topological mapping stack using the construction of homotopy types for topological stacks as developed by Noohi [17–19] and the classical work by Thom [23] on the homotopy and cohomology of mapping spaces.

2 Moduli Stacks of Principal Bundles

In this section we will briefly recall some aspects of the general theory of algebraic principal bundles over algebraic varieties and their moduli stacks. More details can be found, for example, in [6, 10, 12, 16] and [20].

Definition 1 Let S be a scheme and G a group scheme over S. For every scheme X over S, let $G_X = G \times_S X$ be the associated group scheme over X. A *principal G-bundle over X* is a scheme P over X equipped with an action

$$G_X \times_X P \cong G \times_S P \to P$$

of G_X in the category of schemes over X, which is locally trivial in the following sense: there exists a faithful flat covering morphism $U \to X$ and a G_X-equivariant isomorphism

$$U \times_X P \cong U \times_X G_X \cong U \times_S G.$$

Given a morphism of schemes $f : X' \to X$ over X and a principal G-bundle P over X, the fibre product $X' \times_X P$ exists and is a principal G-bundle over X'. We call $X' \times_X P$ the *pullback of P along f*, and denote it by f^*P.

If G is a smooth group scheme over S, any principal G-bundle P over a scheme X over the base S is also smooth as a scheme over X and can actually be trivialised by an étale covering morphism $U \to X$.

Let us recall now the general definition of the moduli stack of principal bundles over an algebraic stack as a Hom-stack (compare [11, 14]).

Definition 2 Let G be a group scheme over a base scheme S and $\mathcal{X}$ be an algebraic stack over the category Sch/S of schemes over S. The moduli stack of principal G-bundles over $\mathcal{X}$ is defined as the Hom-stack

$$\mathcal{B}un_{G,\mathcal{X}} = \mathcal{H}om_S(\mathcal{X}, \mathcal{B}G)$$

where $\mathcal{B}G$ is the classifying stack of G.

In the particular situation that $\mathcal{X}$ is a projective flat scheme X over a Noetherian base scheme S and G is a closed subgroup scheme of GL_n and the fppf-quotient scheme GL_n/G is quasi-projective over S it follows that the moduli stack $\mathcal{B}un_{G,X}$ is actually an algebraic stack which is locally of finite type (see [11]).

In this article we will always work over the field $\mathbb{C}$ of complex numbers and our base scheme will be $S = \mathrm{Spec}(\mathbb{C})$. We also assume that X is a connected smooth complex projective variety and G will be a semisimple complex affine algebraic group. The moduli stack $\mathcal{B}un_{G,X}$ can then also be described via its groupoid of sections as follows: For a given scheme U over $\mathrm{Spec}(\mathbb{C})$, the groupoid $\mathcal{B}un_{G,X}(U)$ of U-valued points of $\mathcal{B}un_{G,X}$ is the groupoid of principal G-bundles over $X \times_{\mathrm{Spec}(\mathbb{C})} U$ and their isomorphisms. Then we have the following (see [4, 5, 14])

Theorem 3 *The moduli stack $\mathcal{B}un_{G,X}$ is a smooth Artin algebraic stack, which is locally of finite type.* □

3 Homotopy Types of Algebraic Stacks

We recollect here some facts about homotopy types of algebraic and topological stacks following Noohi [17–19] (compare also [9]).

In order to analyze homotopy types of complex algebraic stacks we need to understand homotopy types in the underlying topological situation.

Let $\mathcal{X}$ be an algebraic stack over the catgegory **Sch**/$\mathbb{C}$ of schemes over $\mathbb{C}$. If $\mathcal{X}$ is locally of finite type, then there exists an associated topological stack $\mathcal{X}^{top}$ over the category **Top** of topological spaces.

A *topological stack* is basically a stack $\mathcal{X}$ over **Top** which admits an atlas, i. e. a representable morphism $p : X \to \mathcal{X}$ from a topological space X having local sections (see [17] for the general theory of stacks over the site **Top**).

Noohi [17, Prop. 20.2] constructs a well-defined topologification functor $(-)^{top}$: $\mathfrak{AlgSt}/\mathbb{C} \to \mathfrak{TopSt}$ from the 2-category of algebraic stacks over $\mathbb{C}$ to the 2-category of topological stacks. It sends Artin algebraic stacks $\mathcal{X}$, i.e. algebraic stacks with a smooth atlas to topological stacks $\mathcal{X}^{top}$ whose atlas is a local topological fibration. These topological stacks can be considered as the analogs of Artin algebraic stacks in the topological framework. This topologification functor also factors naturally through the 2-category of complex analytic stacks $\mathfrak{AnSt}$ and it basically extends the complex points functor from the category **Sch**/$\mathbb{C}$ of schemes over $\mathbb{C}$ to the category **AnSp** of complex analytic spaces. Here a *complex analytic stack* $\mathcal{X}$ is a stack over the site **AnSp** of complex analytic spaces which admits an atlas, i. e. a complex analytic smooth surjection $X \to \mathcal{X}$ from a complex analytic space X (see also [9]).

Let us now define the homotopy type of a topological stack (see [17], [19]):

Definition 4 A representable morphism $f : \mathcal{X} \to \mathcal{Y}$ between topological stacks is called a *universal weak equivalence* if for any morphism $Y \to \mathcal{Y}$ from a topological space Y, the left vertical morphism in the diagram

$$\begin{array}{ccc} Y \times_{\mathcal{Y}} \mathcal{X} & \longrightarrow & \mathcal{X} \\ \downarrow & & \downarrow f \\ Y & \longrightarrow & \mathcal{Y} \end{array}$$

is a weak homotopy equivalence of topological spaces. A *homotopy type* for a topological stack $\mathcal{X}$ is a CW-complex $\{\mathcal{X}\}$ together with a morphism $\eta : \{\mathcal{X}\} \to \mathcal{X}$ which is a universal weak equivalence.

Every universal weak equivalence is automatically a representable morphism. We could have used also the simplicial space associated to an atlas to define the homotopy type of a topological stack (see [9]). Let $\mathcal{X}$ be a topological stack with atlas $X \to \mathcal{X}$. The associated simplicial topological space $X_\bullet$ is given by the iterated fiber products of the atlas with $X_n = X \times_{\mathcal{X}} \ldots \times_{\mathcal{X}} X$ having $(n+1)$ factors. $X_\bullet$ is simply the nerve of the topological groupoid $[X \times_{\mathcal{X}} X \rightrightarrows X]$ associated to the topological stack $\mathcal{X}$. We have the following relation between the homotopy type of a topological stack $\mathcal{X}$ and the homotopy type of the (thick) geometric realization of $X_\bullet$ (see [17]):

Proposition 5 *Let $\mathcal{X}$ be a topological stack and $X \to \mathcal{X}$ be an atlas. There is a universal weak equivalence $||X_\bullet|| \to \mathcal{X}$, where $||X_\bullet||$ is the (thick) geometric realisation of the simplicial topological space $X_\bullet$.*

The topological space $||X_\bullet||$ is in general *not* a CW-complex, but the (thick) geometric realization of the singular simplicial set $Sing_\bullet(||X_\bullet||)$ is always a CW-complex. In fact, the evaluation map $|Sing_\bullet(||X_\bullet||)| \to ||X_\bullet||$ is then a Serre fibration and a weak homotopy equivalence. Therefore the composition $|Sing_\bullet(||X_\bullet||)| \to \mathcal{X}$ of the evaluation map and the universal weak equivalence above is again a universal weak equivalence and $|Sing_\bullet(||X_\bullet||)|$ is a homotopy type for $\mathcal{X}$ which proves:

Proposition 6 *Any topological stack $\mathcal{X}$ has a homotopy type $\{\mathcal{X}\}$ which is given as a CW-complex.*

Any CW-complex X can be considered as a topological stack and therefore itself has a homotopy type $\{X\}$ which is simply given as the identity map $X \to X$, which is a universal weak equivalence and it follows immediately that $\{X\} \simeq X$.

For the action of a topological group G on a topological space X we have the *quotient stack* $[X/G]$ defined as follows: For every topological space U the groupoid $[X/G](U)$ is the groupoid of all triples (E, π, μ), where $\pi : E \to U$ is a principal G-bundle and $\mu : E \to X$ is a G-equivariant map and the isomorphisms are simply G-bundle isomorphism compatible with the respective equivariant maps.

It follows that $[X/G]$ is a topological stack. An atlas $q : X \to [X/G]$ is given by the triple $(G \times X, pr_X, \rho)$, where $pr_X : G \times X \to X$ is the trivial bundle and

$\rho : G \times X \to X$ the action. This data defines an object in $[X/G](X)$ and via representability a morphism $q : X \to [X/G]$, which in fact is a principal G-bundle.

It is easy to see that the homotopy type $\{[X/G]\}$ of the quotient stack $[X/G]$ is given by the Borel construction $EG \times_G X$, where the universal weak equivalence is the morphism

$$\eta : EG \times_G X \to [X/G]$$

given by the projection map $EG \times X \to X$, which is G-equivariant and the quotient map $EG \times X \to EG \times_G X$ which is a principal G-bundle. In the special case that the action is free the quotient stack $[X/G]$ is isomorphic to the quotient space X/G. If $X = *$ is a point with a trivial action of a topological group G, then it follows that the homotopy type of the *classifying stack* $\mathcal{B}G = [*/G]$ is homotopy equivalent to the classifying space BG, i. e. $\{\mathcal{B}G\} \simeq BG$.

In order to study homotopy types of moduli stacks of principal bundles over complex algebraic varieties we need to understand the homotopy type of mapping stacks. Giving two topological stacks $\mathcal{X}$ and $\mathcal{Y}$ we have the associated mapping stack $\mathcal{M}ap(\mathcal{X}, \mathcal{Y})$, which is basically the topological version of the algebraic Hom-stack given by the groupoid of section as

$$\mathcal{M}ap(\mathcal{X}, \mathcal{Y})(U) = \mathrm{Hom}(U \times \mathcal{X}, \mathcal{Y})$$

with $\mathrm{Hom}(U \times \mathcal{X}, \mathcal{Y})$ the respective groupoid of stack morphisms (see [7, 18]). It turns out that if X is a compact topological space, then the mapping stack $\mathcal{M}ap(X, \mathcal{Y})$ is actually a topological stack [7, Prop. 5.1]. In this particular case, the homotopy type of the mapping stack $\mathcal{M}ap(X, \mathcal{Y})$ is represented by the mapping space $\mathrm{M}ap(X, Y)$, where Y represents the homotopy type of the topological stack $\mathcal{Y}$ (see [18, Cor. 6.5]).

We can now define the (classical) homotopy type of an algebraic stack over the complex numbers via its underlying topological stack.

Definition 7 Let $\mathcal{X}$ be an Artin algebraic stack over $\mathbb{C}$, which is locally of finite type. The *(classical) homotopy type* of $\mathcal{X}$ is defined as the homotopy type of the underlying topological stack $\mathcal{X}^{top}$, i.e. $\{\mathcal{X}\}_{cl} = \{\mathcal{X}^{top}\}$.

Now we can use the machinery of homotopy theory for topological stacks to determine the (classical) homotopy type of moduli stacks of principal G-bundles over fixed smooth complex projective algebraic varieties.

Theorem 8 *Let G be a semisimple affine complex algebraic group and let X be a connected smooth complex projective algebraic variety of special type, i.e. the homotopy type of the moduli stack $\mathcal{B}un_{G,X}$ is given as the homotopy type of the topological mapping stack $\mathcal{M}ap(X, \mathcal{B}G)$. Then the homotopy type of the moduli stack $\mathcal{B}un_{G,X}$ is given as $\{\mathcal{B}un_{G,X}\}_{cl} \simeq \mathrm{M}ap(X, BG)$ where BG is the classifying space of G.*

Proof From Definition 2 and the condition of X being of special type, we see that the homotopy type of $\mathcal{B}un_{G,X}$ is given as the homotopy type of the topological mapping stack

$$\{\mathcal{B}un_{G,X}\}_{cl} = \{\mathcal{M}ap(X, \mathcal{B}G)\}.$$

For any representation of the classifying space BG, the universal bundle EG over BG defines an object in the groupoid $\mathcal{B}G(BG)$ and via representation, a map $BG \to \mathcal{B}G$, which is a weak universal equivalence, as mentioned above.

Now, for any complex projective algebraic variety (and indeed for any compact CW-complex), any map $X \to BG$ induces by composition a map $X \to \mathcal{B}G$, therefore we get a natural morphism

$$Map(X, BG) \to \mathcal{M}ap(X, \mathcal{B}G)$$

which by Noohi [No 1 Cor. 6.5] is again a weak universal equivalence. Therefore the homotopy type of the mapping stack $\mathcal{M}ap(X, \mathcal{B}G)$ is given by the mapping space $Map(X, BG)$, which has the homotopy type of a CW-complex [15]. □

In the special case, that X is a connected smooth complex projective curve, the condition of being special is fulfilled and the previous statement was obtained before by Atitah and Bott [3] using gauge theory and equivariant Morse theory and later also by Teleman [22] using simplicial homotopy theory (compare also [2] and [10]).

4 Rational Homotopy of the Moduli Stack $\mathcal{B}un_{G,X}$

Let G be a semisimple affine complex algebraic group over $\mathbb{C}$. If $T \subset G$ is a maximal torus with Weyl group $W = N(T)/T$, and BG denotes the classifying space of G, then

$$H^*(BG, \mathbb{Q}) \cong H^*(BT, \mathbb{Q})^W \cong \mathbb{Q}[x_1, \ldots, x_n]^W, \tag{1}$$

where $T \cong \mathbb{G}_m^n$, BT is the classifying space of T and x_i is the first Chern class of the universal line bundle on the i-th factor. The right hand side of Eq. 1 is a polynomial ring on the elementary W-invariant polynomials on the classes x_i. These are Chern classes of the universal bundle EG over BG of degree $2n_i$ and $H^*(BG, \mathbb{Q})$ is a direct sum of Tate-Hodge structures. Let $V = \oplus_i V_i$ be the span of these cohomology classes. Since G is semisimple these numbers are always ≥ 2 and define a map from the classifying space of G

$$BG \to \prod_i K(\mathbb{Z}, 2n_i)$$

to a product of Eilenberg-MacLane spaces $K(\mathbb{Z}, 2n_i)$, which induces a *rational* homotopy equivalence.

Let us recall, that given a connected finite dimensional CW-complex X, one has for the homotopy groups of the mapping space from X into an Eilenberg-MacLane space $K(A, n)$ the following description in terms of cohomology groups (see [1, 21])

$$\begin{aligned} \pi_k (\mathrm{M}ap(X, K(A, n))) &\cong \pi_0 \Omega^k \mathrm{M}ap(X, K(A, n)) \\ &\cong \pi_0 \mathrm{M}ap(X, \Omega^k K(A, n)) \\ &\cong \pi_0 \mathrm{M}ap(X, K(A, n-k)) \\ &\cong \quad H^{n-k}(X, A). \end{aligned} \tag{2}$$

Moreover, for any connected and finite dimensional CW-complex (and in particular any connected complex manifold) we have the following theorem due to Thom [23]

Theorem 9 (Thom) *Let X be a connected finite dimensional CW-complex, then, up to homotopy equivalence, we have*

$$\mathrm{M}ap(X, K(A, n)) \simeq \prod_{\substack{0 \leq q \leq dim_{\mathbb{R}}(X) \\ 0 \leq n-q}} K(H^q(X, A); n-q)$$

□

Therefore, in particular, for any connected finite dimensional CW-complex X, one has a rational homotopy equivalence

$$\mathrm{M}ap(X, BG) \simeq_{\mathbb{Q}} \prod_i \prod_{0 \leq q \leq \dim_{\mathbb{R}} X} K(H^q(X, \mathbb{Z}); 2n_i - q) \tag{3}$$

with the convention $K(A; r) = \{*\}$ for $r < 0$.
Let us look at some instructive examples to illustrate this.

Examples

1. If C is a complex algebraic curve of genus g, then $H^q(X, \mathbb{Z})$ is either $\mathbb{Z}$ (for $q = 0, 2$) or $\mathbb{Z}^g$ (for $q = 1$), and therefore

$$\mathrm{Map}(C, BG) \simeq_{\mathbb{Q}} \prod_i \big(K(\mathbb{Z}, 2n_i) \times K(\mathbb{Z}^g, 2n_i - 1) \times K(\mathbb{Z}, 2n_i - 2)\big)$$

(continued)

2. If $X = S^1$, we get

$$\mathrm{Map}(S^1, BG) \simeq_{\mathbb{Q}} \prod_i (K(\mathbb{Z}, 2n_i) \times K(\mathbb{Z}, 2n_i - 1))$$

3. For $X = \mathbb{P}^k$, we get

$$\mathrm{Map}(\mathbb{P}^k, BG) \simeq_{\mathbb{Q}} \prod_i \prod_{0 \leq q \leq k} K(\mathbb{Z}; 2n_i - 2q)$$

again with the convention that $K(\mathbb{Z}, r) = \{*\}$ for $r < 0$.

A more interesting example is given by smooth projective hypersurfaces

Lemma 10 *Assume $X \subset \mathbb{P}^{k+1}$ is a smooth hypersurface, then the Poincaré series of $\mathrm{M}ap(X, BG)$ is of the form*

$$P_t\left(\mathrm{M}ap(X, BG)\right) = \prod_i \frac{1}{(1 - t^{2n_i - 2m})^d} \times \prod_{0 \leq q < m} \frac{1}{1 - t^{2n_i - 2q}}$$

if $k = 2m$ is even and $\dim H^k(X, \mathbb{Q}) = d$, or

$$P_t\left(\mathrm{M}ap(X, BG)\right) = \prod_i (1 + t^{2n_i - 2m - 1})^d \times \prod_{0 \leq q < m} \frac{1}{1 - t^{2n_i - 2q}}$$

if $k = 2m + 1$ is odd and $\dim H^k(X, \mathbb{Q}) = d$.

Proof By Thom's theorem we have

$$\mathrm{M}ap(X, BG) \simeq_{\mathbb{Q}} \prod_i \prod_{0 \leq q \leq \dim_{\mathbb{R}} X} K(H^q(X, \mathbb{Z}); 2n_i - q).$$

By the weak Lefschetz theorem we have

$$H^j(X, \mathbb{Q}) \cong H^j(\mathbb{P}^{k+1}, \mathbb{Q})$$

for all $0 \leq j < k$. Therefore by Poincaré duality, $H^j(X, \mathbb{Q})$ is either 0 or $\mathbb{Q}$ for $j \neq k$, depending on wether j is odd or even. And so the above rational homotopy decomposition becomes

$$\mathrm{M}ap(X, BG) \simeq_{\mathbb{Q}} \prod_i \left(\prod_{2q \neq k} K(\mathbb{Z}; 2n_i - 2q) \right) \times K(H^k(X, \mathbb{Z}); 2n_i - k).$$

Since the Poincaré series are multiplicative and we have

$$\begin{aligned} P_t\left(K(\mathbb{Z},2r)\right) &= \frac{1}{1-t^{2r}} \\ P_t\left(K(\mathbb{Z},2r-1)\right) &= 1+t^{2r-1}, \end{aligned}$$

we get therefore

$$P_t\left(\mathrm{M}ap(X,BG)\right) = \prod_i \frac{1}{(1-t^{2n_i-2m})^d} \times \prod_{0\le q<m} \frac{1}{1-t^{2n_i-2q}}$$

if $k=2m$ and $\dim H^k(X,\mathbb{Q}) = d$, or

$$P_t\left(\mathrm{M}ap(X,BG)\right) = \prod_i (1+t^{2n_i-2m-1})^d \times \prod_{0\le q<m} \frac{1}{1-t^{2n_i-2q}}$$

if $k=2m+1$ and $\dim H^k(X,\mathbb{Q}) = d$, as claimed. □

As in Atiyah and Bott [3] (see also [2, 22]), we can identify G with the based loop space $\Omega BG = \mathrm{M}ap^*(S^1, BG)$ and up to rational homotopy equivalence get a splitting

$$G \simeq \mathrm{M}ap^*(S^1, BG) \simeq_{\mathbb{Q}} \prod_i K(\mathbb{Z}, 2n_i-1).$$

The universal principal G-bundle over $S^1 \times \mathrm{M}ap^*(S^1, BG)$ induces an evaluation map

$$S^1 \times G \cong S^1 \times \mathrm{M}ap^*(S^1, BG) \to BG.$$

And so, the pullback map followed by the projection and integration along S^1 produces a commutative diagram of the form

$$\begin{array}{ccccc} H^*(BG,\mathbb{Q}) & \longrightarrow & H^*(S^1\times G) & \xrightarrow{\cong} & H^0(S^1,\mathbb{Q})\otimes H^*(G,\mathbb{Q}) \oplus H^1(S^1,\mathbb{Q})\otimes H^{*-1}(G,\mathbb{Q}) \\ & \searrow & & & \downarrow \\ & & & & H^1(S^1,\mathbb{Q})\otimes H^{*-1}(G,\mathbb{Q}) \\ & & & & \downarrow{\scriptstyle \int_{S^1}} \\ & & & & H^{*-1}(G,\mathbb{Q}). \end{array}$$

As before, let V be the span of the Chern classes of the universal principal G-bundle EG. The image of V under the above composition can then be identified with $V[1]$ where, as usual, $V[i]_k = V_{k+i}$ and it is not difficult to see that we get an isomorphism $H^*(G, \mathbb{Q}) \cong \Lambda V[1]$.

5 Cohomology and Hodge Theory of the Moduli Stack $\mathcal{B}un_{G,X}$

Assume that X is given as a connected CW-complex of real dimension n and G a semisimple affine algebraic group over $\mathbb{C}$. Then the previous Lemma 10 can be generalized as follows

Proposition 11 *Let X be a connected CW-complex of real dimension n and G be a semisimple complex algebraic group. If $\dim H^j(X, \mathbb{Q}) = d_j$ for $0 \leq j \leq n$, then the Poincaré polynomial of the mapping space $\mathrm{M}ap(X, BG)$ is*

$$\frac{\prod_{q\ odd} \prod_i (1 + t^{2n_i - q})^{d_q}}{\prod_{q\ even} \prod_i (1 - t^{2n_i - q})^{d_q}}$$

where the second product in the numerator and in the denominator involve only those i's satisfying $0 \leq 2n_i - q$ and $0 \leq q \leq n$. □

Proof By Thom's theorem and according to Eq. (3) we have

$$\begin{aligned} \mathrm{M}ap(X, BG) &\simeq_{\mathbb{Q}} \prod_i \prod_{0 \leq q \leq n} K(H^q(X, \mathbb{Z}); 2n_i - q) \\ &\simeq_{\mathbb{Q}} \prod_i \prod_{0 \leq q \leq n} K(\mathbb{Z}^{d_q}; 2n_i - q) \\ &\simeq_{\mathbb{Q}} \prod_{0 \leq q \leq n} \prod_i K(\mathbb{Z}; 2n_i - q)^{d_q} \end{aligned} \tag{4}$$

up to rational weak equivalence. So the assertion follows at once, since the Poincaré polynomials are multiplicative and we have

$$\begin{aligned} P_t\,(K(\mathbb{Z}, 2r)) &= \frac{1}{1 - t^{2r}} \\ P_t\,(K(\mathbb{Z}, 2r - 1)) &= 1 + t^{2r-1}. \end{aligned}$$

□

Since rationally $BG \simeq_{\mathbb{Q}} \prod_i K(\mathbb{Z}, 2n_i)$, Eq. (4) can be interpreted as a rational homotopy splitting of the form

$$\begin{aligned} \mathrm{M}ap(X, BG) &\simeq_{\mathbb{Q}} BG \times (\Omega BG)^{d_1} \times (\Omega^2 BG)^{d_2} \times \cdots \times (\Omega^n BG)^{d_n} \\ &\simeq_{\mathbb{Q}} BG \times \prod_{j=1}^{n} (\Omega^{2j} BG)^{d_{2j}} \times \prod_{j=1}^{n} (\Omega^{2j-1} BG)^{d_{2j-1}} \end{aligned} \tag{5}$$

As we have seen, any connected component $\Omega_0^j BG$ of $\Omega^j BG$ can be identified with the based space

$$\Omega_0^j BG = \mathrm{M}ap^*(S^j, BG).$$

As before, the universal principal G-bundle over $S^j \times Map^*(S^j, BG)$ induces an evaluation map

$$S^j \times \Omega_0^j BG \cong S^j \times \mathrm{M}ap^*(S^j, BG) \longrightarrow BG.$$

Therefore the pullback map followed by the projection and integration along S^j produces

$$\begin{array}{ccccc} H^*(BG, \mathbb{Q}) & \longrightarrow & H^*(S^j \times \Omega_0^j BG) & \xrightarrow{\cong} & \oplus_i H^i(S^j, \mathbb{Q}) \otimes H^{*-i}(\Omega_0^j BG, \mathbb{Q}) \\ & \searrow & & & \downarrow \\ & & & & H^j(S^j, \mathbb{Q}) \otimes H^{*-j}(\Omega_0^j BG, \mathbb{Q}) \\ & & & & \downarrow \int_{S^j} \\ & & & & H^{*-j}(\Omega_0^j BG, \mathbb{Q}) \end{array}$$

The image of V under this composition can be identified with $V[j]$ and it is not difficult to see that

$$H^*(\Omega_0^{2j} BG^0, \mathbb{Q}) \cong \mathrm{Sym}\, V[2j]$$

and

$$H^*(\Omega_0^{2j-1} BG^0, \mathbb{Q}) \cong \Lambda\, V[2j-1],$$

where G^0 denotes the component of the identity of G. Similarly, given a CW-complex X, the universal principal G-bundle over the product $X \times \mathrm{M}ap^*(X, BG)$ induces a classifying map

$$X \times \mathrm{M}ap^*(X, BG) \longrightarrow BG,$$

therefore via pullback, one gets

$$\begin{array}{ccc} H^*(BG, \mathbb{Q}) & \longrightarrow & \oplus_{2j} H^{2j}(X, \mathbb{Q}) \otimes H^{*-2j}(\mathrm{M}ap^*(X, BG), \mathbb{Q}) \\ & \searrow & \downarrow \\ & & H^{2j}(X, \mathbb{Q}) \otimes H^{*-2j}(\mathrm{M}ap^*(X, BG), \mathbb{Q}) \end{array}$$

or equivalently,

$$H^*(BG, \mathbb{Q}) \otimes H^{2j}(X, \mathbb{Q})^\vee \to H^{*-2j}(\mathrm{M}ap^*(X, BG), \mathbb{Q}).$$

The map above induces maps

$$\mathrm{Sym}\, V[2j] \otimes H^{2j}(X, \mathbb{Q})^\vee \to H^{*-2j}\mathrm{M}ap^*(X, BG), \mathbb{Q})$$
$$\Lambda\, V[2j-1] \otimes H^{2j-1}(X, \mathbb{Q})^\vee \to H^{*-2j+1}(\mathrm{M}ap^*(X, BG), \mathbb{Q})$$

which fit together to produce a map

$$\Psi : \mathrm{Sym}\, V \otimes \left(\bigotimes_{j=1}^{n} (\mathrm{Sym}\, V[2j])^{\otimes d_{2j}} \right) \otimes \left(\bigotimes_{j=1}^{n} (\Lambda V[2j-1])^{\otimes d_{2j-1}} \right) \to H^*(\mathrm{M}ap^*(X, BG), \mathbb{Q}).$$

We have the following theorem

Theorem 12 *Let X be a connected CW-complex of real dimension $2n$ and let G be a semisimple complex algebraic group. Assume that X has no cell of odd dimension and that every even dimensional skeleton X^{2j} of X is connected. If $\dim H^{2j}(X, \mathbb{Q}) = d_j$ for $0 \leq j \leq n$, then the cohomology of any connected component $\mathrm{M}ap^0(X, BG)$ of $\mathrm{M}ap(X, BG)$ is given, by*

$$H^*(\mathrm{M}ap^0(X, BG), \mathbb{Q}) \cong \Big(\bigotimes_{1 \leq j \leq n} (Sym\ V[2j])^{\otimes d_j} \Big)$$

and the isomorphism is induced by the pullback map from the classifying space.

Proof We proceed by induction on n, the assertion being clear for $n = 0$.

If $n \geq 1$, the inclusion $X^{2n-2} \hookrightarrow X$, of the $2n-2$ skeleton on X, produces a cofibration

$$X^{2n-2} \longrightarrow X \longrightarrow X/X^{2n-2} \simeq \bigvee S^{2n}$$

which produces a fibration of base point preserving mapping spaces

$$\prod_{d_n} \Omega_0^{2n} \sim Map^*(\bigvee S^{2n}, BG) \longrightarrow Map^*(X, BG) \longrightarrow Map^*(X^{2n-2}, BG)$$

The above fibration, together with the fibration

$$Map^*(X, BG) \longrightarrow Map(X, BG) \longrightarrow BG$$

yields a trigraded spectral sequence of the following form, where all cohomology groups are rational cohomology groups

$$E_2^{pqr} = H^p(BG) \otimes H^q\left(Map^*(X^{2n-2}, BG)\right) \otimes H^r\left(\prod_{d_n} \Omega_0^{2n}\right) \Rightarrow H^{p+q+r}(Map(X, BG))$$

Observe that the sum of terms on the left is just a sum of $|\pi_1(G)|$ copies of the domain of Ψ (under the current hypothesis, there is no odd cohomology), so induction on n and the equality of Poincare series forces that the spectral sequence collapses and we have $E_2 = E_\infty$ and thus the map Ψ is an isomorphism. □

Remember that any compact stratified set in the sense of Thom, and therefore any compact complex manifold (regarded as a differentiable manifold), can be triangulated (see for example [24] or [13]). Therefore we get the following corollary on the rational cohomology of the moduli stack of principal G-bundles over a smooth complex projective variety

Corollary 13 *Let X be a connected smooth complex projective variety of complex dimension n of special type and let G be a semisimple complex algebraic group. If there are no odd dimensional cells on the cellular decomposition of X as a CW-complex (as is the case for H/P for any semisimple group H and any parabolic subgroup $P \subset H$), and $\dim H^{2j}(X, \mathbb{Q}) = d_j$ for $0 \leq j \leq n$, then the rational cohomology of any connected component $\mathcal{B}un^0_{G,X}$ of $\mathcal{B}un_{G,X}$ is given by*

$$H^*(\mathcal{B}un^0_{G,X}, \mathbb{Q}) \cong \Big(\bigotimes_{1 \leq j \leq n} (Sym\, V[2j])^{\otimes d_j} \Big).$$

In particular, it is a direct sum of Hodge-Tate structures.

More generally, without the special condition on the cellular structure we expect the following result

Conjecture 1 Let X be a connected smooth complex projective variety of complex dimension n and G be a semisimple complex algebraic group. If $\dim H^j(X, \mathbb{Q}) = d_j$ for $0 \leq j \leq 2n$, then the rational cohomology of any connected component $\mathcal{B}un^0_{G,X}$ of $\mathcal{B}un_{G,X}$ is given by

$$H^*(\mathcal{B}un^0_{G,X}, \mathbb{Q}) \cong \left(\bigotimes_{j=0}^{n} (\mathrm{Sym}\, V[2j])^{\otimes d_{2j}} \right) \otimes \left(\bigotimes_{j=1}^{n} (\Lambda V[2j-1])^{\otimes d_{2j-1}} \right)$$

and the isomorphism is induced by the pullback map from the classifying space. In particular, it is a direct sum of Hodge-Tate structures.

Acknowledgments Both authors would like to thank Prof. Alexander Schmitt for the kind invitation to speak at the Complex Geometry session as part of the 14th ISAAC Congress 2023 at USP in Ribeirão Preto (Brazil). They also like to thank the local organising committee and the ISAAC board for the great hospitality as well as the referee for the useful comments. The first named author also likes to thank the Centro de Investigación en Matemáticas (CIMAT) for financial support. The second named author thanks the Dipartimento di Matematica 'Felice Casorati' of the University of Pavia for financial support.

References

1. Aguilar, M., Gitler, S., Prieto, C.: Algebraic Topology from a Homotopical Viewpoint. Springer, Heidelberg (2002)
2. Arapura, D., Dhillon, A.: The motive of the moduli stack of G-bundles over the universal curve. Proc. Ind. Acad. Sci. (Math. Sci.) **118**(3), 389–411 (2008)
3. Atiyah, M.F., Bott, R.: The Yang-Mills equations over Riemann surfaces. Phils. Trans. Royal. Soc. Lond. A **308**, 523–615 (1982)
4. Behrend, K.: The Lefschetz trace formula for the moduli stack of principal bundles, PhD thesis, Berkeley, 96pp. (1990)
5. Behrend, K.: The Lefschetz trace formula for algebraic stacks. Invent. Math. **112**, 127–149 (1993)
6. Behrend, K.: Derived l-adic categories for algebraic stacks. Mem. Amer. Math. Soc. **163**(774), viii+93pp. (2003)
7. Behrend, K.; Ginot, G., Noohi, B., Xu, P.: String topology for stacks. Astérisque **343** SMF (2012)
8. Dhillon, A.: On the Cohomology of moduli of vector bundles and the Tamagawa number of SL_n. Can. J. Math. **58**(5), 1000–1025 (2006)
9. Frediani, P., Neumann, F.: Étale homotopy types of moduli stacks of algebraic curves with symmetries. K-Theory **30**, 315–340 (2003)
10. Gaitsgory, D., Lurie, J.: Weil's Conjecture for Function Fields: Volume I. Annals of Mathematics Studies, vol. 199. Princeton University Press, Princeton (2019)
11. Hall, J., Rydh, D.: General Hilbert stacks and quot schemes. Michigan Math. J. **64**, 335–347 (2015)

12. Heinloth, J., Schmitt, A.H.W.: The cohomology rings of moduli stacks of principal bundles over curves. Doc. Math. **15**, 423–488 (2010)
13. Johnson, F.E.A.: On the triangulation of stratified sets and singular varieties. Trans. Amer. Mat. Soc. **275**(1), 333–343 (1983)
14. Laumon, G., Moret-Bailly, L.: Champs algébriques. Ergebnisse der Mathematik und ihrer Grenzgebiete 3. Folge, Band, vol. 39. Springer, Berlin (2000)
15. Milnor, J.W.: On spaces having the homotopy type of a CW-complex. Trans. Amer. Math. Soc. **90**(2), 272–280 (1959)
16. Neumann, F.: Algebraic Stacks and Moduli of Vector Bundles. Publicaçoes Matematicas. IMPA Research Monographs, vol. 36, 2nd edn. IMPA, Rio de Janeiro (2011)
17. Noohi, B.: Foundations of topological stacks I (2005, preprint). arXiv:math/0503257
18. Noohi, B.: Mapping stacks of topological stacks. J. Reine Angew. Math. **646**, 117–133 (2010)
19. Noohi, B.: Homotopy types of topological stacks. Adv. Math. **230**, 2014–2047 (2012)
20. Sorger, C.: Lectures on Moduli of Principal G-Bundles Over Algebraic Curves. School on Algebraic Geometry (Trieste, 1999), ICTP Lecture Notes, vol. 1, pp. 1–57. Abdus Salam International Centre for Theoretical Physics, Trieste (2000)
21. Switzer, R.M.: Algebraic Topology - Homotopy and Homology. Grundlehren der math. Wiss., vol. 212. Springer, Heidelberg (1975)
22. Teleman, C.: Borel-Weil-Bott theory on the moduli stack of G-bundles on a curve. Invent. Math. **134**, 1–57 (1998)
23. Thom, R.: L'homologie des espaces fonctionnels. Colloque de topologie algébrique, Louvain, 1956, 29–39. Georges Thone, Liège (1957).
24. Whitehead, J.H.C.: On C^1 complexes. Ann. Math. **41**(2), 809–824 (1940)

On Slice Regular Bergman Spaces and Fiber Bundles

José Oscar González-Cervantes

Abstract Recently, fiber bundle theory has been widely used in the study of the slice regular functions and continuing with this line of research, the present work shows that the quaternionic slice regular Bergman space is the base space of a coordinate sphere bundle and some properties of quaternionic slice regular Bergman theory are studied from the point of view of the fiber bundle theory.

1 Introduction

Fiber bundle theory has been commonly used to interpret some physical and mathematical phenomena, see [2, 3, 6, 17–19, 24, 25]. Bergman's theory has been developed in hypercomplex analysis see [4, 5, 8–10, 12, 13, 21–23]. In particular, works [8–10] were the first to introduce the theory of slice regular Bergman spaces. In this paper the fiber bundle theory is used to study the second notion of quaternionic Bergman space introduced in [8] that consists of a family of slice regular Bergman spaces associated to a slice generate by $\{1, \mathbf{i}\}$, where $\mathbf{i} \in \mathbb{S}^2$. For ease of reading, Sect. 2 presents basic concepts of the quaternionic slice regular Bergman space and fiber bundle theory and Sect. 3 presents the main results.

2 Preliminaries

2.1 *Quaternionic Slice Regular Bergman Spaces*

The concepts and results of this subsection are given in [7–11, 14, 15].

The skew-field of quaternions $\mathbb{H}$ consists of $q = x_0 + x_1 e_1 + x_2 e_2 + x_3 e_3$ where $x_0, x_1, x_2, x_3 \in \mathbb{R}$ and the quaternionic units satisfy $e_1^2 = e_2^2 = e_3^2 = -1$, $e_1 e_2 =$

J. O. González-Cervantes (✉)
Departamento de Matemáticas, ESFM-Instituto Politécnico Nacional, Ciudad México, México

M. R. Ebert et al. (eds.), *New Tools in Mathematical Analysis and Applications*, Research Perspectives, https://doi.org/10.1007/978-3-031-77050-0_5

$-e_2e_1 = e_3, e_2e_3 = -e_3e_2 = e_1, e_3e_1 = -e_1e_3 = e_2$. The set $\{e_1, e_2, e_3\}$ is called the standard basis of $\mathbb{R}^3$. The vector part of q is denoted by $\mathbf{q} = x_1e_1 + x_2e_2 + x_3e_3$ and is usually identified with the vector $\mathbf{q} = (x_1, x_2, x_3) \in \mathbb{R}^3$. The real part is $q_0 = x_0$. The quaternionic conjugate of q is $\bar{q} = q_0 - \mathbf{q}$ and its norm is $\|q\| := \sqrt{x_0^2 + x_1^2 + x_2^3 + x_3^2} = \sqrt{q\bar{q}} = \sqrt{\bar{q}q}$. Denote $\mathbb{B}^4 := \{q \in \mathbb{H} \mid \|q\| < 1\}$, $\mathbb{S}^2 := \{\mathbf{q} \in \mathbb{R}^3 \mid \|\mathbf{q}\| = 1\}$, $\mathbb{S}^3 := \{q \in \mathbb{H} \mid \|q\| = 1\}$ and $T := \{(\mathbf{i}, \mathbf{j}) \in \mathbb{S}^2 \times \mathbb{S}^2 \mid \{\mathbf{i}, \mathbf{j}, \mathbf{ij}\}$ and $\{e_1, e_2, e_3\}$ are co-oriented$\}$.

Note that $\mathbf{i}^2 = -1$ for any $\mathbf{i} \in \mathbb{S}^2$, so $\mathbb{C}(\mathbf{i}) := \{x + \mathbf{i}y; \mid x, y \in \mathbb{R}\} \approx \mathbb{C}$ as fields and from the algebra of quaternions we see that $\mathbb{H} = \bigcup_{\mathbf{i}\in\mathbb{S}^2} \mathbb{C}(\mathbf{i})$. The quaternionic rotations that preserve $\mathbb{R}^3$ are given by $q \mapsto vq\bar{v}$ for all $q \in \mathbb{H}$, where $v \in \mathbb{S}^3$, see [20]. Then let us consider $R_v(\mathbf{i}, \mathbf{j}) := (v\mathbf{i}\bar{v}, v\mathbf{j}\bar{v})$ for all $(\mathbf{i}, \mathbf{j}) \in T$.

Definition 1 Let $\Omega \subset \mathbb{H}$ be an open set. A real differentiable function $f : \Omega \to \mathbb{H}$ is called slice regular function on Ω, if

$$\bar{\partial}_{\mathbf{i}} f\mid_{\Omega\cap\mathbb{C}(\mathbf{i})} := \frac{1}{2}\left(\frac{\partial}{\partial x} + \mathbf{i}\frac{\partial}{\partial y}\right) f\mid_{\Omega\cap\mathbb{C}(\mathbf{i})} = 0 \text{ on } \Omega \cap \mathbb{C}(\mathbf{i}),$$

for all $\mathbf{i} \in \mathbb{S}^2$ and the Cullen derivative of f is $f' = \partial_{\mathbf{i}} f\mid_{\Omega\cap\mathbb{C}(\mathbf{i})} = \frac{\partial}{\partial x} f\mid_{\Omega\cap\mathbb{C}(\mathbf{i})} = \partial_x f\mid_{\Omega\cap\mathbb{C}(\mathbf{i})}$. The quaternionic right module of the slice regular functions on Ω is denoted by $\mathcal{SR}(\Omega)$.

Definition 2 A set $U \subset \mathbb{H}$ is called axially symmetric if $U \cap \mathbb{R} \neq \emptyset$ and if $x + \mathbf{i}y \in U$ with $x, y \in \mathbb{R}$ then $\{x + \mathbf{j}y \mid \mathbf{j} \in \mathbb{S}^2\} \subset U$. In addition, a domain $U \subset \mathbb{H}$ is a slice domain, or s-domain, if $U_{\mathbf{i}} := U \cap \mathbb{C}(\mathbf{i})$ is a domain in $\mathbb{C}(\mathbf{i})$ for all $\mathbf{i} \in \mathbb{S}^2$. Given an axially symmetric s-domain $\Omega \subset \mathbb{H}$ denote

$$S_\Omega := \{(x, y) \in \mathbb{R}^2 \mid \text{ there exists } \mathbf{i} \in \mathbb{S}^2 \text{ and } x + \mathbf{i}y \in \Omega\}$$

The following theorem presents two important results in the slice regular function theory.

Theorem 1 *Let $\Omega \subset \mathbb{H}$ be an axially symmetric s-domain. Given $f \in \mathcal{SR}(\Omega)$ and $(\mathbf{i}, \mathbf{j}) \in T$.*

1. *(Splitting Lemma) There exist $f_1, f_2 \in Hol(\Omega_{\mathbf{i}})$ such that $f_{|\Omega_{\mathbf{i}}} = f_1 + f_2\mathbf{j}$ on $\Omega_{\mathbf{i}}$.*
2. *(Representation Formula) For every $q = x + \mathbf{I}_q y \in \Omega$ where $x, y \in \mathbb{R}$ and $\mathbf{I}_q \in \mathbb{S}^2$ we have*

$$f(x + \mathbf{I}_q y) = \frac{1}{2}[f(x + \mathbf{i}y) + f(x - \mathbf{i}y)] + \frac{1}{2}\mathbf{I}_q\mathbf{i}[f(x - \mathbf{i}y) - f(x + \mathbf{i}y)].$$

The above results help us directly introduce the following operators and their properties.

Definition 3 Let $\Omega \subset \mathbb{H}$ be an axially symmetric s-domain and $(\mathbf{i},\mathbf{j}) \in T$. Define $Q_{\mathbf{i},\mathbf{j}} : \mathcal{SR}(\Omega) \to \mathrm{Hol}(\Omega_{\mathbf{i}}) + \mathrm{Hol}(\Omega_{\mathbf{i}})\mathbf{j}$ and $P_{\mathbf{i},\mathbf{j}} : \mathrm{Hol}(\Omega_{\mathbf{i}}) + \mathrm{Hol}(\Omega_{\mathbf{i}})\mathbf{j} \to \mathcal{SR}(\Omega)$ as follows: $Q_{\mathbf{i},\mathbf{j}}[f] = f\mid_{\Omega_{\mathbf{i}}} = f_1 + f_2\mathbf{j}$ for all $f \in \mathcal{SR}(\Omega)$ where $f_1, f_2 \in \mathrm{Hol}(\Omega_{\mathbf{i}})$, and $P_{\mathbf{i},\mathbf{j}}[g](q) = \frac{1}{2}\left[(1+\mathbf{I}_q\mathbf{i})g(x-y\mathbf{i}) + (1-\mathbf{I}_q\mathbf{i})g(x+y\mathbf{i})\right]$, for all $g \in \mathrm{Hol}(\Omega_{\mathbf{i}}) + \mathrm{Hol}(\Omega_{\mathbf{i}})\mathbf{j}$ where $q = x + \mathbf{I}_q y \in \Omega$, respectively. In addition, $P_{\mathbf{i},\mathbf{j}} \circ Q_{\mathbf{i},\mathbf{j}} = \mathcal{I}_{\mathcal{SR}(\Omega)}$ and $Q_{\mathbf{i},\mathbf{j}} \circ P_{\mathbf{i},\mathbf{j}} = \mathcal{I}_{\mathrm{Hol}(\Omega_{\mathbf{i}})+\mathrm{Hol}(\Omega_{\mathbf{i}})\mathbf{j}}$, where $\mathcal{I}_{\mathcal{SR}(\Omega)}$ and $\mathcal{I}_{\mathrm{Hol}(\Omega_{\mathbf{i}})+\mathrm{Hol}(\Omega_{\mathbf{i}})\mathbf{j}}$ are identity operators.

On the other hand, given $f \in \mathcal{SR}(\Omega)$ and $(\mathbf{i},\mathbf{j}) \in T$, from direct computation we see that

$$D_1[f,\mathbf{i},\mathbf{j}] := \frac{Q_{\mathbf{i},\mathbf{j}}[f] + \overline{Q_{\mathbf{i},\mathbf{j}}[f]}}{2}, \qquad D_2[f,\mathbf{i},\mathbf{j}] := -\frac{Q_{\mathbf{i},\mathbf{j}}[f]\mathbf{i} + \overline{Q_{\mathbf{i},\mathbf{j}}[f]\mathbf{i}}}{2},$$

$$D_3[f,\mathbf{i},\mathbf{j}] := -\frac{Q_{\mathbf{i},\mathbf{j}}[f]\mathbf{j} + \overline{Q_{\mathbf{i},\mathbf{j}}[f]\mathbf{j}}}{2}, \qquad D_4[f,\mathbf{i},\mathbf{j}] := \frac{Q_{\mathbf{i},\mathbf{j}}[f]\mathbf{ji} + \overline{Q_{\mathbf{i},\mathbf{j}}[f]\mathbf{ji}}}{2}$$

are the real components of $Q_{\mathbf{i},\mathbf{j}}[f]$ in terms of base $\{1,\mathbf{i},\mathbf{j},\mathbf{ij}\}$. Therefore, $Q_{\mathbf{i},\mathbf{j}}[g] = D_1[g,\mathbf{i},\mathbf{j}] + D_2[g,\mathbf{i},\mathbf{j}]\mathbf{i} + D_3[g,\mathbf{i},\mathbf{j}]\mathbf{j} + D_4[g,\mathbf{i},\mathbf{j}]\mathbf{ij}$.

Definition 4 Given $f, g \in \mathcal{SR}(\Omega)$ and $(\mathbf{i},\mathbf{j}) \in T$ define the $\bullet_{\mathbf{i},\mathbf{j}}-$product: $f \bullet_{\mathbf{i},\mathbf{j}} g := P_{\mathbf{i},\mathbf{j}}[\, f_1g_1 + f_2g_2\mathbf{j}\,]$, where $f_1, f_2, g_1, g_2 \in \mathrm{Hol}(\Omega_{\mathbf{i}})$ are such that $Q_{\mathbf{i},\mathbf{j}}[f] = f_1 + f_2\mathbf{j}$ and $Q_{\mathbf{i},\mathbf{j}}[g] = g_1 + g_2\mathbf{j}$.

In particular, if $\Omega = \mathbb{B}^4(0,1)$ then there exists two sequences of quaterinons $(a_n)_{n\geq 0}$ and $(b_n)_{n\geq 0}$ such that $f(q) = \sum_{n=0}^{\infty} q^n a_n$, $g(q) = \sum_{n=0}^{\infty} q^n b_n$ then their $*-$product is $f * g(q) := \sum_{n=0}^{\infty} q^n \sum_{k=0}^{n} a_k b_{n-k}$ for all $q \in \mathbb{B}^4(0,1)$.

Definition 5 Let $\Omega \subseteq \mathbb{H}$ be an axially symmetric s-domain. We introduce a family of quaternionic slice regular Bergman spaces. Given $\mathbf{i} \in \mathbb{S}^2$ denote

$$\mathcal{A}_{\mathbf{i}}(\Omega) := \left\{ f \in \mathcal{SR}(\Omega) \mid \|f\|^2_{\mathcal{A}_{\mathbf{i}}(\Omega)} := \int_{\Omega_{\mathbf{i}}} |f_{|_{\Omega_{\mathbf{i}}}}|^2 d\sigma_{\mathbf{i}} < \infty \right\}$$

equipped with the norm $\|f\|_{\mathcal{A}_{\mathbf{i}}(\Omega)} := \left(\int_{\Omega_{\mathbf{i}}} |f_{|_{\Omega_{\mathbf{i}}}}|^2 d\sigma_{\mathbf{i}}\right)^{\frac{1}{2}}$, $\forall f \in \mathcal{A}_{\mathbf{i}}(\Omega)$, and with the scalar product $\langle f, g\rangle_{\mathcal{A}_{\mathbf{i}}(\Omega)} = \int_{\Omega_{\mathbf{i}}} \overline{f\mid_{\Omega_{\mathbf{i}}}}\, g\mid_{\Omega_{\mathbf{i}}} d\sigma_{\mathbf{i}}$, $\forall f, g \in \mathcal{A}_{\mathbf{i}}(\Omega)$, where $d\sigma_{\mathbf{i}}$ denotes the element of area in the complex plane $\mathbb{C}(\mathbf{i})$.

Remark 1 Given $\mathbf{i}, \mathbf{k} \in \mathbb{S}^2$ and $f \in \mathcal{A}_{\mathbf{k}}(\Omega)$. Then Representation Formula implies that $|f(x+y\mathbf{i})|^2 \leq 2\left[|f(x+y\mathbf{k})|^2 + |f(x-y\mathbf{k})|^2\right]$ and the integration on S_Ω gives us $\int_{\Omega_{\mathbf{i}}} |f_{\Omega_{\mathbf{i}}}|^2 d\sigma_{\mathbf{i}} \leq 2\|f\|^2_{\mathcal{A}_{\mathbf{k}}(\Omega)}$, i.e., $\|f\|_{\mathcal{A}_{\mathbf{i}}(\Omega)} \leq \sqrt{2}\|f\|_{\mathcal{A}_{\mathbf{k}}(\Omega)}$. Therefore, $\mathcal{A}_{\mathbf{i}}(\Omega)$ and $\mathcal{A}_{\mathbf{k}}(\Omega)$ are equal functions sets and have equivalent norms.

The following theorem was proved in [8].

Theorem 2 *Let Ω be a bounded axially symmetric s-domain and $\mathbf{i} \in \mathbb{S}^2$. Then $\left(\mathcal{A}_{\mathbf{i}}(\Omega), \langle\cdot,\cdot\rangle_{\mathcal{A}_{\mathbf{i}}(\Omega)}\right)$ is a quaternionic right Hilbert module. In addition, given $q \in \Omega$ let ϕ_q the evaluation functional at point q on $\mathcal{A}_{\mathbf{i}}(\Omega)$. Then there exists* *an unique function* $K_q(\mathbf{i}) \in \mathcal{A}_\mathbf{i}(\Omega)$ *such that* $\phi_q[f] = \langle K_q(\mathbf{i}), f\rangle_{\mathcal{A}_\mathbf{i}(\Omega)}$ *for all* $f \in \mathcal{A}_\mathbf{i}(\Omega)$*, due to the evaluation functional on* $\mathcal{A}_\mathbf{i}(\Omega)$ *is a bounded quaternionic right-linear functional and T. Riesz representation theorem for quaternionic right Hilbert modules.*

Remark 2 If we denote $\mathcal{K}_{\Omega_\mathbf{i}}(q, \cdot) := \bar{K}_q(\mathbf{i})$ then the slice regular Bergman kernel associated to $\Omega_\mathbf{i}$ is given by $\mathcal{K}_{\Omega_\mathbf{i}}(\cdot, \cdot) : \Omega_\mathbf{i} \times \Omega_\mathbf{i} \to \mathbb{H}$. Representation Formula shows that $\mathcal{K}_\Omega : \Omega \times \Omega \to \mathbb{H}$ satisfies:

$$\mathcal{K}_\Omega(x + y\mathbf{k}, r) := \frac{1}{2}(1 - \mathbf{ki})\mathcal{K}_{\Omega_\mathbf{i}}(x + y\mathbf{i}, r) + \frac{1}{2}(1 + \mathbf{ki})\mathcal{K}_{\Omega_\mathbf{i}}(x + y\mathbf{i}, r),$$

for all $\mathbf{k} \in \mathbb{S}^2$ and $r \in \Omega_\mathbf{i}$.

In this work we will focus on the slice regular Bergman space associated with the quaternionic unit ball $\mathbb{B}^4$ to simplify the notation and computations. If $\Omega = \mathbb{B}^4$ then $\int_{\mathbb{B}^4_\mathbf{i}} |f_{|\mathbb{B}^4_\mathbf{i}}|^2 d\sigma_\mathbf{i} = \int_D |f(x+\mathbf{i}y)|^2 d\mu_{x,y}$ and $\langle f, g\rangle_{\mathcal{A}_\mathbf{i}(\mathbb{B}^4)} = \int_D \overline{f(x+\mathbf{i}y)}g(x + \mathbf{i}y)d\mu_{x,y}$, for all $f, g \in \mathcal{A}_\mathbf{i}(\mathbb{B}^4)$, where $D := \{(x, y) \in \mathbb{R}^2 \mid x^2 + y^2 < 1\}$ and $d\mu_{x,y}$ is the area differential in $\mathbb{R}^2$.

Definition 6 Define $L := \{f : \mathbb{B}^4 \to \mathbb{H} \mid \int_D \|f(x + \mathbf{i}y)\|^2 d\mu_{x,y} < \infty, \quad \forall \mathbf{i} \in \mathbb{S}^2\}$. Given $\mathbf{i} \in \mathbb{S}^2$. The slice regular Bergman projection $B_\mathbf{i} : L \to \mathcal{A}_\mathbf{i}(\mathbb{B}^4)$ is $B_\mathbf{i}[f](q) = \int_{\mathbb{B}^4_\mathbf{i}} \mathcal{K}_{\mathbb{B}^4}(q, z)g(z)d\sigma_\mathbf{i}$ for all $q \in \mathbb{B}^4$, where $\mathcal{K}_{\mathbb{B}^4}$ is the slice regular Bergman Kernel of $\mathcal{A}_\mathbf{i}(\mathbb{B}^4)$.

And, in addition, we introduce the Toeplitz operators in $\mathcal{A}_\mathbf{i}(\mathbb{B}^4)$ extending the concept of Toeplitz operator in the complex Bergman theory given in [1]. Consider $\alpha \in C(\overline{\mathbb{B}^4_\mathbf{i}}, \mathbb{H})$ and as the quaternionic product is non conmutative we have two kinds of Toepliz operators associated to the slice regular Bergman space $\mathcal{A}_\mathbf{i}(\mathbb{B}^4)$. The left-Toeplitz operator, or only Toeplitz operator, with symbol α is given by $T^\mathbf{i}_\alpha f = B_\mathbf{i}[\alpha f]$ for all $f \in L$, i.e.,

$$(T^\mathbf{i}_\alpha f)(q) = \int_{\mathbb{B}^4_\mathbf{i}} \mathcal{K}_{\mathbb{B}^4}(q, z)\alpha(z)f(z)d\sigma_\mathbf{i}, \quad \forall q \in \mathbb{B}^4$$

and the the right-Toeplitz operator with symbol α is given by $T^\mathbf{i}_{r,\alpha} f = B_\mathbf{i}[f\alpha]$ for all $f \in L$.

2.2 On Fiber Bundle Theory

The following basic concepts about fiber bundle theory are given in [6].

Definition 7 A fiber bundle $(X, \mathbf{P}, B, F)$ consists of Hausdorff spaces X, B, F and a topological action group K on F as a group of homeomorphisms. The continuous surjective map $\mathbf{P} : X \to B$ is the bundle projection and each element of B has a neighborhood $U \subset B$ and a homeomorphism called trivialization $\varphi : U \times F \to \mathbf{P}^{-1}(U)$ such that $\mathbf{P} \circ \varphi(x, y) = x$ for all $x \in U$ and $y \in F$. In addition, the family of all trivializations Φ satisfies the following:

1. If $\varphi : U \times F \to \mathbf{P}^{-1}(U)$ belongs to Φ and $V \subset U$ then $\varphi_{|V \times F}$ belongs to Φ.
2. If $\varphi, \phi \in \Phi$ over $U \subset B$ then there exists a map $\psi : U \to K$ such that $\phi(x, y) = \varphi(x, \psi(u)(y))$ for all $x \in U$ and $y \in F$.
3. Φ is a maximal family.

A continuous map $S : B \to X$ is a section of $(X, \mathbf{P}, B, F)$, if $\mathbf{P} \circ S(x) = x$ for all $x \in B$. An arbitrary map $\mathfrak{K} : A \to B$ generates a fiber bundle $(\mathfrak{K}^*(X), \mathbf{P}', A, F)$ called the induced bundle from $\mathbf{P}$ by $\mathfrak{K}$, or pullback bundle, where $\mathfrak{K}^*(X) := \{(a, x) \in A \times X \mid \mathbf{P}(x) = \mathfrak{K}(a)\}$ and $\mathbf{P}'(a, x) = a$ for all $(a, x) \in \mathfrak{K}^*(X)$. If F and K are the sphere and a subgroup of the orthogonal group of an Euclidean space respectively, then $(X, \mathbf{P}, B, F)$ is called sphere bundle. In addition, $(X, \mathbf{P}, B, F)$ is called a coordinate bundle if the mappings $\varphi_{v,x} : F \to \mathbf{P}^{-1}(x)$ given by $\varphi_{v,x}(y) = \varphi_v(x, y)$ where φ_v is a trivialization and $(x, y) \in U \times F$ define the transition functions $g_{v,w}(x) = \varphi_{v,x}^{-1} \circ \varphi_{w,x}$ that satisfy: $g_{v,w} g_{w,l} = g_{v,l}$, $g_{v,v} = e$ and $[g_{v,w}]^{-1} = g_{w,v}$, where e is the identity element.

Let $(X_1, \mathbf{P}_1, B_1, F_1)$ and $(X_2, \mathbf{P}_2, B_2, F_2)$ be two fiber bundles then a morphism $\Gamma : (X_1, \mathbf{P}_1, B_1, F_1) \to (X_2, \mathbf{P}_2, B_2, F_2)$ is a pair of continuous maps $\Gamma_1 : X_1 \to X_2$ and $\Gamma_2 : B_1 \to B_2$ such that $\mathbf{P}_2 \circ \Gamma_1 = \Gamma_2 \circ \mathbf{P_1} : X_1 \to B_2$. In addition, Γ is a isomorphism if there exists a morphism $\Gamma^{-1} : (X_2, \mathbf{P}_2, B_2, F_2) \to (X_1, \mathbf{P}_1, B_1, F_1)$ such that $\Gamma \circ \Gamma^{-1}$ and $\Gamma^{-1} \circ \Gamma$ are the identity morphisms.

3 Main Results

We will define the Hausdorff spaces and the continuous mappings required to explain the quaternionic slice regular Bergman space in terms of the fiber bundle theory.

Definition 8 The set $HL^2(D)$ is formed by pairs of conjugated harmonic functions (a, b) on D such that

$$\|a\|_{L^2}^2 := \int_D |a(x, y)|^2 d\mu_{x,y} = \int_D |a|^2 d\mu, \quad \|b\|_{L^2}^2 := \int_D |b|^2 d\mu < \infty.$$

Set $\mathcal{HL}(D) := \left\{ \left(\begin{pmatrix} a & b \\ c & d \end{pmatrix}, (\mathbf{k}, \mathbf{l}) \right) \mid (a, b), (c, d) \in HL^2(D),\ (\mathbf{k}, \mathbf{l}) \in T \right\}$ equipped with the following metric

$$d\left(\left(\begin{pmatrix} a & b \\ c & d \end{pmatrix}, (\mathbf{k}, \mathbf{l}) \right), \left(\begin{pmatrix} r & s \\ t & u \end{pmatrix}, (\mathbf{m}, \mathbf{n}) \right) \right)$$

$$:= \|a - r\|_{L^2} + \|b - s\|_{L^2} + \|c - t\|_{L^2} + \|d - u\|_{L^2} + \|(\mathbf{k}, \mathbf{l}) - (\mathbf{m}, \mathbf{n})\|_{\mathbb{R}^6},$$

for all $\left(\begin{pmatrix} a & b \\ c & d \end{pmatrix}, (\mathbf{k}, \mathbf{l})\right), \left(\begin{pmatrix} r & s \\ t & u \end{pmatrix}, (\mathbf{m}, \mathbf{n})\right) \in \mathcal{HL}(D)$.

Definition 9 Given $\mathbf{i} \in \mathbb{S}^2$ consider the metric induced by $\| \cdot \|_{\mathcal{A}_{\mathbf{i}}(\mathbb{B}^4)}$ on $\mathcal{A}_{\mathbf{i}}(\mathbb{B}^4)$. In addition, $\mathcal{A}_{\mathbf{i}}(\mathbb{B}^4) \times T$ is equipped with the following metric: $\rho_{\mathbf{i}}\,(f, (\mathbf{k}, \mathbf{l})), (g, (\mathbf{m}, \mathbf{n}))) = \|f - g\|_{\mathcal{A}_{\mathbf{i}}(\mathbb{B}^4)} + \|(\mathbf{k}, \mathbf{l}) - (\mathbf{m}, \mathbf{n})\|_{\mathbb{R}^6}$, for all $(f, (\mathbf{k}, \mathbf{l})), (g, (\mathbf{m}, \mathbf{n})) \in \mathcal{A}_{\mathbf{i}}(\mathbb{B}^4) \times T$.

Remark 3 Given $\left(\begin{pmatrix} a & b \\ c & d \end{pmatrix}, (\mathbf{k}, \mathbf{l})\right) \in \mathcal{HL}(D)$. From Representation Theorem, Splitting Lemma and Remark 1 we have that $P_{\mathbf{k},\mathbf{l}}[a + b\mathbf{k} + c\mathbf{l} + d\mathbf{kl}] \in \mathcal{SR}(\mathbb{B}^4)$ and $\| P_{\mathbf{k},\mathbf{l}}[a + b\mathbf{k} + c\mathbf{l} + d\mathbf{kl}]\|^2_{\mathcal{A}_{\mathbf{i}}(\mathbb{B}^4)} \leq 2\left(\|a\|^2_{L^2} + \|b\|^2_{L^2} + \|c\|^2_{L^2} + \|d\|^2_{L^2}\right)$.

Definition 10 Given $\mathbf{i} \in \mathbb{S}^2$ and using the previous remark the operator $\mathcal{P}_{\mathbb{B}^4} : \mathcal{HL}(D) \to \mathcal{A}_{\mathbf{i}}(\mathbb{B}^4)$ given by $\mathcal{P}_{\mathbb{B}^4}\left(\begin{pmatrix} a & b \\ c & d \end{pmatrix}, (\mathbf{k}, \mathbf{l})\right) := P_{\mathbf{k},\mathbf{l}}[a + b\mathbf{k} + c\mathbf{l} + d\mathbf{kl}]$ is well defined. Given $(\mathbf{k}, \mathbf{l}) \in T$ denote

$$\mathbf{S}_{\mathbf{k},\mathbf{l}}[f] := \left(\begin{pmatrix} D_1[f, \mathbf{k}, \mathbf{l}] & D_2[f, \mathbf{k}, \mathbf{l}] \\ D_3[f, \mathbf{k}, \mathbf{l}] & D_4[f, \mathbf{k}, \mathbf{l}] \end{pmatrix}, (\mathbf{k}, \mathbf{l})\right),$$

for all $f \in \mathcal{A}_{\mathbf{i}}(\mathbb{B}^4)$. In addition, let $U \subset \mathcal{A}_{\mathbf{i}}(\mathbb{B}^4)$ be a neighborhood and $u \in \mathbb{S}^3$ define $\varphi_u[f, (\mathbf{k}, \mathbf{l})] := \left(\begin{pmatrix} D_1[f, u\mathbf{k}\bar{u}, u\mathbf{l}\bar{u}] & D_2[f, u\mathbf{k}\bar{u}, u\mathbf{l}\bar{u}] \\ D_3[f, u\mathbf{k}\bar{u}, u\mathbf{l}\bar{u}] & D_4[f, u\mathbf{k}\bar{u}, u\mathbf{l}\bar{u}] \end{pmatrix}, (u\mathbf{k}\bar{u}, u\mathbf{l}\bar{u})\right)$, for all $(f, (\mathbf{k}, \mathbf{l})) \in U \times T$.

Proposition 1 *Given* $\left(\begin{pmatrix} a & b \\ c & d \end{pmatrix}, (\mathbf{i}, \mathbf{j})\right), \left(\begin{pmatrix} r & s \\ t & u \end{pmatrix}, (\mathbf{k}, \mathbf{l})\right) \in \mathcal{HL}(D)$ *we have that*

$$\begin{aligned} &\| P_{\mathbf{i},\mathbf{j}}[a + b\mathbf{i} + c\mathbf{j} + d\mathbf{ij}] - P_{\mathbf{k},\mathbf{l}}[r + s\mathbf{k} + t\mathbf{l} + u\mathbf{kl}]\|_{\mathcal{A}_{\mathbf{i}}(\mathbb{B}^4)} \\ &\leq 8\left(1 + \|r\|_{L^2} + \|s\|_{L^2} + \|t\|_{L^2} + \|u\|_{L^2}\right) \\ &\times d\left(\left(\begin{pmatrix} a & b \\ c & d \end{pmatrix}, (\mathbf{i}, \mathbf{j})\right), \left(\begin{pmatrix} r & s \\ t & u \end{pmatrix}, (\mathbf{k}, \mathbf{l})\right)\right). \end{aligned}$$

Proof From Representation Theorem and Splitting Lemma we have that

$$\| P_{\mathbf{i},\mathbf{j}}[a + b\mathbf{i} + c\mathbf{j} + d\mathbf{ij}] - P_{\mathbf{k},\mathbf{l}}[r + s\mathbf{k} + t\mathbf{l} + u\mathbf{kl}]\|_{\mathcal{A}_{\mathbf{i}}(\mathbb{B}^4)}$$

$$\leq \| \frac{1}{2}(1 + \mathbf{ii})\,(a + b\mathbf{i} + c\mathbf{j} + d\mathbf{ij}) \circ \zeta - \frac{1}{2}(1 + \mathbf{ik})\,(r + s\mathbf{k} + t\mathbf{l} + u\mathbf{kl}) \circ \zeta \,\|_{L^2(\mathbb{B}^4_{\mathbf{i}}, \mathbb{H})}$$

$$
\begin{aligned}
&+\|\frac{1}{2}(1-\mathbf{ii})\,(a+b\mathbf{i}+c\mathbf{j}+d\mathbf{ij})-\frac{1}{2}(1-\mathbf{ik})\,(r+s\mathbf{k}+t\mathbf{l}+u\mathbf{kl})\,\|_{L^2(\mathbb{B}^4_{\mathbf{i}},\mathbb{H})}\\
&\leq\frac{1}{2}\left\{\|a-r\|_{L^2}+\|b-s\|_{L^2}+\|c-t\|_{L^2}+\|d-u\|_{L^2}\right\}\\
&\quad+\frac{1}{2}\left\{\|s\|_{L^2}\,\|\mathbf{i}-\mathbf{k}\|_{\mathbb{H}}++\|t\|_{L^2}\,\|\mathbf{j}-\mathbf{l}\|_{\mathbb{H}}+\|u\|_{L^2}\,\|\mathbf{ij}-\mathbf{kl}\|_{\mathbb{H}}\right\}\\
&\quad+\frac{1}{2}\left\{\|a-r\|_{L^2}+\|b-s\|_{L^2}+\|c-t\|_{L^2}+\|d-u\|_{L^2}\right\}\\
&\quad+\frac{1}{2}\left\{\|r\|_{L^2}\,\|\mathbf{i}-\mathbf{k}\|_{\mathbb{H}}+\|s\|_{L^2}\,\|\mathbf{ii}-\mathbf{kk}\|_{\mathbb{H}}+\|t\|_{L^2}\,\|\mathbf{ij}-\mathbf{kl}\|_{\mathbb{H}}\right.\\
&\qquad\left.+\|u\|_{L^2}\,\|\mathbf{iij}-\mathbf{kkl}\|_{\mathbb{H}}\right\}\\
&\quad+\frac{1}{2}\left\{\|a-r\|_{L^2}+\|b-s\|_{L^2}+\|c-t\|_{L^2}+\|d-u\|_{L^2}\right\}\\
&\quad+\frac{1}{2}\left\{\|s\|_{L^2}\,\|\mathbf{i}-\mathbf{k}\|_{\mathbb{H}}+\|t\|_{L^2}\,\|\mathbf{j}-\mathbf{l}\|_{\mathbb{H}}+\|u\|_{L^2}\,\|\mathbf{ij}-\mathbf{kl}\|_{\mathbb{H}}\right\}\\
&\quad+\frac{1}{2}\left\{\|r-a\|_{L^2}+\|s-b\|_{L^2}+\|t-c\|_{L^2}+\|u-d\|_{L^2}\}\right\}\\
&\quad+\frac{1}{2}\left\{\|r\|_{L^2}\,\|\mathbf{k}-\mathbf{i}\|_{\mathbb{H}}+\|s\|_{L^2}\,\|\mathbf{kk}-\mathbf{ii}\|_{\mathbb{H}}+\|t\|_{L^2}\,\|\mathbf{kl}-\mathbf{ij}\|_{\mathbb{H}}\right.\\
&\qquad\left.+\|u\|_{L^2}\,\|\mathbf{kkl}-\mathbf{iij}\|_{\mathbb{H}}\right\}\\
&\quad\leq 8\left(1+\|r\|_{L^2}+\|s\|_{L^2}+\|t\|_{L^2}+\|u\|_{L^2}\right)\\
&\qquad\times d\left(\left(\begin{pmatrix} a & b\\ c & d\end{pmatrix},(\mathbf{i},\mathbf{j})\right),\left(\begin{pmatrix} r & s\\ t & u\end{pmatrix},(\mathbf{k},\mathbf{l})\right)\right),
\end{aligned}
$$

where $\zeta(q)=\bar{q}$ for all $q\in\mathbb{H}$ and the relationships $\|a\circ\zeta\|_{L^2(\mathbb{B}^4_{\mathbf{i}},\mathbb{H})}=\|a\|_{L^2(\mathbb{B}^4_{\mathbf{i}},\mathbb{H})}=\|a\|_{L^2}$ and $\|\mathbf{kl}-\mathbf{ij}\|_{\mathbb{H}}\leq 2\left(\|\mathbf{k}-\mathbf{i}\|_{\mathbb{H}}+\|\mathbf{l}-\mathbf{j}\|_{\mathbb{H}}\right)\leq 4\|(\mathbf{i},\mathbf{j})-(\mathbf{k},\mathbf{l})\|_{\mathbb{R}^6}$ were used.

Corollary 1 *$\mathcal{P}_{\mathbb{B}^4}:\mathcal{HL}(D)\to\mathcal{A}_{\mathbf{i}}(\mathbb{B}^4)$ is a continuous surjective operator.*

Proposition 2 *If $(f,(\mathbf{i},\mathbf{j})),(g,(\mathbf{k},\mathbf{l}))\in\mathcal{A}_{\mathbf{i}}(\mathbb{B}^4)\times T$ then*

$$
\begin{aligned}
&d\left(\left(\begin{pmatrix} D_1[f,\mathbf{i},\mathbf{j}] & D_2[f,\mathbf{i},\mathbf{j}]\\ D_3[f,\mathbf{i},\mathbf{j}] & D_4[f,\mathbf{i},\mathbf{j}]\end{pmatrix},(\mathbf{i},\mathbf{j})\right),\left(\begin{pmatrix} D_1[g,\mathbf{k},\mathbf{l}] & D_2[g,\mathbf{k},\mathbf{l}]\\ D_3[g,\mathbf{k},\mathbf{l}] & D_4[g,\mathbf{k},\mathbf{l}]\end{pmatrix},(\mathbf{k},\mathbf{l})\right)\right)\\
&\leq 4\left(1+(1+\sqrt{2})\|g\|_{\mathcal{A}_{\mathbf{i}}(\mathbb{B}^4)}\right)\,\rho_{\mathbf{i}}\left((f,(\mathbf{i},\mathbf{j})),(g,(\mathbf{k},\mathbf{l}))\right).
\end{aligned}
$$

Proof From Representation Formula, Splitting Lemma and similar computations to the previous proof we can conclude that

$$d\left(\left(\begin{pmatrix} D_1[f,\mathbf{i},\mathbf{j}] & D_2[f,\mathbf{i},\mathbf{j}] \\ D_3[f,\mathbf{i},\mathbf{j}] & D_4[f,\mathbf{i},\mathbf{j}] \end{pmatrix}, (\mathbf{i},\mathbf{j})\right), \left(\begin{pmatrix} D_1[g,\mathbf{k},\mathbf{l}] & D_2[g,\mathbf{k},\mathbf{l}] \\ D_3[g,\mathbf{k},\mathbf{l}] & D_4[g,\mathbf{k},\mathbf{l}] \end{pmatrix}, (\mathbf{k},\mathbf{l})\right)\right)$$

$$\leq 4\left[\left(\int_D \|f(x+\mathbf{i}y) - g(x+\mathbf{i}y)\|^2 d\mu_{x,y}\right)^{\frac{1}{2}}\right.$$

$$+\left(\int_D \|g(x+\mathbf{i}y) - g(x+\mathbf{k}y)\|^2 d\mu_{x,y}\right)^{\frac{1}{2}}$$

$$\left.+\left(\int_D \|g(x+\mathbf{i}y)\|^2 d\mu_{x,y}\right)^{\frac{1}{2}} \|(\mathbf{i},\mathbf{j}) - (\mathbf{k},\mathbf{l})\|_{\mathbb{R}^6}\right] + \|(\mathbf{i},\mathbf{j}) - (\mathbf{k},\mathbf{l})\|_{\mathbb{B}^6}$$

$$\leq 4(1 + \|g\|_{\mathcal{A}_{\mathbf{i}}(D)})\, \rho_{\mathbf{i}}\left((f,(\mathbf{i},\mathbf{j})),(g,(\mathbf{k},\mathbf{l}))\right)$$

$$+4\left(\int_D \|g(x+\mathbf{i}y) - g(x+\mathbf{k}y)\|^2 d\mu_{x,y}\right)^{\frac{1}{2}}.$$

Given $(x, y) \in D$ from Remark 1 and Representation Formula we get

$$g(x+\mathbf{i}y) - g(x+\mathbf{k}y) = \frac{1}{2}\left[(\mathbf{i}-\mathbf{k})\mathbf{k}g(x+\mathbf{k}y) + (\mathbf{i}-\mathbf{k})\mathbf{k}g(x-\mathbf{k}y)\right].$$

Therefore, $\left(\int_D \|g(x+\mathbf{i}y) - g(x+\mathbf{k}y)\|^2 d\mu_{x,y}\right)^{\frac{1}{2}} \leq \sqrt{2}\|\mathbf{i}-\mathbf{k}\|_{\mathbb{H}}\|g\|_{\mathcal{A}_{\mathbf{i}}(\mathbb{B}^4)}$. and the two previous inequalities give us the main result.

Corollary 2 *Given* $(\mathbf{i},\mathbf{j}) \in T$ *and* $u \in \mathbb{S}^3$ *then* $\mathbf{S}_{\mathbf{i},\mathbf{j}}$ *and of* φ_u *are continuous operators.*

Proposition 3 *Consider* $(\mathbf{i},\mathbf{j}) \in T$. *Then* $(\mathcal{HL}(D), \mathcal{P}_{\mathbb{B}^4}, \mathcal{A}_{\mathbf{i}}(\mathbb{B}^4), T)$ *is a coordinate sphere bundle where* $K = \{R_v \mid v \in \mathbb{S}^3\}$ *is the structure group,* $\Phi = \{\varphi_v \mid v \in \mathbb{S}^3\}$ *is the family of trivializations and the operator* $\mathbf{S}_{\mathbf{i},\mathbf{j}}$ *is a section.*

Proof From Corollary 1 we see that $\mathcal{P}_{\mathbb{B}^4} : \mathcal{HL}(D) \to \mathcal{A}_{\mathbf{i}}(\mathbb{B}^4)$ is a continuous surjective operator and Corollary 2 shows that $\mathbf{S}_{\mathbf{i},\mathbf{j}}$ and φ_v are continuous operators. The set $K = \{R_v \mid v \in \mathbb{S}^3\}$ equipped with the composition is a group of homeomorphisms on T and $\{\varphi_v \mid v \in \mathbb{S}^3\}$ is the family of trivializations, see [16, Proposition 3.6] since our functions are slice regular functions. The transition functions are given by

$$g_{v,w}(f) = \varphi_{v,f}^{-1} \circ \varphi_{w,f}(\mathbf{i},\mathbf{j}) = \varphi_{v,f}^{-1} \circ \varphi_v(f, R_{\bar{v}w}(\mathbf{i},\mathbf{j})) = R_{\bar{v}w}(\mathbf{i},\mathbf{j}),$$

where $v, w \in \mathbb{S}^3$, $f \in \mathcal{A}_{\mathbf{i}}(\mathbb{B}^4)$ and $(\mathbf{i}, \mathbf{j}) \in T$. From direct computations we see that

$$\mathcal{P}_{\mathbb{B}^4} \circ \mathbf{S}_{\mathbf{i},\mathbf{j}}[f] = \mathcal{P}_{\mathbb{B}^4}\left(\begin{pmatrix} D_1[f, \mathbf{i}, \mathbf{j}] & D_2[f, \mathbf{i}, \mathbf{j}] \\ D_3[f, \mathbf{i}, \mathbf{j}] & D_4[f, \mathbf{i}, \mathbf{j}] \end{pmatrix}, (\mathbf{i}, \mathbf{j})\right) = f,$$

i.e., $\mathbf{S}_{\mathbf{i},\mathbf{j}}$ is a section of $(\mathcal{HL}(D), \mathcal{P}_{\mathbb{B}^4}, \mathcal{A}_{\mathbf{i}}(\mathbb{B}^4), T)$.

Remark 4 The operations, norm and quaternionic inner product induced on $\mathcal{HL}(D)$ by $\mathcal{A}_{\mathbf{i}}(\mathbb{B}^4)$ can be used to study the orthogonality, reproducing property of the Bergman kernel, the behavior of Toepliz operators among others facts from the point of view fiber bundle theory.

Proposition 4 *Given* $\mathbf{i}, \mathbf{k} \in \mathbb{S}^2$ *then* $(\mathcal{HL}(D), \mathcal{P}_{\mathbb{B}^4}, \mathcal{A}_{\mathbf{i}}(\mathbb{B}^4), T)$ *and* $(\mathcal{HL}(D), \mathcal{P}_{\mathbb{B}^4}, \mathcal{A}_{\mathbf{k}}(\mathbb{B}^4), T)$ *are isomorphic sphere bundles.*

Proof The isomorphism is $(\mathcal{I}, I)$, where $\mathcal{I} : \mathcal{HL}(D) \to \mathcal{HL}(D)$ and $I : \mathcal{A}_{\mathbf{i}}(\mathbb{B}^4) \to \mathcal{A}_{\mathbf{k}}(\mathbb{B}^4)$ are the identity mappings.

Remark 5 In a profoundly similar way to the computations presented in [16, Proposition 3.11] given $A, B \in \mathcal{HL}(D)$ we can define $A+B$, $A \bullet_{\mathbf{i},\mathbf{j}} B$, $\mathcal{D}(A)$, $\mathcal{R}_v(A)$ and $A * B$ to see some algebraic properties of $\mathcal{P}_{\mathbb{B}^4}$ such as $\mathcal{P}_{\mathbb{B}^4}(A + B) = \mathcal{P}_{\mathbb{B}^4}(A) + \mathcal{P}_{\mathbb{B}^4}(B)$, $\mathcal{P}_{\mathbb{B}^4}(A \bullet_{\mathbf{i},\mathbf{j}} B) = \mathcal{P}_{\mathbb{B}^4}(A) \bullet_{\mathbf{i},\mathbf{j}} \mathcal{P}_{\mathbb{B}^4}(B)$, $(\mathcal{P}_{\mathbb{B}^4}(A))' = \mathcal{P}_{\mathbb{B}^4}(\mathcal{D}(A))$, $\mathcal{R}_v(A) = P_{v\mathbf{i}\bar{v}, v\mathbf{j}\bar{v}}[\, v Q_{\mathbf{i},\mathbf{j}}[\mathcal{P}_{\mathbb{B}^4}(A)]\bar{v}\,]$ and $\mathcal{P}_{\mathbb{B}^4}(A * B) = \mathcal{P}_{\mathbb{B}^4}(A) * \mathcal{P}_{\mathbb{B}^4}(B)$.

On the other hand, the slice regular Bergman projection $B_{\mathbf{i}} : L \to \mathcal{A}_{\mathbf{i}}(\mathbb{B}^4)$ induces the pullback bundle $(B_{\mathbf{i}}^*(\mathcal{HL}(D)), \mathcal{P}'_{\mathbb{B}^4}, L, T)$, where the total space

$$B_{\mathbf{i}}^*(\mathcal{HL}(D)) := \{(f, A) \in L \times \mathcal{HL}(D) \mid \mathcal{P}_{\mathbb{B}^4}(A) = B_{\mathbf{i}}(f)\}$$

and $\mathcal{P}'_{\mathbb{B}^4}(f, A) = f$ for all $(f, A) \in B_{\mathbf{i}}^*(\mathcal{HL}(D))$. Note that $(f, A) \in L \times \mathcal{HL}(D)$ iff A under $\mathcal{P}_{\mathbb{B}^4}$ is assigned to slice regular part of f.

In addition, Toeplitz operator associated to $\mathcal{A}_{\mathbf{i}}(\mathbb{B}^4)$ are deeply related with pullback bundles. Given $\alpha \in C(\overline{\mathbb{B}_{\mathbf{i}}^4}, \mathbb{H})$ we see that the left- and right-Toeplitz operator with symbol α, induce the pullback bundles $((T_\alpha^{\mathbf{i}})^*(\mathcal{HL}(D)), \mathcal{P}'_{\mathbb{B}^4}, L, T)$ and $((T_{r,\alpha}^{\mathbf{i}})^*(\mathcal{HL}(D)), \mathcal{P}'_{\mathbb{B}^4}, L, T)$ respectively, where

$$(T_\alpha^{\mathbf{i}})^*(\mathcal{HL}(D)) := \{(f, A) \in L \times \mathcal{HL}(D) \mid \mathcal{P}_{\mathbb{B}^4}(A) = T_\alpha(f)\},$$

$$(T_{r,\alpha}^{\mathbf{i}})^*(\mathcal{HL}(D)) := \{(f, A) \in L \times \mathcal{HL}(D) \mid \mathcal{P}_{\mathbb{B}^4}(A) = T_{r,\alpha}(f)\},$$

$\mathcal{P}'_{\mathbb{B}^4}(f, A) = f$ for all $(f, A) \in (T_\alpha^{\mathbf{i}})^*(\mathcal{HL}(D))$ and $\mathcal{P}'_{\mathbb{B}^4}(f, A) = f$ for all $(f, A) \in (T_{r,\alpha}^{\mathbf{i}})^*(\mathcal{HL}(D))$.

Acknowledgments The author would like to thank the Institute. Politécnico Nacional (grant number SIP20241638) and CONAHCYT for partial support.

References

1. Axler, S., Conway, J.B., McDonald, G.: Toeplitz operators on Bergman spaces. Can. J. Math. **34**(2), 466–483 (1982)
2. Bernstein, H.J., Philips, A.: Fiber bundles and quantum theory. Sci. Amer. **245**(1), 122–137 (1981)
3. Bleecker, D.: Gauge Theory and Variational Principles. Dover Books on Physics Dover Books on Mathemtics. Courier Corporation, North Chelmsford (2005)
4. Brackx, F., Delanghe, R.: Hypercomplex function theory and Hilbert modules with reproducing kernel. Proc. Amer. Math. Soc. **37**, 545–576 (1978)
5. Brackx, F., Delanghe, R., Sommen, F.: Clifford Analysis. Pitman Research Notes in Mathematics, vol. 76. Pitman, London (1982)
6. Bredon, G.E.: Topology and Geometry. Springer, Berlin (1913)
7. Colombo, F., Gentili, G., Sabadini, I., Struppa, D.C.: Extension results for slice regular functions of a quaternionic variable. Adv. Math. **222**, 1793–1808 (2009)
8. Colombo, F., González-Cervantes, J.O., Luna-Elizarrarás, M.E., Sabadini, I., Shapiro, M.: On two Approaches to the Bergman Theory for Slice Regular Functions. Advances in Hypercomplex Analysis. Springer-Indam Series, vol. 1. Springer, Berlin (2012)
9. Colombo, F., González-Cervantes, J.O., Sabadini, I.: On slice biregular functions and isomorphisms of Bergman spaces. Compl. Var. Ell. Equ. **57**, 825–839 (2012)
10. Colombo, F., González-Cervantes, J.O., Sabadini, I.: The C-property for slice regular functions and applications to the Bergman space. Compl. Var. Ell. Equ. **58**, 1355–1372 (2013)
11. Colombo, F., Sabadini, I., Struppa, D.C.: Entire slice regular functions. Springer Briefs in Mathematics. Springer, Berlin (2016)
12. Constales, D.: The Bergman and Szegö kernels for separately monogenic functions. Zeit. Anal. Anwen. **9**, 97–103 (1990)
13. Delanghe, R.: On Hilbert modules with reproducing kernel. In: Functional and Theoretical Methods: Partial Differential Equations, Proceedings of the International Symposium. Darmstadt, 1976. Lecture Notes in Mathematics, vol. 561, pp. 158–170 (1976)
14. Gentili, G., Struppa, D.C.: A new theory of regular functions of a quaternionic variable. Adv. Math. **216**, 279–301 (2007)
15. Gentili, G., Stoppato, C., Struppa, D.C.: Regular Functions of a Quaternionic Variable. Springer Monographs in Mathematics. Springer, Heidelberg (2013)
16. González-Cervantes, J.O.: A fiber bundle over the quaternionic slice regular functions. Adv. Appl. Clifford Algebras **31**, 55 (2021). https://doi.org/10.1007/s00006-021-01158-z
17. González-Cervantes, J.O.: Quaternionic slice regular functions with some sphere bundles. Complex Var. Elliptic Equ. **67**(12), 3036–3047 (2022). https://doi.org/10.1080/17476933.2021.1971658
18. González-Cervantes, J.O.: On fiber bundles and quaternionic slice regular functions. Complex Anal. Oper. Theory **16**, 72 (2022). https://doi.org/10.1007/s11785-022-01253-4
19. González-Cervantes, J.O.: An extension of slice regular functions in terms of fiber bundle theory. Adv. Appl. Clifford Algebras **34**, 5 (2024). https://doi.org/10.1007/s00006-023-01309-4
20. Heidrich, R., Jank, G.: On iteration of quaternionic Möbius transformation. Compl. Var. Theory Appl. **29**, 313–318 (1996)
21. Shapiro, M., Vasilevski, N.: On the Bergman kernel function in hyperholomorphic analysis. Acta Appl. Math. **46**, 1–27 (1977)
22. Shapiro, M., Vasilevski, N.: On the Bergman kernel function in Clifford analysis. In: Bracks, F. et al. (eds.) Clifford Analysis and Their Applications in Mathematical Physics. Proceedings of the Third Conference, Deinze, Belgium, 1993. Fundamental Theories of Physics, vol. 55, pp. 183–192. Kluver Academic Publisher, Dordrecht (1993)

23. Shapiro, M., Vasilevski, N.: On the Bergman kernel functions in quaternionic analysis. Russian Math. Izv. VUZov **42**(2), 81–85 (1998)
24. Steenrod, N.: The Topology of Fibre Bundles. Princeton University Press, Princeton (1951)
25. Weatherall, J.O.: Fiber bundles, Yang-Mills theory, and general relativity. Synthese **193**, 2389–2425 (2016)

Part II
Complex Variables and Potential Theory

Inequalities for the Inner Radii of Domains Containing an Arbitrary Ellipse Points and Infinity

Iryna Denega and Yaroslav Zabolotnyi

Abstract An extremal problem for the maximum of product of the inner radii on a system of $n + 1$ mutually non-overlapping multiply connected domains D_k containing the points a_k, $k = 1, \ldots, n$, located on an arbitrary ellipse $\frac{x^2}{d^2} + \frac{y^2}{t^2} = 1$, where $d^2 - t^2 = 1$, and infinity is from the D_{n+1} is considered.

1 Preliminaries and Basic Definitions

In geometric function theory of a complex variable, a prominent place is occupied by the theory of extremal problems on the classes of univalent and multivalent functions [1–29]. This work is devoted to the study of extremal problems on the classes of analytic functions without common values. The symmetrization method plays an important role in solving the problems of geometric function theory of a complex variable. In early works [1–9] symmetrization methods required individual adaptation to each problem and this caused significant difficulties. In [10] it was developed the separating transformation method, which improved the possibilities of researching various problems. By this method authors considered the problems on extremal decomposition of the complex plane when the points are located on the unit circle [10, 15, 19] or on the radial systems [17–26].

In [27] we have considered an extremal problem on the maximum of product of the inner radii on a system of n mutually non-overlapping multiply connected domains D_k containing the points a_k, $k = 1, ..., n$, located on an arbitrary ellipse $\frac{x^2}{d^2} + \frac{y^2}{t^2} = 1$ for which $d^2 - t^2 = 1$. This work is an addendum to the paper [27], where we consider an extremal problem for the maximum of product of the inner radii on a system of $n + 1$ mutually non-overlapping multiply connected domains D_k, $k = 1, ..., n+1$, containing the points located on an arbitrary ellipse and $a_{n+1} = \infty \in D_{n+1}$.

I. Denega (✉) · Y. Zabolotnyi
Institute of mathematics of the National Academy of Sciences of Ukraine, Kyiv-4, Ukraine

M. R. Ebert et al. (eds.), *New Tools in Mathematical Analysis and Applications*, Research Perspectives, https://doi.org/10.1007/978-3-031-77050-0_6

Let $\mathbb{N}$, $\mathbb{R}$ be the sets of natural and real numbers, respectively, $\mathbb{C}$ be the complex plane, $\overline{\mathbb{C}} = \mathbb{C}\bigcup\{\infty\}$ be its one point compactification, U be the open unit disk in $\mathbb{C}$, $\mathbb{R}^+ = (0, \infty)$.

Definition 1 A function $g_D(z, a)$ which is continuous in $\overline{\mathbb{C}}$, harmonic in $D\backslash\{a\}$ apart from z, vanishes outside D, and in the neighborhood of a has the following asymptotic expansion

$$g_D(z, a) = -\ln|z - a| + \gamma + o(1), \quad z \to a,$$

(if $a = \infty$, then $g_D(z, \infty) = \ln|z| + \gamma + o(1)$, $z \to \infty$) is called the (classical) Green function of the domain D with pole at $a \in D$ [7, 8, 17, 19].

A domain D which carries a classical Green function will be said to be admissible. It is known that if each boundary point of D is the end-point of some Jordan arc lying in the exterior of D, then D is an admissible domain [12].

Definition 2 The inner radius $r(D, a)$ of the domain D with respect to a point a is the quantity e^{γ}.

Since for simply connected domains the inner radius coincides with the conformal radius [1, 7], if a function f maps the domain D conformally and univalently onto a domain $f(B)$, then

$$r(D, a)|f'(a)| = r(f(D), f(a)) \tag{1}$$

for each $a \in D$.

The inner radius increases monotonically with the growth of the domain: if $D \subset D'$ then

$$r(D, a) \leqslant r(D', a), \quad a \in B.$$

Let G be a domain in extended complex plane $\overline{\mathbb{C}}_z$.

Definition 3 By a quadratic differential in G we mean the expression

$$Q(z)dz^2, \tag{2}$$

where $Q(z)$ is a meromorphic function in G.

Definition 4 A finite point $z_0 \in G$ is called a zero or a pole of order n of the differential (2) if it is a zero or a pole, respectively, of the function $Q(z)$.

Definition 5 A circle domain for quadratic differential $Q(z)dz^2$ is called simply connected domain G, containing a unique double pole of the quadratic differential $Q(z)dz^2$ in the point $w = a \in G$, such that at univalent conformal mapping $w = f(z)$ $(f(a) = 0)$ the domain G onto the unit circle, the identity holds

$$Q(z)dz^2 \equiv -k\frac{dw^2}{w^2}, \quad k \in \mathbb{R}^+.$$

Thanks to the separating transformation method [10, 17, 19], it was possible for the first time to obtain an accurate estimate of the complex functional on families of mutually non-overlapping domains and for all values of the natural parameter n. Thus the following statement is valid.

Theorem 1 ([10]) *For any different points a_k, $k = 0, 1, \dots, n$ $(n \geqslant 2)$, $a_0 = 0$, $|a_k| = 1$, and any pairwise non-overlapping domains D_k, $a_k \in D_k \subset \overline{\mathbb{C}}$, $k = 0, 1, \dots, n$, $a_0 = 0 \in D_0$, the following inequality holds*

$$\prod_{k=0}^{n} r(D_k, a_k) \leqslant \frac{4^{\frac{1}{n}+n} n^n}{(n^2-1)^{\frac{1}{n}+n}} \left(\frac{n-1}{n+1}\right)^2. \tag{3}$$

In addition, if the domains D_k, $k = 0, 1, \dots, n$, are admissible, then equality holds in (3) *if and only if the D_k and a_k are the circle domains and poles of the quadratic differential*

$$Q(w)dw^2 = -\frac{(n^2-1)w^n + 1}{w^2(w^n-1)^2}\, dw^2.$$

The statement of the following problem is borrowed from Bakhtina's papers (see [17]):

$$\prod_{k=0}^{n} r(D_k, a_k) \longrightarrow \sup;$$

$a_0 = 0$, $|a_k| = 1$, $k = 1, \dots, n$, $a_k \in D_k$, $k = 0, 1, \dots, n$, $D_k \cap D_l = \emptyset$, $k \neq l$, $k, l = 0, 1, \dots, n$, $D_k = \{z : 1/\overline{z} \in D_k\}$, $k = 1, \dots, n$.

Theorem 2 ([17]) *Let a_k, $k = 1, \dots, n$ $(n \geqslant 2)$, be arbitrary different points on the circle $|z| = 1$, let $a_0 = 0$, and let D_k, $k = 0, 1, \dots, n$, be pairwise non-overlapping domains in $\overline{\mathbb{C}}$ such that $a_k \in D_k$, $k = 0, 1, \dots, n$, $D_0 \subset U$, and the D_k with $k = 1, \dots, n$, are mirror-symmetric relative to the circle $|z| = 1$. Then*

$$\prod_{k=0}^{n} r(D_k, a_k) \leqslant \frac{2^{2n+\frac{1}{n}}}{(n^2-2)^{\frac{1}{n}+\frac{n}{2}}} \left(\frac{n-\sqrt{2}}{n+\sqrt{2}}\right)^{\sqrt{2}}.$$

In addition, if the D_k, $k = 0, 1, \ldots, n$, *are admissible domains, then equality holds here if and only if the* D_k *and* a_k *are the circle domains and poles of the quadratic differential*

$$Q(w)dw^2 = -\frac{w^{2n} + (2n^2 - 2)w^n + 1}{w^2(w^n - 1)^2}dw^2. \tag{4}$$

Theorem 2 is the consequence of the following result.

Theorem 3 ([19]) *For any different points* a_k, $k = 1, \ldots, n$ $(n \geqslant 2)$, *on the circle* $|z| = 1$, *for* $a_0 = 0$, $a_{n+1} = \infty$ *and for any pairwise non-overlapping domains* D_k, $a_k \in D_k \subset \overline{\mathbb{C}}$, $k = 0, 1, \ldots, n+1$, $a_0 = 0 \in D_0$, $a_{n+1} = \infty \in D_{n+1}$, *the inequality holds*

$$\sqrt{r\left(D_0, a_0\right) r\left(D_{n+1}, a_{n+1}\right)} \prod_{k=1}^{n} r\left(D_k, a_k\right) \leqslant \frac{2^{2n+\frac{1}{n}}}{(n^2-2)^{\frac{1}{n}+\frac{n}{2}}} \left(\frac{n-\sqrt{2}}{n+\sqrt{2}}\right)^{\sqrt{2}}.$$

In addition, if the D_k, $k = 0, 1, \ldots, n+1$, *are admissible domains, then equality holds here if and only if the* D_k *are the circle domains and the* a_k *are the poles of the quadratic differential (4).*

The following theorem (see also [15]) is a significant strengthening of the result of Theorem 3.

Theorem 4 ([15]) *Let* $n \in \mathbb{N}$, $n \geqslant 2$, $\gamma \in \left(0, \frac{n^2}{8}\right]$. *Then for any different points* a_k, $k = 1, \ldots, n$, *on the circle* $|z| = 1$, *for* $a_0 = 0$, $a_{n+1} = \infty$ *and for any pairwise non-overlapping simply connected domains* D_k, $a_k \in D_k \subset \overline{\mathbb{C}}$, $k = 0, 1, \ldots, n+1$, $a_0 = 0 \in D_0$, $a_{n+1} = \infty \in D_{n+1}$, *the inequality holds*

$$\left(r\left(D_0, a_0\right) r\left(D_{n+1}, a_{n+1}\right)\right)^{\gamma} \prod_{k=1}^{n} r\left(D_k, a_k\right)$$

$$\leqslant \left(r\left(D_0^{(0)}, a_0\right) r\left(D_{n+1}^{(0)}, a_{n+1}\right)\right)^{\gamma} \prod_{k=1}^{n} r\left(D_k^{(0)}, a_k^{(0)}\right),$$

where $a_k^{(0)} = e^{2\pi i k/n}$, $k = 1, \ldots, n$, *and* $D_0^{(0)}$, $D_\infty^{(0)}$, $D_k^{(0)}$, $k = 1, \ldots, n$, *are the circle domains of the quadratic differential*

$$Q(w)dw^2 = -\frac{\gamma w^{2n} + (n^2 - 2\gamma)w^n + \gamma}{w^2(w^n - 1)^2}dw^2.$$

Combining the results of papers [23, 26, 28], we obtain the next statement.

Theorem 5 ([28]) *Let $n \in \mathbb{N}$, $n \geqslant 2$, $\gamma \in \mathbb{R}^+$. Then, for any fixed system of different points $A_n = \{a_k\}_{k=1}^n \in \mathbb{C}\backslash\{0\}$ and for any collection of mutually non-overlapping domains D_0, D_∞, D_k, $a_0 = 0 \in D_0 \subset \overline{\mathbb{C}}$, $\infty \in D_\infty \subset \overline{\mathbb{C}}$, $a_k \in D_k \subset \overline{\mathbb{C}}$, $k = 1, ..., n$, the following inequalities hold:*

(1) if $0 < \gamma \leqslant \frac{n+2}{2}$, then

$$(r(D_0, 0)\, r(D_\infty, \infty))^\gamma \prod_{k=1}^{n} r(D_k, a_k) \leqslant (n+1)^{-\gamma\frac{n+1}{n+2}} \prod_{k=1}^{n} |a_k|^{\frac{2\gamma}{n+2}}$$

$$\times \left((n-1)^{-\frac{n}{4}} \left(\prod_{1\leqslant k<p\leqslant n} |a_k - a_p| \right)^{\frac{2}{n-1}} \right)^{1-\frac{2\gamma}{n+2}},$$

(2) if $\gamma > \frac{n+2}{2}$, then

$$(r(D_0, 0)\, r(D_\infty, \infty))^\gamma \prod_{k=1}^{n} r(D_k, a_k) \leqslant (n+1)^{-\frac{n+1}{2}} \prod_{k=1}^{n} |a_k|.$$

2 Main Results

Let $M = \left\{ z = x + iy : \frac{x^2}{d^2} + \frac{y^2}{t^2} = 1 \right\}$ be an ellipse, where $d^2 - t^2 = 1$ and let $d^* = d - \sqrt{d^2-1}$. Then the following theorem holds.

Theorem 6 *Let $n \in \mathbb{N}$, $n \geqslant 3$, be some natural number; $A_n = \{a_k\}_{k=0}^n$ be any set of different fixed points of the complex plane such that $a_0 = \infty$, $a_k \in M$, $k = 1, ..., n$. Then, for any collection of mutually non-overlapping domains B_k, $k = 0, 1, ..., n$, such that $a_k \in B_k \subset \overline{\mathbb{C}} \setminus [-1; 1]$, $k = 0, 1, ..., n$, the following inequality holds*

$$\sqrt{r(B_0, \infty)} \prod_{k=1}^{n} r(B_k, a_k)$$

$$\leqslant \frac{\sqrt{2}(d^*)^n 2^{2n+\frac{1}{n}}}{(n^2-2)^{\frac{1}{n}+\frac{n}{2}}} \left(\frac{n-\sqrt{2}}{n+\sqrt{2}} \right)^{\sqrt{2}} \prod_{k=1}^{n} \left| \frac{\sqrt{a_k^2-1}}{a_k - \sqrt{a_k^2-1}} \right|. \tag{5}$$

Proof Let B_k^* and a_k^* be the images, respectively, of the domain B_k and the point a_k, $k = 0, 1, ..., n$, by the mapping

$$w = \frac{1}{d^*}\left(z - \sqrt{z^2 - 1}\right). \tag{6}$$

We consider the branch of the root for which $\sqrt{1} = 1$. Note that w maps the ellipse M onto the unit circle, therefore $|a_k^*| = 1$, $k = 1, ..., n$, and $a_0^* = 0$. Also note that $\bigcup\limits_{k=0}^{n} B_k^* \subset D$, where $D = \left\{w : |w| < \frac{1}{d^*}\right\}$. Let us determine how the inner radius of the domain changes after mapping w. We first consider the case $k = 1, ..., n$. Taking into account the invariance of the Green function under conformal and univalent mappings, we get

$$g_{B_k}(z, a_k) = g_{B_k^*}(w, a_k^*) = \ln \frac{1}{|w - a_k^*|} + \ln r(B_k^*, a_k^*) + o(1). \tag{7}$$

Let us now evaluate the expression $\ln \frac{1}{|w-a_k^*|}$. By using the mapping (6) we obtain

$$\ln \frac{1}{|w - a_k^*|} = \ln \left| \frac{d^*}{z - \sqrt{z^2 - 1} - a_k + \sqrt{a_k^2 - 1}} \right|.$$

It is easily seen that

$$\ln \frac{1}{|w - a_k^*|} = \ln \frac{1}{|z - a_k|} + \ln \left| \frac{d^*}{1 - \frac{\sqrt{z^2-1}-\sqrt{a_k^2-1}}{z-a_k}} \right|.$$

Then from here by multiplying and dividing the fraction in the denominator on the value $(\sqrt{z^2 - 1} + \sqrt{a_k^2 - 1})$ we get

$$\ln \frac{1}{|w - a_k^*|} = \ln \frac{1}{|z - a_k|} + \ln \left| \frac{d^*}{1 - \frac{(\sqrt{z^2-1}-\sqrt{a_k^2-1})(\sqrt{z^2-1}+\sqrt{a_k^2-1})}{(z-a_k)(\sqrt{z^2-1}+\sqrt{a_k^2-1})}} \right| =$$

$$= \ln \frac{1}{|z - a_k|} + \ln \left| \frac{d^*}{1 - \frac{z+a_k}{\sqrt{z^2-1}+\sqrt{a_k^2-1}}} \right|.$$

By performing simple transformations it follows that

$$\ln \frac{1}{|w - a_k^*|} = \ln \frac{1}{|z - a_k|} + \ln d^* \left| \frac{\sqrt{z^2 - 1} + \sqrt{a_k^2 - 1}}{\sqrt{z^2 - 1} + \sqrt{a_k^2 - 1} - z - a_k} \right| =$$

$$= \ln \frac{1}{|z - a_k|} + \ln d^* \left| \frac{\sqrt{a_k^2 - 1}}{a_k - \sqrt{a_k^2 - 1}} \right| + \ln \left| \frac{\sqrt{\frac{z^2-1}{a_k^2-1}} + 1}{1 + \frac{\sqrt{z^2-1}-z}{\sqrt{a_k^2-1}-a_k}} \right|.$$

Substituting this expression in (7) and taking into account that

$$\ln \left| \frac{\sqrt{\frac{z^2-1}{a_k^2-1}} + 1}{1 + \frac{\sqrt{z^2-1}-z}{\sqrt{a_k^2-1}-a_k}} \right| \to 0 \quad \text{at} \quad z \to a_k,$$

we obtain the equality

$$g_{B_k}(z, a_k) = \ln \frac{1}{|z - a_k|} + \ln d^* \left| \frac{\sqrt{a_k^2 - 1}}{a_k - \sqrt{a_k^2 - 1}} \right| r(B_k^*, a_k^*) + o(1).$$

From here there follows that

$$r(B_k, a_k) = d^* \left| \frac{\sqrt{a_k^2 - 1}}{a_k - \sqrt{a_k^2 - 1}} \right| r(B_k^*, a_k^*). \tag{8}$$

Now we prove that

$$r(B_0, \infty) = 2d^* r(B_0^*, 0). \tag{9}$$

From above-posed considerations

$$g_{B_0}(z, \infty) = g_{B_0^*}(w, 0) = \ln \frac{1}{|w|} + \ln r(B_0^*, 0) + o(1) =$$

$$= \ln \frac{d^*}{\left| z - \sqrt{z^2 - 1} \right|} + \ln r(B_0^*, 0) + o(1).$$

And we deduce

$$g_{B_0}(z,\infty)=\ln d^*|z+\sqrt{z^2-1}|+\ln r(B_0^*,0)+o(1)=$$

$$=\ln|z|+\ln 2d^*+\ln\left|\frac{1+\sqrt{1-\dfrac{1}{z^2}}}{2}\right|+\ln r(B_0^*,0)+o(1).$$

Thus the following equality holds

$$g_{B_0}(z,\infty)=\ln|z|+\ln 2d^* r(B_0^*,0)+o(1),$$

which proves the equality (9). Using the equalities (8) and (9), we obtain that

$$\sqrt{r\left(B_0,\infty\right)}\prod_{k=1}^{n} r\left(B_k,a_k\right)$$

$$=\sqrt{2}(d^*)^{\frac{1}{2}+n}\prod_{k=1}^{n}\left|\frac{\sqrt{a_k^2-1}}{a_k-\sqrt{a_k^2-1}}\right|\sqrt{r\left(B_0^*,0\right)}\prod_{k=1}^{n} r(B_k^*,a_k^*). \tag{10}$$

Denote $B_{n+1}^*=\left\{w:|w|>\dfrac{1}{d^*}\right\}$. It's clear that

$$r\left(B_{n+1}^*,\infty\right)=d^*.$$

Therefore, the following equality holds

$$\sqrt{r\left(B_0^*,0\right)}\prod_{k=1}^{n} r(B_k^*,a_k^*)=\frac{1}{\sqrt{d^*}}\sqrt{r\left(B_0^*,0\right)r\left(B_{n+1}^*,\infty\right)}\prod_{k=1}^{n} r(B_k^*,a_k^*). \tag{11}$$

The domains B_k^*, $k=0,...,n+1$, form a system of mutually non-overlapping domains for which all conditions of Theorem 6 of the paper [10] are correct. Thus the inequality holds

$$\sqrt{r\left(B_0^*,0\right)r\left(B_{n+1}^*,\infty\right)}\prod_{k=1}^{n} r(B_k^*,a_k^*)\leqslant\frac{2^{2n+\frac{1}{n}}}{(n^2-2)^{\frac{1}{n}+\frac{n}{2}}}\left(\frac{n-\sqrt{2}}{n+\sqrt{2}}\right)^{\sqrt{2}}.$$

Substituting the given inequality successively into the equality (10) and (11), we obtain the inequality (5). □

Denote

$$P = \exp\left(\frac{4\gamma}{n}\log\frac{\sqrt{\gamma}}{n} - n\log\frac{2}{n} - 2n\left(\frac{1}{2}-\frac{\sqrt{\gamma}}{n}\right)^2\log\left(\frac{1}{2}-\frac{\sqrt{\gamma}}{n}\right)\right)$$
$$\times\exp\left(-2n\left(\frac{1}{2}+\frac{\sqrt{\gamma}}{n}\right)^2\log\left(\frac{1}{2}+\frac{\sqrt{\gamma}}{n}\right)\right).$$

Theorem 7 *Let $n \in \mathbb{N}$, $n \geqslant 3$, be some natural number; γ be some positive real number and $\gamma \leqslant \frac{n^2}{8}$; $A_n = \{a_k\}_{k=0}^n$ be any set of different fixed points of the complex plane such that $a_0 = \infty$, $a_k \in M$, $k = 1, ..., n$. Then, for any collection of mutually non-overlapping simply connected domains B_k, $k = 0, 1, ..., n$, such that $a_k \in B_k \subset \overline{\mathbb{C}} \setminus [-1; 1]$, $k = 0, 1, ..., n$, the following inequality holds*

$$r^{\gamma}(B_0,\infty)\prod_{k=1}^{n} r(B_k,a_k) \leqslant 2^{\gamma}(d^*)^n \prod_{k=1}^{n}\left|\frac{\sqrt{a_k^2-1}}{a_k-\sqrt{a_k^2-1}}\right| \cdot P. \tag{12}$$

Proof As in the previous theorem, let B_k^* and a_k^* be the images, respectively, of the domain B_k and the point a_k, $k = 0, 1, ..., n$, by the mapping (6). Using the equality (1), for this mapping, domains B_k and points a_k, we get the equality

$$r(B_k,a_k) = d^*\left|\frac{\sqrt{a_k^2-1}}{a_k-\sqrt{a_k^2-1}}\right| r(B_k^*,a_k^*) \quad \text{for} \quad k = 1, ..., n$$

and the equality

$$r(B_0,\infty) = 2d^* r(B_0^*,0).$$

Hence

$$r^{\gamma}(B_0,\infty)\prod_{k=1}^{n} r(B_k,a_k) \tag{13}$$

$$= 2^{\gamma}(d^*)^{\gamma+n}\prod_{k=1}^{n}\left|\frac{\sqrt{a_k^2-1}}{a_k-\sqrt{a_k^2-1}}\right| r^{\gamma}\left(B_0^*,0\right)\prod_{k=1}^{n} r(B_k^*,a_k^*).$$

Then taking above-posed considerations from Theorem 6 into account, we have the next equality

$$r^{\gamma}\left(B_0^*,0\right)\prod_{k=1}^{n} r(B_k^*,a_k^*)=\frac{\left(r\left(B_0^*,0\right)r\left(B_{n+1}^*,\infty\right)\right)^{\gamma}}{(d^*)^{\gamma}}\prod_{k=1}^{n} r(B_k^*,a_k^*). \tag{14}$$

The domains B_k^*, $k=0,1,\ldots,n+1$, form a system of mutually non-overlapping domains for which all conditions of Theorem 2 of the paper [15] are correct. Therefore, the inequality holds

$$\left(r\left(B_0^*,0\right)r\left(B_{n+1}^*,\infty\right)\right)^{\gamma}\prod_{k=1}^{n} r(B_k^*,a_k^*)\leqslant P.$$

Substituting the given inequality successively into the equality (14) and (13), we obtain the inequality (12). □

Theorem 8 *Let $n\in\mathbb{N}$, $n\geqslant 3$, be some natural number; γ be some positive real number and $\gamma\in\mathbb{R}^{+}$; $A_n=\{a_k\}_{k=0}^{n}$ be any set of different fixed points of the complex plane such that $a_0=\infty$, $a_k\in M$, $k=1,\ldots,n$. Then, for any collection of mutually non-overlapping domains B_k, $k=0,1,\ldots,n$, such that $a_k\in B_k\subset\overline{\mathbb{C}}\setminus[-1;1]$, $k=0,1,\ldots,n$, the following inequality holds*

$$r^{\gamma}\left(B_0,\infty\right)\prod_{k=1}^{n} r\left(B_k,a_k\right)\leqslant 2^{\gamma}(d^*)^{n}\prod_{k=1}^{n}\left|\frac{\sqrt{a_k^2-1}}{a_k-\sqrt{a_k^2-1}}\right|\cdot T, \tag{15}$$

where

$$T=(n+1)^{-\gamma\frac{n+1}{n+2}}\left((n-1)^{-\frac{n}{4}}\left(\prod_{1\leqslant k<p\leqslant n}|a_k-a_p|\right)^{\frac{2}{n-1}}\right)^{1-\frac{2\gamma}{n+2}}\prod_{k=1}^{n}|a_k|^{\frac{2\gamma}{n+2}}.$$

The proof of Theorem 8 is similar to that of Theorem 6, taking Theorem 4 [28] and Theorem 2.1 [26] into account, so we have chosen to omit the analogous details.

Remark 1 Using some linear transformation $w=pz+z_0$ we can transform an arbitrary ellipse $\frac{x-x_0}{d_0^2}+\frac{y-y_0}{t_0^2}=1$ on the complex plane onto an ellipse of the form

$$\frac{x^2}{d^2}+\frac{y^2}{t^2}=1\quad\text{for which}\quad d^2-t^2=1.$$

Moreover, the inner radii of respective domains in this transformation will be treated as $|p|:1$. Therefore, in order to obtain an estimate of the product of the inner radii of non-overlapping domains containing points of an arbitrary ellipse and infinity, it is necessary to transform it onto the ellipse M by an appropriate linear transformation and apply Theorem 6.

References

1. Lavrentiev, M.A.: On the theory of conformal mappings. Trans. Sci. Inst. An. USSR. **5**, 159–245 (1934)
2. Polya, G., Szego, G.: Isoperimetric Inequalities in Mathematical Physics. Princeton University Press, Princeton (1951)
3. Jenkins, J.A.: Univalent Functions and Conformal Mappings. Springer, Berlin (1962)
4. Tamrazov, P.M.: On the general coefficient theorem. Math. USSR-Sb. **1**(1), 49–59 (1967)
5. Tamrazov, P.M.: Some extremal problems in conformal mapping. Math. USSR-Sb. **2**(1), 85–109 (1967)
6. Tamrazov, P.M.: Extremal conformal mappings and poles of quadratic differentials. Izv. Akad. Nauk SSSR. Ser. Mat. **32**, 1033–1043 (1968)
7. Goluzin, G.M.: Geometric Theory of Functions of a Complex Variable. American Mathematical Society, Providence (1969)
8. Lebedev, N.A.: The Area Principle in the Theory of Univalent Functions. Science, Moscow (1975)
9. Duren, P.: Univalent Functions. Springer, Heidelberg (1983)
10. Dubinin, V.N.: A separating transformation of domains and problems on extremal decomposition. J. Soviet Math. **53**(3), 252–263 (1991)
11. Hayman, W.K.: Multivalent Functions. Cambridge University Press, Cambridge (1994)
12. Ransford, T.: Potential Theory in the Complex Plane. Cambridge University Press, Cambridge (1995)
13. Tamrazov, P.M.: Moduli and extremal metrics on nonorientable and twisted Riemannian manifolds. Ukr. Math. J. **50**(10), 1586–1598 (1998)
14. Bergweiler, W.: On the number of critical points in parabolic basins. Ergod. Th. Dynam. Syst. **22**, 655–669 (2002)
15. Kuz'mina, G.V.: Problems of extremal decomposition of the Riemann sphere. J. Math. Sci. **118**(1), 4880–4894 (2003)
16. Bakhtin, A.K., Targonskii, A.L.: Extremal problems and quadratic differential. Nonlinear Oscillations **8**, 296–301 (2005)
17. Bakhtin, A.K., Bakhtina, G.P., Zelinskii, Y.B.: Topologico-algebraic structures and geometric methods in complex analysis. Zb. Prats Inst. of Math. of NASU, Kyiv **73**, 308pp. (2008)
18. Bakhtin, A.K., Targonskii, A.L.: Generalized (n, d)-ray systems of points and inequalities for nonoverlapping domains and open sets. Ukr. Math. J. **63**, 999–1012 (2011)
19. Dubinin, V.N.: Condenser Capacities and Symmetrization in Geometric Function Theory. Birkhäuser/Springer, Basel (2014)
20. Denega, I.V., Zabolotnii, Y.V.: Estimates of products of inner radii of non-overlapping domains in the complex plane. Complex Var. Elliptic Equ. **11**, 1611–1618 (2017)
21. Bakhtin, A.: Extremal decomposition of the complex plane with restrictions for free poles. J. Math. Sci. **231** 1–15 (2018)
22. Bakhtin, A.: Separating transformation and extremal problems on nonoverlapping simply connected domains. J. Math. Sci. **234**, 1–13 (2018)
23. Bakhtin, A.K., Denega, I.V.: Extremal decomposition of the complex plane with free poles II. J. Math. Sci. **246**(5), 602–616 (2020)
24. Bakhtin, A.K.: A problem of extreme decomposition of the complex plane with free poles. Ukr. Math. J. **71**(10), 1485–1509 (2020)

25. Denega, I.: Extremal decomposition of the complex plane for n-radial system of points. Azerbaijan J. Math. (Special issue dedicated to the 67th birth anniversary of Prof. M. Mursaleen) **11**, 64–74 (2021)
26. Bakhtin, A.K., Zabolotnii, Y.V.: Estimates of the products on inner radii for multiconnected domains. Ukr. Math. J. **73**(1), 6–21 (2021)
27. Zabolotnii, Y.V., Denega, I.V.: An extremal problem on non-overlapping domains containing ellipse points. Eurasian Math. J. **12**(4), 82–91 (2021)
28. Bakhtin, A.K., Denega, I.V.: Generalized M.A. Lavrentiev's inequality. J. Math. Sci. **262**(2), 138–153 (2022)
29. Denega, I.V., Zabolotnii, Y.V.: Application of upper estimates for products of inner radii to distortion theorems for univalent functions. Mat. Stud. **60**(2), 138–144 (2023)

On the Carleman-Bers-Vekua Equation

G. Giorgadze, G. Makatsaria, and N. Manjavidze

Abstract In the present work the Carleman-Bers-Vekua equations on the complex plane with the regular coefficients are considered. We have introduced special classes of solutions of these equations, which we call generalized meromorphic functions, and we study them both from the point of view of the pure theory of functions, as well as from the point of view of the analysis of boundary value problems of the theory of functions. Sufficiently important information about the behavior (hence the structure) of the generalized meromorphic function in the neighborhood of the point at infinity has been determined. Based on these results, it is possible to correctly pose natural boundary value problems for generalized meromorphic functions and carry out their (in some sense) complete analysis.

1 Generalized Meromorphic Functions

We will use the following terms and notations. $\mathbb{C}$ denotes a complex plane of the variable $z = x + iy$.

Let a and b be the regular coefficients, i.e. $a, b \in L_{p,2}(\mathbb{C})$, $p > 2$. This class was introduced by I. Vekua [4], for a special purpose. We say that the function $f \in L_{p,2}(G)$, $p > 2$, if

$$\iint_G |f(z)|^p \, dz < \infty, \iint_G \left|\frac{1}{z} f(\frac{1}{z})\right|^p dz < \infty, G = \{|z| \leq 1\}.$$

G. Giorgadze
Ivane Javakhishvili Tbilisi State University, Tbilisi, Georgia
e-mail: gia.giorgadze@tsu.ge

G. Makatsaria
Vladimer Chavchanidze Institute of Cybernetics of the Georgian Technical University, Tbilisi, Georgia

N. Manjavidze (✉)
Ilia State University, Tbilisi, Georgia
e-mail: ninomanjavidze@iliauni.edu.ge

M. R. Ebert et al. (eds.), *New Tools in Mathematical Analysis and Applications*, Research Perspectives, https://doi.org/10.1007/978-3-031-77050-0_7

The generalized analytic functions are interpreted in different ways. We define it according to [4]. The generalized analytic functions are the solutions of the equation, so called Carleman-Bers-Vekua equation

$$\frac{\partial w}{\partial \bar{z}} + aw + b\bar{w} = 0 \tag{1}$$

where the derivative is understood in Sobolev sense. The generalized solutions of this equation are called the generalized analytic functions [4]. General representation of generalized analytic functions by means of analytic functions is the strongest tool in order to investigate the solutions of the equation $(C - B - V)$. In particular, there exists the formula

$$w(z) = \Phi(z)\, e^{\Omega(z)}, \tag{2}$$

where

$$\Omega(z) = \frac{1}{\pi} \iint_{\mathbb{C}} \left(A(\zeta) + B(\zeta) \frac{\overline{w(\zeta)}}{w(\zeta)} \right) \frac{dC_\zeta}{\zeta - z} \tag{3}$$

By these formulas the bijective relationship between the generalized analytic functions and the classical analytic functions is established. Φ is called the analytic divisor of the analytic function w.

It is essential that the general solution is given by means of an analytic function having the function $\Omega(z)$ as exponent. When the coefficients a and b are regular, many properties of the analytic functions are easily transferred to the solution, but in case of the singular coefficients, more effort is needed. Every bounded solution of the equation $(C - B - V)$ with regular coefficients have the same property as the entire function of a complex variable.

Let $D \subset \mathbb{C}$ be a discrete set $D = \{z_1, z_2, \ldots, z_n, \ldots\}$. A function w is called generalized meromorphic function with D singularities, if:

1. w is a generalized regular solution of the Eq. (1) in $\mathbb{C} \setminus D$;
2. $w(z) \to \infty, z \to z_k, (k = 1, 2, \ldots)$.

Recall that the research of the generalized analytic functions is based on Vekua's Theorem and the formulas (2),(3) hold. Below the following problem, which we conventionally call Problem A, is of essential importance.

Problem A Given a number $\nu \in R$, $\nu > 0$, a continuous function $H : R \to R$ with the period 2π, and nonnegative integer $N \in N_0 = \{0, 1, 2, \ldots\}$. Find a generalized analytic function w such that

$$\log^+ |w(z)| = O\left(|z|^\nu H(\varphi) + N \log^+ |z|\right), z \to \infty. \tag{4}$$

The following theorem holds.

Theorem 1 *Let H be a function such that for some numbers $\varphi_0 \in R, n \in N$, we have*

$$H\left(\varphi_k\right) \leq 0, k = 0, 1, \ldots, n,$$

where $\varphi_k = \varphi_0 + \frac{2\pi}{n}k$ and $\nu < \frac{n}{2}$. Then the solution $w(z)$ of the Problem A in the class of generalized meromorphic functions, is a generalized rational function [4] whose order of pole at infinity doesn't exceed n. Therefore, the function Φ in (2) will turn out to be a rational function whose pole order at infinity doesn't exceed n.

Remark 1 If $\nu = 0, n = 0$, H is arbitrary function, and the coefficients a and b are zeros almost everywhere, then Theorem 1 gives us the Liouville classical theorem [2, 3, 5].

Remark 2 As was mentioned above, Problem A has many applications in the investigation of the equation of type (1), with singular coefficients, but this is another subject, and we will not touch it here.

Remark 3 Theorem 1 allows us to correctly pose the boundary value problems for the Eq. (1) and to make in some sense their complete investigation.

Theorem 2 *If the generalized analytic function $w(z)$ satisfies (4), where*

$$\int_0^{2\pi} H\left(\varphi\right) d\varphi < 0,$$

then it is a trivial function.

It should be mentioned that the obtained results are new for the classical meromorphic functions. All the above results are essentially based on the iterative transformation of the meromorphic functions studied by the authors. The obtained results continue the research started in the monograph [1].

2 Boundary Value Problems

Consider the Riemann-Hilbert boundary value problem

$$Re\left\{\overline{\lambda\left(t\right)}w\left(t\right)\right\} = \gamma\left(t\right), t \in \Gamma, \qquad \text{(R - H)}$$

for the Carleman-Bers-Vekua equation

$$\partial_{\bar{z}}w + A\left(z\right)w + B\left(z\right)\bar{w} = 0 \qquad \text{(C - B - V)}$$

with the polar singularities in the finite, $m+1$-connected domain G of the complex plane with sufficiently smooth boundary provided that the given functions $\lambda\left(t\right)$ and $\gamma\left(t\right)$ are the Hölder continuous functions.

Let G be a bounded complex domain with the boundary Γ consisting of closed nonintersecting Liapunov smooth $\Gamma_0, \Gamma_1, \ldots, \Gamma_m$ contours and Γ_0 encloses all the rest. Let G^* be some finite subset of the set G

$$G^* = \{z_1, z_2, \ldots, z_N\}, N \geq 1. \tag{5}$$

Consider the equation $(C - B - V)$ in the domain G, provided that the coefficient $B(z) \in L_p(G)$, $p > 2$ and the coefficient $A(z)$ admits the following representation

$$A(z) = \overline{g(z)} + \sum_{k=1}^{N} \frac{A_k(z)}{|z - z_k|^{\nu_k}},$$

where the function $g(z)$ is holomorphic in $G \setminus G^*$ and has continuous boundary value on Γ. The function $A_k(z)$ admits the following representation

$$A_k(z) = a_k(z) \exp\{i n_k arg(z - z_k)\},$$

where

$$\frac{a_k(z) - \lambda_k}{|z - z_k|^{\nu_k}} \in L_p(G), p > 2.$$

The constants λ_k, ν_k, n_k are correspondingly complex, positive, and entire numbers for every $k = 1, 2, \ldots, N$.

Under the solution of the equation $(C - B - V)$ is understood the continuous generalized solution in $G \setminus G^*$. Denote by $\mathcal{R}(A, B, G \setminus G^*)$ the set of all possible such solutions.

Everywhere below the fulfillment of the following condition

$$\lambda_k \neq 0, |n_k - 1| > 2(\nu_k - 1) > 0, k = 1, 2, \ldots, N \tag{6}$$

is assumed.

As seen from (5) the coefficient $A(z)$ has the polar singularities of the form $|z - z_k|^{-\nu_k}$ and the singularities of the function $\overline{g(z)}$ in the points z_k. Below it will be established that the structure of the solutions of the Carleman-Bers-Vekua equation depends on the relationship between the parameters λ_k, ν_k, n_k, and if the required conditions between them are violated then generally speaking the assumptions proved below are not valid.

Below we will need the following notations:

$$q_k = Res_{z=z_k} g(z), k = 1, 2, \ldots, N; \; Q_k = \frac{1}{2\pi i} \int_{\Gamma_k} g(t)\, dt, k = 1, 2, \ldots, m.$$

Introduce an auxiliary function given by the formula:

$$f(z) = \int_{\Gamma_{\zeta_0,z}} \overline{g(t)dt} - \sum_{k=1}^{m} \overline{Q_k} \log(\overline{z} - \overline{\tau_k}) - \sum_{k=1}^{N} \overline{q_k} \log(\overline{z} - \overline{z_k}),$$

in the domain $G \setminus G^*$ where ζ_0 is some fixed point in $G \setminus G^*$; $\Gamma_{\zeta_0,z}$ is a smooth contour connecting the points ζ_0 and z and lying in $G \setminus G^*$; τ_k is an arbitrary fixed point inside the contour Γ_k, $k = 1, 2, \ldots, m$. Consider also the function

$$F(z) = \Lambda(z) \exp\{2iImf(z)\} \chi(z), z \in G \setminus G^*,$$

where

$$\Lambda(z) = \prod_{k=1}^{N} (z - z_k)^{-[2Req_k]}$$

$$\chi(z) = \exp\left\{2\sum_{k=1}^{m} Q_k \log|z - \tau_k| + 2\sum_{k=1}^{N} q_k \log|z - z_k|\right\}.$$

It follows from the conditions (6) that $2 - \nu_k - n_k \neq 0, k = 1, 2, \ldots, N$ and therefore by the formulas

$$\delta_k^* = \frac{2\lambda_k}{2 - \nu_k - n_k}, k = 1, 2, \ldots, N$$

the definite non-zero numbers are given. Assume

$$R(z) \equiv \sum_{k=1}^{N} \frac{\delta_k^*}{|z - z_k|^{\nu_{k-1}}} \exp\{i(n_k - 1) arg(z - z_k)\},$$

$$\Psi(z) \equiv F(z) \exp\{R(z)\}.$$

Consider the following Carleman-Bers-Vekua equation

$$\frac{\partial w_*}{\partial \acute{z}} + A_*(z) w_* + B_*(z) \overline{w_*} = 0, \tag{7}$$

where

$$A_*(z) = \sum_{k=1}^{N} \frac{a_k(z) - \lambda_k}{|z - z_k|^{\nu_k}} e^{in_k arg(z - z_k)},$$

$$B_*(z) = \frac{B(z)\Psi(z)}{\overline{\Psi(z)}}.$$

It is easy to see, that $A_*(z)$, $B_*(z) \in L_p(G)$, $p > 2$ and hence (7) is the regular Carleman-Bers-Vekua equation.
The following preparation theorem takes place.

Theorem 3 *When $w_* \in R(A_*, B_*, G \setminus G^*)$, $w \in R(A, B, G \setminus G^*)$, the relation*

$$w_*(z) = \Psi(z)\, w(z),\, z \in G \setminus G^* \tag{8}$$

is a bijection between the classes $R(A, B, G \setminus G^*)$ and $R(A_*, B_*, G \setminus G^*)$.

Let $\delta = (\delta_1, \delta_2, \ldots, \delta_N)$ and $\sigma = (\sigma_1, \sigma_2, \ldots, \sigma_N)$ are given N-dimensional vectors with the non-negative components. Denote by $\Omega_0[\delta, \sigma]$ the class of all possible functions from the set $R(A, B, G \setminus G^*)$ satisfying the condition

$$w(z) = O\left(\exp\left\{\delta_k |z - z_k|^{-\sigma_k}\right\}\right),\, z \to z_k,\, k = 1, 2, \ldots, N.$$

Denote by $\overline{\Omega_0}[\delta, \sigma]$ the class of all possible functions from the set $\Omega_0[\delta, \sigma]$ admitting the continuous extension in $\overline{G} \setminus G^*$. By δ^* and ν^* the following vectors are denoted

$$\delta^* \equiv \left(\left|\delta_1^*\right|, \left|\delta_2^*\right|, \ldots, \left|\delta_N^*\right|\right),\, \nu^* \equiv (\nu_1 - 1, \nu_2 - 1, \ldots, \nu_N - 1).$$

The following theorem is true.

Theorem 4 *If for some value k of the index the inequality $\delta_k < \left|\delta_k^*\right|$ is fulfilled then the class $\Omega_0\left[\delta, \nu^*\right]$ is a trivial class (i.e., it contains only zero functions).*

Theorem 4 implies that if for some value k of the index the inequality $\sigma_k < \nu_k - 1$ is fulfilled then for every vector δ the class $\Omega_0[\delta, \sigma]$ is a trivial class. Therefore, if $\sigma_k = \nu_k - 1, k = 1, 2, \ldots, N$, and for some k_0 on of the inequalities $\delta_{k_0} < \left|\delta_{k_0}^*\right|$ or $\sigma_{k_0} < \nu_{k_0} - 1$ is fulfilled, then the class $\Omega_0[\delta, \sigma]$ is a trivial class.
It is natural to investigate the class $\Omega_0[\delta, \sigma]$. The following theorem gives us the representation of the solution of this class.

Theorem 5 *Relation (8) establishes the bijection between the pairs of classes*

$$\Omega_0\left[\delta^*, \nu^*\right] \textit{and } R(A_*, B_*, G),$$

$$\overline{\Omega_0}\left[\delta^*, \nu^*\right] \textit{and } R(A_*, B_*, G) \cap C\left(\overline{G}\right)$$

To pose correctly the Riemann-Hilbert boundary value problem it is clear from the aforementioned results that it is sufficient to require from the solution of the equation $(C - B - V)$ the fulfillment of the asymptotic condition of the form

$$w(z) = O\left(\exp\left\{\left|\delta_k^*\right| |z - z_k|^{-\nu_k+1}\right\}\right), z \to z_k, k = 1, 2 \ldots, N \tag{*}$$

Consider the following boundary value problem: on the boundary Γ the Hölder continuous functions $\lambda(t)$ and $\gamma(t)$ are given, $\gamma(t)$ is a real function and $|\lambda(t)| = 1$.

Find the function $w(z) \in \overline{\Omega_0}[\delta, \sigma]$ satisfying the boundary equation

$$Re\left\{\overline{\lambda(t)} w(t)\right\} = \gamma(t), t \in \Gamma. \tag{9}$$

From the Theorem 4 it follows that in case when $\gamma(t) \neq 0$ the problem (9) isn't solvable in the class $\overline{\Omega_0}[\delta, \sigma]$ if $\sigma_k = \nu_k - 1$ for some k, or if $\sigma_k = \nu_k - 1, k = 1, 2, \ldots, N$, but $\delta_k < \left|\delta_k^*\right|$ for some k.

Let $\delta_k \neq 0, \sigma_k \geq \nu_k - 1, k = 1, 2, \ldots, N$. Denote by $\mathcal{H}$ the set of all possible values k for which $\delta_k < \left|\delta_k^*\right|$.

The following theorem takes place.

Theorem 6 *The homogeneous boundary value problem (9) ($\gamma(t) = 0$) in the class $\overline{\Omega_0}[\delta, \sigma]$ has infinite number of linearly independent solutions if and only if one of the conditions is fulfilled:*

1. $\mathcal{H} = \emptyset$; $\sum_{k=1}^{N}(\delta_k + \sigma_k) > \sum_{k=1}^{N}\left(\left|\delta_k^*\right| + \nu_k - 1\right)$;
2. $\mathcal{H} = \emptyset$; $\sigma_k > \nu_k - 1, k \in \mathcal{H}$.

Proof Let the condition 1 be fulfilled. Then for all $k = 1, 2, \ldots, N$, the following inequalities hold

$$\delta_k \geq \left|\delta_k^*\right| + \nu_k - 1,$$

and there exists at least one $k = k_0$ so that the strict inequality is fulfilled

$$\delta_{k_0} + \sigma_{k_0} > \left|\delta_{k_0}^*\right| + \nu_{k_0} - 1.$$

The last inequality implies that one of the inequalities $\delta_{k_0} \geq \left|\delta_{k_0}^*\right|$ or $\sigma_{k_0} \geq \nu_{k_0} - 1$, is also strict. Let $\delta_{k_0} > \left|\delta_{k_0}^*\right|$ and let us fix an arbitrary number $S \geq 0$. Consider the solutions $w_*(z)$ from the class $R(A_*, B_*, G \setminus G^*)$ which are representable in the form

$$w_*(z) = \frac{\Phi(z)}{\left(z - z_{k_0}\right)^s} \exp(\omega(z)), \tag{10}$$

where $\Phi(z)$ is holomorphic in G and continuous in $\overline{G}$ function. It is easy to see that every function of the form (10) defines the solution $w(z)$ of the equation $(C - B - V)$

of the class $\overline{\Omega_0}[\delta, \sigma]$ by means of the relation (8). Further on, we can see that by the relation

$$w_*(z) = \frac{\omega_0(z)}{\left(z - z_{k_0}\right)^s} \tag{11}$$

the bijective correspondence between the class of all functions of the form (10) and the class

$$R\left(A_*, B_*, \left(\frac{z - z_{k_0}}{\overline{z} - \overline{z_{k_0}}}\right)^s, G\right) \cap C\left(\overline{G}\right)$$

of the solutions of the equation

$$\frac{\partial \omega_0}{\partial \overline{z}} + A_*(z)\,\omega_0 + B_*\left(\frac{z - z_{k_0}}{\overline{z} - \overline{z_{k_0}}}\right)^s \overline{\omega_0} = 0 \tag{12}$$

is established.

Parallel with the problem (9) consider the following boundary value problem: Find the generalized solution of the problem (12) continuous in G and satisfying the boundary condition

$$Re\left\{\frac{\overline{\lambda(t)}}{\left(t - z_{k_0}\right)^s \Psi(t)}\omega_0(t)\right\} = 0,\ t \in \Gamma. \tag{13}$$

By the formulas (11)and (8), every system of linearly independent solutions of the problem (13) defines the system of linearly independent solutions of homogeneous problem (9). On the other hand the number of linearly independent solutions of the problem (13) l, satisfies the inequality

$$l \geq 2ind\left(\frac{\lambda(t)}{\left(\overline{t} - \overline{z_{k_0}}\right)^s \overline{\Psi(t)}}\right) - m + 1,$$

by virtue of which we get

$$l \geq 2\left(ind\lambda(t) + S + \sum_{k=1}^{N}[2Req_k]\right) - m + 1.$$

From here it follows that if an appropriate choice of S, the number l will be arbitrarily large and therefore the homogeneous problem (9) has infinite number of linearly independent solutions. One can prove similarly that the set of linearly independent solutions of the homogeneous problem (9) is infinite in case

$$\sigma_{k_0} > \nu_{k_0} - 1, \delta_{k_0} = \left|\delta^*_{k_0}\right|.$$

Hence we obtain that if the condition 1 of theorem is fulfilled, then the homogeneous problem (9) has the infinite number of linearly independent solutions. Assume now the condition 2 holds. Then based on the relation

$$\overline{\Omega_0}\left[\delta^{(1)}, \sigma^{(1)}\right] \subset \overline{\Omega_0}\left[\delta^{(2)}, \sigma^{(2)}\right]$$

which follows directly from the conditions

$$\delta_k^{(2)} \neq 0, k = 1, 2, \ldots, N; \sigma_k^{(2)} > \sigma_k^{(1)}, k \in \mathcal{H}.$$

We get that the homogeneous problem (9) has infinite number of linearly independent solutions. The sufficiency of one of the conditions 1,2 is proved.
Let us prove the necessity.

Let the homogeneous problem (9) has infinite number of linearly independent solutions and let $\mathcal{H} \neq \emptyset$. Then for every $k \in \mathcal{H}$ we have $\sigma_k > \nu_k - 1$. Indeed, if for at least one $k_0 \in \mathcal{H}$ we have $\sigma_{k_0} > \nu_{k_0} - 1$, then the class $\overline{\Omega_0}[\delta, \sigma]$ consists of only zero elements. Thus we get a contradiction. Hence, when $\mathcal{H} \neq \emptyset$ the condition 2 is fulfilled.
Let now $\mathcal{H} = \emptyset$. Then the following is true

$$\sum_{k=1}^{N} (\delta_k - \sigma_k) > \sum_{k=1}^{N} \left(\left|\delta_k^*\right| + \nu_k - 1\right). \tag{14}$$

Indeed, otherwise we would have

$$\sum_{k=1}^{N} (\delta_k + \sigma_k) = \sum_{k=1}^{N} \left(\left|\delta_k^*\right| + \nu_k - 1\right) \tag{15}$$

and therefore $\delta_k = \left|\delta_k^*\right|, \sigma_k = \nu_k - 1, k = 1, 2, \ldots, N$. Hence, we obtain that the homogeneous problem (9) has infinite number of linearly independent solutions in the class $\overline{\Omega_0}\left[\delta^*, \nu^*\right]$, but the problem has finite number of linearly independent solutions. Indeed, using the relation (8) the bijective correspondence is established between the solutions of the problem (9) and the following boundary value problem: find the generalized solution of the equation

$$\frac{\partial w_*}{\partial \overline{z}} + A_* (z)\, w_* (t) + B_* \overline{w_*} = 0 \tag{16}$$

continuous in the domain $\overline{G}$ satisfying the boundary condition

$$Re\left\{\frac{\overline{\lambda(t)}}{\Psi(t)}w_*(t)\right\}=\gamma(t),t\in\Gamma. \tag{17}$$

The homogeneous problem (16) and (17) ($\gamma(t)=0$) based on [3] has finite number of linearly independent solutions. Because of why the homogeneous problem (9) has finite number of linearly independent solutions in the class $\overline{\Omega_0}\left[\delta^*,\nu^*\right]$. Therefore the condition (15) is not fulfilled and thus, (14) holds.
Theorem 6 is completely proved. □

Consider the boundary value problem: find the generalized solution continuous in the class $\overline{G}$ of the equation

$$\frac{\partial w_*^{'}}{\partial\overline{z}}-A_*(z)\,w_*^{'}(t)-\overline{B_*(z)\overline{w_*^{'}}}=0$$

satisfying the boundary condition

$$Re\left\{\frac{\lambda(t)}{\overline{\Psi(t)}}t^{'}(s)\,w_*^{'}(t)\right\}=0,t\in\Gamma. \tag{18}$$

It is easy to see that the number of linearly independent solutions of the problem (18) l_*, is finite and it is clear that for the problem (16) to be solvable it is necessary and sufficient the validity of the following equation

$$\int_\Gamma\frac{\lambda(t)}{\overline{\Psi(t)}}\gamma(t)\,w_*^{'}(t)\,dt=0 \tag{19}$$

for every solution of the problem (18).
Based on above obtained results the following theorem becomes evident.

Theorem 7 *The homogeneous problem (9) in the class* $\overline{\Omega_0}\left[\delta^*,\nu^*\right]$ *has finite number of linearly independent solutions and the non-homogeneous problem is solvable if and only if the condition (19) is fulfilled.*

Let l be a number of linearly independent solutions of the homogeneous problem (9). By means of the evident equality

$$ind\left(\frac{1}{\overline{\Psi(t)}}\right)=\sum_{k=1}^{N}[2Req_k]$$

it follows the validity of the next theorem.

Theorem 8 *The following condition*

$$l - l_* = 2n + 2\sum_{k=1}^{N} [2Req_k] - m + 1$$

takes place, where q_k is a residue of the function $q(z)$ at a point z_k.

From the results obtained above in particular we get

Theorem 9 *For the problem (9) to be Noetherian in the class $\overline{\Omega_0}[\delta, \sigma]$ it is necessary and sufficient the fulfillment of the condition*

$$\delta = \delta^*, \sigma = \nu^*.$$

Remark 4 We do not give the proofs of some theorems formulated above, because they are too extensive to fit the publishing format.

Acknowledgments This work was supported by grant no. FR 22-354 from the Shota Rustaveli National Science Foundation.

References

1. Akhalaia, G., Giorgadze, G., Jikia, V., Makatsaria, G., Manjavidze, N.: Elliptic Systems on Riemann Surfaces, vol. 13, pp. 1–154. Tbilisi International Center of Mathematics and Informatics, Tbilisi (2012)
2. Goncharov, A.L., Klimentov, S.B., Usmanov, Z.D.: The Analog of Liouville theorem for one class of Cauchy-Riemann type system with singular coefficients. Vladikavkaz. Math. J. **7**(4), 7–16 (2005)
3. Makatsaria, G.: Liouville-type theorems for generalized analytic functions. Bull. Acad. Sci. Georgian SSR **113**(3), 485–488 (1984)
4. Vekua, I.: Generalized Analytic Functions. Pergamon, Oxford (1962)
5. Vinogradov, V.S.: Liouville's theorem for generalized analytic functions. Dokl. Akad. Nauk. SSSR **183**(3), 500–508 (1968)

On Asymptotics of Ring Q-Homeomorphisms with Respect to p-Modulus Near the Origin

Ruslan Salimov and Bogdan Klishchuk

Abstract We consider the class of ring Q-homeomorphisms with respect to p-modulus in $\mathbb{R}^n$ with $p > n$, and obtain lower bounds for limsups of the distance distortions under such mappings. These estimates can be treated as Hölder's continuity of the inverses near the origin. The sharpness is illustrated by example.

1 Introduction

In this paper, we study the distortion of distances under ring Q-homeomorphisms with respect to p-modulus in $\mathbb{R}^n$ with $p > n$. For such mappings we derive lower bounds for upper limits.

Let us recall some definitions, see [1]. Let Γ be a family of curves γ in $\mathbb{R}^n$, $n \geqslant 2$. A Borel measurable function $\rho : \mathbb{R}^n \to [0, \infty]$ is called *admissible* for Γ, (abbr. $\rho \in \operatorname{adm} \Gamma$), if

$$\int_{\gamma} \rho(x)\, ds \geqslant 1$$

for all locally rectifiable $\gamma \in \Gamma$. Let $p \in (1, \infty)$.

The quantity

$$M_p(\Gamma) = \inf_{\rho \in \operatorname{adm} \Gamma} \int_{\mathbb{R}^n} \rho^p(x)\, dm(x)$$

is called *p–modulus* of the family Γ. Here $dm(x)$ corresponds to the Lebesgue measure in $\mathbb{R}^n$.

R. Salimov · B. Klishchuk (✉)
Institute of Mathematics of NAS of Ukraine, Kyiv-4, Ukraine

M. R. Ebert et al. (eds.), *New Tools in Mathematical Analysis and Applications*, Research Perspectives, https://doi.org/10.1007/978-3-031-77050-0_8

Let $Q : D \to [0, \infty]$ be a measurable function. Recall that a homeomorphism $f : D \to \mathbb{R}^n$ is said to be a *Q-homeomorphism with respect to p-modulus* if

$$\mathrm{M}_p(f\Gamma) \leqslant \int\limits_D Q(x)\, \rho^p(x)\, dm(x)$$

for every family Γ of paths in D and every admissible function ρ for Γ.

For arbitrary sets E, F and G of $\mathbb{R}^n$ we denote by $\Delta(E, F, G)$ a set of all continuous curves $\gamma : [a, b] \to \mathbb{R}^n$, that connect E and F in G, i.e., such that $\gamma(a) \in E$, $\gamma(b) \in F$ and $\gamma(t) \in G$ for $a < t < b$.

Let D be a domain in $\mathbb{R}^n$, $n \geqslant 2$, $x_0 \in D$ and $d_0 = \operatorname{dist}(x_0, \partial D)$. Set

$$\mathbb{A}(x_0, r_1, r_2) = \{x \in \mathbb{R}^n : r_1 < |x - x_0| < r_2\}\,,$$

$$B(x_0, r) = \{x \in \mathbb{R}^n : |x - x_0| < r\}\,,$$

$$S_i = S(x_0, r_i) = \{x \in \mathbb{R}^n : |x - x_0| = r_i\}\,, \quad i = 1, 2\,.$$

The following concept generalizes and localizes the concept of a Q-homeomorphism with respect to p-modulus.

Let a function $Q : D \to [0, \infty]$ be Lebesgue measurable. We say that a homeomorphism $f : D \to \mathbb{R}^n$ is a *ring Q-homeomorphism with respect to p-modulus* at $x_0 \in D$, if the relation

$$M_p(\Delta(fS_1, fS_2, fD)) \leqslant \int\limits_{\mathbb{A}} Q(x)\, \eta^p(|x - x_0|)\, dm(x)$$

holds for any ring $\mathbb{A} = \mathbb{A}(x_0, r_1, r_2)$, $0 < r_1 < r_2 < d_0$, $d_0 = \operatorname{dist}(x_0, \partial D)$, and for any measurable function $\eta : (r_1, r_2) \to [0, \infty]$ such that

$$\int\limits_{r_1}^{r_2} \eta(r)\, dr = 1\,.$$

The theory of Q-homeomorphisms for $p = n$ was developed in [2–6], whereas for $1 < p < n$ and $p > n$ see [7–14] and [15–25], respectively.

Denote by ω_{n-1} the area of the unit sphere $\mathbb{S}^{n-1} = \{x \in \mathbb{R}^n : |x| = 1\}$ in $\mathbb{R}^n$ and by

$$q_{x_0}(r) = \frac{1}{\omega_{n-1}\, r^{n-1}} \int\limits_{S(x_0,r)} Q(x)\, d\mathcal{A}$$

the integral mean of Q over the sphere $S(x_0, r) = \{x \in \mathbb{R}^n : |x - x_0| = r\}$, here $d\mathcal{A}$ is the element of the surface area.

Now we formulate a criterion which guarantees for a homeomorphism to be a ring Q-homeomorphism with respect to p-modulus for $p > 1$ in $\mathbb{R}^n$, $n \geqslant 2$, see Theorem 2.3 in [12].

Proposition 1 *Let D be a domain in $\mathbb{R}^n$, $n \geqslant 2$, $x_0 \in D$, and let $Q : D \to [0, \infty]$ be a Lebesgue measurable function such that $q_{x_0}(r) \neq \infty$ for a.e. $r \in (0, d_0)$, $d_0 = \operatorname{dist}(x_0, \partial D)$. A homeomorphism $f : D \to \mathbb{R}^n$ is a ring Q-homeomorphism with respect to p-modulus at x_0 if and only if the inequality*

$$M_p\left(\Delta(fS_1, fS_2, f\mathbb{A})\right) \leqslant \frac{\omega_{n-1}}{\left(\int\limits_{r_1}^{r_2} \frac{dr}{r^{\frac{n-1}{p-1}} q_{x_0}^{\frac{1}{p-1}}(r)}\right)^{p-1}}$$

holds for any $0 < r_1 < r_2 < d_0$.

We also need the following statement [15, Thm. 1].

Theorem 1 *Let D be a bounded domain in $\mathbb{R}^n$, $n \geqslant 2$ and let $f : D \to \mathbb{R}^n$ be a ring Q-homeomorphism with respect to p-modulus at a point $x_0 \in D$ with $p > n$. Assume that the function Q satisfies the growth condition*

$$q_{x_0}(t) \leqslant q_0\, t^{-\alpha}, \; q_0 \in (0, \infty), \; \alpha \in [0, \infty),$$

for a.e. $t \in (0, d_0)$, $d_0 = \operatorname{dist}(x_0, \partial D)$. Then for all $r \in (0, d_0)$, the estimate

$$m(fB(x_0, r)) \geqslant \Omega_n \left(\frac{p-n}{\alpha + p - n}\right)^{\frac{n(p-1)}{p-n}} q_0^{\frac{n}{n-p}}\, r^{\frac{n(\alpha+p-n)}{p-n}},$$

holds, where Ω_n is a volume of the unit ball in $\mathbb{R}^n$.

An upper bound for the area of a disc image under quasiconformal mappings was established by M.A. Lavrent'ev in his monograph [26]. In [27, Proposition 3.7], the Lavrentiev's inequality was refined in terms of the angular dilatation. Also in [28] and [29], there were obtained upper bounds for the area distortion for the ring and lower Q-homeomorphisms. V. Kruglikov has obtained a bound for the measure of the image of a ball for mappings quasiconformal in the mean in $\mathbb{R}^n$ (see Lemma 9 in [30]). In [12], there was established an upper bound for the measure of a ball image under ring Q-homeomorphisms with respect to p-modulus ($1 < p \leqslant n$). In [15, 16] there were obtained the lower bounds for the measure of a ball image as $p > n$. Note that the corresponding problems for the planar case has been studied in [17–19].

2 Main Results

In the following theorem, we provide a lower bound for upper limits of distance ratios under ring Q-homeomorphisms with respect to p-modulus with $p > n$.

Theorem 2 *Let $f : \mathbb{B}^n \to \mathbb{R}^n$ be a ring Q-homeomorphism with respect to p-modulus at a point $x_0 = 0$ with $p > n$ and $f(0) = 0$. Assume that the function Q satisfies the condition*

$$q_{x_0}(t) \leqslant q_0\, t^{-\alpha},\ q_0 \in (0, \infty)\,,\ \alpha \in [0, \infty),$$

for a.e. $t \in (0, \varepsilon_0)$, $\varepsilon_0 \in (0, 1)$. Then

$$\limsup_{x \to 0} \frac{|f(x)|}{|x|^{\frac{\alpha+p-n}{p-n}}} \geqslant \left(\frac{p-n}{\alpha+p-n}\right)^{\frac{p-1}{p-n}} q_0^{\frac{1}{n-p}}\,. \tag{1}$$

Proof We denote by $L_f(r) = \max\limits_{|x|=r} |f(x)|$. Since $f(0) = 0$, $\Omega_n L_f^n(r) \geqslant m(fB(0,r))$. Hence,

$$L_f(r) \geqslant \left(\frac{m(fB(0,r))}{\Omega_n}\right)^{\frac{1}{n}}.$$

Then, by Theorem 1, we have

$$L_f(r) \geqslant \left(\frac{m(fB(0,r))}{\Omega_n}\right)^{\frac{1}{n}} \geqslant \left(\frac{p-n}{\alpha+p-n}\right)^{\frac{p-1}{p-n}} q_0^{\frac{1}{n-p}}\, r^{\frac{\alpha+p-n}{p-n}}$$

for any $r \in [0, \varepsilon_0)$.

Thus, we obtain

$$\limsup_{x \to 0} \frac{|f(x)|}{|x|^{\frac{\alpha+p-n}{p-n}}} = \limsup_{r \to 0} \frac{L_f(r)}{r^{\frac{\alpha+p-n}{p-n}}} \geqslant \left(\frac{p-n}{\alpha+p-n}\right)^{\frac{p-1}{p-n}} q_0^{\frac{1}{n-p}}\,.$$

This completes the proof of Theorem 2. □

Assuming a stronger growth condition on the majorant Q we deduce

Corollary 1 *Let $f : \mathbb{B}^n \to \mathbb{R}^n$ be a ring Q-homeomorphism with respect to p-modulus at a point $x_0 = 0$ with $p > n$ and $f(0) = 0$. Assume that the function Q satisfies the condition*

$$Q(x) \leqslant K\, |x|^{-\alpha},\ K \in (0, \infty)\,,\ \alpha \in [0, \infty),$$

for a.e. $x \in B(0, \varepsilon_0)$, $\varepsilon_0 \in (0, 1)$. *Then*

$$\limsup_{x \to 0} \frac{|f(x)|}{|x|^{\frac{\alpha+p-n}{p-n}}} \geqslant \left(\frac{p-n}{\alpha+p-n} \right)^{\frac{p-1}{p-n}} K^{\frac{1}{n-p}} .$$

Letting $\alpha = 0$ in Theorem 2, we obtain the following statement.

Corollary 2 *Let* $f : \mathbb{B}^n \to \mathbb{R}^n$ *be a ring* Q*-homeomorphism with respect to* p*-modulus at a point* $x_0 = 0$ *with* $p > n$ *and* $f(0) = 0$. *Assume that the function* Q *satisfies the condition*

$$q_{x_0}(t) \leqslant q_0, \;\; q_0 \in (0, \infty),$$

for a.e. $t \in (0, \varepsilon_0)$, $\varepsilon_0 \in (0, 1)$. *Then*

$$\limsup_{x \to 0} \frac{|f(x)|}{|x|} \geqslant q_0^{\frac{1}{n-p}} .$$

Similarly to Corollary 1, we get

Corollary 3 *Let* $f : \mathbb{B}^n \to \mathbb{R}^n$ *be a ring* Q*-homeomorphism with respect to* p*-modulus at a point* $x_0 = 0$ *with* $p > n$ *and* $f(0) = 0$. *If the function* $Q(x) \leqslant K$, *for a.e.* $x \in \mathbb{B}^n$, $K \in (0, \infty)$, *then*

$$\limsup_{x \to 0} \frac{|f(x)|}{|x|} \geqslant K^{\frac{1}{n-p}} .$$

Now we verify the sharpness of the bound in our main result.

Example Let $p > n$ and $q_0 > 0$. Define $f_0 : \mathbb{B}^n \to \mathbb{R}^n$ by

$$f_0(x) = \begin{cases} q_0^{\frac{1}{n-p}} \left(\frac{p-n}{\alpha+p-n} \right)^{\frac{p-1}{p-n}} |x|^{\frac{\alpha+p-n}{p-n}} \frac{x}{|x|}, & x \neq 0 \\ 0, & x = 0 . \end{cases}$$

It can be easily seen that the estimate (1) is sharp, and it becomes an equality under f_0.

Let us show that f_0 is a ring Q-homeomorphism with respect to p-modulus with $Q(x) = q_0 \, |x|^{-\alpha}$ at $x_0 = 0$ with $p > n$. Clearly, $q_{x_0}(t) = q_0 \, t^{-\alpha}$. Consider a ring $\mathbb{A}(0, r_1, r_2)$, $0 < r_1 < r_2 < 1$. Note that the mapping f_0 maps the ring $\mathbb{A}(0, r_1, r_2)$ onto the ring $\widetilde{\mathbb{A}}(0, \widetilde{r}_1, \widetilde{r}_2)$, where

$$\widetilde{r}_i = q_0^{\frac{1}{n-p}} \left(\frac{p-n}{\alpha+p-n} \right)^{\frac{p-1}{p-n}} r_i^{\frac{\alpha+p-n}{p-n}}, \quad i = 1, 2.$$

Denote by Γ a set of all curves that join the spheres $S_1 = S(0, r_1)$ and $S_2 = S(0, r_2)$ in the ring $\mathbb{A} = \mathbb{A}(0, r_1, r_2)$. Then one can calculate the p-modulus of the family of curves $f_0\Gamma$ in an implicit form:

$$M_p(f_0\Gamma) = \omega_{n-1} \left(\frac{p-n}{p-1}\right)^{p-1} \left(\widetilde{r}_2^{\frac{p-n}{p-1}} - \widetilde{r}_1^{\frac{p-n}{p-1}}\right)^{1-p}$$

(see, e.g., (2) in [31]).

Substituting the values $\widetilde{r}_1$ and $\widetilde{r}_2$, defined above, in the previous equality one gets

$$M_p(f_0\Gamma) = \omega_{n-1}\, q_0 \left(\frac{\alpha+p-n}{p-1}\right)^{p-1} \left(r_2^{\frac{\alpha+p-n}{p-1}} - r_1^{\frac{\alpha+p-n}{p-1}}\right)^{1-p}.$$

Note that the last equality can be written as

$$M_p(f_0\Gamma) = \frac{\omega_{n-1}}{\left(\int\limits_{r_1}^{r_2} \frac{dt}{t^{\frac{n-1}{p-1}}\, q_{x_0}^{\frac{1}{p-1}}(t)}\right)^{p-1}},$$

where $q_{x_0}(t) = q_0\, t^{-\alpha}$ and $x_0 = 0$.

Hence, by Proposition 1, the homeomorphism f_0 is a ring Q-homeomorphism with respect to p-modulus for $p > n$ with $Q(x) = q_0\, |x|^{-\alpha}$ at origin.

Acknowledgments This work was supported by a grant from the Simons Foundation (1290607, R.R.S., B.A.K.).

References

1. Väisälä, J.: Lectures on n-Dimensional Quasiconformal Mappings. Lecture Notes in Mathematics, vol. 229. Springer, Berlin (1971)
2. Ryazanov, V.I., Sevost'yanov, E.A.: Equicontinuous classes of ring Q-homeomorphisms. Siberian Math. J. **48**(6), 1093–1105 (2007)
3. Martio, O., Ryazanov, V., Srebro, U., Yakubov, E.: Q-homeomorphisms. Contemp. Math. **364**, 193–203 (2004)
4. Martio, O., Ryazanov, V., Srebro, U., Yakubov, E.: On Q-homeomorphisms. Ann. Acad. Sci. Fenn. Ser. A1. Math. **30**(1), 49–69 (2005)
5. Martio, O., Ryazanov, V., Srebro, U., Yakubov, E.: Moduli in Modern Mapping Theory. New York: Springer Science + Business Media, LLC (2009)
6. Salimov, R.: ACL and differentiability of a generalization of quasiconformal maps. Izvestiya: Math. **72**(5), 977–984 (2008)
7. Golberg, A.: Differential properties of (α, Q)-homeomorphisms. In: Further Progress in Analysis, Proceedings of the Sixth ISAAC Congress, pp. 218–228 (2009)
8. Golberg, A.: Integrally quasiconformal mappings in space. Trans. Inst. Math. NAS of Ukraine **7**(2), 53–64 (2010)

9. Golberg, A., Salimov, R.: Logarithmic Holder continuity of ring homeomorphisms with controlled p-module. Complex Variables Ellipt. Equ. **59**(1), 91–98 (2014)
10. Golberg, A., Salimov, R., Sevost'yanov, E.: Distortion estimates under mappings with controlled p-module. Ann. Univ. Bucharest, Ser. Math. **LXIII**, 95–114 (2014)
11. Salimov, R.: On finitely Lipschitz space mappings. Siberian Electron. Math. Rep. **8**, 284–295 (2011)
12. Salimov, R.R.: Estimation of the measure of the image of the ball. Siberian Math. J. **53**(4), 739–747 (2012)
13. Salimov, R.: To a theory of ring Q-homeomorphisms with respect to a p-modulus. J. Math. Sci. **196**(5), 679–692 (2014)
14. Salimov, R.: One property of ring Q-homeomorphisms with respect to a p-module. Ukrainian Math. J. **65**(5), 806–813 (2013)
15. Salimov, R., Klishchuk, B.: An extremal problem for the volume functional. Mat. Stud. **50**(1), 36–43 (2018)
16. Klishchuk, B., Salimov, R.: Lower bounds for the volume of the image of a ball. Ukrainian Math. J. **71**(6), 883–895 (2019)
17. Salimov, R.R., Klishchuk, B.A.: The extremal problem for the area of an image of a disc. Rep. NAS of Ukraine **10**, 22–27 (2016)
18. Klishchuk, B.A., Salimov, R.R.: Lower bounds for the area of the image of a circle. UFA Math. J. **9**(2), 55–61 (2017)
19. Salimov, R.R., Klishchuk, B.A.: Extremal problem for the area of the image of a disk (in Russian). J. Math. Sci. **234**(3), 373–380 (2018)
20. Salimov, R., Klishchuk, B.: On the Behavior at Infinity of One Class of Homeomorphisms. Current Trends in Analysis, its Applications and Computation. Trends in Mathematics. Birkhauser, Cham, pp. 173–180 (2022)
21. Salimov, R., Klishchuk, B.: On the behavior of one class of homeomorphisms at infinity. Ukrainian Math. J. **74**, 1617–1628 (2023)
22. Salimov, R., Klishchuk, B.: On the asymptotic behavior at infinity of ring Q-homeomorphisms with respect to p-modulus. Proc. IAMM NASU **36**(1), 26–35 (2022)
23. Salimov, R., Klishchuk, B.: On the ring Q-homeomorphisms with respect to p-modulus. Trans. Inst. Math. NAS Ukraine **19**(1), 197–209 (2022)
24. Klishchuk, B.: On power-law behavior of some mapping class at infinity. J. Math. Sci. **268**(2), 192–198 (2022)
25. Klishchuk, B., Salimov R., Stefanchuk M.: On the asymptotic behavior at infinity of one mapping class. Proc. Int. Geom. Center **16**(1), 50–58 (2023)
26. Lavrent'ev, M.A.: The Variational Method in Boundary-Value Problems for Systems of Equations of Elliptic Type (in Russian). Izd-vo AN SSSR, Moscow (1962)
27. Bojarski, B., Gutlyanskii, V., Martio, O., Ryazanov, V.: Infinitesimal geometry of quasiconformal and bi-Lipschitz mappings in the plane. EMS Tracts in Mathematics, vol. 19. European Mathematical Society (EMS), Zurich (2013). ISBN: 978-3-03719-122-4
28. Lomako, T.V., Salimov, R.R.: To the theory of extremal problems (in Russian). Trans. Inst. Math. NAS of Ukraine **7**(2), 264–269 (2010)
29. Salimov, R.R.: Lower estimates of p-modulus and mappings of Sobolev's class (in Russian). Algebra i Anal. **26**(6), 143–171 (2014)
30. Kruglikov, V.I.: Capacity of condensers and spatial mappings quasiconformal in the mean (in Russian). Matem. Sborn. **130**(2), 185–206 (1986)
31. Gehring, F.W.: Lipschitz mappings and the p-capacity of ring in n-space. In: Advances in the Theory of Riemann Surfaces (Proceedings of Conference Stonybrook, 1969). Annals of Mathematics Studies, vol. 66, pp. 175–193 (1971)

Approximate Solutions of Some Problems in Thermoelasticity

Flavia Lanzara

Abstract A fast approximation method to 3-dimensional equations in static and quasi-static uncoupled thermoelasticity is proposed. We approximate the density via Gaussian approximating functions introduced in the method approximate approximations. In this way, the action of the integral operators on such functions is presented in a simple analytical form. For densities with separated representation, we derive a tensor product representation of the integral operator which admits efficient cubature procedures. These results were obtained in collaboration with Vladimir Maz'ya and Gunther Schmidt.

1 Introduction

The equations of thermoelasticity describe the elastic and thermal behavior of elastic, heat conductive media, in particular, the reciprocal actions between elastic stresses and temperature differences.

In the static uncoupled thermoelasticity, thermal effects on a body are restricted to strains due to a steady-state temperature distribution. The thermoelastostatic state of a medium, corresponding to the mass force $\mathbf{f} = (f_1, f_2, f_3)$ and the heat source Q, is the couple $(\mathbf{u}, T)$ satisfying the equations

$$\mu\,\Delta\mathbf{u} + (\lambda + \mu)\operatorname{grad}\operatorname{div}\mathbf{u} - \gamma\operatorname{grad}T + \mathbf{f} = 0, \tag{1}$$

$$\Delta T + Q = 0, \tag{2}$$

together with the corresponding boundary conditions (cf., e.g., [1–3]). The scalar function T is the temperature and the vector $\mathbf{u} = (u_1, u_2, u_3)$ is the thermoelastic displacement. We consider the Eqs. (1) and (2) in the whole space $\mathbb{R}^3$. We suppose that f_1, f_2, f_3 and Q belong to the Schwartz space $\mathcal{S}(\mathbb{R}^3)$ of smooth functions

F. Lanzara (✉)
Mathematics Department, Sapienza University of Rome, Rome, Italy
e-mail: flavia.lanzara@uniroma1.it

M. R. Ebert et al. (eds.), *New Tools in Mathematical Analysis and Applications*, Research Perspectives, https://doi.org/10.1007/978-3-031-77050-0_9

whose derivatives (including the function itself) decay at infinity faster than any power.

Uncoupled quasi-static thermoelasticity can be employed when slowly varying thermal and mechanical loads are encountered and dissipative effects can be neglected. The equations are a coupling of the equations of elasticity and the heat equation (cf., e.g., [1, 2])

$$\mu\,\Delta\,\mathbf{u} + (\lambda + \mu)\mathrm{grad}\,\mathrm{div}\,\mathbf{u} - \gamma\,\mathrm{grad}\,T + \mathbf{F} = 0 \tag{3}$$

$$\frac{\partial T}{\partial t} - a\Delta_{\mathbf{x}} T = 0 \tag{4}$$

$$T(\mathbf{x}, 0) = g(\mathbf{x}) \tag{5}$$

for $(\mathbf{x}, t) \in \mathbb{R}^3 \times [0, \infty)$, together with the corresponding initial and boundary conditions. The function $T(\mathbf{x}, t)$ is the temperature, the vector $\mathbf{F}(\mathbf{x}, t) = (F_1(\mathbf{x}, t), F_2(\mathbf{x}, t), F_3(\mathbf{x}, t))$ is the mass force and $\mathbf{u}(\mathbf{x}, t) = (u_1(\mathbf{x}, t), u_2(\mathbf{x}, t), u_3(\mathbf{x}, t))$ is the thermoelastic displacement. We suppose that $g, F_1(\cdot, t), F_2(\cdot, t), F_3(\cdot, t) \in \mathcal{S}(\mathbb{R}^3)$. Here λ and μ are the Lamé constants. The set of quantities μ, γ, a are positive and $3\lambda + 2\mu > 0$.

The goal of this paper is to derive semi-analytic cubature formulas for solutions to (1) and (2) and for solutions to (3)–(5) of an arbitrary high-order which are fast and accurate by using the basis functions introduced in the theory *approximate approximations*. This concept, introduced by V. Maz'ya in [9, 10] and described in detail in the monograph [11], allows to construct efficient high-order cubature formulas for convolution integral operators with singular kernel functions. Consider, for example, the harmonic potential

$$\mathcal{L}f(\mathbf{x}) = \frac{1}{4\pi} \int\limits_{\mathbb{R}^3} \frac{f(\mathbf{y})}{|\mathbf{x} - \mathbf{y}|}\, d\mathbf{y}, \tag{6}$$

which provides the unique solution $T = \mathcal{L}(Q)$ of (2) such that $T(\mathbf{x}) \to 0$ as $|\mathbf{x}| \to \infty$.

The idea is to replace the density f in (6) by a quasi-interpolant of the form

$$\mathcal{M}_{h,\mathcal{D}} f(\mathbf{x}) = \mathcal{D}^{-3/2} \sum_{\mathbf{m} \in \mathbb{Z}^3} f(h\mathbf{m}) \eta\left(\frac{\mathbf{x} - h\mathbf{m}}{h\sqrt{\mathcal{D}}}\right), \tag{7}$$

where h and $\mathcal{D}$ are two positive parameters, and η is a smooth and rapidly decaying function, which satisfies the moment condition of order N

$$\int\limits_{\mathbb{R}^3} \mathbf{x}^\alpha \eta(\mathbf{x})\, d\mathbf{x} = \delta_{0,\alpha}, \quad 0 \le |\alpha| < N\,. \tag{8}$$

It is known that

$$|f(\mathbf{x}) - \mathcal{M}_{h,\mathcal{D}} f(\mathbf{x})| \le c(\sqrt{\mathcal{D}}h)^N \|\nabla_N f\|_{L_\infty} + \sum_{k=0}^{N-1} \varepsilon_k (\sqrt{\mathcal{D}}h)^k \big|\nabla_k f(\mathbf{x})\big|$$

with

$$0 \le \varepsilon_k \le \sum_{\mathbf{m}\in\mathbb{Z}^3\setminus\{0\}} \big|\nabla_k \mathcal{F}\eta(\sqrt{\mathcal{D}}\mathbf{m})\big|, \quad \lim_{\mathcal{D}\to\infty} \sum_{\mathbf{m}\in\mathbb{Z}^3\setminus\{0\}} \big|\nabla_k \mathcal{F}\eta(\sqrt{\mathcal{D}}\mathbf{m})\big| = 0.$$

Hence for any saturation error $\varepsilon > 0$ one can fix the parameter $\mathcal{D}$ such that

$$|f(\mathbf{x}) - \mathcal{M}_{h,\mathcal{D}} f(\mathbf{x})| = \mathcal{O}((\sqrt{\mathcal{D}}h)^N + \varepsilon)\|f\|_{W_\infty^N} . \tag{9}$$

Then the linear combination

$$\mathcal{L}_{h,\mathcal{D}}^{(N)} f(\mathbf{x}) := \mathcal{L}\mathcal{M}_{h,\mathcal{D}} f(\mathbf{x}) = \frac{h^2}{\sqrt{\mathcal{D}}} \sum_{\mathbf{m}\in\mathbb{Z}^3} f(h\mathbf{m})\mathcal{L}\eta\Big(\frac{\mathbf{x}-h\mathbf{m}}{\sqrt{\mathcal{D}}h}\Big) \tag{10}$$

gives rise to new classes of semi-analytic cubature formulas with the following approximation behavior: For any $\varepsilon > 0$ one can fix the parameter $\mathcal{D}$ such that (10) differs in some uniform or L_p—norm from the integral (6) by

$$\mathcal{O}((\sqrt{\mathcal{D}}h)^N + (\sqrt{\mathcal{D}}h)^2\varepsilon) \quad \text{as} \quad h \to 0, \tag{11}$$

where N is determined by (8) and -2 is the order of the pseudodifferential operator $\mathcal{L}$ (cf. [11, Theorem 4.10]). This follows from the general asymptotic error estimate (9). The saturation error ε does not converge to zero as $h \to 0$, however, because of

$$\varepsilon = O\Big(\max_{\mathbf{x}} \Big| \sum_{\mathbf{k}\in\mathbb{Z}^3\setminus\{\mathbf{0}\}} \mathcal{F}\eta(\sqrt{\mathcal{D}}\mathbf{k}) e^{2\pi i(\mathbf{k},\mathbf{x})/h}\Big|\Big), \tag{12}$$

it can be made arbitrarily small if the parameter $\mathcal{D}$ is sufficiently large. Additionally, the harmonic potential maps the fast oscillating saturation term (12) into a function with norm of order $\mathcal{O}(h^2\varepsilon)$, which establishes the estimate (11). Therefore, in numerical computations with $\mathcal{D} \ge 3$, formula (10) behaves like a high-order cubature formula. Therefore to construct high-order cubature formulas for (6), it remains to compute the integral $\mathcal{L}\eta$, analytically or at least efficiently. This is performed by using one-dimensional integral representations. As basis functions

η in (7) we take the tensor products of univariate basis functions

$$\eta_{2M}(\mathbf{x}) = \prod_{\ell=1}^{3} \widetilde{\eta}_{2M}(x_\ell); \quad \widetilde{\eta}_{2M}(x) = \frac{(-1)^{M-1}}{2^{2M-1}\sqrt{\pi}\,(M-1)!} \frac{H_{2M-1}(x)e^{-x^2}}{x} \tag{13}$$

which satisfies the moment condition (8) of order $N = 2M$ (cf. [11, p. 56]), where H_k are the Hermite polynomials

$$H_k(x) = (-1)^k e^{x^2} \left(\frac{d}{dx}\right)^k e^{-x^2}.$$

We have (cf. [4])

$$\mathcal{L}\eta_{2M}(\mathbf{x}) = \frac{1}{4} \int_0^\infty \prod_{\ell=1}^{3} \frac{e^{-x_\ell^2/(1+t)}}{\sqrt{1+t}} P_M(x_\ell, t)dt \tag{14}$$

where

$$P_M(x,t) = \frac{1}{\sqrt{\pi}} \sum_{k=0}^{M-1} \frac{(-1)^k}{k!\,4^k} \frac{1}{(1+t)^k} H_{2k}\Big(\frac{x}{\sqrt{1+t}}\Big). \tag{15}$$

The computation of $\mathcal{L}_{h,\mathcal{D}}^{(2M)} f$ on the uniform grid $h\mathbf{k} = (hk_1 hk_2, hk_3)$ leads to the discrete convolutions

$$\mathcal{L}_{h,\mathcal{D}}^{(2M)} f(h\mathbf{k}) = \frac{h^2}{\sqrt{\mathcal{D}}} \sum_{\mathbf{m}\in\mathbb{Z}^3} f(h\mathbf{m}) \mathcal{L}\eta_{2M}\Big(\frac{\mathbf{k}-\mathbf{m}}{\sqrt{\mathcal{D}}}\Big).$$

A suitable quadrature with nodes τ_p and weights ω_p gives the separate representations of (14)

$$\mathcal{L}\eta_{2M}\Big(\frac{\mathbf{m}}{\sqrt{\mathcal{D}}}\Big) \approx \frac{1}{4} \sum_p \omega_p \tau_p \prod_{\ell=1}^{3} \frac{e^{-m_\ell^2/(\mathcal{D}(1+\tau_p))}}{\sqrt{1+\tau_p}} P_M\Big(\frac{m_\ell}{\sqrt{\mathcal{D}}}, \tau_p\Big), \quad \mathbf{m} \in \mathbb{Z}^3. \tag{16}$$

The representations (16) are very useful if f admits a separated representation that is, within a prescribed accuracy, can be represented as the sum of products of one-dimensional functions

$$f(\mathbf{x}) = \sum_{s=1}^{S} \prod_{\ell=1}^{3} f_\ell^{(s)}(x_\ell) + \mathcal{O}(\varepsilon), \quad \mathbf{x} = (x_1, x_2, x_3). \tag{17}$$

Then the value $\mathcal{L}_{h,\mathcal{D}}^{(2M)} f(h\mathbf{k})$ can be obtained by computing

$$\mathcal{L}_{h,\mathcal{D}}^{(2M)} f(h\mathbf{k}) \approx \frac{1}{4} \sum_{s=1}^{S} \sum_{p} \omega_p \tau_p \prod_{\ell=1}^{3} \sigma_{\ell,s}^{(M)}(k_\ell, \tau_p)$$

with the one-dimensional discrete convolutions

$$\sigma_{\ell,s}^{(M)}(k,\tau) = \sum_{m\in\mathbb{Z}} f_\ell^{(s)}(hm) \frac{e^{-(k_\ell - m_\ell)^2/(\mathcal{D}(1+\tau))}}{\sqrt{1+\tau}} P_M(\frac{k_\ell - m_\ell}{\sqrt{\mathcal{D}}}, \tau).$$

Thus for the computation of the 3−dimensional convolution (6) only one-dimensional operations are used. This gives rise to a fast algorithm.

The outline of the paper is the following. We start in Sect. 2 by describing the fast formulas for the approximation of the solution to (1) and (2). We then consider in Sect. 3 fast cubature formulas for the solution to (3)–(5). The results in Sect. 2 are taken from [7]; those in Sect. 3 are obtained in [8].

2 Static Thermoelasticity

Let $U = (\mathbf{u}, u_4)$, where $\mathbf{u} = (u_1, u_2, u_3)$ is the thermoelastic displacement and $u_4 = T$ the temperature. The matrix of fundamental solutions of Eqs. (1) and (2) is

$$\Phi_{k\ell}(\mathbf{x}) = (1-\delta_{k4})(1-\delta_{\ell 4})\Gamma_{k\ell}(\mathbf{x}) + \frac{\gamma\delta_{\ell 4}(1-\delta_{k\ell})}{8\pi(\lambda+2\mu)}\frac{x_k}{|\mathbf{x}|} + \frac{\delta_{k4}\delta_{\ell 4}}{4\pi}\frac{1}{|\mathbf{x}|}, \ k,\ell = 1,2,3,4$$

(cf. [3, p. 97]) and $\{\Gamma_{k\ell}\}_{k,\ell=1,2,3}$ is the Kelvin fundamental matrix (cf. [3, p. 84])

$$\Gamma_{k\ell}(\mathbf{x}) = \frac{\lambda'\delta_{k\ell}}{8\pi|\mathbf{x}|} + \frac{\mu'}{8\pi}\frac{x_k x_\ell}{|\mathbf{x}|^3}, \quad k,\ell = 1,2,3, \tag{18}$$

with $\lambda' = (\lambda+3\mu)/(\mu(\lambda+2\mu))$, $\mu' = (\lambda+\mu)/(\mu(\lambda+2\mu))$. Then if we set $f_4 = Q$, a solution of (1)–(2) is given by the volume potentials

$$u_k(\mathbf{x}) = \sum_{\ell=1}^{4} \int_{\mathbb{R}^3} \Phi_{k\ell}(\mathbf{x}-\mathbf{y}) f_\ell(\mathbf{y}) d\mathbf{y}, \quad k = 1,2,3,4. \tag{19}$$

The problem of determining T is an independent of $\mathbf{u}$ problem. Indeed, if T is a solution of (2), the obtained value of T when substituted into the system

$$\mu\,\Delta\mathbf{u} + (\lambda+\mu)\mathrm{grad}\,\mathrm{div}\,\mathbf{u} = \gamma\,\mathrm{grad}\,T - \mathbf{f}$$

leads to the problem of elasticity with respect to $\mathbf{u}$.

Set

$$\Psi_k g(\mathbf{x}) = \frac{\gamma}{8\pi(\lambda+2\mu)} \int\limits_{\mathbb{R}^3} \frac{(x_k - y_k)}{|\mathbf{x}-\mathbf{y}|} g(\mathbf{y})d\mathbf{y}, \quad k = 1,2,3. \tag{20}$$

The components of $U = (u_1, u_2, u_3, u_4)$ in (19) can be written as

$$u_k(\mathbf{x}) = \sum_{\ell=1}^{3} \Gamma^{(k,\ell)} f_\ell(\mathbf{x}) + \Psi_k f_4(\mathbf{x}), \quad k = 1,2,3, \quad u_4(\mathbf{x}) = \mathcal{L} f_4(\mathbf{x})$$

where

$$\Gamma^{(k,\ell)} g(\mathbf{x}) = \int\limits_{\mathbb{R}^3} \Gamma_{k\ell}(\mathbf{x}-\mathbf{y}) g(\mathbf{y}) d\mathbf{y}, \quad k,\ell = 1,2,3, \tag{21}$$

Fast cubature formulas of high-order for the harmonic potential $\mathcal{L}$ have been described in the Introduction. In the next Subsections we describe similar formulas for the approximation of $\Gamma^{(k,\ell)}$ and Ψ_k.

Approximation of the Elastic Potential $\Gamma^{(k,\ell)}$ An approximation formula for $\Gamma^{(k,\ell)}$ in (21) can be obtained if the density g is replaced by the approximate quasi-interpolant $\mathcal{M}_{h,\mathcal{D}} g$ in (7). Then the sum

$$\Gamma^{(k,\ell)}(\mathcal{M}_{h,\mathcal{D}} g)(\mathbf{x}) = \frac{h^2}{\sqrt{\mathcal{D}}} \sum_{\mathbf{m}\in\mathbb{Z}^3} g(h\mathbf{m}) \Gamma^{(k,\ell)} \eta \left(\frac{\mathbf{x} - h\mathbf{m}}{h\sqrt{\mathcal{D}}} \right)$$

provides an approximation formula for $\Gamma^{(k,\ell)} g$. Since the mapping properties of the elastic potential are similar to those of the harmonic potential, the estimation of the cubature error $\Gamma^{(k,\ell)} g - \Gamma^{(k,\ell)}(\mathcal{M}_{h,\mathcal{D}} g) = \Gamma^{(k,\ell)}(g - \mathcal{M}_{h,\mathcal{D}} g)$ can be carried out analogously to those of the harmonic potential. If the function η is smooth, of rapid decay and satisfies the moment condition (8), then the approximations converge with the rate $\mathcal{O}((h\sqrt{\mathcal{D}})^N + (\sqrt{\mathcal{D}}h)^2\varepsilon)$ in some uniform or L_p-norm (cf. [11, p. 113]).

Rewriting (18) as (cf. [3, (1.3) p. 84])

$$\Gamma_{k\ell}(\mathbf{x}) = \frac{\delta_{k\ell}}{4\pi\mu|\mathbf{x}|} - \frac{\mu'}{8\pi} \frac{\partial^2}{\partial x_k \partial x_\ell} |\mathbf{x}|$$

then

$$\Gamma^{(k,\ell)} \eta(\mathbf{x}) = \frac{\delta_{kl}}{\mu} \mathcal{L}\eta(\mathbf{x}) - \frac{\mu'}{2} I_{k\ell} \eta(\mathbf{x}) \tag{22}$$

where

$$I_{k\ell}g(\mathbf{x}) = \frac{1}{4\pi}\frac{\partial^2}{\partial x_k \partial x_\ell}\int_{\mathbb{R}^3} |\mathbf{x}-\mathbf{y}| g(\mathbf{y}) d\mathbf{y}. \tag{23}$$

Theorem 1 ([6, Theorems 1 and 2]) *Let $M \geq 1$. The integrals $I_{k\ell}$ in* (23) *applied to η_{2M} in* (13) *admit the following one-dimensional integral representations*

$$I_{\ell\ell}\eta_{2M}(\mathbf{x}) = \frac{1}{4}\int_0^\infty \frac{t\, e^{-\frac{|\mathbf{x}|^2}{1+t}}}{(1+t)^{5/2}} \prod_{j=1}^{3} P_M(x_j,t)dt - \frac{1}{2}\int_0^\infty \frac{t\, e^{-\frac{|\mathbf{x}|^2}{1+t}}}{(1+t)^{7/2}} R_M(x_\ell,t) \prod_{\substack{j=1\\ j\neq \ell}}^{3} P_M(x_j,t)dt,$$

and, for $k \neq \ell$,

$$I_{k\ell}\eta_{2M}(\mathbf{x}) = -\frac{1}{2}\int_0^\infty \frac{t\, e^{-\frac{|\mathbf{x}|^2}{1+t}}}{(1+t)^{7/2}} Q_M(x_\ell,t) Q_M(x_k,t) P_M(x_i,t)dt$$

denoting $i \in \{1,2,3\} \setminus \{k,\ell\}$. Here P_M are given in (15),

$$\begin{aligned} Q_M(x,t) &= x\, P_M(x,t) + \frac{2}{\sqrt{\pi}}\sum_{s=1}^{M-1} \frac{1}{(1+t)^{s-1/2}} \frac{(-1)^{s-1}}{(s-1)!4^s} H_{2s-1}\left(\frac{x}{\sqrt{1+t}}\right); \\ R_M(x,t) &= x\, Q_M(x,t) + \frac{1}{\sqrt{\pi}}\sum_{s=1}^{M-1} \frac{1}{(1+t)^{s-1}} \frac{(-1)^{s-1}}{(s-1)!4^s} H_{2s}\left(\frac{x}{\sqrt{1+t}}\right). \end{aligned} \tag{24}$$

From (22), (14) and Theorem 1, we obtain a one-dimensional integral representation for $\Gamma^{(k,\ell)}\eta_{2M}$. A separated representation of $\Gamma^{(k,\ell)}\eta_{2M}$ is obtained by applying an accurate quadrature rule for this integral.

Approximation of Ψ_k The sum

$$\Psi_k(\mathcal{M}_{h,\mathcal{D}}g)(\mathbf{x}) = h^3 \sum_{\mathbf{m}\in\mathbb{Z}^3} g(h\mathbf{m})(\Psi_k\eta)\left(\frac{\mathbf{x}-h\mathbf{m}}{h\sqrt{\mathcal{D}}}\right), \quad k=1,2,3, \tag{25}$$

provides an approximation formula for $\Psi_k g$ in (20). From the inequality

$$|\Psi_k g(\mathbf{x})| \leq \frac{\gamma}{8\pi(\lambda+2\mu)} ||g||_{L_1}$$

we get

$$|\Psi_k g(\mathbf{x}) - \Psi_k(\mathcal{M}_{h,\mathcal{D}} g)(\mathbf{x})| = |\Psi_k(g - \mathcal{M}_{h,\mathcal{D}} g)(\mathbf{x})| \leq \frac{\gamma}{8\pi(\lambda+2\mu)} ||g - \mathcal{M}_{h,\mathcal{D}} g||_{L_1}.$$

We deduce that, if η is a smooth and rapidly decaying function, and satisfies the moment condition (8) then $|\Psi_k g(\mathbf{x}) - \Psi_k(\mathcal{M}_{h,\mathcal{D}} g)(\mathbf{x})| = \mathcal{O}((\sqrt{\mathcal{D}}h)^N) + \varepsilon$.

Theorem 2 ([7, Theorem 4.1]) *Let $M \geq 1$. Assume as basis functions the tensor product functions* (13). *We have, for $k = 1, 2, 3$*

$$\Psi_k \eta_{2M}(\mathbf{x}) = \frac{\gamma}{8\pi^{3/2}(\lambda+2\mu)} \int_0^\infty Q_M(x_k, t) \prod_{\substack{\ell=1 \\ \ell \neq k}}^{3} P_M(x_j, t) \frac{t\, e^{-\frac{|\mathbf{x}|^2}{(1+t)}}}{(1+t)^{5/2}} dt \tag{26}$$

where P_M and Q_M are given in (15) *and* (24)*, respectively.*

A separated representation of $\Psi_k \eta_{2M}$ is obtained by applying a quadrature rule. Then, if g has a separated representation (17) then we get tensor product formulas for $\Psi_k g$ at the point of a uniform grid.

3 Uncoupled Quasi-Static Thermoelasticity

The Cauchy problem (4) and (5) can be solved by the Poisson integral

$$T(\mathbf{x}, t) = (\mathcal{P}_t g)(\mathbf{x}) := \frac{1}{(4\pi a t)^{3/2}} \int_{\mathbb{R}^3} e^{-\frac{|\mathbf{x}-\mathbf{y}|^2}{4at}} g(\mathbf{y}) d\mathbf{y}. \tag{27}$$

Moreover $\lim_{|\mathbf{x}|\to\infty} T(\mathbf{x}, t) = 0, \quad \forall t > 0$.

Approximation of T We approximate the functions g by the approximate quasi-interpolants $\mathcal{M}_{h,\mathcal{D}} g$ in (7) with the basis functions (13). Then the sum

$$\mathcal{P}_t^{(2M)} g(\mathbf{x}) := (\mathcal{P}_t \mathcal{M}_{h,\mathcal{D}} g)(\mathbf{x}) = \frac{1}{\mathcal{D}^{3/2}} \sum_{\mathbf{m}\in\mathbb{Z}^3} g(h\mathbf{m})\, \mathcal{P}_{\frac{t}{h^2\mathcal{D}}} \eta_{2M} \left(\frac{\mathbf{x} - h\mathbf{m}}{h\sqrt{\mathcal{D}}} \right)$$

provides an approximation of $\mathcal{P}_t\, g(\mathbf{x})$ with the error $\mathcal{O}((h\sqrt{\mathcal{D}})^{2M} + \varepsilon)$. Since the Poisson integral is a smoothing operator, one can prove that also the saturation error tends to 0 as $h \to 0$ and the approximate solution $\mathcal{P}_t^{(2M)} g$ converges for any fixed $t > 0$ with order $\mathcal{O}(h^{2M})$ to $\mathcal{P}_t\, g$ (cf. [11, Theorem 6.1]).

Theorem 3 ([5, Theorem 3.1]) *Let $M \geq 1$. The Poisson integral applied to the function η_{2M} in (13) can be written as*

$$\mathcal{P}_t \eta_{2M}(\mathbf{x}) = e^{-\frac{|\mathbf{x}|^2}{1+4at}} \prod_{\ell=1}^{3} \frac{P_M(x_\ell, 4at)}{\sqrt{1+4at}}$$

where $P_M(x,t)$ is defined in (15).

From Theorem 3, the sum

$$\mathcal{P}_t^{(2M)} g(\mathbf{x}) = \frac{1}{\mathcal{D}^{3/2}} \sum_{\mathbf{m}\in\mathbb{Z}^3} g(h\mathbf{m}) e^{-\frac{|\mathbf{x}-h\mathbf{m}|^2}{h^2\mathcal{D}+4at}} \prod_{\ell=1}^{3} \frac{P_M(\frac{x_\ell - hm_\ell}{h\sqrt{\mathcal{D}}}, 4a\frac{t}{h^2\mathcal{D}})}{\sqrt{1+4at/(h^2\mathcal{D})}} \tag{28}$$

is a semi-analytic cubature formula for $\mathcal{P}_t g$. The computation of (28) is very efficient if the functions g allow the separated representation (17). Then, at the point of a uniform grid $\{h\mathbf{j}\}$,

$$\mathcal{P}_t g(h\mathbf{j}) \approx \mathcal{P}_t^{(2M)} g(h\mathbf{j}) \approx \sum_{s=1}^{S} \prod_{\ell=1}^{3} S_{\ell,s}^{(M)}(j_\ell, \frac{4at}{h^2\mathcal{D}})$$

where $S_{\ell,s}^{(M)}$ denote the one-dimensional convolutions

$$S_{\ell,s}^{(M)}(j,t) = \frac{1}{\sqrt{\mathcal{D}}} \sum_{m\in\mathbb{Z}} g_\ell^{(s)}(hm) \frac{e^{-\frac{(j-m)^2}{\mathcal{D}(1+t)}}}{\sqrt{1+t}} P_M(\frac{j-m}{\sqrt{\mathcal{D}}}, t)\,.$$

Approximation of u When the temperature field T is known, the displacement field $\mathbf{u}(\mathbf{x},t) = (u_1(\mathbf{x},t), u_2(\mathbf{x},t), u_3(\mathbf{x},t))$ is obtained by solving (3) where the gradient of T is treated as a body force. The field $\mathbf{u}$ with $\lim_{|\mathbf{x}|\to\infty} |\mathbf{u}(\mathbf{x},t)| = 0$, $t > 0$, can be represented by means of the Kelvin fundamental matrix $\{\Gamma_{k\ell}\}_{k,\ell=1,2,3}$ as follows

$$u_k(\mathbf{x},t) = \sum_{\ell=1}^{3} \int_{\mathbb{R}^3} \Gamma_{k\ell}(\mathbf{x}-\mathbf{y})(F_\ell(\mathbf{y},t) - \gamma \frac{\partial}{\partial y_\ell} T(\mathbf{y},t)) d\mathbf{y}, \quad k = 1,2,3.$$

Fast formulas of high-order for the approximation of $\Gamma^{(k,\ell)} F_\ell$ have been considered in Sect. 2. Integrating by parts the integral representation

$$u_k^{(2)}(\mathbf{x},t) = -\gamma \sum_{\ell=1}^{3} \int_{\mathbb{R}^3} \Gamma_{k\ell}(\mathbf{x}-\mathbf{y}) \frac{\partial}{\partial y_\ell} T(\mathbf{y},t) d\mathbf{y}, \quad k = 1,2,3.$$

and using the relation (cf. [3, p. 84])

$$\sum_{\ell=1}^{3} \frac{\partial}{\partial x_\ell} \Gamma_{k\ell}(\mathbf{x}) = \frac{1}{4\pi(\lambda+2\mu)} \frac{\partial}{\partial x_k} \frac{1}{|\mathbf{x}|}$$

we obtain

$$u_k^{(2)}(\mathbf{x},t) = -\frac{c_{\gamma,\lambda+2\mu}}{4\pi} \frac{\partial}{\partial x_k} \int_{\mathbb{R}^3} \frac{T(\mathbf{y},t)}{|\mathbf{x}-\mathbf{y}|} d\mathbf{y} \quad \text{with} \quad c_{\gamma,\lambda+2\mu} = \frac{\gamma}{\lambda+2\mu}, \ k=1,2,3.$$

Since $T(\mathbf{y},t) = \mathcal{P}_t g(\mathbf{y})$, we use the representation (27) to get

$$u_k^{(2)}(\mathbf{x},t) = -\frac{c_{\gamma,\lambda+2\mu}}{4\pi^{3/2}} \frac{1}{(4at)^{1/2}} \frac{\partial}{\partial x_k} \int_0^\infty \frac{d\tau}{(1+\tau)^{3/2}} \int_{\mathbb{R}^3} e^{-\frac{|\frac{\mathbf{x}-\mathbf{z}}{\sqrt{4at}}|^2}{1+\tau}} g(\mathbf{z}) d\mathbf{z}.$$

Now we replace g by $\mathcal{M}_{h,\mathcal{D}} g$ with the basis functions (13) and we set

$$(\mathcal{N}_{h,\mathcal{D}} g)_k(\mathbf{x},t) := -\frac{c_{\gamma,\lambda+2\mu}}{4\pi^{3/2}} \frac{h^3}{(4at)^{1/2}} \sum_{\mathbf{m}\in\mathbb{Z}^3} g(h\mathbf{m})$$

$$\times \frac{\partial}{\partial x_k} \int_0^\infty \frac{d\tau}{(1+\tau)^{3/2}} \int_{\mathbb{R}^3} e^{-\frac{|r_{\mathbf{m}}(\mathbf{x})-\mathbf{w}|^2}{(1+\tau)\frac{4at}{h^2\mathcal{D}}}} \eta_{2M}(\mathbf{w})\, d\mathbf{w}$$

with $r_{\mathbf{m}}(\mathbf{x}) = (\mathbf{x} - h\mathbf{m})/(h\sqrt{\mathcal{D}})$. In [8, Theorem 3.1] we proved that, for fixed $t > 0$, the approximations $\mathcal{N}_{h,\mathcal{D}} g(\cdot,t)$ converge with the rate h^{2M} in some L_p−norm to $\mathbf{u}^{(2)}(\cdot,t)$. Moreover

$$(\mathcal{N}_{h,\mathcal{D}} g)_k(\mathbf{x},t) = \frac{c_{\gamma,\lambda+2\mu}}{\pi^{3/2} h\, \mathcal{D}^2} a\,t \sum_{\mathbf{m}\in\mathbb{Z}^3} g(h\mathbf{m})$$

$$\times \int_0^\infty \prod_{\substack{j=1\\ j\neq k}}^{3} P_M\left(\frac{x_j - hm_j}{h\sqrt{\mathcal{D}}}, (1+\tau)\frac{4at}{h^2\mathcal{D}}\right) \times \tag{29}$$

$$\mathcal{R}_M\left(\frac{x_k - hm_k}{h\sqrt{\mathcal{D}}}, (1+\tau)\frac{4at}{h^2\mathcal{D}}\right) e^{-\frac{|\mathbf{x}-h\mathbf{m}|^2}{h^2\mathcal{D}+(1+\tau)4at}} \frac{(h\sqrt{\mathcal{D}})^3}{(h^2\mathcal{D}+(1+\tau)4at)^{3/2}} d\tau,$$

where

$$\mathcal{R}_M(x,\Lambda) = \frac{2x}{1+\Lambda} P_M(x,\Lambda) + 2\sqrt{1+\Lambda}\mathcal{A}_M(x,\Lambda)$$

with

$$\mathcal{A}_M(x,\Lambda) = 2\sum_{s=1}^{M-1} \frac{1}{(1+\Lambda)^{s-1/2}} \frac{(-1)^{s-1}}{(s-1)!4^s} H_{2s-1}\left(\frac{x}{\sqrt{1+\Lambda}}\right)$$

and P_M defined in (15). The quadrature of the one-dimensional integrals which appears in (29) leads to a separated representation.

So the problem is reduced to find efficient quadrature formulas for the integrals over the half-line. It is known that the classical trapezoidal rule is exponentially converging for certain classes of integrands, for example rapidly decaying functions on the real line. As first observed in [12], if the integrand has doubly exponential decay, then the trapezoidal rule is optimal with respect to the economy of the number of nodes. Therefore we make special substitutions which transform the one-dimensional integrals on the half-line to integrals over $\mathbb{R}$ with integrands decaying doubly exponentially. Then we apply the trapezoidal rule. Numerical tests showing that these formulas are efficient and provide approximations of order $\mathcal{O}(h^{2M})$, $M = 1, 2, 3, 4$, are given in [7] and [8].

References

1. Gaul, L., Wagner, M., Kö, M.: Boundary Element Methods for Engineers and Scientists. An Introductory Course with Advanced Topics. Springer, Berlin (2003)
2. Kozlov, V.A., Maz'ya, V.G., Fomin, A.V.: Uniqueness of the solution to an inverse thermoelasticity problem. Comput. Math. Math. Phys. **49**, 525–531 (2009)
3. Kupradze, V.D., Gegelia, T.G., Basheleishvili, M.O., Burchuladze, T.V.: Three-dimensional problems of the mathematical theory of elasticity and thermoelasticity. Kupradze, V.D. (ed.) North-Holland Series in Applied Mathematics and Mechanics, vol. 25. North-Holland, Amsterdam (1979)
4. Lanzara, F., Maz'ya, V., Schmidt, G.: On the fast computation of high dimensional volume potentials. Math. Comput. **80**, 887–904 (2011)
5. Lanzara, F., Maz'ya, V., Schmidt, G.: Approximation of solutions to multidimensional parabolic equations by approximate approximations. Appl. Comput. Harmon. Anal. **41**, 749–767 (2016)
6. Lanzara, F., Maz'ya, V., Schmidt, G.: Fast computation of elastic and hydrodynamic potentials using approximate approximations. Anal. Math. Phys. **10**, 81 (2020)
7. Lanzara, F., Maz'ya, V., Schmidt, G.: Approximation of solutions to equations in static thermoelasticity. J. Math. Sci. **268**, 422–434 (2022)
8. Lanzara, F., Maz'ya, V., Schmidt, G.: Approximation of uncoupled quasi-static thermoelsticity solutions based on gaussians. J. Math. Fluid Mech. **25**, 44 (2023)
9. Maz'ya, V.: A new approximation method and its applications to the calculation of volume potentials. Boundary point method. In: DFG-Kolloqium des DFG-Forschungsschwerpunktes "Randelementmethoden" (1991)
10. Maz'ya, V.: Approximate approximations. In: Whiteman, J.R. (ed.) The Mathematics of Finite Elements and Applications. Highlights 1993 (1994)
11. Maz'ya, V., Schmidt, G.: Approximate Approximations, vol. 141. American Mathematical Society, Providence (2007)
12. Takahasi, H., Mori, M.: Doubly exponential formulas for numerical integration. Publ. RIMS Kyoto Univ. **9**, 721–741 (1974)

On Local Hölder and Lipschitz Continuities of Solutions to Nonlinear Beltrami Equation

Igor Petkov, Ruslan Salimov, and Mariia Stefanchuk

Abstract A sufficient condition providing Hölder's continuity for the regular solutions of Sobolev class $W^{1,1}_{\rm loc}$ to a nonlinear Beltrami equation is established.

1 Introduction

In this paper we continue the studying of distortion estimates of Hölder and Lipschitz types for regular solutions to a nonlinear Beltrami equation involving the p-modular method; see [1–3].

Let G be a domain in the complex plane $\mathbb{C}$, i.e., a connected open subset of $\mathbb{C}$, and let $\mu : G \to \mathbb{C}$ be a measurable function with $|\mu(z)| < 1$ a.e. (almost everywhere) in G. The *Beltrami equation* is a partial linear equation

$$f_{\bar{z}} = \mu(z)\, f_z\,, \tag{1}$$

where $f_{\bar{z}} = (f_x + i f_y)/2$, $f_z = (f_x - i f_y)/2$, $z = x + iy$, and f_x and f_y are partial derivatives of f in x and y, correspondingly. The function μ is called a *complex coefficient*, and

$$K_\mu(z) = \frac{1 + |\mu(z)|}{1 - |\mu(z)|}$$

is said to be a *dilatation quotient* of the Eq. (1). The Beltrami equation (1) is called *degenerate* if $\operatorname{ess\,sup} K_\mu(z) = \infty$. The existence of homeomorphic $W^{1,1}_{\rm loc}$ solutions to various degenerate Beltrami equations was recently established, see, e.g., related references in the recent monographs [4] and [5], and in the surveys [6] and [7].

I. Petkov
Admiral Makarov National University of Shipbuilding, Mykolaiv, Ukraine

R. Salimov · M. Stefanchuk (✉)
Institute of Mathematics of NAS of Ukraine, Kyiv-4, Ukraine

M. R. Ebert et al. (eds.), *New Tools in Mathematical Analysis and Applications*, Research Perspectives, https://doi.org/10.1007/978-3-031-77050-0_10

Let $\sigma : G \to \mathbb{C}$ be a measurable function and $m \geqslant 0$. We consider the following equation written in the polar coordinates (r, θ), $z = z_0 + re^{i\theta}$, $z_0 \in G$:

$$f_r = \sigma(re^{i\theta}) \, |f_\theta|^m \, f_\theta, \tag{2}$$

where f_r and f_θ are the partial derivatives of f by r and θ, respectively. Applying the relations between these derivatives and the formal derivatives

$$rf_r = (z - z_0) f_z + \overline{(z - z_0)} f_{\bar{z}}, \qquad f_\theta = i((z - z_0) f_z - \overline{(z - z_0)} f_{\bar{z}}),$$

see, e.g. the formulas (21.51) in [8], one can rewrite the Eq. (2) in the Cartesian form:

$$f_{\bar{z}} = \left(\frac{z - z_0}{|z - z_0|}\right)^2 \frac{\sigma(z) \, N_m(f, z, z_0) - 1}{\sigma(z) N_m(f, z, z_0) + 1} f_z, \tag{3}$$

where $N_m(f, z, z_0) = i \, |z - z_0| \, |(z - z_0) f_z - \overline{(z - z_0)} f_{\bar{z}}|^m$.

Under $m = 0$, the Eq. (3) turns to be the standard linear Beltrami equation (1) with the complex coefficient

$$\mu(z) = \left(\frac{z - z_0}{|z - z_0|}\right)^2 \frac{i\sigma(z) \, |z - z_0| - 1}{i\sigma(z) \, |z - z_0| + 1}.$$

Picking $m = 0$ and $\sigma = -i/|z - z_0|$ in (3), we arrive at the classical Cauchy-Riemann system. For $m > 0$ the Eq. (3) provides a partial case of the general nonlinear system of equations (7.33) given in [8], Sect. 7.7. Later on we assume that $m > 0$.

A homeomorphism $f : G \to \mathbb{C}$ is called *regular at a point* $z_0 \in G$, if f has the total differential at this point and its Jacobian $J_f = |f_z|^2 - |f_{\bar{z}}|^2$ does not vanish, cf. I. 1.6 in [9]. A homeomorphism f of Sobolev class $W^{1,1}_{\text{loc}}$ is called *regular*, if $J_f > 0$ a.e. Note that any homeomorphism of $W^{1,1}_{\text{loc}}$ has the total differential a.e.

By a *regular homeomorphic solution of the Eq. (3)* we call a regular homeomorphism $f : G \to \mathbb{C}$, which satisfies (3) a.e. in G.

2 Preliminaries

Let $z_0 \in \mathbb{C}$. Later on we use the following notations

$$B(z_0, r) = \{z \in \mathbb{C} : |z - z_0| < r\}, \quad \mathbb{B} = \{z \in \mathbb{C} : |z| < 1\}$$

and

$$C(z_0, r) = \{z \in \mathbb{C} : |z - z_0| = r\}, \quad \mathbb{A}(z_0, r_1, r_2) = \{z \in \mathbb{C} : r_1 < |z - z_0| < r_2\}.$$

Let us recall some definitions, see [10]. A path γ in $\mathbb{C}$ is a continuous mapping $\gamma : \Delta \to \mathbb{C}$, where Δ is an interval in $\mathbb{R}$. Its locus $\gamma(\Delta)$ is denoted by $|\gamma|$. Given a family of paths Γ in $\mathbb{C}$, a Borel function $\rho : \mathbb{C} \to [0, \infty]$ is called *admissible* for Γ, abbr. $\rho \in adm\, \Gamma$, if

$$\int_\gamma \rho(z)|dz| \geqslant 1$$

for each $\gamma \in \Gamma$.

Let $p \geqslant 1$. Recall also that the *p-modulus* of Γ is the quantity

$$\mathrm{M}_p(\Gamma) = \inf_{\rho \in adm\, \Gamma} \int_{\mathbb{C}} \rho^p(z) dm(z),$$

where $dm(z)$ corresponds to the Lebesgue measure in $\mathbb{C}$.

Later on, given sets E_1, E_2 and E in $\mathbb{C}$, $\Delta(E_1, E_2; E)$ denotes a family of all paths $\gamma : [a, b] \to \mathbb{C}$ joining E_1 and E_2 in E, i.e. $\gamma(a) \in E_1$, $\gamma(b) \in E_2$ and $\gamma(t) \in E$ for all $t \in (a, b)$.

Here a condenser is a pair $\mathcal{E} = (A, C)$, where $A \subset \mathbb{C}$ is open and C is a nonempty compact set contained in A. A condenser $\mathcal{E}$ is a ringlike condenser if $A \setminus C$ is a ring, i.e., $A \setminus C$ is a domain whose complement $\overline{\mathbb{C}} \setminus (A \setminus C)$ has exactly two components, where $\overline{\mathbb{C}} = \mathbb{C} \cup \{\infty\}$. A condenser $\mathcal{E} = (A, C)$ is said to be in a domain G if $A \subset G$.

The following statement is immediate.

Proposition 1 *If $f : G \to \mathbb{C}$ is homeomorphism and $\mathcal{E} = (A, C)$ is a condenser in G, then $f\mathcal{E} = (fA, fC)$ is a condenser in fG.*

Let $\mathcal{E} = (A, C)$ be a condenser. Denote by $\mathcal{C}_0(A)$ a set of continuous functions $u : A \to \mathbb{R}^1$ with compact support. Let $W_0(\mathcal{E}) = W_0(A, C)$ be a family of nonnegative functions $u : A \to \mathbb{R}^1$ such that 1) $u \in \mathcal{C}_0(A)$, 2) $u(z) \geqslant 1$ for $z \in C$, 3) u belongs to the class ACL and

$$|\nabla u| = \sqrt{\left(\frac{\partial u}{\partial x}\right)^2 + \left(\frac{\partial u}{\partial y}\right)^2}.$$

For $p \geqslant 1$, the quantity

$$\operatorname{cap}_p \mathcal{E} = \operatorname{cap}_p (A, C) = \inf_{u \in W_0(\mathcal{E})} \int_A |\nabla u|^p \, dm(z)$$

is called *p-capacity* of the condenser $\mathcal{E}$.

It is known that for $p > 1$,

$$\operatorname{cap}_p \mathcal{E} = \operatorname{cap}_p (A, C) = M_p(\Delta(\partial A, \partial C; A \setminus C)), \tag{4}$$

see Theorem 1 in [11].

The following estimate holds; see Proposition 6 in [12].

Proposition 2 *Suppose that $\mathcal{E} = (A, C)$ is a condenser such that C is connected and $p > 1$. Then*

$$\operatorname{cap}_p \mathcal{E} = \operatorname{cap}_p (A, C) \geqslant \nu_p \frac{d^p(C)}{|A|^{p-1}},$$

where ν_p is a positive constant that depends only on p, $d(C)$ is a diameter of C and $|A|$ is the Lebesgue measure of A in $\mathbb{C}$.

The following statement is true, see, e.g., Sec. 1.4 in [13].

Proposition 3 *Suppose that $\mathcal{E} = (A, C)$ is a condenser, $1 < p < 2$. Then*

$$\operatorname{cap}_p \mathcal{E} = \operatorname{cap}_p (A, C) \geqslant \alpha_p |C|^{\frac{2-p}{2}},$$

where α_p is a positive constant that depends only on p.

Motivated by the ring definition of quasiconformality in [14], the following notion was introduced in [15] (cf. also [16, 17] and Chapters 7 and 11 in [5]).

Let G be a domain in $\mathbb{C}$, $z_0 \in G$, and $Q : G \to [0, \infty]$ be a measurable function. A homeomorphism $f : G \to \mathbb{C}$ is called a *ring Q−homeomorphism with respect to p-modulus at $z_0 \in G$* if

$$M_p(\Delta(fC_1, fC_2; fG)) \leqslant \int\limits_{\mathbb{A}} Q(z) \cdot \eta^p(|z - z_0|)\, dm(z)$$

for every ring

$$\mathbb{A} = \mathbb{A}(z_0, r_1, r_2) = \{z \in \mathbb{C} : r_1 < |z - z_0| < r_2\}, \quad 0 < r_1 < r_2 < dist\,(z_0, \partial G),$$

and every measurable function $\eta : (r_1, r_2) \to [0, \infty]$ such that

$$\int\limits_{r_1}^{r_2} \eta(r)\, dr = 1,$$

where $C_1 = \{z \in \mathbb{C} : |z - z_0| = r_1\}$ and $C_2 = \{z \in \mathbb{C} : |z - z_0| = r_2\}$.

Denote by $q_{z_0}(r) = \frac{1}{2\pi r} \int\limits_{C(z_0,r)} Q(z)|dz|$ the integral mean over the circle $C(z_0, r) = \{z \in \mathbb{C} : |z - z_0| = r\}$, $r \in (0, \mathrm{dist}(z_0, \partial G))$.

Now we formulate a criterion, which guarantees that a homeomorphism is a ring Q-homeomorphism with respect to p-modulus for $p > 1$ in $\mathbb{C}$, see Theorem 1 in [15].

Proposition 4 *Let G be a domain in $\mathbb{C}$ and $Q : G \to [0, \infty]$ be a Lebesgue measurable function such that $q_{z_0}(r) \neq \infty$ for a.a. $r \in (0, d_0)$, $d_0 = \mathrm{dist}(z_0, \partial G)$. A homeomorphism $f : G \to \mathbb{C}$ is a ring Q-homeomorphism with respect to p-modulus at a point $z_0 \in G$ if and only if the inequality*

$$\mathrm{M}_p(\Delta(fC_1, fC_2; f\mathbb{A})) \leqslant \frac{2\pi}{\left(\int\limits_{r_1}^{r_2} \frac{dr}{r^{\frac{1}{p-1}} q_{z_0}^{\frac{1}{p-1}}(r)}\right)^{p-1}}$$

holds for any r_1 and r_2 such that $0 < r_1 < r_2 < d_0$.

The following theorem gives a sufficient condition for regular homeomorphic solutions to the nonlinear Beltrami equation (3) of Sobolev class $W^{1,1}_{\mathrm{loc}}$ to be ring Q-homeomorphisms with respect to p-modulus, see Theorem 2.2 in [18].

Theorem 1 *Let $f : G \to \mathbb{C}$ be a regular homeomorphic solution of the Eq. (3) which belongs to Sobolev class $W^{1,1}_{\mathrm{loc}}$. Suppose that $\int\limits_{C(z_0,r)} \frac{|dz|}{\left(\mathrm{Im}\,\overline{\sigma(z)}\right)^{\frac{1}{m+1}}} \neq \infty$ for a.a. $r \in (0, d_0)$, $d_0 = \mathrm{dist}(z_0, \partial G)$. Then f is a ring Q-homeomorphism with respect to p-modulus at the point z_0 with $Q = \frac{1}{|z-z_0|\left(\mathrm{Im}\,\overline{\sigma(z)}\right)^{\frac{1}{m+1}}}$ and $p = \frac{m+2}{m+1}$.*

Combining Theorem 1 and Proposition 4, we have the following statement.

Lemma 1 *Let $f : G \to \mathbb{C}$ be a regular homeomorphic solution of the Eq. (3) which belongs to Sobolev class $W^{1,1}_{\mathrm{loc}}$. Suppose that $\int\limits_{C(z_0,r)} \frac{|dz|}{\left(\mathrm{Im}\,\overline{\sigma(z)}\right)^{\frac{1}{m+1}}} \neq \infty$ for a.a. $r \in (0, d_0)$, $d_0 = \mathrm{dist}(z_0, \partial G)$. Then*

$$\mathrm{M}_{\frac{m+2}{m+1}}(\Delta(fC_1, fC_2; f\mathbb{A})) \leqslant \frac{2\pi}{\left(\int\limits_{r_1}^{r_2} \frac{dr}{K_{m,\sigma}(z_0,r)}\right)^{\frac{1}{m+1}}}$$

holds for every r_1 and r_2 such that $0 < r_1 < r_2 < d_0$, where $K_{m,\sigma}(z_0, r) = \left(\frac{1}{2\pi r} \int\limits_{C(z_0,r)} \frac{|dz|}{\left(\mathrm{Im}\,\overline{\sigma(z)}\right)^{\frac{1}{m+1}}}\right)^{m+1}$, $C_j = \{z \in \mathbb{C} : |z - z_0| = r_j\}$, $j = 1, 2$, and $\mathbb{A} = \mathbb{A}(z_0, r_1, r_2)$.

Lemma 2 *Let $f\colon G \to \mathbb{C}$ be a regular homeomorphic solution of the Eq.* (3) *which belongs to Sobolev class $W^{1,1}_{\mathrm{loc}}$, $\beta > \frac{m+2}{m+1}$ and $C_{z_0} > 0$. If*

$$\int\limits_{C(z_0,r)} \frac{|dz|}{\left(\operatorname{Im}\overline{\sigma(z)}\right)^{\frac{1}{m+1}}} \leqslant C_{z_0} r^{\beta} \tag{5}$$

for a.a. $r \in (0, d_0)$, $d_0 = \operatorname{dist}(z_0, \partial G)$, then

$$\operatorname{cap}_{\frac{m+2}{m+1}}\left(fB(z_0, r_2), f\overline{B(z_0, r_1)}\right) \leqslant C_{z_0}\left(\frac{\beta(m+1)-m-2}{r_1^{m+2-\beta(m+1)} - r_2^{m+2-\beta(m+1)}}\right)^{\frac{1}{m+1}} \tag{6}$$

for every r_1 and r_2 such that $0 < r_1 < r_2 < d_0$.

Proof Consider a ring $\mathbb{A} = \mathbb{A}(z_0, r_1, r_2)$, $0 < r_1 < r_2 < d_0$, $d_0 = \operatorname{dist}(z_0, \partial G)$.

Since $\left(fB(z_0, r_2), f\overline{B(z_0, r_1)}\right)$ is a ringlike condenser in $\mathbb{C}$, according to (4),

$$\operatorname{cap}_{\frac{m+2}{m+1}}\left(fB(z_0, r_2), f\overline{B(z_0, r_1)}\right) = \mathrm{M}_{\frac{m+2}{m+1}}(\Delta(fC_1, fC_2; f\mathbb{A})),$$

where $C_i = \{z \in \mathbb{C} : |z - z_0| = r_i\}$, $i = 1, 2$.

Thus, by Lemma 1, we obtain

$$\operatorname{cap}_{\frac{m+2}{m+1}}\left(fB(z_0, r_2), f\overline{B(z_0, r_1)}\right) \leqslant \frac{2\pi}{\left(\int\limits_{r_1}^{r_2} \frac{dr}{K_{m,\sigma}(z_0,r)}\right)^{\frac{1}{m+1}}},$$

where $K_{m,\sigma}(z_0, r) = \left(\frac{1}{2\pi r}\int\limits_{C(z_0,r)} \frac{|dz|}{\left(\operatorname{Im}\overline{\sigma(z)}\right)^{\frac{1}{m+1}}}\right)^{m+1}$. Therefore, in view of the condition (5), we come to the estimate (6). □

3 The Main Results

The following theorem provides a sufficient condition for the regular solutions of Sobolev class $W^{1,1}_{\mathrm{loc}}$ to the nonlinear Beltrami equation (3) to be locally Hölder continuous at a point z_0.

Theorem 2 *Let $f\colon G \to \mathbb{C}$ be a regular homeomorphic solution of the Eq.* (3) *which belongs to Sobolev class $W^{1,1}_{\mathrm{loc}}$, $\beta > \frac{m+2}{m+1}$ and $C_{z_0} > 0$. If*

$$\int\limits_{C(z_0,r)} \frac{|dz|}{\left(\operatorname{Im}\overline{\sigma(z)}\right)^{\frac{1}{m+1}}} \leqslant C_{z_0} r^{\beta} \tag{7}$$

for a.a. $r \in (0, d_0)$, $d_0 = \operatorname{dist}(z_0, \partial G)$, then

$$|f(z) - f(z_0)| \leqslant \nu_0 C_{z_0}^{\frac{m+1}{m}} |z - z_0|^{\frac{\beta(m+1)-m-2}{m}} \tag{8}$$

for every $z \in B(z_0, \frac{d_0}{4})$, where ν_0 is a positive constant which depends only on m and β.

Proof We put $\varepsilon = |z - z_0|$, where $\varepsilon \in (0, \frac{d_0}{4})$, $d_0 = \operatorname{dist}(z_0, \partial G)$. Consider the condenser $\left(B(z_0, 2\varepsilon), \overline{B(z_0, \varepsilon)}\right)$. By Lemma 2, we have

$$\operatorname{cap}_{\frac{m+2}{m+1}}\left(fB(z_0, 2\varepsilon), f\overline{B(z_0, \varepsilon)}\right) \leqslant \nu_1 C_{z_0} \varepsilon^{\frac{\beta(m+1)-m-2}{m+1}}, \tag{9}$$

where ν_1 is a positive constant which depends only on m and β.

On the other hand, by Proposition 2,

$$\operatorname{cap}_{\frac{m+2}{m+1}}\left(fB(z_0, 2\varepsilon), f\overline{B(z_0, \varepsilon)}\right) \geqslant \nu_2 \frac{\left(d\left(f\overline{B(z_0, \varepsilon)}\right)\right)^{\frac{m+2}{m+1}}}{|fB(z_0, 2\varepsilon)|^{\frac{1}{m+1}}}, \tag{10}$$

where ν_2 is a positive constant which depends only on m.

Combining (9) and (10), we obtain

$$d\left(f\overline{B(z_0, \varepsilon)}\right) \leqslant \nu_3 C_{z_0}^{\frac{m+1}{m+2}} \varepsilon^{\frac{\beta(m+1)-m-2}{m+2}} |fB(z_0, 2\varepsilon)|^{\frac{1}{m+2}}, \tag{11}$$

where ν_3 is a positive constant which depends only on m and β.

Now consider the condenser $\left(B(z_0, 4\varepsilon), \overline{B(z_0, 2\varepsilon)}\right)$. By Lemma 2, we get an estimate

$$\operatorname{cap}_{\frac{m+2}{m+1}}\left(fB(z_0, 4\varepsilon), f\overline{B(z_0, 2\varepsilon)}\right) \leqslant \nu_4 C_{z_0} \varepsilon^{\frac{\beta(m+1)-m-2}{m+1}}, \tag{12}$$

where ν_4 is a positive constant which depends only on m and β.

On the other hand, by Proposition 3, we have

$$\operatorname{cap}_{\frac{m+2}{m+1}}\left(f B(z_0, 4\varepsilon), f\overline{B(z_0, 2\varepsilon)}\right) \geqslant \nu_5 \left| f\overline{B(z_0, 2\varepsilon)}\right|^{\frac{m}{2(m+1)}}, \tag{13}$$

where ν_5 is a positive constant which depends only on m.

Combining (12) and (13), we obtain

$$\left| f\overline{B(z_0, 2\varepsilon)}\right| \leqslant \nu_6\, C_{z_0}^{\frac{2(m+1)}{m}}\, \varepsilon^{\frac{2(\beta(m+1)-m-2)}{m}}, \tag{14}$$

where ν_6 is a positive constant which depends only on m and β.

Finally, combining (11) and (14), we come to the following estimate

$$d\left(f\overline{B(z_0, \varepsilon)}\right) \leqslant \nu_0\, C_{z_0}^{\frac{m+1}{m}}\, \varepsilon^{\frac{\beta(m+1)-m-2}{m}},$$

where ν_0 is a positive constant which depends only on m and β.

Now the desired estimate (8) follows from the obvious inequality

$$d\left(f\overline{B(z_0, \varepsilon)}\right) \geqslant |f(z) - f(z_0)|.$$

The theorem is proved. □

The following example shows the sharpness of power in (8).

Example 1 Let $\beta > \frac{m+2}{m+1}$. Consider the equation

$$f_r = -i r^{(m+1)(1-\beta)} |f_\theta|^m f_\theta \tag{15}$$

in the unit disk $\mathbb{B}$.

Let $f = C_0\, r^{\frac{\beta(m+1)-m-2}{m}} e^{i\theta}$, where $C_0 = \left(\frac{\beta(m+1)-m-2}{m}\right)^{\frac{1}{m}}$. Obviously, the mapping f belongs to the Sobolev class $W^{1,1}_{\mathrm{loc}}(\mathbb{B})$. The partial derivatives of f with respect to r and θ are

$$f_r = C_0^{m+1} r^{\frac{(m+1)(\beta-2)}{m}} e^{i\theta}, \quad f_\theta = i\, C_0\, r^{\frac{\beta(m+1)-m-2}{m}} e^{i\theta}.$$

Now we show that the mapping $f = C_0\, r^{\frac{\beta(m+1)-m-2}{m}} e^{i\theta}$ is a solution of Eq. (15). Clearly,

$$\sigma = \frac{f_r}{f_\theta |f_\theta|^m} = -i\, r^{(m+1)(1-\beta)}.$$

Thus, (7) holds, since

$$\int\limits_{C(0,r)} \frac{|dz|}{\left(\mathrm{Im}\,\overline{\sigma(z)}\right)^{\frac{1}{m+1}}} = 2\pi\, r^{\beta},$$

where $C(0,r) = \{z \in \mathbb{C} : |z| = r\}$.

On the other hand, $|f(z)| = C_0|z|^{\frac{\beta(m+1)-m-2}{m}}$.

Choosing $\beta = 2$ in Theorem 2, we derive the Lipschitz continuity.

Corollary 1 *Let $f : G \to \mathbb{C}$ be a regular homeomorphic solution of the Eq.* (3) *which belongs to Sobolev class $W^{1,1}_{\mathrm{loc}}$ and $C_{z_0} > 0$. If*

$$\int\limits_{C(z_0,r)} \frac{|dz|}{\left(\mathrm{Im}\,\overline{\sigma(z)}\right)^{\frac{1}{m+1}}} \leqslant C_{z_0} r^2$$

for a.a. $r \in (0, d_0)$, $d_0 = \mathrm{dist}(z_0, \partial G)$, then

$$|f(z) - f(z_0)| \leqslant \nu_0 C_{z_0}^{\frac{m+1}{m}} |z - z_0|$$

for every $z \in B(z_0, \frac{d_0}{4})$, where ν_0 is a positive constant which depends only on m.

Assuming the following growth bound for $\sigma(z)$, one gets

Corollary 2 *Let $f : G \to \mathbb{C}$ be a regular homeomorphic solution of the Eq.* (3) *which belongs to Sobolev class $W^{1,1}_{\mathrm{loc}}$, $\sigma_0 > 0$ and $\alpha > 1$. If*

$$\mathrm{Im}\,\overline{\sigma(z)} \geqslant \frac{\sigma_0}{|z - z_0|^{\alpha}} \tag{16}$$

for a.a. $z \in B(z_0, d_0)$, $d_0 = \mathrm{dist}(z_0, \partial G)$, then

$$|f(z) - f(z_0)| \leqslant \nu_0\, \sigma_0^{-\frac{1}{m}} |z - z_0|^{\frac{\alpha-1}{m}} \tag{17}$$

for every $z \in B(z_0, \frac{d_0}{4})$, where ν_0 is a positive constant which depends only on m and α.

Proof Indeed, in view of the condition (16), we have

$$\int\limits_{C(z_0,r)} \frac{|dz|}{\left(\mathrm{Im}\,\overline{\sigma(z)}\right)^{\frac{1}{m+1}}} \leqslant \frac{2\pi}{\sigma_0^{\frac{1}{m+1}}}\, r^{1+\frac{\alpha}{m+1}}$$

for a.a. $r \in (0, d_0)$, $d_0 = \mathrm{dist}(z_0, \partial G)$.

Next, applying Theorem 2 with the parameters $C_{z_0} = \frac{2\pi}{\sigma_0^{\frac{1}{m+1}}}$ and $\beta = 1 + \frac{\alpha}{m+1}$, we get the estimate (17). □

Choosing $\alpha = m+1$ in Corollary 2, we again obtain the following Lipschitzness.

Corollary 3 *Let $f\colon G \to \mathbb{C}$ be a regular homeomorphic solution of the Eq.* (3) *which belongs to Sobolev class $W^{1,1}_{\rm loc}$ and $\sigma_0 > 0$. If*

$$\mathrm{Im}\,\overline{\sigma(z)} \geqslant \frac{\sigma_0}{|z - z_0|^{m+1}}$$

for a.a. $z \in B(z_0, d_0)$, $d_0 = \mathrm{dist}(z_0, \partial G)$, then

$$|f(z) - f(z_0)| \leqslant \nu_0\, \sigma_0^{-\frac{1}{m}} |z - z_0|$$

for every $z \in B(z_0, \frac{d_0}{4})$, where ν_0 is a positive constant which depends only on m.

Acknowledgments This work was supported by a grant from the Simons Foundation (1290607, R.R.S., M.V.S.).

References

1. Golberg, A., Salimov, R., Sevost'yanov, E.: Distortion estimates under mappings with controlled p-module. Ann. Univ. Buchar. Math. Ser. **5(LXIII)**(1), 95–114 (2014)
2. Golberg, A., Salimov, R., Stefanchuk, M.: Asymptotic dilation of regular homeomorphisms. Complex Anal. Oper. Theory **13**(6), 2813–2827 (2019)
3. Golberg, A., Salimov, R.: Nonlinear Beltrami equation. Complex Var. Ellipt. Equ. **65**(1), 6–21 (2020)
4. Gutlyanskii, V., Ryazanov, V., Srebro, U., Yakubov, E.: The Beltrami Equation: A Geometric Approach. Developments in Mathematics, vol. 26. Springer, New York (2012)
5. Martio, O., Ryazanov, V., Srebro, U., Yakubov, E.: Moduli in Modern Mapping Theory. Springer Monographs in Mathematics. Springer, New York (2009)
6. Gutlyanskii, V., Ryazanov, V., Srebro, U., Yakubov, E.: On recent advances in the Beltrami equations. Ukr. Mat. Visn. **7**(4), 467–515 (2010); reprinted in J. Math. Sci. (N.Y.) **175** (2011), no. 4, 413–449
7. Srebro, U., Yakubov, E.: The beltrami equation. In: Handbook in Complex Analysis: Geometric Function Theory, vol. 2, pp. 555–597. Elseiver, Amsterdam (2005)
8. Astala, K., Iwaniec, T., Martin, G.: Elliptic Partial Differential Equations and Quasiconformal Mappings in the Plane. Princeton Mathematical Series, vol. 48. Princeton University Press, Princeton (2009)
9. Lehto, O., Virtanen, K.: Quasiconformal Mappings in the Plane, 2nd edn. Springer, New York (1973)
10. Väisälä, J.: Lectures on n-Dimensional Quasiconformal Mappings. Lecture Notes in Mathematics, vol. 229. Springer, Berlin (1971)
11. Shlyk, V.A.: On the Equality Between p-Capacity and p-Modulus (in Russian). Sibirsk. Mat. Zh. **34**(6), 216–221 (1993); transl. in Siberian Math. J. **34** (1993) no. 6, 1196–1200

12. Kruglikov, V.I.: Capacity of condensers and spatial mappings quasiconformal in the mean. Mat. Sb. **130**(172), (1986); (**2**, no. 6), 185–206 [Math. USSR-Sb. **58** (1987), 185–205]
13. Maz'ya, V.G.: Classes of domains, measures and capacities in the theory of spaces of differentiable functions (in Russian). In Itogi Nauki Tekh., Ser. Sovrem. Probl. Mat., Fundam. Napravleniya Vol. 26: Analysis III. Spaces of Differentiable Functions (VINITI, Moscow, 1988), 159–228 [Encycl. Math. Sci. 26, 141–211 (1991)]
14. Gehring, F.W.: Rings and quasiconformal mappings in space. Trans. Amer. Math. Soc. **103**, 353–393 (1962)
15. Salimov, R.: On finitely Lipschitz space mappings. Sib. Electron. Math. Rep. **8**(1), 284–295 (2011)
16. Ryazanov, V., Srebro, U., Yakubov, E.: On ring solutions of Beltrami equation. J. d'Anal. Math. **96**, 117–150 (2005)
17. Ryazanov, V., Sevost'yanov, E.: Equicontinuous classes of ring Q-homeomorphisms. Sib. Mat. Zh. **48**(6), 1361–1376 (2007). MR2397516
18. Salimov, R., Stefanchuk, M.: Finite Lipschitzness of regular solutions to nonlinear Beltrami equation. Complex Var. Ellipt. Equ. **69**, 913–923 (2024). Published Online 12.01.2023. https://doi.org/10.1080/17476933.2023.2166498

A Piecewise Continuous Schwarz-Type Boundary-Value Problem in an Angle for Monogenic Functions in a Commutative Biharmonic Algebra

Serhii Gryshchuk and Sergiy Plaksa

Abstract We consider a piecewise continuous Schwarz-type boundary-value problem in an angle for monogenic functions in a commutative biharmonic algebra. This problem is reduced to a system of integral equations on the positive semiaxis. It is shown that, on each segment of this semiaxis, the set of solutions of this system coincides with the set of solutions of a certain system of Fredholm integral equations.

1 Biharmonic Monogenic Functions in Biharmonic Algebra

We say that an associative commutative two-dimensional algebra $\mathbb{B}$ with unit e over the field of complex numbers $\mathbb{C}$ is *biharmonic* (this notion is proposed in [1]) if in $\mathbb{B}$ there exists a *biharmonic* basis, i.e., a basis $\{e_1, e_2\}$ satisfying the conditions

$$(e_1^2 + e_2^2)^2 = 0, \qquad e_1^2 + e_2^2 \neq 0\,. \tag{1}$$

V. F. Kovalev and I. P. Mel'nichenko [1] found a multiplication table for a biharmonic basis $\{e_1, e_2\}$:

$$e_1 = e, \qquad e_2^2 = e_1 + 2ie_2,$$

where i is the imaginary complex unit.

In the paper [2] I. P. Mel'nichenko proved that there exists the unique biharmonic algebra $\mathbb{B}$ with a non-biharmonic basis $\{e, \rho\}$ for which

$$\rho = 2e_1 + 2ie_2$$

S. Gryshchuk · S. Plaksa (✉)
Institute of Mathematics of the National Academy of Sciences of Ukraine, Kyiv-4, Ukraine

M. R. Ebert et al. (eds.), *New Tools in Mathematical Analysis and Applications*, Research Perspectives, https://doi.org/10.1007/978-3-031-77050-0_11

and $\rho^2 = 0$. Note that the algebra $\mathbb{B}$ is isomorphic to four-dimensional over the field of real numbers $\mathbb{R}$ algebras considered by A. Douglis [3] and L. Sobrero [4].

We now define the Euclidean norm $\|a\| := \sqrt{|z_1|^2 + |z_2|^2}$ in the algebra $\mathbb{B}$, where $a = z_1 e_1 + z_2 e_2$ and $z_1, z_2 \in \mathbb{C}$.

Consider a *biharmonic plane* $\mu_{e_1,e_2} := \{\zeta = x\, e_1 + y\, e_2 : x, y \in \mathbb{R}\}$ which is a linear span over the field $\mathbb{R}$ of the elements of biharmonic basis $\{e_1, e_2\}$.

With a domain D we associate the congruent domain $D_\zeta := \{\zeta = x e_1 + y e_2 : (x, y) \in D\}$ in the biharmonic plane μ_{e_1,e_2}. Its boundaries are denoted by ∂D, ∂D_ζ, respectively. Let $\overline{D_\zeta}$ (or $\overline{D}$) be the closure of domain D_ζ (or D, respectively).

Everywhere in what follows, $\zeta = x\, e_1 + y\, e_2$, $z = x + iy$, where $(x, y) \in D$, and $\zeta_0 = x_0 e_1 + y_0 e_2$, $z_0 = x_0 + i y_0$, where $(x_0, y_0) \in \partial D$.

Any function $\Phi : D_\zeta \longrightarrow \mathbb{B}$ has an expansion

$$\Phi(\zeta) = U_1(x, y)\, e_1 + U_2(x, y)\, i e_1 + U_3(x, y)\, e_2 + U_4(x, y)\, i e_2\,, \qquad (2)$$

where $U_l : D \longrightarrow \mathbb{R}$, $l = \overline{1, 4}$, are real-valued component-functions. We shall use the following notation: $\mathrm{U}_l\,[\Phi] := U_l$, $l = \overline{1, 4}$.

We say that a function $\Phi : D_\zeta \longrightarrow \mathbb{B}$ is *monogenic* in a domain D_ζ if it has at every point $\zeta \in D_\zeta$ the classical derivative

$$\Phi'(\zeta) := \lim_{h \to 0,\, h \in \mu_{e_1,e_2}} \big(\Phi(\zeta + h) - \Phi(\zeta)\big)\, h^{-1}.$$

Every monogenic function $\Phi : D_\zeta \longrightarrow \mathbb{B}$ has the derivative of any order in D_ζ (cf., e.g., [5, 6]) and satisfies the biharmonic equation due to the equalities

$$\left(\frac{\partial^4}{\partial x^4} + 2\frac{\partial^4}{\partial x^2 \partial y^2} + \frac{\partial^4}{\partial y^4}\right) \Phi(\zeta) = \Phi^{(4)}(\zeta)\, (e_1^2 + e_2^2)^2 = 0 \quad \forall\, \zeta \in D_\zeta$$

and the conditions (1). In addition, every function $U(x, y)$ biharmonic in a simply connected domain D is the first component $U_1 \equiv U$ in the expression (2) of a certain monogenic function $\Phi : D_\zeta \longrightarrow \mathbb{B}$ and, moreover, all such functions Φ are found in an explicit form (cf., e.g., [5, 6]).

An example of monogenic function in both domains D_ζ and $\mu_{e_1,e_2} \setminus \overline{D_\zeta}$ is the *biharmonic Cauchy type integral*

$$\mathcal{B}[\varphi](\zeta) = \frac{1}{2\pi i} \int\limits_{\partial D_\zeta} \varphi(\tau)(\tau - \zeta)^{-1} d\tau$$

with a continuous density $\varphi : \partial D_\zeta \longrightarrow \mathbb{B}$.

In what follows, the integrals over the unbounded contour $\Gamma \in \mu_{e_1,e_2}$ are always understood in a sense of the Cauchy principal value, i.e.,

$$\int_{\Gamma} g(\tau,\cdot)\,d\tau := \lim_{N\to+\infty} \int_{\{\tau\in\Gamma:\|\tau\|\le N\}} g(\tau,\cdot)\,d\tau.$$

2 A Schwarz-Type BVP for Monogenic Functions

Consider a boundary value problem on finding a function $\Phi\colon D_\zeta \longrightarrow \mathbb{B}$ which is monogenic in a domain D_ζ when limiting values of the first and the fourth component-functions in (2) are given on the boundary ∂D_ζ, i.e., the following boundary conditions are satisfied:

$$U_1(x_0,y_0) = u_1(\zeta_0)\,, \quad U_4(x_0,y_0) = u_4(\zeta_0) \qquad \forall\,\zeta_0 \in \partial D_\zeta$$

where

$$U_j(x_0,y_0) = \lim_{\zeta\to\zeta_0,\zeta\in D_\zeta} \mathrm{U}_j\left[\Phi\left(\zeta\right)\right], \qquad j\in\{1,4\},$$

and u_1, u_4 are given continuous functions.

If D_ζ is unbounded then one demands additionally the existence of finite limits

$$\lim_{\|\zeta\|\to\infty,\,\zeta\in D_\zeta} \mathrm{U}_j\left[\Phi(\zeta)\right], \qquad j\in\{1,4\},$$

as well as the assumption that every given function u_j, $j \in \{1,4\}$, has a finite limit

$$u_j(\infty) := \lim_{\|\zeta\|\to\infty,\,\zeta\in\partial D_\zeta} u_j(\zeta)$$

in the case where the boundary ∂D_ζ is unbounded.

We shall call such a problem by the (1-4)-problem.

V. F. Kovalev [7] considered similar Schwartz-type boundary value problems for monogenic functions (while the limiting values of a pair of component-functions in (2) are given on ∂D_ζ) with additional assumptions that the sought-for function $\Phi : \overline{D_\zeta} \longrightarrow \mathbb{B}$ is continuous in $\overline{D_\zeta}$ and has the limit (if D_ζ is unbounded)

$$\lim_{\|\zeta\|\to\infty,\,\zeta\in D_\zeta} \Phi(\zeta) =: \Phi(\infty) \in \mathbb{B}.$$

He named such problems as *biharmonic Schwarz problems* owing to their analogy with the classic Schwarz problem on finding an analytic function of a complex variable when values of its real part are given on the boundary of domain.

We call a (1-4)-problem with these Kovalov's additional assumptions as (1-4)-*problems in the sense of Kovalev*. More detailed information about relation between the (1-4)-problems in both senses were considered in [8]. Note, that in the papers [9–11] we interpret the (1-4)-problem as the (1-4)-problem in the sense of Kovalev.

In [12], we investigated the (1-4)-problem for the case where D_ζ is either a half-plane or a unit disk in the biharmonic plane. Its solutions were found in explicit form with using of some integrals analogous to the classic Schwarz integral.

In [10], using a hypercomplex analog of the Cauchy type integral, we reduced the (1-4)-problem in the sense of Kovalev to a system of integral equations and established sufficient conditions under which this system has the Fredholm property. The similar is done for the (1-4)-problem in [8, 13]. It made for the case where the boundary of domain belongs to a class being wider than the class of Lyapunov curves that was usually required in the plane elasticity theory (cf., e.g., [14–18]).

3 A Piecewise Continuous Schwarz-Type BVP for Monogenic Functions in an Angle

The theory of boundary-value problems for elliptic equations in domains with piecewise continuous boundaries was developed by numerous authors (see, e.g., [19–22] and the references therein). In these works, the authors studied the problem of existence and uniqueness of solutions to the boundary-value problems for elliptic operators in weight spaces and analyzed the asymptotic properties of solutions in neighborhoods of the angular points, the influence of angular points on the smoothness of solutions, etc.

In [10, 11], a relation between (1-4)-problem and boundary value problems of the plane elasticity theory is considered. Namely, there is considered a problem on finding an elastic equilibrium for isotropic body occupying D with given limiting values of partial derivatives $\frac{\partial u}{\partial x}$, $\frac{\partial v}{\partial y}$ for the displacements $u = u(x, y)$, $v = v(x, y)$ on the boundary ∂D. In particular, it is shown in [8, 11] that such a problem is reduced to the (1-4)-problem (see also [10]).

A new method for the solution of a biharmonic problem based on the relationship between biharmonic and monogenic functions in the algebra $\mathbb{B}$ and on representations of solutions by hypercomplex integrals similar to the classical Schwarz integrals and Cauchy-type integrals was developed in the papers [8–13].

In [23], such a hypercomplex method intended for solving biharmonic problems is applied in the case of a specific unbounded angular domain

$$D := \{(x, y) \in \mathbb{R} \times \mathbb{R} : x = r\cos\theta, y = r\sin\theta, 0 < \theta < \alpha, 0 < r < \infty\} \quad (3)$$

with fixed $\alpha \in (0; 2\pi)$, which is an angle in the Cartesian plane xOy with vertex at the origin of coordinates. These biharmonic problems are associated with a Schwarz-type boundary value problem for monogenic functions in the angle D_ζ

when the limiting values of two component-functions U_1 and U_3 in (2) are given on the set $\partial D \setminus \{(0,0)\}$ as piecewise continuous functions for which singularities are permitted at the vertex of the angle.

The aim of our present research is to solve a similar (1-4)-problem.

Let us formulate this *piecewise continuous* (1-4)*-problem for monogenic functions in the angle* D_ζ associated with the angle (3).

Consider a boundary value problem on finding a monogenic function $\Phi\colon D_\zeta \longrightarrow \mathbb{B}$ for which the limit values of the components $\mathrm{U}_j\,[\Phi(\zeta)]$, $j \in \{1,4\}$, in decomposition (2) satisfy the boundary conditions

$$\lim_{\zeta\to\zeta_0,\zeta\in D_\zeta} \mathrm{U}_j\,[\Phi(\zeta)] = u_j(\zeta_0) \qquad \forall\,\zeta_0 \in \partial D_\zeta \setminus \{0\}, \qquad j \in \{1,4\}, \tag{4}$$

where u_1, u_4 are given continuous functions on the set $\partial D \setminus \{(0,0)\}$ with the following asymptotic behaviors at the origin and at the infinity:

$$\left|u_j(\zeta_0)\right| \le c\ \|\zeta_0\|^{-\beta_0}, \qquad \sqrt{x_0^2+y_0^2} \equiv \|\zeta_0\| \to 0, \qquad (x_0,y_0) \in \partial D\,, \tag{5}$$

$$\left|u_j(\zeta_0)\right| \le c\ \|\zeta_0\|^{-\beta_\infty}, \qquad \|\zeta_0\| \to \infty, \quad (x_0,y_0) \in \partial D\,, \ j \in \{1,4\}, \tag{6}$$

where a constant c is independent of (x_0, y_0); $\beta_0, \beta_\infty \in (0,1)$ are fixed numbers.

In addition, according to estimates (5) and (6), the required function Φ must satisfy the similar inequalities

$$\left|\mathrm{U}_j\,[\Phi(\zeta)]\right| \le c\,\|\zeta\|^{-\beta_0}, \qquad \zeta \to 0, \quad \zeta \in D_\zeta\,, \tag{7}$$

$$\left|\mathrm{U}_j\,[\Phi(\zeta)]\right| \le c\,\|\zeta\|^{-\beta_\infty}, \qquad \|\zeta\| \to \infty, \quad \zeta \in D_\zeta\,, \tag{8}$$

for $j \in \{1,4\}$, where the constant c is independent of ζ.

We seek the solutions of this (1-4)-problem in the class of functions that can be represented in the form of the biharmonic Cauchy-type integral

$$\Phi(\zeta) = \mathcal{B}[\varphi](\zeta) \quad \forall\,\zeta \in D_\zeta\,, \tag{9}$$

where

$$\varphi(\zeta) = \varphi_1(\zeta)\,e_1 + \varphi_4(\zeta)\,ie_2 \quad \forall\,\zeta \in \partial D_\zeta \tag{10}$$

and the functions $\varphi_j\colon \partial D_\zeta \setminus \{0\} \longrightarrow \mathbb{R}$, $j \in \{1,4\}$, are continuous on $\partial D_\zeta \setminus \{0\}$. The conditions on the behavior of the functions φ_1 and φ_4 at infinity and in a neighborhood of zero are imposed in what follows.

4 Auxiliary Statements

Denote $k_1 := \cos\alpha$, $k_2 := \sin\alpha$. We introduce open rays $R_+ := \{xe_1 : x \in (0, +\infty)\}$ and $L_\alpha := \{rk_1e_1 + rk_2e_2 : r \in (0, +\infty)\}$, which are parts of the boundary ∂D_ζ.

We consider a function of the type $\upsilon\colon R_+ \longrightarrow \mathbb{R}$ as a function of real variable and preserve the same notation for it, i.e., $\upsilon(x) \equiv \upsilon(xe_1)$ for $x > 0$.

For a function of the type $\upsilon\colon L_\alpha \longrightarrow \mathbb{R}$ defined on the ray L_α, we introduce a function of real variable $\widetilde{\upsilon}(r) := \upsilon\left(r(k_1e_1 + k_2e_2)\right)$ for $r > 0$.

For $s_1, s \in (0, \infty)$, we denote

$$q(s_1, s) := \left(s^2 - 2k_1 s s_1 + s_1^2\right)^2 \equiv \left((k_1 s - s_1)^2 + k_2^2 s^2\right)^2.$$

Let $E \subset \mathbb{R}$. For a function of the type $\upsilon_*\colon E \longrightarrow \mathbb{R}$ continuous on the set E, we consider its modulus of continuity on E:

$$\omega_E(\upsilon_*, \varepsilon) = \sup_{t_1, t_2 \in E\,:\,|t_1 - t_2| \le \varepsilon} |\upsilon_*(t_1) - \upsilon_*(t_2)|\,.$$

Lemma 1 ([23]) *Suppose that a function $\upsilon\colon \partial D_\zeta \setminus \{0\} \longrightarrow \mathbb{R}$ is continuous on $\partial D_\zeta \setminus \{0\}$ and satisfies the inequalities*

$$\left|\upsilon(\tau)\right| \le c\,\|\tau\|^{-\beta_0}, \qquad \tau \to 0, \quad \tau \in \partial D_\zeta \setminus \{0\}, \tag{11}$$

$$\left|\upsilon(\tau)\right| \le c\,\|\tau\|^{-\beta_\infty}, \qquad \|\tau\| \to \infty, \quad \tau \in \partial D_\zeta, \tag{12}$$

where $\beta_0, \beta_\infty \in (0, 1)$ and c is a constant independent of τ. Then the components $\mathrm{U}_1\left[\mathcal{B}[\upsilon]\right]$, $\mathrm{U}_3\left[\mathcal{B}[\upsilon]\right]$, $\mathrm{U}_4\left[\mathcal{B}[\upsilon]\right]$ *of the integral $\mathcal{B}[\upsilon]$ are continuously extended onto $\partial D_\zeta \setminus \{0\}$ from the domain D_ζ. Moreover, for all $\zeta_0 \in R_+$, the equalities*

$$\lim_{\zeta \to \zeta_0,\, \zeta \in D_\zeta} \mathrm{U}_1\left[\mathcal{B}[\upsilon](\zeta)\right] = \frac{1}{2}\,\upsilon\left(\|\zeta_0\|\right) + \frac{k_2^3\|\zeta_0\|}{\pi}\int\limits_0^\infty \frac{s^2\,\widetilde{\upsilon}(s)}{q\left(\|\zeta_0\|, s\right)}\,ds\,, \tag{13}$$

$$\lim_{\zeta \to \zeta_0,\, \zeta \in D_\zeta} \mathrm{U}_3\left[\mathcal{B}[\upsilon](\zeta)\right] = -\frac{k_2^2\|\zeta_0\|}{\pi}\int\limits_0^\infty \frac{s\left(k_1 s - \|\zeta_0\|\right)\widetilde{\upsilon}(s)}{q\left(\|\zeta_0\|, s\right)}\,ds\,, \tag{14}$$

$$\lim_{\zeta \to \zeta_0,\, \zeta \in D_\zeta} \mathrm{U}_4\left[\mathcal{B}[\upsilon](\zeta)\right] = \frac{k_2\|\zeta_0\|}{2\pi}\int\limits_0^\infty \frac{\left(k_2^2 s^2 - \left(k_1 s - \|\zeta_0\|\right)^2\right)\widetilde{\upsilon}(s)}{q\left(\|\zeta_0\|, s\right)}\,ds\,, \tag{15}$$

are true and, for all $\zeta_0 \in L_\alpha$, the following equalities are true:

$$\lim_{\zeta\to\zeta_0,\,\zeta\in D_\zeta} \mathrm{U}_1\left[\mathcal{B}[\upsilon](\zeta)\right] = \frac{1}{2}\,\widetilde{\upsilon}\,(\|\zeta_0\|) + \frac{k_2^3\|\zeta_0\|^3}{\pi}\int\limits_0^\infty \frac{\upsilon(s)}{q\big(\|\zeta_0\|, s\big)}\,ds\,, \tag{16}$$

$$\lim_{\zeta\to\zeta_0,\,\zeta\in D_\zeta} \mathrm{U}_3\left[\mathcal{B}[\upsilon](\zeta)\right] = \frac{k_2^2\|\zeta_0\|^2}{\pi}\int\limits_0^\infty \frac{\big(s - k_1\|\zeta_0\|\big)\upsilon(s)}{q\big(\|\zeta_0\|, s\big)}\,ds\,, \tag{17}$$

$$\lim_{\zeta\to\zeta_0,\,\zeta\in D_\zeta} \mathrm{U}_4\left[\mathcal{B}[\upsilon](\zeta)\right] = \frac{k_2\|\zeta_0\|}{2\pi}\int\limits_0^\infty \frac{\Big(k_2^2\|\zeta_0\|^2 - \big(s - k_1\|\zeta_0\|\big)^2\Big)\upsilon(s)}{q\big(\|\zeta_0\|, s\big)}\,ds. \tag{18}$$

Lemma 2 ([23]) *Under the conditions of Lemma 1, the components* $\mathrm{U}_1\left[\mathcal{B}[\upsilon]\right]$, $\mathrm{U}_3\left[\mathcal{B}[\upsilon]\right]$, $\mathrm{U}_4\left[\mathcal{B}[\upsilon]\right]$ *of the integral* $\mathcal{B}[\upsilon]$ *satisfy inequalities of the form* (7), (8).

5 Reducing the Piecewise Continuous (1-4)-Problem to a System of Integral Equations

We apply relations (13)–(18) to the required solution (9) of the piecewise continuous (1-4)-problem, where the function φ has the form (10) and the functions $\varphi_j\colon \partial D_\zeta \setminus \{0\} \longrightarrow \mathbb{R}$, $j \in \{1, 4\}$, satisfy the conditions imposed on the function υ in Lemma 1. By using the equalities

$$\mathrm{U}_1\left[ie_2\mathcal{B}[\varphi_4](\zeta)\right] = -\mathrm{U}_4\left[\mathcal{B}[\varphi_4](\zeta)\right],$$

$$\mathrm{U}_4\left[ie_2\mathcal{B}[\varphi_4](\zeta)\right] = \mathrm{U}_1\left[\mathcal{B}[\varphi_4](\zeta)\right] - 2\mathrm{U}_4\left[\mathcal{B}[\varphi_4](\zeta)\right]$$

and the notation $\xi := \|\zeta_0\|$ we reduce the boundary conditions (4) to the following system of integral equations for the functions φ_1 and φ_4 for all $\xi > 0$:

$$\frac{1}{2}\,\varphi_1(\xi) + \frac{k_2\,\xi}{2\pi}\int\limits_0^\infty \frac{2k_2^2s^2\,\widetilde{\varphi}_1(s) + \big((k_1s - \xi)^2 - k_2^2s^2\big)\,\widetilde{\varphi}_4(s)}{q(\xi, s)}\,ds = u_1\,(\xi)\,,$$

$$\frac{1}{2}\,\varphi_4(\xi) + \frac{k_2\,\xi}{2\pi}\int\limits_0^\infty \frac{\big(k_2^2s^2 - (k_1s - \xi)^2\big)\,\widetilde{\varphi}_1(s) + 2\,(k_1s - \xi)^2\,\widetilde{\varphi}_4(s)}{q(\xi, s)}\,ds = u_4\,(\xi)\,,$$

$$\frac{1}{2}\,\widetilde{\varphi}_1(\xi)+\frac{k_2\,\xi}{2\pi}\int\limits_0^{\infty}\frac{2k_2^2\,\xi^2\,\varphi_1(s)-\left(k_2^2\xi^2-(s-k_1\xi)^2\right)\varphi_4(s)}{q(\xi,s)}\,ds=\widetilde{u}_1\,(\xi)\,,$$

$$\frac{1}{2}\,\widetilde{\varphi}_4(\xi)+\frac{k_2\,\xi}{2\pi}\int\limits_0^{\infty}\frac{\left(k_2^2\,\xi^2-(s-k_1\xi)^2\right)\varphi_1(s)+2\,(s-k_1\xi\,)^2\,\varphi_4(s)}{q(\xi,s)}\,ds=\widetilde{u}_4\,(\xi)\,. \tag{19}$$

It follows from Lemmas 1 and 2 that each solution of system (19), which gives the functions φ_1 and φ_4 from equality (10) by the formulas

$$\varphi_j(\tau)=\begin{cases}\varphi_j(\|\tau\|) & \text{for}\quad \tau\in R_+\,,\\ \widetilde{\varphi}_j(\|\tau\|) & \text{for}\quad \tau\in L_\alpha\,,\end{cases}\qquad j\in\{1,4\},$$

such that these functions satisfy the conditions of the form (11) and (12), generates the solution of the piecewise continuous (1-4)-problem by relation (9).

To rewrite system (19) in the vector form, we introduce the required notation. Let $\mathbf{a}(s,\cdot)$ be a four-dimensional column vector whose components are real-valued functions $a_j(s,\cdot)$, $j=\overline{1,4}$, defined for all $s>0$. Similarly, $\int\limits_0^{\infty}\mathbf{a}(s,\cdot)\,ds$ denotes a four-dimensional column vector whose components are integrals $\int\limits_0^{\infty}a_j(s,\cdot)\,ds$, $j=\overline{1,4}$, in the case where these integrals exist.

By $\mathbf{L}^{\infty}_{[0,\infty)}$ we denote a Banach space of four-dimensional column vectors $\mathbf{b}(s)$ whose components $b_j(s)$, $j=\overline{1,4}$, are essentially bounded functions on the set $[0,\infty)$ with the norm

$$\|\mathbf{b}\|_{\mathbf{L}^{\infty}_{[0,\infty)}}:=\max_{j=\overline{1,4}}\{\operatorname*{ess\ sup}_{s\in[0,\infty)}|b_j(s)|\}\,.$$

By $\mathbf{C}_{[0,\infty)}$ we denote a subspace of the space $\mathbf{L}^{\infty}_{[0,\infty)}$ formed by the column vectors $\mathbf{b}(s)$ whose components $b_j(s)$, $j=\overline{1,4}$, are continuous functions on the set $[0,\infty)$ with finite limits as $s\to\infty$.

We introduce a column vector $\boldsymbol{\varphi}(\xi)$, $\xi>0$, whose components are the required functions $\varphi_1(\xi)$, $\varphi_4(\xi)$, $\widetilde{\varphi}_1(\xi)$, $\widetilde{\varphi}_4(\xi)$, and a column vector $\mathbf{u}(\xi)$, $\xi>0$, whose components are the functions $u_1(\xi)$, $u_4(\xi)$, $\widetilde{u}_1(\xi)$ and $\widetilde{u}_4(\xi)$ given in the system (19).

Assume that the vector functions $\boldsymbol{\varphi}$ and $\mathbf{u}$ have the form

$$\boldsymbol{\varphi}(\xi)=\frac{(\xi+1)^{\beta_0-\beta_\infty}}{\xi^{\beta_0}}\,\mathbf{g}(\xi),\qquad \mathbf{u}(\xi)=\frac{1}{2}\,\frac{(\xi+1)^{\beta_0-\beta_\infty}}{\xi^{\beta_0}}\,\mathbf{v}(\xi)\,,$$

where the vector functions

$$\mathbf{g}(\xi) := \begin{pmatrix} g_1(\xi) \\ g_4(\xi) \\ \widetilde{g_1}(\xi) \\ \widetilde{g_4}(\xi) \end{pmatrix}, \qquad \mathbf{v}(\xi) := \begin{pmatrix} v_1(\xi) \\ v_4(\xi) \\ \widetilde{v_1}(\xi) \\ \widetilde{v_3}(\xi) \end{pmatrix}$$

belong to the space $\mathbf{L}^{\infty}_{[0,\infty)}$.

By $0_{2\times 2}$ we denote the 2×2 null matrix and introduce the following functional matrix:

$$a(\xi, s) := \begin{pmatrix} 2k_2^2 s^2 & \left((\xi - k_1 s)^2 - k_2^2 s^2\right) \\ -\left((\xi - k_1 s)^2 - k_2^2 s^2\right) & 2\,(k_1 s - \xi)^2 \end{pmatrix}.$$

Finally, we consider a (4×4)-dimensional functional matrix and, by using the introduced 2×2 matrices, represent it in the block form as follows:

$$A(\xi, s) := \begin{pmatrix} 0_{2\times 2} & a\,(\xi, s) \\ a\,(s, \xi) & 0_{2\times 2} \end{pmatrix}.$$

Thus, under the assumptions imposed on the vector functions $\boldsymbol{\varphi}$, $\mathbf{u}$, we can rewrite system (19) in the vector form:

$$\mathcal{I}[\mathbf{g}](\xi) + \mathcal{A}[\mathbf{g}](\xi) = \mathbf{v}(\xi), \qquad \xi > 0, \tag{20}$$

where $\mathcal{I}$ is the identity operator,

$$\mathcal{A}[\mathbf{g}](\xi) := \frac{k_2\,\xi^{\,1+\beta_0}\,(\xi + 1)^{\beta_\infty - \beta_0}}{\pi} \int\limits_0^\infty A(\xi, s)\,\mathbf{g}(s) \frac{ds}{s^{\beta_0}\,(s + 1)^{\beta_\infty - \beta_0}\,q(\xi, s)}, \tag{21}$$

and equality (20) holds almost everywhere on $(0, \infty)$ with respect to the Lebesgue measure.

By analyzing Eq. (20) for $\xi \in [a, b]$, we rewrite operator (21) in the form of a sum of three operators as follows:

$$\mathcal{A}[\mathbf{g}](\xi) = \frac{k_2\,\xi^{\,1+\beta_0}\,(\xi + 1)^{\beta_\infty - \beta_0}}{\pi} \int\limits_0^\delta A(\xi, s)\mathbf{g}(s) \frac{ds}{s^{\beta_0}\,(s + 1)^{\beta_\infty - \beta_0}\,q(\xi, s)} +$$

$$+ \frac{k_2\,\xi^{\,1+\beta_0}\,(\xi + 1)^{\beta_\infty - \beta_0}}{\pi} \int\limits_\delta^N A(\xi, s)\mathbf{g}(s) \frac{ds}{s^{\beta_0}\,(s + 1)^{\beta_\infty - \beta_0}\,q(\xi, s)} +$$

$$+\frac{k_2\,\xi^{1+\beta_0}\,(\xi+1)^{\beta_\infty-\beta_0}}{\pi}\int\limits_N^\infty A(\xi,s)\mathbf{g}(s)\frac{ds}{s^{\beta_0}\,(s+1)^{\beta_\infty-\beta_0}\,q(\xi,s)}=:$$

$$=:\mathcal{A}_\delta[\mathbf{g}](\xi)+\mathcal{A}_0[\mathbf{g}](\xi)+\mathcal{A}_N[\mathbf{g}](\xi)\,,$$

where the positive numbers δ, N are chosen such that, for all $\xi\in[a,b]$, the inequalities

$$\|\mathcal{A}_\delta[\mathbf{g}](\xi)\|_{\mathbf{L}^\infty_{[0,\infty)}}<\frac{1}{2}\,\|\mathbf{g}\|_{\mathbf{L}^\infty_{[0,\infty)}}\,,\qquad \|\mathcal{A}_N[\mathbf{g}](\xi)\|_{\mathbf{L}^\infty_{[0,\infty)}}<\frac{1}{2}\,\|\mathbf{g}\|_{\mathbf{L}^\infty_{[0,\infty)}}$$

are true.

We introduce the operators

$$\widehat{\mathcal{A}}_\delta[\mathbf{g}](\xi):=\begin{cases}\mathcal{A}_\delta[\mathbf{g}](a) & \text{for}\quad 0\le\xi<a\,,\\ \mathcal{A}_\delta[\mathbf{g}](\xi) & \text{for}\quad a\le\xi\le b\,,\\ \mathcal{A}_\delta[\mathbf{g}](b) & \text{for}\quad \xi\ge b\,,\end{cases}$$

$$\widehat{\mathcal{A}}_N[\mathbf{g}](\xi):=\begin{cases}\mathcal{A}_N[\mathbf{g}](a) & \text{for}\quad 0\le\xi<a\,,\\ \mathcal{A}_N[\mathbf{g}](\xi) & \text{for}\quad a\le\xi\le b\,,\\ \mathcal{A}_N[\mathbf{g}](b) & \text{for}\quad \xi>b\,,\end{cases}$$

$$\widehat{\mathcal{A}}[\mathbf{g}](\xi):=\begin{cases}\mathcal{A}_0[\mathbf{g}](a) & \text{for}\quad 0\le\xi<a\,,\\ \mathcal{A}_0[\mathbf{g}](\xi) & \text{for}\quad a\le\xi\le b\,,\\ \mathcal{A}_0[\mathbf{g}](b) & \text{for}\quad \xi\ge b\,.\end{cases}$$

It is clear that, for all $\xi\ge 0$, the operators $\widehat{\mathcal{A}}_\delta$ and $\widehat{\mathcal{A}}_N$ satisfy the estimates

$$\left\|\widehat{\mathcal{A}}_\delta[\mathbf{g}](\xi)\right\|_{\mathbf{L}^\infty_{[0,\infty)}}<\frac{1}{2}\,\|\mathbf{g}\|_{\mathbf{L}^\infty_{[0,\infty)}}\,,\qquad \left\|\widehat{\mathcal{A}}_N[\mathbf{g}](\xi)\right\|_{\mathbf{L}^\infty_{[0,\infty)}}<\frac{1}{2}\,\|\mathbf{g}\|_{\mathbf{L}^\infty_{[0,\infty)}}\,,\tag{22}$$

and the operator $\widehat{\mathcal{A}}$ acts from the space $\mathbf{L}^\infty_{[0,\infty)}$ into the space $\mathbf{C}_{[0,\infty)}$. In this case, by the Ascoli–Arzela theorem, the operator $\widehat{\mathcal{A}}\colon \mathbf{L}^\infty_{[0,\infty)}\longrightarrow\mathbf{C}_{[0,\infty)}$ is compact and, hence, the operator $\widehat{\mathcal{A}}$ is also compact in the space $\mathbf{L}^\infty_{[0,\infty)}$.

Parallel with Eq. (20), we consider the equation

$$\mathcal{I}[\mathbf{g}](\xi)+\widehat{\mathcal{A}}_\delta[\mathbf{g}](\xi)+\widehat{\mathcal{A}}_N[\mathbf{g}](\xi)+\widehat{\mathcal{A}}[\mathbf{g}](\xi)=\mathbf{v}(\xi),\qquad \xi\ge 0\,.\tag{23}$$

It is clear that each solution of Eq. (20) satisfies Eq. (23) for $\xi\in[a,b]$. Conversely, each solution of Eq. (23) satisfies Eq. (20) for $\xi\in[a,b]$.

In the theory of operators in Banach spaces, a *Noetherian* operator is defined as an operator with the closed image whose kernel and cokernel are finite-dimensional.

The difference between the dimensions of kernel and cokernel is called the *index* of the operator. A Noetherian operator with index zero is called a *Fredholm* operator (see, e.g., [24]).

Theorem 1 *The operator* $\mathcal{I} + \widehat{\mathcal{A}}_\delta + \widehat{\mathcal{A}}_N + \widehat{\mathcal{A}}$ *is a Fredholm operator in the space* $\mathbf{L}^\infty_{[0,\infty)}$.

Proof It follows from estimates (22) that the operator $B := \mathcal{I} + \widehat{\mathcal{A}}_\delta + \widehat{\mathcal{A}}_N$ possesses the inverse operator in the space $\mathbf{L}^\infty_{[0,\infty)}$. Hence, B is a Fredholm operator in the space $\mathbf{L}^\infty_{[0,\infty)}$. Thus, by the theorem on stability of the Noetherian property and the index of the operator under its compact perturbation (see, e.g., [24]), $B + \widehat{\mathcal{A}}$ is a Fredholm operator in the space $\mathbf{L}^\infty_{[0,\infty)}$. □

Thus, on any segment $[a, b]$, where $a > 0$, the set of solutions of Eq. (20) from the space $\mathbf{L}^\infty_{[0,\infty)}$ coincides with the set of solutions of the Fredholm equation (23) in this space.

Acknowledgments This work was supported by a grant from the Simons Foundation (1030291, S.V.G.) for the first author. The second author acknowledges the support of the UNIPD Scholar Rescue Fund and of INdAM.

References

1. Kovalev, V.F., Mel'nichenko, I.P.: Biharmonic functions on the biharmonic plane (in Russian). Reports Acad. Sci. USSR Ser. A **8**, 25–27 (1981)
2. Mel'nichenko, I.P.: Biharmonic bases in algebras of the second rank. Ukr. Math. J. **38**(2), 224–226 (1986)
3. Douglis, A.: A function-theoretic approach to elliptic systems of equations in two variables. Commun. Pure Appl. Math. **6**(2), 259–289 (1953)
4. Sobrero, L.: Nuovo metodo per lo studio dei problemi di elasticità, con applicazione al problema della piastra forata. Ricerche di Ingegneria. **13**(2), 255–264 (1934)
5. Grishchuk, S.V., Plaksa, S.A.: Monogenic functions in a biharmonic algebra (in Russian). Ukr. Mat. Zh. **61**(12), 1587–1596 (2009); English transl. (Springer) in Ukr. Math. J. **61**, No. 12, 1865–1876 (2009)
6. Gryshchuk, S.V., Plaksa, S.A.: Basic Properties of Monogenic Functions in a Biharmonic Plane. In: Agranovskii, M.L., Ben-Artzi, M., Galloway, G., Karp, L., Maz'ya, V. (eds.) Complex Analysis and Dynamical Systems V. Contemporary Mathematics, vol. 591, pp. 127–134. American Mathematical Society, Providence (2013)
7. Kovalev, V.F.: Biharmonic Schwarz problem (in Russian). Preprint No. 86.16. Institute of Mathematics, Academy of Sciences USSR, Institute of Mathematical Publications, Kiev (1986)
8. Gryshchuk, S.V., Plaksa, S.A.: Schwartz-type boundary value problems for monogenic functions in a biharmonic algebra. In: Rogosin, S., Çelebi, A. (eds.) Analysis as a Life, Dedicated to Heinrich Begehr on the Occasion of His 80th Birthday. Trends in Mathematics, pp. 193–211. Birkhäuser, Basel (2019)
9. Gryshchuk, S.V., Plaksa, S.A.: A Schwartz-Type Boundary Value Problem in a Biharmonic Plane. Lobachevskii J. Math. **38**(3), 435–442 (2017)

10. Gryshchuk, S.V., Plaksa, S.A.: Reduction of a Schwartz-type boundary value problem for biharmonic monogenic functions to Fredholm integral equations. Open Math. **15**(1), 374–381 (2017)
11. Gryshchuk, S.V.: $\mathbb{B}$-valued monogenic functions and their applications to boundary value problems in displacements of 2-D elasticity. In: S.V. Rogosin, M.V. Dubatovskaya (eds.), Analytic Methods of Analysis and Differential Equations: AMADE 2015, vol. 45, pp. 37–47. Cambridge Scientic, Cambridge (2016). ISBN 978-1-908106-56-8 (paperback)
12. Gryshchuk, S.V., Plaksa, S.A.: Schwartz-type boundary value problems for canonical domains in a biharmonic plane. J. Math. Sci. **259**(1), 37–52 (2021)
13. Gryshchuk, S.V., Plaksa, S.A.: A hypercomplex method for solving boundary value problems for biharmonic functions. In: Hošková-Mayerová, Š., Flaut, C., Maturo, F. (eds.), Algorithms as a Basis of Modern Applied Mathematics. Studies in Fuzziness and Soft Computing, vol. 404, pp. 231–255. Springer, Cham (2021)
14. Mikhlin, S.G.: The plane problem of the theory of elasticity (in Russian). In: Transmission Institute of Seismology, Academy of Sciences USSR, vol. 65. Academy of Sciences USSR Publishing, Moscow (1935)
15. Muskhelishvili, N.I.: Some basic problems of the mathematical theory of elasticity. In: Fundamental Equations, Plane Theory of Elasticity, Torsion and Bending. English transl. from the 4th Russian edition by Radok R.M. Noordhoff International Publishing, Leiden (1977)
16. Lurie, A.I.: Theory of Elasticity. English translation by Belyaev A. Springer, Berlin (2005)
17. Mikhlin, S.G., Morozov, N.F., Paukshto, M.V.: The Integral Equations of the Theory of Elasticity/TEUBNER-TEXTE zur Mathematik Band 135. Translation from Russian by Rainer Radok. Ed. by Herbert Gajewski. Springer Fachmedien Wiesbaden GmbH, Stuttgart (1995)
18. Lu, J.: Complex Variable Methods in Plane Elasticity. Series in Pure Mathematics, vol. 22. World Scientific, Singapore (1995)
19. Kondrat'ev, V. A.: Boundary-value problems for elliptic equations in domains with conic or angular points (in Russian). Tr. Mosk. Mat. Obshch. **16**, 209–292 (1967)
20. Kondrat'ev, V.A., Oleinik, O.A.: Boundary-value problems for partial differential equations in nonsmooth domains (in Russian). Usp. Mat. Nauk. **38**(2), 3–76 (1983)
21. Kufner, A., Sändig, A.M.: Some Applications of Weighted Sobolev Spaces (TEUBNER-TEXTE zur Mathematik, vol. 100). BSB BG Teubner Verlagsgesellschaft, Leipzig (1987)
22. Maz'ya, V., Nazarov, S., Plamenevskij, B.: Asymptotic theory of elliptic boundary value problems in singularly perturbed domains. In: Vol. 1 (Operator Theory Advances and Applications), vol. 111. Springer, Berlin (2000)
23. Gryshchuk, S.V., Plaksa, S.A.: Biharmonic problem for an angle and monogenic functions. Ukr. Math. J. **74**(11), 1686–1700 (2023)
24. Krein, S.G.: Linear Equations in Banach Spaces. Birkhäuser/Springer, Boston (1982)

On Distortion Estimates of Mappings with the Poletsky Condition in Domains with Poincare Inequality

Oleksandr Dovhopiatyi and Evgeny Sevost'yanov

Abstract The article is devoted to the study of mappings that distort the modulus of families of paths according to the Poletsky inequality type. At the boundary points of the domain, we have obtained an estimate of the distance distortion for such mappings provided that their characteristic has finite integral averages over the balls, and the image domain of mappings is Ahlfors regular and such that the Poincare inequality holds. The manuscript considers the cases of homeomorphisms and mappings with branching.

1 Introduction

It is well known that, quasiconformal mappings and mappings with finite distortion satisfy Hölder inequality under certain conditions. The problem of the fulfillment of this inequality has been investigated both at inner and boundary points of the domain by various authors, see, e.g., [2, 7, 8, 11] and [14]. In our works, this problem was studied for Sobolev and Orlicz-Sobolev classes, see e.g. [12, 13] and [15]. Note that we previously considered mappings of the unit ball onto itself, and therefore in this manuscript we expand both the class of mappings under study and the class of domains between which these mappings act. Our immediate task is to consider mappings that satisfy the Poletsky inequality with respect to p-modulus. As for domains, we are mainly interested in the case when these maps act in a domain with the $(1; p)$-Poincare inequality. Here we may see four natural cases: when the mappings are homeomorphisms and their domain of definition is good; when mappings have branch points and their domain of definition is good. The remaining two cases correspond to bad boundaries under the same assumptions

O. Dovhopiatyi
Zhytomyr Ivan Franko State University, Zhytomyr, Ukraine

E. Sevost'yanov (✉)
Zhytomyr Ivan Franko State University, Zhytomyr, Ukraine

Institute of Applied Mathematics and Mechanics of NAS of Ukraine, Slov'yans'k, Ukraine

M. R. Ebert et al. (eds.), *New Tools in Mathematical Analysis and Applications*, Research Perspectives, https://doi.org/10.1007/978-3-031-77050-0_12

on the mappings. Below we use the following notations: $B(x_0, r) = \{x \in \mathbb{R}^n : |x - x_0| < r\}$, $\mathbb{B}^n = B(0, 1)$, $S(x_0, r) = \{x \in \mathbb{R}^n : |x - x_0| = r\}$, $\mathbb{S}^{n-1} = S(0, 1)$, $\Omega_n = m(\mathbb{B}^n)$, $\omega_{n-1} = \mathcal{H}^{n-1}(\mathbb{S}^{n-1})$. Let m be a Lebesgue measure in $\mathbb{R}^n$, let $\mathcal{H}^{n-1}$ be $(n-1)$-measured Hausdorff measure,

$$A(x_0, r_1, r_2) := \left\{x \in \mathbb{R}^n : r_1 < |x - x_0| < r_2\right\},$$

and $M_p(\Gamma)$ denotes *p-modulus of* Γ (see [16]). Given sets E, $F \subset \overline{\mathbb{R}^n}$ and a domain $D \subset \mathbb{R}^n$ denote by $\Gamma(E, F, D)$ the family of all paths $\gamma : [a, b] \to \overline{\mathbb{R}^n}$ such that $\gamma(a) \in E, \gamma(b) \in F$ and $\gamma(t) \in D$ for $t \in (a, b)$. Given a family Γ of paths $\gamma : [a, b] \to D$ in D, by $f(\Gamma)$ we denote the family of paths $\{(f \circ \gamma) : [a, b] \to f(D), \gamma \in \Gamma\}$. (If $\gamma : [a, b] \to \overline{D}$, $\gamma(t) \in D$ if $a < t < b$, then under $f(\gamma)$ we understand the path $f(\gamma(t))$, $t \in (a, b)$. Similarly we are reasoning for the corresponding families Γ of paths γ, whose end points do not belong to D, but belong to the closure of D).

Let $Q : \mathbb{R}^n \to [0, \infty]$ be a Lebesgue measurable function equal to zero outside D. Consider the following concept, see [11, Section 7.6]. We say that a mapping $f : D \to \overline{\mathbb{R}^n}$ is a *ring Q-mapping at the point* $x_0 \in \overline{D}$ *with respect to p-modulus,* $x_0 \neq \infty$, $p \geqslant 1$, if for some $r_0 = r(x_0) > 0$ and arbitrary $0 < r_1 < r_2 < r_0$ the condition

$$M_p(f(\Gamma(S(x_0, r_1),\ S(x_0, r_2),\ D))) \leqslant \int\limits_{A} Q(x) \cdot \eta^p(|x - x_0|)\, dm(x)\,, \tag{1}$$

is fulfilled, where $\eta : (r_1, r_2) \to [0, \infty]$ is arbitrary Lebesgue measurable function that satisfies inequality

$$\int\limits_{r_1}^{r_2} \eta(r)\, dr \geqslant 1\,. \tag{2}$$

According to [5, Section 7.22], we say that a Borel function $\rho \colon D \to [0, \infty]$ is an *upper gradient* of the function $u \colon D \to \mathbb{R}$, if the relation $|u(x) - u(y)| \leqslant \int\limits_{\gamma} \rho\, |dx|$ holds for all rectifiable paths γ, joining the points x and $y \in D$. Let $u_B := \frac{1}{m(B)} \int\limits_{B} u(x)\, dm(x)$. Given $p \geqslant 1$, we say that D supports $(1; p)$-*Poincare inequality,* if there are constants $C \geqslant 1$ and $\tau > 0$ such that, the relation

$$\frac{1}{m(B)} \int\limits_{B} |u(x) - u_B|\, dm(x) \leqslant C \cdot (\operatorname{diam} B) \left(\frac{1}{m(\tau B)} \int\limits_{\tau B} \rho^p(x)\, dm(x)\right)^{1/p}$$

holds for any ball $B \subset D$, an arbitrary bounded continuous function $u \colon D \to \mathbb{R}$ and any upper gradient ρ of u. A domain D is called *Ahlfors regular,* if there is

$C \geqslant 1$ such that the inequality $\frac{1}{C}R^n \leqslant m(B(x_0, R) \cap D) \leqslant CR^n$ holds for any $x_0 \in D$ and an arbitrary $R < \operatorname{diam} D$. Given sets $A, B \subset \mathbb{R}^n$, we put, as usual, $\operatorname{diam} A = \sup\limits_{x,y \in A} |x - y|$, $\operatorname{dist}(A, B) = \inf\limits_{x \in A, y \in B} |x - y|$. Sometimes instead of $\operatorname{diam} A$ and $\operatorname{dist}(A, B)$ we also write $d(A)$ and $d(A, B)$, respectively.

Given numbers $\delta > 0$ and $p \geqslant 1$, domains $D, D' \subset \mathbb{R}^n$, $n \geqslant 2$, $x_0 \in \partial D$, a continuum $A \subset D$ and given a function $Q : D \to [0, \infty]$ denote by $\mathfrak{F}^{p,x_0}_{Q,A,\delta}(D, D')$ the family of all ring Q-homeomorphisms f of D onto D' at the point x_0 with respect to p-modulus, which satisfy the condition $\operatorname{diam}(f(A)) \geqslant \delta$. The following statement is valid.

Theorem 1 *Let $x_0 \in \partial D$, $x_0 \neq \infty$, $n - 1 < p \leqslant n$, D' be an Ahlfors regular bounded domain with $(1; p)$-Poincare inequality. Assume that, the following conditions hold: 1) there exists $r_0' = r_0'(x_0) > 0$ such that the set $B(x_0, r) \cap D$ is connected for all $0 < r < r_0'$; 2) there is $0 < C(x_0) < \infty$ such that $\limsup\limits_{\varepsilon \to 0} \frac{1}{\Omega_n \cdot \varepsilon^n} \int\limits_{B(x_0,\varepsilon) \cap D} Q(x)\,dm(x) \leqslant C$. Then there exist $0 < \varepsilon_0 = \varepsilon_0(x_0, D', n, p, \delta, A) < 1$ and $\widetilde{C} = \widetilde{C}(p, n, C, D') > 0$ such that the inequality*

$$|f(x) - f(y)| \leqslant \widetilde{C} \cdot \max\left\{\frac{1}{\log^{n-1}\frac{\varepsilon_0}{|x-x_0|}}, \frac{1}{\log^{n-1}\frac{\varepsilon_0}{|y-x_0|}}\right\} \tag{3}$$

holds for all $x, y \in B(x_0, \varepsilon_0^2) \cap D$ and all $f \in \mathfrak{F}^{p,x_0}_{Q,A,\delta}(D, D')$.

Corollary 1 *Under the conditions of Theorem 1, any $f \in \mathfrak{F}^{p,x_0}_{Q,A,\delta}(D, D')$ has a continuous extension to x_0 and*

$$|f(x) - f(x_0)| \leqslant \widetilde{C} \cdot \frac{1}{\log^{n-1}\frac{\varepsilon_0}{|x-x_0|}}. \tag{4}$$

Remark 1 In particular, condition 1) of Theorem 1 is fulfilled if D is a convex domain.

Some analog of Theorem 1 is also holds for mappings with branching, see below. Recall that, a mapping $f : D \to \overline{\mathbb{R}^n}$ between domains $D \subset \mathbb{R}^n$ and $D' \subset \overline{\mathbb{R}^n}$ is called *closed* if $C(f, \partial D) \subset \partial D'$, where, as usual, $C(f, \partial D)$ is the cluster set of f at ∂D.

Later, in the extended Euclidean space $\overline{\mathbb{R}^n} = \mathbb{R}^n \cup \{\infty\}$ we use the *spherical (chordal) metric* $h(x, y) = |\pi(x) - \pi(y)|$, where π is the stereographic projection of $\overline{\mathbb{R}^n}$ onto the sphere $S^n(\frac{1}{2}e_{n+1}, \frac{1}{2})$ in $\mathbb{R}^{n+1}$. Namely,

$$h(x, \infty) = \frac{1}{\sqrt{1 + |x|^2}},$$

$$h(x, y) = \frac{|x - y|}{\sqrt{1 + |x|^2}\sqrt{1 + |y|^2}}, \quad x \neq \infty \neq y \tag{5}$$

(see, e.g., [16, Definition 12.1]). Further, for the sets $A, B \subset \overline{\mathbb{R}^n}$ we set $h(A, B) = \inf\limits_{x\in A, y\in B} h(x, y)$, $h(A) = \sup\limits_{x,y\in A} h(x, y)$, where h is the chordal distance defined in (5). Given $\delta > 0$, domains $D, D' \subset \mathbb{R}^n$, $n \geqslant 2$, a point $x_0 \in \partial D$ and a Lebesgue measurable function $Q : D \to [0, \infty]$ we denote by $\mathfrak{R}^{p,x_0}_{Q,\delta}(D, D')$ the family of all open, discrete and closed ring Q-mappings $f : D \to D'$ with respect to p-modulus at the point x_0 for which there is a continuum $K_f \subset D'$ such that $\operatorname{diam}(K_f) \geqslant \delta$ and $h(f^{-1}(K_f), \partial D) \geqslant \delta > 0$. The following is true.

Theorem 2 *Let $x_0 \in \partial D$, $x_0 \neq \infty$, $n - 1 < p \leqslant n$, let D' be an Ahlfors regular bounded domain supporting $(1; p)$-by Poincare inequality. Assume that, the following conditions hold: 1) there exists $r_0' = r_0'(x_0) > 0$ such that the set $B(x_0, r) \cap D$ is connected for all $0 < r < r_0'$; 2) there is $0 < C = C(x_0) < \infty$ such that $\limsup\limits_{\varepsilon\to 0} \frac{1}{\Omega_n \varepsilon^n} \int\limits_{B(x_0,\varepsilon)\cap D} Q(x)\, dm(x) \leqslant C$. Then there exist $0 < \varepsilon_0 = \varepsilon_0(x_0, D', n, p, \delta, A) < 1$ and $\widetilde{C} = \widetilde{C}(p, n, C, D') > 0$ such that the inequality*

$$|f(x) - f(y)| \leqslant \widetilde{C} \cdot \max\left\{\frac{1}{\log^{n-1}\frac{\varepsilon_0}{|x-x_0|}}, \frac{1}{\log^{n-1}\frac{\varepsilon_0}{|y-x_0|}}\right\}$$

holds for any $x, y \in B(x_0, \varepsilon_0^2) \cap D$ and all $f \in \mathfrak{R}^{p,x_0}_{Q,\delta}(D, D')$.

Corollary 2 *Under the conditions of Theorem 2, the mapping f has a continuous extension to x_0 and*

$$|f(x) - f(x_0)| \leqslant \widetilde{C} \cdot \frac{1}{\log^{n-1}\frac{\varepsilon_0}{|x-x_0|}}.$$

Remark 2 In particular, all bounded convex domains D' are Ahlfors regular domains with $(1; p)$-Poincare inequality for any $n - 1 < p \leqslant n$.

Indeed, bounded convex domains are John domains (see [9, Remark 2.4]) and hence uniform domains in the sense of Martio-Sarvas (see [9, note 2.13(c)]). Such domains are also quasiextremal distance (QED) domains by Gering-Martio. In other words, there exists a constant $1 \leqslant A_0^* < \infty$ such that

$$M(\Gamma(E, F, \mathbb{R}^n)) \leqslant A_0^* \cdot M(\Gamma(E, F, D)) \tag{6}$$

(see [3, Lemma 2.18]). Since $\mathbb{R}^n$ is a Loewner space ([5, Theorem 8.1]), due to (6) bounded convex domains are also Loewner as metric spaces. But then by [5, Proposition 8.19] bounded convex domains are also Ahlfors regular. Finally, by Theorem 10.5 in [4] bounded convex domains satisfy the $(1; 1)$-Poincare inequality.

Therefore, by Hölder inequality, they also satisfy $(1; p)$-Poincare inequality for any $p > 1$.

2 Proof of Theorem 1

Set $\varepsilon_0 = \min\{r_0', d(x_0, A), 1\}$, in addition, let r_0' and A are number and a continuum from conditions of the theorem, respectively. Let $x, y \in B(x_0, \varepsilon_0^2)$ and let $f \in \mathfrak{F}^{p,x_0}_{Q,A,\delta}(D, D')$. Then $x, y \in B(x_0, \varepsilon_0)$. Without loss of restriction, we may consider that $|x - x_0| \geqslant |y - x_0|$. By the definition of $\varepsilon_0 > 0$,

$$A \subset D \setminus B(x_0, \varepsilon_0)\,. \tag{7}$$

Let $\varepsilon_1 > 0$ and $\varepsilon_2 > 0$ be such that the closed balls $\overline{B(x, \varepsilon_1)}$ and $\overline{B(y, \varepsilon_2)}$ belong to D. Let $x_* \in B(x, \varepsilon_1) \cap B(x_0, |x - x_0|)$ and $y_* \in B(y, \varepsilon_2) \cap B(x_0, |x - x_0|)$. Since $B(x_0, r) \cap D$ is connected for any $0 < r < r_0'$, the points x_* and y_* may be joined by a path γ_* which completely belongs to $\overline{B(x_0, |x - x_0|)}$ and belong to D. In turn, the sets $\overline{B(x_0, |x - x_0|)} \cap B(x, \varepsilon_1)$ and $\overline{B(x_0, |x - x_0|)} \cap B(x, \varepsilon_2)$ are convex as intersection of two convex sets, so x and x_* may be joined by a segment γ_1 in $\overline{B(x_0, |x - x_0|)} \cap B(x, \varepsilon_1) \subset \overline{B(x_0, |x - x_0|)} \cap D$, and the points y and y_* may be joined by a segment γ_2 in $\overline{B(x_0, |x - x_0|)} \cap B(y, \varepsilon_2) \subset \overline{B(x_0, |x - x_0|)} \cap D$. Let K be a path which is sequentially union of the paths γ_1, γ_* and γ_2. Now, K joins the points x and y in $\overline{B(x_0, |x - x_0|)} \cap D$.

Let $z, w \in f(A) \subset D'$ and $u, v \in A$ be such that

$$\operatorname{diam} f(A) = |z - w| = |f(u) - f(v)| \geqslant \delta\,. \tag{8}$$

Note that, $f(K)$ and $f(A)$ are continua as a continuous image of the continua K and A under the mapping f, respectively. We fix arbitrarily $y_0 \in D'$. Since D' is bounded, there exists $R_0 > 0$ such that $D' \subset B(y_0, R_0) := B_{R_0}$, see Fig. 1. In this case, by Adamowicz and Shanmugalingam [1, Proposition 4.7]

$$M_p(\Gamma(f(K), f(A), D')) \geqslant \frac{1}{M} \cdot \frac{\min\{\operatorname{diam} f(K), \operatorname{diam} f(A)\}}{R_0^{1+p-n}}\,. \tag{9}$$

Observe that,

$$\Gamma(|K|, A, D) > \Gamma(S(x_0, |x - x_0|), S(x_0, \varepsilon_0), D)\,. \tag{10}$$

Indeed, let $\gamma \in \Gamma(|K|, A, D)$, $\gamma : [0, 1] \to D$, $\gamma(0) \in |K|$ and $\gamma(1) \in A$. Since $|K| \subset B(x_0, |x - x_0|)$, due to (7), we obtain that $A \subset D \setminus B(x_0, |x - x_0|)$. Then $|\gamma| \cap B(x_0, |x - x_0|) \neq \varnothing \neq |\gamma| \cap (D \setminus B(x_0, |x - x_0|))$. By Kuratowski [6, Theorem 1.I.5.46] there exists $t_1 \in (0, 1)$ such that $\gamma(t_1) \in S(x_0, |x - x_0|)$. Consider

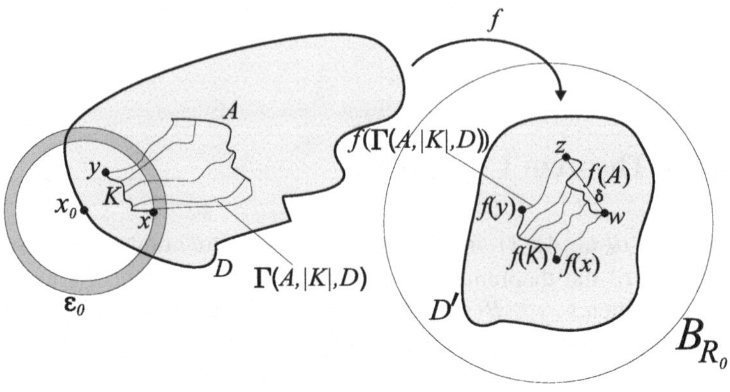

Fig. 1 To the proof of Theorem 1

the path $\gamma_1 := \gamma|_{[t_1,1]}$. Recall that $|K| \subset B(x_0, \varepsilon_0)$, in addition, due to (7) $A \subset D \setminus B(x_0, \varepsilon_0)$. Then $\gamma_1 \cap B(x_0, \varepsilon_0) \neq \varnothing \neq |\gamma| \cap (D \setminus B(x_0, \varepsilon_0))$. By Kuratowski [6, theorem 1.I.5.46] there exists $t_2 \in (0, t_1)$ such that $\gamma_1(t_2) = \gamma(t_2) \in S(x_0, \varepsilon_0)$. Put $\gamma_2 := \gamma|_{[t_1,t_2]}$. Then γ_2 is a subpath of γ and $\gamma_2 \in \Gamma(S(x_0, |x - x_0|), S(x_0, \varepsilon_0), D)$. This proves (10). In in this case, by the minorizaion principle of the modulus of families of paths (see, e.g., [16, Theorem 6.4]), by (1) and (10) we obtain that

$$\begin{aligned} M_p(\Gamma(f(|K|), f(A), D')) &= M_p(f(\Gamma(|K|, A, D))) \\ &\leqslant M_p(f(\Gamma(S(x_0, |x - x_0|), S(x_0, \varepsilon_0), D))) \qquad (11) \\ &\leqslant \int\limits_A Q(y) \cdot \eta^p(|y - x_0|)\, dm(y)\,, \end{aligned}$$

where η is an arbitrary Lebesgue measurable function satisfying (2) for $r_1 = |x - x_0|$ and $r_2 = \varepsilon_0$. Set $\eta(t) := \begin{cases} \dfrac{1}{\left(\log \frac{\varepsilon_0}{|x-x_0|}\right)^{n/p} t^{n/p}}, & t \in (|x - x_0|, \varepsilon_0)\,, \\ 0\,, & t \notin (|x - x_0|, \varepsilon_0). \end{cases}$ Observe that, $\int\limits_{|x-x_0|}^{\varepsilon_0} \frac{dt}{t \log \frac{\varepsilon_0}{|x-x_0|}} = \frac{1}{\log \frac{\varepsilon_0}{|x-x_0|}} \cdot \int\limits_{|x-x_0|}^{\varepsilon_0} \frac{dt}{t} = 1$. Then, by Hölder inequality, we obtain that

$$1 = \int\limits_{|x-x_0|}^{\varepsilon_0} \frac{dt}{t \log \frac{\varepsilon_0}{|x-x_0|}} \leqslant \left(\int\limits_{|x-x_0|}^{\varepsilon_0} \frac{dt}{t^{\frac{n}{p}} \cdot \left(\log \frac{\varepsilon_0}{|x-x_0|}\right)^{\frac{n}{p}}} \right)^{\frac{p}{n}} \cdot (\varepsilon_0 - |x - x_0|)^{\frac{n-p}{n}}$$

$$\leqslant \left(\int\limits_{|x-x_0|}^{\varepsilon_0} \frac{dt}{t^{\frac{n}{p}} \cdot \left(\log \frac{\varepsilon_0}{|x-x_0|}\right)^{\frac{n}{p}}}\right)^{\frac{p}{n}} = \left(\int\limits_{|x-x_0|}^{\varepsilon_0} \eta(t)\,dt\right)^{\frac{p}{n}}. \tag{12}$$

By (12), the function η satisfies the condition (2) for $r_1 = |x - x_0|$ and $r_2 = \varepsilon_0$. In this case, by (11) we obtain that

$$M_p(\Gamma(f(|K|), f(A), D'))$$

$$\leqslant \frac{1}{\log^n \frac{\varepsilon_0}{|x-x_0|}} \int\limits_{A(x_0,|x-x_0|,\varepsilon_0)} \frac{Q(y)}{|y-x_0|^n}\,dm(y)\,. \tag{13}$$

Since $|x - x_0| < \varepsilon_0^2$,

$$\log \frac{1}{|x-x_0|} < 2\log \frac{\varepsilon_0}{|x-x_0|}\,. \tag{14}$$

Due to [15, Lemma 3.1] and by (13) and (14) we obtain that

$$M_p(\Gamma(f(|K|), f(A), D')) \leqslant \frac{2}{\log^{n-1} \frac{\varepsilon_0}{|x-x_0|}} \frac{C\Omega_n 2^n}{\log 2}\,. \tag{15}$$

Uniting (9) and (15), we have

$$\frac{1}{M} \cdot \frac{\min\{\operatorname{diam} f(K), \operatorname{diam} f(A)\}}{R_0^{1+p-n}} \leqslant \frac{2^{n+1}}{\log^{n-1} \frac{\varepsilon_0}{|x-x_0|}} \frac{C\Omega_n}{\log 2}\,.$$

Since $\omega_{n-1} = n\Omega_n$, the latter relation may be rewritten as

$$\frac{1}{M} \cdot \frac{\min\{\operatorname{diam} f(K), \operatorname{diam} f(A)\}}{R_0^{1+p-n}} \leqslant \frac{\omega_{n-1}}{\log^{n-1} \frac{\varepsilon_0}{|x-x_0|}} \frac{C2^{n+1}}{n\log 2}\,. \tag{16}$$

By (8), $\min\{\operatorname{diam} f(K), \operatorname{diam} f(A)\} \geqslant \min\{\operatorname{diam} f(K), \delta\}$. Then, from (16) we obtain that

$$\min\{\operatorname{diam} f(K), \delta\} \leqslant \frac{\omega_{n-1} M R_0^{1+p-n}}{\log^{n-1} \frac{\varepsilon_0}{|x-x_0|}} \cdot \frac{C2^{n+1}}{n\log 2}\,. \tag{17}$$

Observe that, $\frac{\omega_{n-1}MR_0^{1+p-n}}{\log^{n-1}\frac{\varepsilon_0}{|x-x_0|}}\cdot\frac{C2^{n+1}}{n\log 2}\rightarrow 0$ as $x\rightarrow x_0$. Then there is $0<\sigma=\sigma(x_0, M, R_0, n, p, \delta)$ such that

$$\frac{\omega_{n-1}MR_0^{1+p-n}}{\log^{n-1}\frac{\varepsilon_0}{|x-x_0|}}\cdot\frac{C2^{n+1}}{n\log 2}<\delta,\qquad \forall\, x\in B(x_0,\sigma)\,. \tag{18}$$

Let $|x-x_0|<\sigma$, by (17) and (18), we obtain that

$$|f(x)-f(y)|\leqslant \operatorname{diam} f(K)\leqslant \frac{1}{\log^{n-1}\frac{\varepsilon_0}{|x-x_0|}}\cdot\frac{\omega_{n-1}MR_0^{1+p-n}C2^{n+1}}{n\log 2}\,.$$

Observe that, by the triangle inequality $|f(x)-f(y)|\leqslant 2R_0$, thus for $|x-x_0|\geqslant\sigma$ we have that

$$\frac{1}{\log^{n-1}\frac{\varepsilon_0}{|x-x_0|}}\geqslant\frac{1}{\log^{n-1}\frac{\varepsilon_0}{\sigma}}:=P_0\,. \tag{19}$$

Therefore,

$$|f(x)-f(y)|\leqslant 2R_0\leqslant\frac{2R_0}{P_0}\cdot\frac{1}{\log^{n-1}\frac{\varepsilon_0}{|x-x_0|}}\,. \tag{20}$$

Set $\widetilde{C}:=\max\left\{\frac{\omega_{n-1}MR_0^{1+p-n}C2^{n+1}}{n\log 2}, \frac{2R_0}{P_0}\right\}$. The constant $\widetilde{C}$ depends only on p, n, C and D', because M and R_0 are defined by D'. The proof is complete. □

Proof of Corollary 1 If f has no a limit as $x\rightarrow x_0$, we may construct at least two sequences $x_m\rightarrow x_0$ and $y_m\rightarrow x_0$, $m\rightarrow\infty$, such that $|f(x_m)-f(y_m)|\geqslant\delta>0$ for some positive $\delta>0$ and all $m=1,2,\ldots$. But this contradicts with (3). Thus, the limits of f as $x\rightarrow x_0$ exists. To prove (4) it remains to go to the limit as $y\rightarrow x_0$ in (3). □

3 Proof of Theorem 2

Put $\varepsilon_0=\min\{r_0',\delta,1\}$. In addition, let r_0' be a number from the conditions of the theorem and δ is a number from the definition of the class $\mathfrak{R}^{p,x_0}_{Q,\delta}(D,D')$. We put $\widetilde{\varepsilon_0}:=\varepsilon_0^2$. Let $x,y\in B(x_0,\widetilde{\varepsilon_0})$ and $f\in\mathfrak{R}^{p,x_0}_{Q,\delta}(D,D')$. Then $x,y\in B(x_0,\varepsilon_0)$. Without loss of generalization, we may assume that $|x-x_0|\geqslant|y-x_0|$. Let $K_f\subset D'$ be a continuum such that $\operatorname{diam}(K_f)\geqslant\delta$ and $h(f^{-1}(K_f),\partial D)\geqslant\delta>0$ (it exists by definition of the class $\mathfrak{R}^{p,x_0}_{Q,\delta}(D,D')$). By the definition of ε_0,

$$f^{-1}(K_f)\subset D\setminus B(x_0,\varepsilon_0)\,.$$

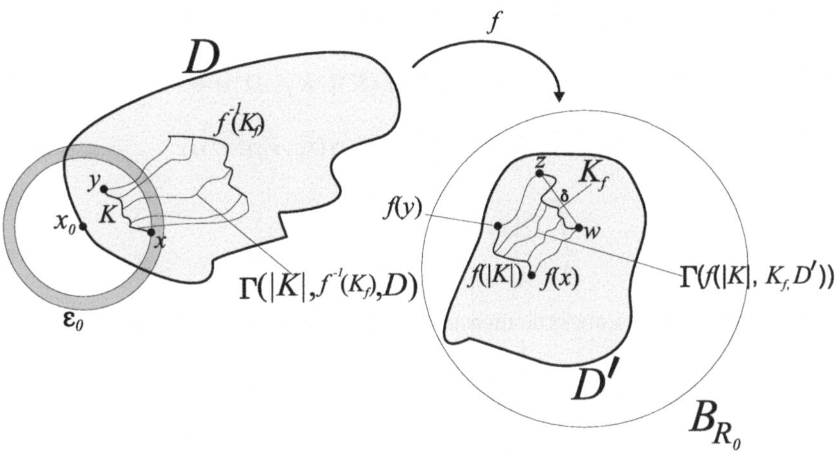

Fig. 2 To the proof of Theorem 2

Since the points $S(x_0, r) \cap D$, $0 < r < r_0'$, are accessible from the domain D using some path γ and the set $B(x_0, r) \cap D$ is connected for $0 < r < r_0'$, points x and y may be joined K, which completely belongs to the ball $\overline{B(x_0, |x - x_0|)}$ and also belongs to D, see Fig. 2. Let $z, w \in K_f \subset D'$ be such that

$$\operatorname{diam} K_f = |z - w| \geqslant \delta . \tag{21}$$

The further proof is very similar to proof of Theorem 1. Namely, note that $f(K)$ is a continuum as a continuous image of the continuum K under the mapping f. Fix $y_0 \in D'$. Then, since D' is bounded, there exists $R_0 > 0$ such that $D' \subset B(y_0, R_0) := B_{R_0}$. In this case, by Adamowicz and Shanmugalingam [1, Proposition 4.7]

$$M_p(\Gamma(f(K), K_f, D')) \geqslant \frac{1}{M} \cdot \frac{\min\{\operatorname{diam} f(K), \operatorname{diam} K_f\}}{R_0^{1+p-n}} . \tag{22}$$

Let Γ^* be a family $\gamma : [0, 1) \to D$ of all of maximal f-lifting of paths $\gamma' : [0, 1] \to D'$ in $\Gamma = \Gamma(|f(K)|, K_f, D')$ starting in $|K|$. Such liftings exist due to [10, Lemma 3.12], cf. [17, Lemma 3.7]. By the same Lemmas, due to the closeness of the mapping f, any path $\gamma \in \Gamma^*$ has a continuous extension $\gamma : [0, 1] \to D$ to the point $b = 1$. Then $\gamma(1) \in f^{-1}(K_f)$, that is, $\Gamma^* \subset \Gamma(|K|, f^{-1}(K_f), D)$.

Reasoning similarly to the proof of the relation (10), it is possible prove that

$$\Gamma(|K|, f^{-1}(K_f), D) > \Gamma(S(x_0, |x - x_0|), S(x_0, \varepsilon_0), D) . \tag{23}$$

Note that, $f(\Gamma^*) = \Gamma = \Gamma(|f(K)|, K_f, D')$. In this case, by the minorization principle of the modulus (see, e.g., [16, Theorem 6.4]), due to (23) and (1) we obtain that

$$M_p(f(\Gamma^*)) = M_p(\Gamma(|f(K)|, K_f, D')) \leqslant$$
$$\leqslant M_p(f(\Gamma(S(x_0, |x-x_0|), S(x_0, \varepsilon_0), D))) \tag{24}$$
$$\leqslant \int\limits_A Q(y)\cdot \eta^p(|y-x_0|)\, dm(y)\,,$$

where η is an arbitrary Lebesgue measurable function that satisfies ratio (2) for $r_1 = |x-x_0|$, $r_2 = \varepsilon_0$. Put $\eta(t) := \begin{cases} \frac{1}{\left(\log \frac{\varepsilon_0}{|x-x_0|}\right)^{n/p} t^{n/p}}\,, & t\in(|x-x_0|, \varepsilon_0)\,, \\ 0\,, & t\notin (|x-x_0|, \varepsilon_0). \end{cases}$ By (12), η satisfies condition (2) when $r_1 = |x-x_0|$ and $r_2 = \varepsilon_0$. Then from (24) it follows that

$$M_p(\Gamma(|f(K)|, K_f, D')) \leqslant \frac{1}{\log^n \frac{\varepsilon_0}{|x-x_0|}} \int\limits_{A(x_0, |x-x_0|, \varepsilon_0)} \frac{Q(y)}{|y-x_0|^p}\, dm(y)\,. \tag{25}$$

Since $|x-x_0| < \varepsilon_0^2$, (14) holds. By choosing ε_0, by Ryazanov et al. [15, Lemma 3.1] and by (25) we obtain that

$$M_p(\Gamma(f(|K|), K_f, D')) \leqslant \frac{2}{\log^{n-1} \frac{\varepsilon_0}{|x-x_0|}} \frac{C\Omega_n 2^n}{\log 2}\,. \tag{26}$$

Combining (22) and (26), we obtain that

$$\frac{1}{M}\cdot \frac{\min\{\operatorname{diam} f(K), \operatorname{diam} K_f\}}{R_0^{1+p-n}} \leqslant \frac{2}{\log^{n-1} \frac{\varepsilon_0}{|x-x_0|}} \frac{C\Omega_n 2^n}{\log 2}\,.$$

Since $\omega_{n-1} = n\Omega_n$, the last relation may be rewritten as

$$\frac{1}{M}\cdot \frac{\min\{\operatorname{diam} f(K), \operatorname{diam} K_f\}}{R_0^{1+p-n}} \leqslant \frac{\omega_{n-1}}{\log^{n-1} \frac{\varepsilon_0}{|x-x_0|}} \frac{C 2^{n+1}}{n\log 2}\,. \tag{27}$$

By (21), $\min\{\operatorname{diam} f(K), \operatorname{diam} K_f\} \geqslant \min\{\operatorname{diam} f(K), \delta\}$. Then by (27) we obtain that

$$\min\{\operatorname{diam} f(K), \delta\} \leqslant \frac{\omega_{n-1} M R_0^{1+p-n}}{\log^{n-1} \frac{\varepsilon_0}{|x-x_0|}} \cdot \frac{C 2^{n+1}}{n\log 2}\,. \tag{28}$$

Arguing similarly to (18) and (19), by (28) we obtain that

$$|f(x) - f(y)| \leqslant \operatorname{diam} f(K) \leqslant \widetilde{C} \cdot \frac{1}{\log^{n-1} \frac{\varepsilon_0}{|x-x_0|}}, \tag{29}$$

where $\widetilde{C} := \max\left\{\frac{\omega_{n-1} M R_0^{1+p-n} C 2^{n+1}}{n \log 2}, \frac{2R_0}{P_0}\right\}$. The constant $\widetilde{C}$ depends only on p, n, C and D', because M and R_0 are totally defined by D'. The proof is complete. □

Proof of Corollary 2 is similar to proof of Corollary 1. □

References

1. Adamowicz, T., & Shanmugalingam, N.: Non-conformal Loewner type estimates for modulus of curve families. Ann. Acad. Sci. Fenn. Math. **35**, 609–626 (2010)
2. Arsenović, M., Mateljević, M.: On the Hölder continuity of ring Q-homeomorphisms. Georgian Math. J. **29**, 805–811 (2022)
3. Gehring, F.W., Martio, O.: Quasiextremal distance domains and extension of quasiconformal mappings. J. Anal. Math. **45**, 181–206 (1985)
4. Hajlasz, P., Koskela, P.: Sobolev met Poincare. Mem. Amer. Math. Soc. **145**(688), 1–101 (2000)
5. Heinonen, J.: Lectures on Analysis on Metric Spaces. Springer, New York (2001)
6. Kuratowski, K.: Topology, v. 2. Academic Press, New York (1968)
7. Martio, O., Näkki, R.: Boundary Hölder continuity and quasiconformal mappings. J. Lond. Math. Soc. **44**, 339–350 (1991)
8. Martio, O., Rickman, S., Väisälä, J.: Distortion and singularities of quasiregular mappings. Ann. Acad. Sci. Fenn. Ser. A1 **465**, 1–13 (1970)
9. Martio, O., Sarvas, J.: Injectivity theorems in plane and space. Ann. Acad. Sci. Fenn. Ser. A1 Math. **4**, 384–401 (1978/1979)
10. Martio, O., Rickman, S., Väisälä, J.: Topological and metric properties of quasiregular mappings. Ann. Acad. Sci. Fenn. Ser. A1 **488**, 1–31 (1971)
11. Martio, O., Ryazanov, V., Srebro, U., Yakubov, E.: Moduli in Modern Mapping Theory. Springer Monographs in Mathematics. Springer, Berlin (2009)
12. Mateljević, M., Sevost'yanov, E.: On the behavior of Orlicz-Sobolev mappings with branching on the unit sphere. J. Math. Sci. **270**, 467–499 (2023)
13. Mateljević, M., Salimov, R., Sevost'yanov, E.: Hölder and lipschitz continuity in orlicz-sobolev classes, distortion and harmonic mappings. Filomat **36**, 5359–5390 (2022)
14. Ryazanov, V.I., Sevost'yanov, E.A.: Toward the theory of ring Q-homeomorphisms. Israel J. Math. **168**, 101–118 (2008)
15. Ryazanov, V.I., Salimov, R.R., Sevost'yanov, E.A.: On the Hölder property of mappings in domains and on boundaries. J. Math. Sci. **246**(1), 60–74 (2020)
16. Väisälä, J.: Lectures on n-Dimensional Quasiconformal Mappings. Lecture Notes in Mathematics, vol. 229. Springer, Berlin (1971)
17. Vuorinen, M.: Exceptional sets and boundary behavior of quasiregular mappings in $n-$space. Ann. Acad. Sci. Fenn. Ser. AI Math. Dissertattiones **11**, 1–44 (1976)

Part III
Function Spaces and Their Applications

Well-Posedness for a System of Nonlinear Schrödinger Equations with Derivative Nonlinearity via the Energy Method

Hiroyuki Hirayama, Shinya Kinoshita, and Mamoru Okamoto

Abstract We consider the Cauchy problem for a system of nonlinear Schrödinger equations with derivative nonlinearity. By using the energy method, we prove the local well-posedness in the Sobolev space for this system. The energy structure of this system depends on coefficients in linear terms and nonlinear terms. In particular, we give a condition for the coefficients in nonlinear terms that the local well-posedness holds regardless of the coefficients of linear terms.

1 Introduction

We consider the Cauchy problem of the system of nonlinear Schrödinger equations:

$$\begin{cases} i\partial_t u + \alpha\Delta u = p(\nabla \cdot w)v, & t > 0,\ x \in \mathbb{R}^d, \\ i\partial_t v + \beta\Delta v = q(\nabla \cdot \overline{w})u, & t > 0,\ x \in \mathbb{R}^d, \\ i\partial_t w + \gamma\Delta w = r\nabla(u \cdot \overline{v}), & t > 0,\ x \in \mathbb{R}^d, \\ (u, v, w)|_{t=0} = (u_0, v_0, w_0), & x \in \mathbb{R}^d, \end{cases} \tag{1.1}$$

where $\alpha, \beta, \gamma, p, q, r \in \mathbb{R}\backslash\{0\}$, $d \in \mathbb{N}$, and the unknown function (u, v, w) is $\mathbb{C}^d \times \mathbb{C}^d \times \mathbb{C}^d$-valued. The operators Δ and ∇ denote the Laplacian and the nabla

H. Hirayama
Faculty of Education, University of Miyazaki, Miyazaki, Japan
e-mail: h.hirayama@cc.miyazaki-u.ac.jp

S. Kinoshita (✉)
Department of Mathematics, Institute of Science Tokyo, Tokyo, Japan
e-mail: kinoshita@math.titech.ac.jp

M. Okamoto
Department of Mathematics, Graduate School of Science, Osaka University, Osaka, Japan
e-mail: okamoto@math.sci.osaka-u.ac.jp

M. R. Ebert et al. (eds.), *New Tools in Mathematical Analysis and Applications*, Research Perspectives, https://doi.org/10.1007/978-3-031-77050-0_13

on $\mathbb{R}^d$, respectively. Namely,

$$\Delta = \sum_{l=1}^{d} \partial_l^2, \quad \nabla = (\partial_1, \cdots, \partial_d),$$

where ∂_l denotes the partial differential operator with respect to the l-th component of the spatial variable $x \in \mathbb{R}^d$. The initial data is given in the Sobolev space

$$\mathcal{H}^s(\mathbb{R}^d) := H^s(\mathbb{R}^d)^d \times H^s(\mathbb{R}^d)^d \times H^s(\mathbb{R}^d)^d$$

equipped with the norm

$$\begin{aligned} \|(u_1, u_2, u_3)\|_{\mathcal{H}^s} &:= \left(\|u_1\|_{H^s}^2 + \|u_2\|_{H^s}^2 + \|u_3\|_{H^s}^2\right)^{\frac{1}{2}} \\ &= \left\{\sum_{j=1}^{d} \left(\|u_{1,j}\|_{H^s}^2 + \|u_{2,j}\|_{H^s}^2 + \|u_{3,j}\|_{H^s}^2\right)\right\}^{\frac{1}{2}}, \end{aligned}$$

where $u_{k,j}$ denotes the j-th component of $u_k \in H^s(\mathbb{R}^d)^d$. When $(p, q, r) = (-1, -1, 1)$, then (1.1) is known as a model of laser-plasma interaction [1]. The well-posedness of (1.1) with $(p, q, r) = (-1, -1, 1)$ is well studied in previous works [2–6]. In particular, the following is proved by using the iteration argument.

Theorem 1.1 ([2, 3, 5]) *Let $d \in \mathbb{N}$ and $(p, q, r) = (-1, -1, 1)$.*

(i) If $(\alpha - \gamma)(\beta + \gamma) \neq 0$, then (1.1) is locally well-posed in $\mathcal{H}^s(\mathbb{R}^d)$ for $s \geq 1$ and $s > \frac{d}{2} - 1$.
(ii) If $(\alpha - \gamma)(\beta + \gamma) = 0$, then the data-to-solution map of (1.1) is not C^2 in $\mathcal{H}^s(\mathbb{R}^d)$ for any $s \in \mathbb{R}$.

□

Moreover, the following is proved by using the energy method.

Theorem 1.2 ([6]) *Let $d \in \mathbb{N}$, $s > \frac{d}{2} + 3$, and $(p, q, r) = (-1, -1, 1)$.*

(i) If $\beta + \gamma \neq 0$, then (1.1) is locally well-posed for small initial data in $\mathcal{H}^s(\mathbb{R}^d)$.[1]
(ii) If $\beta\gamma > 0$, then (1.1) is locally well-posed in $\mathcal{H}^s(\mathbb{R}^d)$.

□

In [6], the authors construct a modified energy with respect to $(\Delta u, \Delta v, \Delta w)$ or $(\partial_t u, \partial_t v, \partial_t w)$. This is a reason why we need the regularity $\frac{d}{2} + 3 = \left(\frac{d}{2} + 1\right) + 2$ in Theorem 1.2.

[1] Theorem 1.1 in [6] says the global well-posedness for small initial data. However, the maximum time of existence can not be take to infinity even if the initial data are small. See Section 4 in [6]. Thus, their result is the local well-posedness for small initial data.

Remark 1.3 When $\beta + \gamma \neq 0$ and $(p, q, r) = (-1, -1, 1)$, the system (1.1) has an energy structure. However, we can not construct a modified energy for $\beta + \gamma = 0$ and $(p, q, r) = (-1, -1, 1)$. Moreover, Theorem 1.1 (ii) implies that an iteration argument fails to show the well-posedness for (1.1) in this case. In particular, the well-posedness for (1.1) is not obtained when $\beta + \gamma = 0$ and $(p, q, r) = (-1, -1, 1)$. □

By putting

$$u_1 := \overline{u}, \quad u_2 := v, \quad u_3 := w, \qquad \varphi_1 := \overline{u_0}, \quad \varphi_2 := v_0, \quad \varphi_3 := w_0,$$
$$\sigma_1 := -\alpha, \quad \sigma_2 := \beta, \quad \sigma_3 := \gamma, \qquad \lambda_1 := -p, \quad \lambda_2 := q, \quad \lambda_3 := r,$$

and

$$\begin{aligned} &U := (u_1, u_2, u_3), \qquad \Phi := (\varphi_1, \varphi_2, \varphi_3), \\ &N_1(U) := \lambda_1(\nabla \cdot \overline{u_3})\overline{u_2}, \quad N_2(U) := \lambda_2(\nabla \cdot \overline{u_3})\overline{u_1}, \quad N_3(U) := \lambda_3\nabla(\overline{u_1} \cdot \overline{u_2}), \end{aligned} \tag{1.2}$$

the system (1.1) can be rewritten as follows:

$$\begin{cases} i\partial_t u_k + \sigma_k \Delta u_k = N_k(U), \quad k = 1, 2, 3, \\ U|_{t=0} = \Phi. \end{cases} \tag{1.3}$$

From here, we consider (1.3) instead of (1.1). Throughout this paper, we assume that $d \in \mathbb{N}$ and $\sigma_1, \sigma_2, \sigma_3, \lambda_1, \lambda_2, \lambda_3$ are nonzero real numbers. The main result in the present paper is the following.

Theorem 1.4 *Let $s > \frac{d}{2} + 1$. If λ_1, λ_2, and λ_3 have the same sign,* (1.3) *is locally well-posed in $\mathcal{H}^s(\mathbb{R}^d)$.* □

To obtain the well-posedness of (1.3), we will use the energy method (without a modified energy). The system (1.3) has good energy structure when λ_1, λ_2, and λ_3 have the same sign, even if $\beta + \gamma = 0$. Note that $(p, q, r) = (-1, -1, 1)$ corresponds to $(\lambda_1, \lambda_2, \lambda_3) = (1, -1, 1)$. Moreover, our argument works when

$$\mu_1 := \lambda_2\lambda_3, \quad \mu_2 := \lambda_1\lambda_3, \quad \mu_3 := \lambda_1\lambda_2 \tag{1.4}$$

have the same sign. See (2.7) below. However, this condition is equivalent to that λ_1, λ_2, and λ_3 have the same sign, which is also equivalent to that μ_1, μ_2, and μ_3 are positive.

In the remaining this paper, we assume that λ_1, λ_2, and λ_3 have the same sign as in Theorem 1.4. Then, it follows from (1.4) that

$$\mu_1, \mu_2, \mu_3 > 0. \tag{1.5}$$

Remark 1.5 The result (i) in Theorem 1.1 is proved by using the iteration argument. Because the nonlinear estimates using in the iteration argument hold regardless of the values of p, q, r, the same result as in (i) of Theorem 1.1 holds for any $p, q, r \in \mathbb{R} \setminus \{0\}$. Therefore, Theorem 1.4 with $(\sigma_1 + \sigma_3)(\sigma_2 + \sigma_3) \neq 0$ is not new result. But, the result in the case $(\sigma_1 + \sigma_3)(\sigma_2 + \sigma_3) = 0$ is new because the well-posedness in $\mathcal{H}^s(\mathbb{R}^d)$ for $s > \frac{d}{2} + 1$ is not obtained in Theorem 1.1. □

We define the operator J^s as

$$\mathcal{F}[J^s f](\xi) = (1 + |\xi|^{2s})^{\frac{1}{2}} \mathcal{F}[f](\xi),$$

where $\mathcal{F}[f]$ is the Fourier transform of f. For any $\mathbb{C}^d$-valued function $f = (f_1, \cdots, f_d)$, we define the norms $\|f\|_{L_x^2}$ and $\|\nabla f\|_{L_x^2}$ by

$$\|f\|_{L_x^2}^2 = \sum_{j=1}^{d} \|f_j\|_{L_x^2}^2 = \sum_{j=1}^{d} \int_{\mathbb{R}^d} |f_j(x)|^2 dx,$$

$$\|\nabla f\|_{L_x^2}^2 = \sum_{j=1}^{d} \|\nabla f_j\|_{L_x^2}^2 = \sum_{j=1}^{d} \sum_{l=1}^{d} \|\partial_l f_j\|_{L_x^2}^2,$$

respectively. For $\mathbb{C}^d$-valued functions $f = (f_1, \cdots, f_d)$ and $g = (g_1, \cdots, g_d)$, we define the inner product $(f, g)_{L_x^2}$ by

$$(f, g)_{L_x^2} = \sum_{j=1}^{d} (f_j, g_j)_{L_x^2} = \sum_{j=1}^{d} \int_{\mathbb{R}^d} f_j(x) \overline{g_j(x)} dx.$$

2 Energy Estimates

First, we consider the regularized system

$$\begin{cases} \partial_t u_k = (\epsilon + i\sigma_k)\Delta u_k - iN_k(U), \quad k = 1, 2, 3, \\ U|_{t=0} = \Phi, \end{cases} \tag{2.1}$$

where $\epsilon > 0$ and U and N_k are defined in (1.2). By the standard iteration argument, we obtain the following.

Theorem 2.1 *Let* $s > \frac{d}{2} + 1$. *Then, for any* $\epsilon > 0$, (2.1) *is locally well-posed in* $\mathcal{H}^s(\mathbb{R}^d)$. □

Remark 2.2 The time T of existence for the solution to (2.1) depends only on d, s, ϵ, and $\|\Phi\|_{\mathcal{H}^s}$. In particular, if the forward maximal time $T_{\max}$ of existence is finite, then $\lim_{t\nearrow T_{\max}} \|U(t)\|_{\mathcal{H}^s} = \infty$.

For the solution U^ϵ to (2.1), we will construct the solution to (1.3) as $\lim_{\epsilon\to+0} U^\epsilon$. To prove the convergence of U^ϵ, we use the energy estimate.

Proposition 2.3 *Let* $s_0 > \frac{d}{2} + 1$, $s > 0$, *and* $U^\epsilon = (u_1^\epsilon, u_2^\epsilon, u_3^\epsilon) \in C([0,T]; \mathcal{H}^s(\mathbb{R}^d))$ *be the solution to (2.1). It holds that*

$$\left|\sum_{k=1}^{3} \mu_k \left(\frac{d}{dt}\|J^s u_k^\epsilon(t)\|_{L_x^2}^2 + 2\epsilon\|\nabla J^s u_k^\epsilon(t)\|_{L_x^2}^2\right)\right| \lesssim \|U^\epsilon(t)\|_{\mathcal{H}^s}^2 \|U^\epsilon(t)\|_{\mathcal{H}^{s_0}} \tag{2.2}$$

for $\epsilon > 0$, *where the implicit constant does not depend on* ϵ. *Here,* μ_1, μ_2, *and* μ_3 *are defined in* (1.4).

Proof In this proof, we write U and u_k instead of U^ϵ and u_k^ϵ, respectively. By (2.1), we have

$$\begin{aligned}
\frac{d}{dt}\|J^s u_k\|_{L_x^2}^2 &= 2\mathrm{Re}\int_{\mathbb{R}^d} J^s\overline{u_k}\cdot J^s\partial_t u_k dx \\
&= 2\mathrm{Re}\int_{\mathbb{R}^d} J^s\overline{u_k}\cdot J^s\{(\epsilon + i\sigma_k)\Delta u_k - iN_k(U)\}\,dx \\
&= -2\epsilon\|\nabla J^s u_k\|_{L_x^2}^2 + 2\mathrm{Im}\big(J^s N_k(U), J^s u_k\big)_{L_x^2}.
\end{aligned}$$

It follows from (1.2) and (1.4) that

$$\begin{aligned}
&\sum_{k=1}^{3} \mu_k\big(J^s N_k(U), J^s u_k\big)_{L_x^2} \\
&= \lambda_1\lambda_2\lambda_3\Big\{\big(J^s((\nabla\cdot\overline{u_3})\overline{u_2}), J^s u_1\big)_{L_x^2} + \big(J^s((\nabla\cdot\overline{u_3})\overline{u_1}), J^s u_2\big)_{L_x^2} \\
&\qquad + \big(J^s\nabla(\overline{u_1}\cdot\overline{u_2}), J^s u_3\big)_{L_x^2}\Big\} \\
&= \lambda_1\lambda_2\lambda_3\Big\{\big(J^s\overline{u_1}, J^s((\nabla\cdot u_3)u_2)\big)_{L_x^2} + \big(J^s\overline{u_2}, J^s((\nabla\cdot u_3)u_1)\big)_{L_x^2} \\
&\qquad + \big(J^s\nabla(\overline{u_1}\cdot\overline{u_2}), J^s u_3\big)_{L_x^2}\Big\}.
\end{aligned}$$

We note that

$$
\begin{aligned}
&\left(J^s\overline{u_k}, J^s((\nabla\cdot u_3)u_m)\right)_{L^2_x}\\
&=\left(J^s\overline{u_k}, (\nabla\cdot J^s u_3)u_m\right)_{L^2_x}+\left(J^s\overline{u_k}, J^s((\nabla\cdot u_3)u_m)-(\nabla\cdot J^s u_3)u_m\right)_{L^2_x}\\
&=\left((J^s\overline{u_k})\cdot\overline{u_m}, \nabla\cdot J^s u_3\right)_{L^2_x}+\left(J^s\overline{u_k}, J^s((\nabla\cdot u_3)u_m)-(\nabla\cdot J^s u_3)u_m\right)_{L^2_x}
\end{aligned}
$$

for $(k, m) \in \{(1, 2), (2, 1)\}$ and

$$\left(J^s\nabla(\overline{u_1}\cdot\overline{u_2}), J^s u_3\right)_{L^2_x}=-\left(J^s(\overline{u_1}\cdot\overline{u_2}), \nabla\cdot J^s u_3\right)_{L^2_x}.$$

Therefore, we have

$$\sum_{k=1}^{3}\mu_k\left(J^s N_k(U), J^s u_k\right)_{L^2_x}=\lambda_1\lambda_2\lambda_3(I_1+I_2+I_3),$$

where

$$
\begin{aligned}
I_1&=\left((J^s\overline{u_1})\cdot\overline{u_2}+\overline{u_1}\cdot J^s\overline{u_2}-J^s(\overline{u_1}\cdot\overline{u_2}), \nabla\cdot J^s u_3\right)_{L^2_x},\\
I_2&=\left(J^s\overline{u_1}, J^s((\nabla\cdot u_3)u_2)-(\nabla\cdot J^s u_3)u_2\right)_{L^2_x},\\
I_3&=\left(J^s\overline{u_2}, J^s((\nabla\cdot u_3)u_1)-(\nabla\cdot J^s u_3)u_1\right)_{L^2_x}.
\end{aligned}
$$

It suffices to prove that

$$|I_k| \lesssim \|U\|^2_{\mathcal{H}^s}\|U\|_{\mathcal{H}^{s_0}} \tag{2.3}$$

for $s > 0$, $s_0 > \frac{d}{2} + 1$, and $k = 1, 2, 3$.

Estimate for I_1 We put

$$M(u_1, u_2) := (J^s u_1)\cdot u_2 + u_1\cdot J^s u_2 - J^s(u_1\cdot u_2).$$

Then, the integral by parts and the Cauchy-Schwarz inequality yields that

$$|I_1| \le \|\nabla M(u_1, u_2)\|_{L^2_x}\|J^s u_3\|_{L^2_x}. \tag{2.4}$$

Note that

$$\begin{aligned}\|\nabla M(u_1,u_2)\|_{L^2_x} &\le \sum_{l=1}^{d}\left(\|M(\partial_l u_1,u_2)\|_{L^2_x}+\|M(u_1,\partial_l u_2)\|_{L^2_x}\right)\\ &=\sum_{l=1}^{d}\left(\|M(u_2,\partial_l u_1)\|_{L^2_x}+\|M(u_1,\partial_l u_2)\|_{L^2_x}\right).\end{aligned}$$

We use the following estimate.

Lemma 2.4 (Lemma X.1 in [7]) *Let* $s>0$ *and* $1<p<\infty$. *Then, it holds that*

$$\|J^s(fg)-fJ^sg\|_{L^p}\lesssim\|\nabla f\|_{L^\infty}\|J^{s-1}g\|_{L^p}+\|J^sf\|_{L^p}\|g\|_{L^\infty}.$$

□

By Lemma 2.4 and the Hölder inequality, we get

$$\|J^s(fg)-fJ^sg-(J^sf)g\|_{L^p}\lesssim\|\nabla f\|_{L^\infty}\|J^{s-1}g\|_{L^p}+\|J^sf\|_{L^p}\|g\|_{L^\infty}. \tag{2.5}$$

By using (2.5) as $p=2$ and the Sobolev inequality, we have

$$\begin{aligned}\|M(u_1,\partial_l u_2)\|_{L^2_x} &\lesssim \|\nabla u_1\|_{L^\infty_x}\|J^{s-1}\partial_l u_2\|_{L^2_x}+\|J^s u_1\|_{L^2_x}\|\partial_l u_2\|_{L^\infty_x}\\ &\lesssim \|u_1\|_{H^{s_0}}\|u_2\|_{H^s}+\|u_1\|_{H^s}\|u_2\|_{H^{s_0}}\end{aligned} \tag{2.6}$$

for $s>0$ and $s_0>\frac{d}{2}+1$. We can obtain the same estimate for $M(u_2,\partial_l u_1)$.

As a result, by (2.4) and (2.6), we have (2.3) with $k=1$.

Estimate for I_2 *and* I_3 We only consider for I_2 because we can treat I_3 by the same way. By using the Cauchy-Schwarz inequality and applying Lemma 2.4 as $p=2$, $f=u_{2,j}$, $g=\nabla\cdot u_3$, we have

$$\begin{aligned}|I_2| &\lesssim \sum_{j=1}^{d}\|J^s u_{1,j}\|_{L^2_x}\|J^s(u_{2,j}(\nabla\cdot u_3))-u_{2,j}J^s(\nabla\cdot u_3)\|_{L^2_x}\\ &\lesssim \sum_{j=1}^{d}\|J^s u_{1,j}\|_{L^2_x}(\|\nabla u_{2,j}\|_{L^\infty_x}\|J^{s-1}\nabla\cdot u_3\|_{L^2_x}+\|J^s u_{2,j}\|_{L^2_x}\|\nabla\cdot u_3\|_{L^\infty_x})\\ &\lesssim \|U\|^2_{\mathcal{H}^s}\|U\|_{\mathcal{H}^{s_0}}\end{aligned}$$

for $s>0$ and $s_0>\frac{d}{2}+1$. We thus obtain (2.3) with $k=2$. □

Remark 2.5 We can also obtain (2.2) for $s = 0$, because (1.2) and (1.4) yield that

$$\sum_{k=1}^{3} \mu_k \big(N_k(U), u_k\big)_{L_x^2} = \lambda_1\lambda_2\lambda_3 \int_{\mathbb{R}^d} (\overline{u_1} \cdot \overline{u_2})(\nabla \cdot \overline{u_3}) dx.$$

By using the Hölder inequality and the Sobolev inequality, we get (2.2) for $s = 0$. □

For $s \geq 0$, we define the energy functional E_s as

$$E_s(U) := \sum_{k=1}^{3} \mu_k \| J^s u_k \|_{L_x^2}^2. \tag{2.7}$$

Recall that we have (1.5). We obtain the following.

Corollary 2.6 *Let* $s > \frac{d}{2} + 1$ *and* $\Phi \in \mathcal{H}^s(\mathbb{R}^d)$. *Let* T_* *satisfy*

$$0 < T_* \ll \|\Phi\|_{H^s}^{-1}. \tag{2.8}$$

Then, for $\epsilon > 0$, *the solution* $U^\epsilon = (u_1^\epsilon, u_2^\epsilon, u_3^\epsilon)$ *to (2.1) exists in* $C([0, T_*]; \mathcal{H}^s(\mathbb{R}^d))$ *and satisfies*

$$\sup_{0 \leq t \leq T_*} E_s(U^\epsilon(t)) \leq 4 E_s(\Phi).$$

Proof Proposition 2.3 and (1.5) imply that

$$\begin{aligned}
\frac{d}{dt} E_s(U^\epsilon(t)) &\leq \sum_{k=1}^{3} \mu_k \Big(\frac{d}{dt} \| J^s u_k^\epsilon \|_{L_x^2}^2 + 2\epsilon \| J^{s+1} u_k^\epsilon(t) \|_{L_x^2}^2 \Big) \\
&\leq \Big| \sum_{k=1}^{3} \mu_k \Big(\frac{d}{dt} \| J^s u_k^\epsilon \|_{L_x^2}^2 + 2\epsilon \| J^{s+1} u_k^\epsilon(t) \|_{L_x^2}^2 \Big) \Big| \\
&\leq C_s E_s(U^\epsilon(t))^{\frac{3}{2}}
\end{aligned}$$

for $s > \frac{d}{2}+1$, where $C_s > 0$ is a constant independent of ϵ. This ordinary differential inequality yields that

$$E_s(U^\epsilon(t)) \leq \Big(-\frac{C_s}{2} t + E_s(\Phi)^{-\frac{1}{2}} \Big)^{-2} < 4 E_s(\Phi)$$

provided that $0 < t < \frac{1}{C_s} E_s(\Phi)^{-\frac{1}{2}}$. With Remark 2.2, the maximal time of existence for U^ϵ is larger than $\frac{1}{C_s} E_s(\Phi)^{-\frac{1}{2}}$. Moreover, the desired bound follows from $E_s(\Phi) \sim \|\Phi\|_{H^s}^2$. □

3 Existence and Uniqueness of Solution

First, we prove that U^ϵ satisfies Cauchy's condition.

Proposition 3.1 *Let $s > \frac{d}{2} + 1$, $\Phi \in \mathcal{H}^s(\mathbb{R}^d)$, and T_* be as in* (2.8). *Then, the solution $U^\epsilon = (u_1^\epsilon, u_2^\epsilon, u_3^\epsilon) \in C([0, T_*]; \mathcal{H}^s(\mathbb{R}^d))$ to (2.1) satisfies*

$$\sup_{0 \le t \le T_*} E_0(U^\epsilon(t) - U^\delta(t)) \lesssim |\epsilon - \delta|$$

for $\epsilon, \delta > 0$.

Proof Let $\epsilon, \delta > 0$ and $0 < t < T_*$. By (2.7) and (2.1), we have

$$\begin{aligned}
\frac{d}{dt} E_0(U^\epsilon - U^\delta) &= \sum_{k=1}^{3} \mu_k \frac{d}{dt} \|u_k^\epsilon - u_k^\delta\|_{L_x^2}^2 \\
&= 2 \sum_{k=1}^{3} \mu_k \mathrm{Re} \int_{\mathbb{R}^d} (\overline{u_k^\epsilon} - \overline{u_k^\delta}) \cdot \partial_t (u_k^\epsilon - u_k^\delta) dx \\
&= 2 \sum_{k=1}^{3} \mu_k \mathrm{Re} \int_{\mathbb{R}^d} (\overline{u_k^\epsilon} - \overline{u_k^\delta}) \\
&\quad \cdot \left\{ (\epsilon + i\sigma_k) \Delta u_k^\epsilon - (\delta + i\sigma_k) \Delta u_k^\delta - i(N(U^\epsilon) - N(U^\delta)) \right\} dx \\
&= 2 \sum_{k=1}^{3} \mu_k \mathrm{Re} \int_{\mathbb{R}^d} (\overline{u_k^\epsilon} - \overline{u_k^\delta}) \cdot (\epsilon \Delta u_k^\epsilon - \delta \Delta u_k^\delta) dx \\
&\quad - 2 \sum_{k=1}^{3} \mu_k \sigma_k \mathrm{Im} \int_{\mathbb{R}^d} (\overline{u_k^\epsilon} - \overline{u_k^\delta}) \cdot (\Delta u_k^\epsilon - \Delta u_k^\delta) dx \\
&\quad + 2 \sum_{k=1}^{3} \mu_k \mathrm{Im} \int_{\mathbb{R}^d} (\overline{u_k^\epsilon} - \overline{u_k^\delta}) \cdot (N_k(U^\epsilon) - N_k(U^\delta)) dx \\
&=: 2\mathrm{Re}(I) - 2\mathrm{Im}(II) + 2\mathrm{Im}(III).
\end{aligned}$$

Estimate for I By the integral by parts, we obtain

$$\begin{aligned}
&\int_{\mathbb{R}^d} (\overline{u_k^\epsilon} - \overline{u_k^\delta}) \cdot (\epsilon \Delta u_k^\epsilon - \delta \Delta u_k^\delta) dx \\
&= -\sum_{j=1}^{d} \int_{\mathbb{R}^d} (\nabla \overline{u_{k,j}^\epsilon} - \nabla \overline{u_{k,j}^\delta}) \cdot (\epsilon \nabla u_{k,j}^\epsilon - \delta \nabla u_{k,j}^\delta) dx \\
&= -\epsilon \sum_{j=1}^{d} \int_{\mathbb{R}^d} |\nabla u_{k,j}^\epsilon - \nabla u_{k,j}^\delta|^2 dx - (\epsilon - \delta) \sum_{j=1}^{d} \int_{\mathbb{R}^d} (\nabla \overline{u_{k,j}^\epsilon} - \nabla \overline{u_{k,j}^\delta}) \cdot \nabla u_{k,j}^\delta dx.
\end{aligned}$$

Since the first part on the right hand side is negative, we can neglect it. Thus, (1.5), (2.7), and Corollary 2.6 yield that

$$\mathrm{Re}(I) \le |\epsilon - \delta| \sum_{k=1}^{3} \mu_k \|\nabla (u_k^\epsilon - u_k^\delta)\|_{L_x^2} \|\nabla u_k^\delta\|_{L_x^2} \lesssim |\epsilon - \delta| E_1(\Phi).$$

Estimate for II By the integral by parts, we obtain

$$\int_{\mathbb{R}^d} (\overline{u_k^\epsilon} - \overline{u_k^\delta}) \cdot (\Delta u_k^\epsilon - \Delta u_k^\delta) dx = -\|\nabla (u_k^\epsilon - u_k^\delta)\|_{L_x^2}^2.$$

Therefore, we have $\mathrm{Im}(II) = 0$.

Estimate for III By (1.5), (1.2), and (1.4), we obtain

$$\begin{aligned}
III = \lambda_1\lambda_2\lambda_3 \mathrm{Im} \Bigg(&\int_{\mathbb{R}^d} (\overline{u_1^\epsilon} - \overline{u_1^\delta}) \cdot \left\{ (\nabla \cdot \overline{u_3^\epsilon}) \overline{u_2^\epsilon} - (\nabla \cdot \overline{u_3^\delta}) \overline{u_2^\delta} \right\} dx \\
&+ \int_{\mathbb{R}^d} (\overline{u_2^\epsilon} - \overline{u_2^\delta}) \cdot \left\{ (\nabla \cdot \overline{u_3^\epsilon}) \overline{u_1^\epsilon} - (\nabla \cdot \overline{u_3^\delta}) \overline{u_1^\delta} \right\} dx \\
&- \int_{\mathbb{R}^d} \left\{ \nabla \cdot (\overline{u_3^\epsilon} - \overline{u_3^\delta}) \right\} (\overline{u_1^\epsilon} \cdot \overline{u_2^\epsilon} - \overline{u_1^\delta} \cdot \overline{u_2^\delta}) dx \Bigg) \\
= \lambda_1\lambda_2\lambda_3 \mathrm{Im} &\int_{\mathbb{R}^d} \left\{ (\overline{u_1^\epsilon} - \overline{u_1^\delta}) \cdot (\overline{u_2^\epsilon} - \overline{u_2^\delta}) \right\} \left\{ \nabla \cdot (\overline{u_3^\epsilon} + \overline{u_3^\delta}) \right\} dx.
\end{aligned}$$

It follows from (2.7) and Corollary 2.6 that

$$\begin{aligned}
|III| &\lesssim \|u_1^\epsilon - u_1^\delta\|_{L_x^2} \|u_2^\epsilon - u_2^\delta\|_{L_x^2} (\|\nabla \cdot u_3^\epsilon\|_{L_x^2} + \|\nabla \cdot u_3^\delta\|_{L_x^2}) \\
&\lesssim E_1(\Phi)^{\frac{1}{2}} \|U^\epsilon - U^\delta\|_{L_x^2}^2.
\end{aligned}$$

As a result, we obtain

$$\frac{d}{dt}E_0(U^\epsilon - U^\delta) \lesssim |\epsilon - \delta| E_1(\Phi) + E_1(\Phi)^{\frac{1}{2}} E_0(U^\epsilon - U^\delta).$$

By Gronwall's inequality, there exists $C = C(\|\Phi\|_{\mathcal{H}^1}) > 0$ such that

$$E_0(U^\epsilon(t) - U^\delta(t)) \leq C|\epsilon - \delta|$$

for any $t \in [0, T_*]$. □

From Proposition 3.1 and (2.7), there exists a limit $U \in C([0, T_*]; \mathcal{H}^0(\mathbb{R}^d)$ such that

$$\lim_{\epsilon \to +0} \sup_{t\in[0,T_*]} \|U^\epsilon(t) - U(t)\|_{L_x^2} = 0. \tag{3.1}$$

Proposition 3.2 *Let $s > \frac{d}{2} + 1$, $\Phi \in \mathcal{H}^s(\mathbb{R}^d)$, and T_* be as in* (2.8)*. Then, the limit $U \in C([0, T_*]; \mathcal{H}^0(\mathbb{R}^d)$ in* (3.1) *satisfies $U \in L^\infty([0, T_*]; \mathcal{H}^s(\mathbb{R}^d))$ and*

$$\sup_{t\in[0,T_*]} \|U(t)\|_{\mathcal{H}^s} \lesssim \|\Phi\|_{\mathcal{H}^s}.$$

Proof For $F = (f_1, f_2, f_3)$ and $G = (g_1, g_2, g_3)$, we define an inner product $B^s(F, G)$ as

$$B^s(F, G) := \sum_{k=1}^{3} \mu_k (f_k, g_k)_{H^s} = \sum_{k=1}^{3} \mu_k (J^s f_k, J^s g_k)_{L_x^2}.$$

For any $\Psi \in \mathcal{H}^s(\mathbb{R}^d)$ and $\eta > 0$, there exists $\Psi_\eta \in \mathcal{H}^\infty(\mathbb{R}^d)$ such that

$$\|\Psi - \Psi_\eta\|_{\mathcal{H}^s} < \eta.$$

Let U^ϵ and U^δ be as in Proposition 3.1. Then, it follows from Corollary 2.6 and Proposition 3.1 that

$$\begin{aligned}
|B^s(U^\epsilon - U^\delta, \Psi)| &\leq |B^s(U^\epsilon - U^\delta, \Psi - \Psi_\eta)| + |B^s(U^\epsilon - U^\delta, \Psi_\eta)| \\
&\leq E_s(U^\epsilon - U^\delta)^{\frac{1}{2}} \|\Psi - \Psi_\eta\|_{H^s} + E_0(U^\epsilon - U^\delta)^{\frac{1}{2}} \|J^{2s}\Psi_\eta\|_{L_x^2} \\
&\lesssim E_s(\Phi)^{\frac{1}{2}} \eta + |\epsilon - \delta| \|\Psi_\eta\|_{\mathcal{H}^{2s}}.
\end{aligned}$$

For any $\eta > 0$, we have

$$\limsup_{\epsilon,\delta \to +0} \sup_{t\in[0,T_*]} |B^s(U^\epsilon(t) - U^\delta(t), \Psi)| \lesssim E_s(\Phi)^{\frac{1}{2}} \eta.$$

Since $\eta > 0$ is arbitrary, this implies that

$$\lim_{\epsilon,\delta\to+0} \sup_{t\in[0,T_*]} B^s(U^\epsilon(t) - U^\delta(t), \Psi) = 0$$

and

$$\lim_{\epsilon\to+0} B^s(U^\epsilon(t), \Psi) = B^s(U(t), \Psi).$$

With Corollary 2.6 and (2.7), we obtain that

$$\begin{aligned}\|U(t)\|_{\mathcal{H}^s} &\lesssim \sup_{\|\Psi\|_{\mathcal{H}^s}=1} |B^s(U(t), \Psi)| \\ &= \sup_{\|\Psi\|_{\mathcal{H}^s}=1} \lim_{\epsilon\to+0} |B^s(U^\epsilon(t), \Psi)_{\mathcal{H}^s}| \lesssim \lim_{\epsilon\to+0} \|U^\epsilon(t)\|_{\mathcal{H}^s} \lesssim \|\Phi\|_{\mathcal{H}^s}\end{aligned}$$

for $t \in [0, T_*]$. This means that $U \in L^\infty([0, T_*]; \mathcal{H}^s(\mathbb{R}^d))$. □

The proof in Proposition 3.2 implies that the limit U in (3.1) belongs to $C_w([0, T_*]; \mathcal{H}^s(\mathbb{R}^d))$, where $C_w([0, T_*]; \mathcal{H}^s(\mathbb{R}^d))$ denotes the set of all $\mathcal{H}^s(\mathbb{R}^d)$-valued weakly continuous functions. By considering the corresponding integral equation to (2.1) and taking the limit, $U = (u_1, u_2, u_3)$ satisfies

$$u_k(t) = \varphi_k + i\sigma_k\Delta \int_0^t u_k(\tau)d\tau - i\int_0^t N_k(U(\tau))d\tau$$

in $H^{s-2}(\mathbb{R}^d)$ for $t \in [0, T_*]$ and $k = 1, 2, 3$. Hence, U is $\mathcal{H}^{s-2}(\mathbb{R}^d)$-valued absolutely continuous, namely, $AC([0, T_*]; \mathcal{H}^{s-2}(\mathbb{R}^d))$. Therefore, U is a solution to (1.3) in the sense that U satisfies the first equation in (1.3) in $H^{s-2}(\mathbb{R}^d)$ for almost every $t \in [0, T_*]$. This shows the existence of the solution to (1.3).

Next, we consider the uniqueness.

Proposition 3.3 *Let $s > \frac{d}{2} + 1$ and $\Phi, \Psi \in \mathcal{H}^s(\mathbb{R}^d)$. Let $U = (u_1, u_2, u_3)$, $V = (v_1, v_2, v_3)$ be solutions to (1.3) with $U(0) = \Phi$ and $V(0) = \Psi$ such that*

$$U, V \in C([0, T]; \mathcal{H}^0(\mathbb{R}^d)) \cap C_w([0, T]; \mathcal{H}^s(\mathbb{R}^d)) \cap AC([0, T]; \mathcal{H}^{s-2}(\mathbb{R}^d)).$$

Then, it holds that

$$\|U(t) - V(t)\|_{L_x^2} \leq e^{-C\int_0^t(\|U(\tau)\|_{H^s}+\|V(\tau)\|_{H^s})d\tau} \|\Phi - \Psi\|_{L_x^2}$$

for all $t \in [0, T]$, where $C = C(s) > 0$ is a constant.

Proof Note that $E_0(W)$ is absolutely continuous on $[0, T]$. We put $W := U - V$ and $w_k := u_k - v_k$ $(k = 1, 2, 3)$. By (2.7), we have

$$\frac{d}{dt}E_0(W) = \sum_{k=1}^{3}\mu_k \frac{d}{dt}\|w_k\|_{L_x^2}^2 = 2\mathrm{Re}\sum_{k=1}^{3}\mu_k\int_{\mathbb{R}^d}\overline{w_k}\cdot\partial_t w_k dx$$

for almost every $t \in (0, T)$. It follows from (1.3) that

$$\begin{aligned}\int_{\mathbb{R}^d}\overline{w_k}\cdot\partial_t w_k dx &= \int_{\mathbb{R}^d}\overline{w_k}\cdot(i\sigma_k\Delta w_k - i(N_k(U) - N_k(V)))dx\\ &= -i\sigma_k\|\nabla w_k\|_{L_x^2}^2 - i\int_{\mathbb{R}^d}\overline{w_k}\cdot(N_k(U) - N_k(V))dx.\end{aligned}$$

From (1.2) and (1.4), we have

$$\begin{aligned}&\mathrm{Re}\sum_{k=1}^{3}\mu_k\int_{\mathbb{R}}\overline{w_k}\cdot\partial_t w_k dx\\ &= \mathrm{Im}\left(\mu_1\int_{\mathbb{R}^d}\overline{w_1}\cdot(N_1(U) - N_1(V))dx + \mu_2\int_{\mathbb{R}^d}\overline{w_2}\cdot(N_2(U) - N_2(V))dx\right.\\ &\qquad\left. + \mu_3\int_{\mathbb{R}^d}\overline{w_3}\cdot(N_3(U) - N_3(V))dx\right)\\ &= \lambda_1\lambda_2\lambda_3\mathrm{Im}\left(\int_{\mathbb{R}}\overline{w_1}\cdot\{(\nabla\cdot\overline{u_3})\overline{u_2} - (\nabla\cdot\overline{v_3})\overline{v_2}\}\,dx\right.\\ &\qquad + \int_{\mathbb{R}^d}\overline{w_2}\cdot\{(\nabla\cdot\overline{u_3})\overline{u_1} - (\nabla\cdot\overline{v_3})\overline{v_1}\}\,dx\\ &\qquad\left. - \int_{\mathbb{R}^d}(\nabla\cdot\overline{w_3})(\overline{u_1}\cdot\overline{u_2} - \overline{v_1}\cdot\overline{v_2})dx\right).\end{aligned}$$

Because it holds

$$\begin{aligned}&\overline{w_1}\cdot\{(\nabla\cdot\overline{u_3})\overline{u_2}) - (\nabla\cdot\overline{v_3})\overline{v_2}\} + \overline{w_2}\cdot\{(\nabla\cdot\overline{u_3})\overline{u_1} - (\nabla\cdot\overline{v_3})\overline{v_1}\}\\ &\quad - (\nabla\cdot\overline{w_3})(\overline{u_1}\cdot\overline{u_2} - \overline{v_1}\cdot\overline{v_2})\\ &= (\overline{w_1}\cdot\overline{w_2})(\nabla\cdot\overline{u_3} + \nabla\cdot\overline{v_3}),\end{aligned}$$

we obtain

$$\left|\mathrm{Re}\sum_{k=1}^{3}\mu_k\int_{\mathbb{R}^d}\overline{w_k}\cdot\partial_t w_k dx\right| \le |\lambda_1\lambda_2\lambda_3|\|w_1\|_{L_x^2}\|w_2\|_{L_x^2}(\|\nabla\cdot u_3\|_{L_x^\infty} + \|\nabla\cdot v_3\|_{L_x^\infty}).$$

By the Sobolev inequality, we get

$$\begin{aligned}\frac{d}{dt}E_0(W(t)) &\leq 2|\lambda_1\lambda_2\lambda_3|\|W(t)\|_{L_x^2}^2(\|U(t)\|_{H^s}+\|V(t)\|_{H^s})\\ &\lesssim E_0(W(t))(\|U(t)\|_{H^s}+\|V(t)\|_{H^s})\end{aligned}$$

for $s > \frac{d}{2}+1$ and almost every $t \in (0,T)$. As a result, we have

$$\frac{d}{dt}(e^{-C\int_0^t(\|U(\tau)\|_{H^s}+\|V(\tau)\|_{H^s})d\tau}E_0(W(t))) \leq 0$$

for almost every $t \in (0,T)$ and some constant $C = C(s) > 0$, which concludes the proof. □

By Proposition 3.3, if U and V are solutions to (1.3) with the same initial data $\Phi \in \mathcal{H}^s(\mathbb{R}^d)$, then we have $U(t) = V(t)$.

4 Continuous Dependence of the Solution on Initial Data

In this section, we prove the continuous dependence of the solution on initial data by the Bona-Smith approximation.

Proposition 4.1 *Let $s > \frac{d}{2}+1$, $\Phi \in \mathcal{H}^s(\mathbb{R}^d)$, and T_* be as in* (2.8). *Assume that $\{\Phi_\eta\}_{0<\eta\leq 1} \subset \mathcal{H}^\infty(\mathbb{R}^d)$ satisfies*

$$\lim_{\eta\to 0}\|\Phi_\eta-\Phi\|_{\mathcal{H}^s}=0,\quad \|\Phi-\Phi_\eta\|_{\mathcal{H}^{s-l}}\lesssim \eta^l\|\Phi\|_{\mathcal{H}^s},\quad \|\Phi_\eta\|_{\mathcal{H}^{s+a}}\lesssim \eta^{-a}\|\Phi\|_{\mathcal{H}^s} \tag{4.1}$$

for $0 \leq l \leq s$ and $a \geq 0$. For $0 < \eta \leq 1$, $U_\eta = (u_{\eta,1}, u_{\eta,2}, u_{\eta,3})$ denotes the solution to (1.3) on $[0, T_*]$ with $U_\eta(0) = \Phi_\eta$. Then, it holds that

$$\|U_\eta - U_\delta\|_{L^\infty([0,T_*];\mathcal{H}^s)} \lesssim \eta^{\frac{2s(2s-d-2)}{2s-d}} + \|\Phi_\eta-\Phi_\delta\|_{\mathcal{H}^s}$$

for any $0 < \delta \leq \eta \leq 1$, where the implicit constant does not depend on η, δ.

Proof We put $W = U_\eta - U_\delta$ and $w_k = u_{\eta,k} - u_{\delta,k}$ for $k = 1, 2, 3$. By (1.3), we have

$$\begin{aligned}\frac{d}{dt}\|J^s w_k(t)\|_{L_x^2}^2 &= 2\mathrm{Re}\int_{\mathbb{R}^d} J^s\overline{w_k(t)}\cdot J^s\partial_t w_k(t)dx\\ &= 2\mathrm{Re}\int_{\mathbb{R}^d} J^s\overline{w_k(t)}\cdot J^s\{i\sigma_k\Delta w_k(t)\end{aligned}$$

$$-i(N_k(U_\eta(t)) - N_k(U_\delta(t)))\Big\} dx$$

$$= 2\mathrm{Im} \int_{\mathbb{R}^d} J^s \overline{w_k(t)} \cdot J^s(N_k(U_\eta(t)) - N_k(U_\delta(t)))dx.$$

It follows from (1.2) that

$$\begin{aligned}
N_1(U_\eta) - N_1(U_\delta) &= \lambda_1\{(\nabla \cdot \overline{u_{\eta,3}})\overline{u_{\eta,2}} - (\nabla \cdot \overline{u_{\delta,3}})\overline{u_{\delta,2}}\} \\
&= \lambda_1\{(\nabla \cdot \overline{u_{\eta,3}})\overline{w_2} + (\nabla \cdot \overline{w_3})(\overline{u_{\eta,2}} - \overline{w_2})\}, \\
N_2(U_\eta) - N_2(U_\delta) &= \lambda_2\{(\nabla \cdot \overline{u_{\eta,3}})\overline{u_{\eta,1}} - (\nabla \cdot \overline{u_{\delta,3}})\overline{u_{\delta,1}}\} \\
&= \lambda_2\{(\nabla \cdot \overline{u_{\eta,3}})\overline{w_1} + (\nabla \cdot \overline{w_3})(\overline{u_{\eta,1}} - \overline{w_1})\}, \\
N_3(U_\eta) - N_3(U_\delta) &= \lambda_3\{\nabla(\overline{u_{\eta,1}} \cdot \overline{u_{\eta,2}} - \overline{u_{\delta,1}} \cdot \overline{u_{\delta,2}})\} \\
&= \lambda_3\{\nabla(\overline{u_{\eta,1}} \cdot \overline{w_2} + \overline{w_1} \cdot (\overline{u_{\eta,2}} - \overline{w_2}))\}.
\end{aligned}$$

Therefore, we obtain

$$\begin{aligned}
&\int_{\mathbb{R}^d} J^s \overline{w_1} \cdot J^s(N_1(U_\eta) - N_1(U_\delta))dx \\
&= \lambda_1 \left\{ \big(J^s\overline{w_1}, (J^s\nabla \cdot u_{\eta,3})w_2\big)_{L^2_x} + \big(J^s\overline{w_1}, (J^s\nabla \cdot w_3)(u_{\eta,2} - w_2)\big)_{L^2_x} \right\} \\
&\quad + \lambda_1\big(J^s\overline{w_1}, J^s\big((\nabla \cdot u_{\eta,3})w_2\big) - (J^s\nabla \cdot u_{\eta,3})w_2\big)_{L^2_x} \\
&\quad + \lambda_1\big(J^s\overline{w_1}, J^s\big((\nabla \cdot w_3)(u_{\eta,2} - w_2)\big) - (J^s\nabla \cdot w_3)(u_{\eta,2} - w_2)\big)_{L^2_x} \\
&=: A_1 + B_{1,1} + B_{1,2},
\end{aligned}$$

$$\begin{aligned}
&\int_{\mathbb{R}^d} J^s \overline{w_2} \cdot J^s(N_2(U_\eta) - N_2(U_\delta))dx \\
&= \lambda_2 \left\{ \big(J^s\overline{w_2}, (J^s\nabla \cdot u_{\eta,3})w_1\big)_{L^2_x} + \big(J^s\overline{w_2}, (J^s\nabla \cdot w_3)(u_{\eta,1} - w_1)\big)_{L^2_x} \right\} \\
&\quad + \lambda_2\big(J^s\overline{w_2}, J^s\big((\nabla \cdot u_{\eta,3})w_1\big) - (J^s\nabla \cdot u_{\eta,3})w_1\big)_{L^2_x} \\
&\quad + \lambda_2\big(J^s\overline{w_2}, J^s\big((\nabla \cdot w_3)(u_{\eta,1} - w_2)\big) - (J^s\nabla \cdot w_3)(u_{\eta,1} - w_2)\big)_{L^2_x} \\
&=: A_2 + B_{2,1} + B_{2,2},
\end{aligned}$$

$$\begin{aligned}
&\int_{\mathbb{R}^d} J^s \overline{w_3} \cdot J^s(N_3(U_\eta) - N_3(U_\delta))dx \\
&= -\lambda_3\big(J^s(\overline{u_{\eta,1}} \cdot \overline{w_2} + \overline{w_1} \cdot \overline{u_{\eta,2}} - \overline{w_1} \cdot \overline{w_2}), J^s\nabla \cdot w_3\big)_{L^2_x} \\
&= -\lambda_3 \Big\{ \big((J^s\overline{u_{\eta,1}}) \cdot \overline{w_2} + \overline{u_{\eta,1}} \cdot J^s\overline{w_2}, J^s\nabla \cdot w_3\big)_{L^2_x}
\end{aligned}$$

$$
\begin{aligned}
&+\big((J^s\overline{w_1})\cdot\overline{u_{\eta,2}}+\overline{w_1}\cdot J^s\overline{u_{\eta,2}},\,J^s\nabla\cdot w_3\big)_{L^2_x}\\
&-\big((J^s\overline{w_1})\cdot\overline{w_2}+\overline{w_1}\cdot J^s\overline{w_2},\,J^s\nabla\cdot w_3\big)_{L^2_x}\Big\}\\
&+\lambda_3\big((J^s\overline{u_{\eta,1}})\cdot\overline{w_2}+\overline{u_{\eta,1}}\cdot J^s\overline{w_2}-J^s(\overline{u_{\eta,1}}\cdot\overline{w_2}),\,J^s\nabla\cdot w_3\big)\\
&+\lambda_3\big((J^s\overline{w_1})\cdot\overline{u_{\eta,2}}+\overline{w_1}\cdot J^s\overline{u_{\eta,2}}-J^s(\overline{w_1}\cdot\overline{u_{\eta,2}}),\,J^s\nabla\cdot w_3\big)\\
&-\lambda_3\big((J^s\overline{w_1})\cdot\overline{w_2}+\overline{w_1}\cdot J^s\overline{w_2}-J^s(\overline{w_1}\cdot\overline{w_2}),\,J^s\nabla\cdot w_3\big)\\
=:&\ A_3+B_{3,1}+B_{3,2}+B_{3,3}.
\end{aligned}
$$

By the same calculation in "Estimate for I_2 and I_3" in Proposition 2.3, we have

$$
|B_{k,l}|\lesssim\|W\|_{\mathcal{H}^s}^2\|U_\eta\|_{\mathcal{H}^s}+\|W\|_{\mathcal{H}^s}^2\|U_\eta-W\|_{\mathcal{H}^s}\le(\|U_\eta\|_{\mathcal{H}^s}+\|U_\delta\|_{\mathcal{H}^s})\|W\|_{\mathcal{H}^s}^2
$$

for $s>\frac{d}{2}+1$ and $(k,l)\in\{(1,1),(1,2),(2,1),(2,2)\}$. Similarly, the same argument as in "Estimate for I_1" in Proposition 2.3 yields that

$$
|B_{3,1}|+|B_{3,2}|+|B_{3,3}|\lesssim\|W\|_{\mathcal{H}^s}^2\|U_\eta\|_{\mathcal{H}^s}+\|W\|_{\mathcal{H}^s}^3\le(\|U_\eta\|_{\mathcal{H}^s}+\|U_\delta\|_{\mathcal{H}^s})\|W\|_{\mathcal{H}^s}^2
$$

for $s>\frac{d}{2}+1$. Because Proposition 3.2 and (4.1) with $a=0$ yield that

$$
\|U_\eta(t)\|_{\mathcal{H}^s}\lesssim\|\Phi_\eta\|_{\mathcal{H}^s}\lesssim\|\Phi\|_{\mathcal{H}^s} \tag{4.2}
$$

for any $0\le t\le T_*$ and $0<\eta\le 1$, it holds that

$$
|B_{k,l}|\lesssim\|W\|_{H^s}^2
$$

for any (k,l), where the implicit constant depends on $\|\Phi\|_{\mathcal{H}^s}$ (which is independent of η,δ).

On the other hand, direct calculation shows that

$$
\begin{aligned}
&\mu_1A_1+\mu_2A_2+\mu_3A_3\\
&=\lambda_1\lambda_2\lambda_3\Big\{\big(J^s\overline{w_1}\cdot\overline{w_2}+\overline{w_1}\cdot J^s\overline{w_2},\,J^s\nabla\cdot u_{\eta,3}\big)_{L^2_x}\\
&\qquad-\big(J^s\overline{u_{\eta,1}}\cdot\overline{w_2}+\overline{w_1}\cdot J^s\overline{u_{\eta,2}},\,J^s\nabla\cdot w_3\big)_{L^2_x}\Big\}\\
&=\lambda_1\lambda_2\lambda_3\Big\{\big(J^s\overline{w_1}\cdot\overline{w_2}+\overline{w_1}\cdot J^s\overline{w_2},\,J^s\nabla\cdot u_{\eta,3}\big)_{L^2_x}\\
&\qquad+\sum_{l=1}^{d}\big(J^s\partial_l\overline{u_{\eta,1}}\cdot\overline{w_2}+\overline{w_1}\cdot J^s\partial_l\overline{u_{\eta,2}},\,J^sw_3\big)_{L^2_x}\Big\}
\end{aligned}
$$

$$+ \lambda_1\lambda_2\lambda_3 \sum_{l=1}^{d} \left(J^s \overline{u_{\eta,1}} \cdot \partial_l \overline{w_2} + \partial_l \overline{w_1} \cdot J^s \overline{u_{\eta,2}}, J^s w_3 \right)_{L_x^2}$$

$$=: C_1 + C_2.$$

By using the Hölder inequality and (4.2), we obtain

$$|C_2| \lesssim \|W\|_{\mathcal{H}^s}^2,$$

where implicit constant depends on $\|\Phi\|_{\mathcal{H}^s}$. We show the estimate for C_1. By the Hölder inequality and the Gagliardo-Nirenberg inequality, we have

$$|C_1| \lesssim \|W\|_{\mathcal{H}^s} \|W\|_{L_x^\infty} \|U_\eta\|_{\mathcal{H}^{s+1}} \lesssim \|W\|_{\mathcal{H}^s}^{1+\frac{d}{2s}} \|W\|_{L_x^2}^{1-\frac{d}{2s}} \|U_\eta\|_{\mathcal{H}^{s+1}}$$

for $s > \frac{d}{2}$. We note that

$$\|U_\eta\|_{\mathcal{H}^{s+1}} \lesssim \|\Phi_\eta\|_{\mathcal{H}^{s+1}} \lesssim \eta^{-1} \|\Phi\|_{\mathcal{H}^s}$$

holds by Proposition 3.2 and (4.1) with $a = 1$. Furthermore, by Proposition 3.3 and the second estimate in (4.1) with $l = s$, we get

$$\|W\|_{L_x^2} = \|U_\eta - U_\delta\|_{L_x^2} \lesssim \|\Phi_\eta - \Phi_\delta\|_{L_x^2} \le \|\Phi_\eta - \Phi\|_{L_x^2} + \|\Phi - \Phi_\delta\|_{L_x^2} \lesssim \eta^s \|\Phi\|_{\mathcal{H}^s}$$

for $0 < \delta \le \eta \le 1$. Therefore, we obtain

$$|C_1| \lesssim \eta^{s-\frac{d}{2}-1} \|W\|_{\mathcal{H}^s}^{1+\frac{d}{2s}},$$

where the implicit constant depends on $\|\Phi\|_{H^s}$. We put $\theta = 1 + \frac{d}{2s}$ (< 2). Young's inequality implies that

$$\eta^{s-\frac{d}{2}-1} \|W\|_{\mathcal{H}^s}^{1+\frac{d}{2s}} = (\eta^{\frac{s(2s-d-2)}{2s-d}})^{2-\theta} \|W\|_{\mathcal{H}^s}^{\theta} \lesssim \eta^{\frac{2s(2s-d-2)}{2s-d}} + \|W\|_{\mathcal{H}^s}^2.$$

As a result, for $s > \frac{d}{2} + 1$ and $0 < t < T_*$, we obtain

$$\frac{d}{dt} E_s(W(t)) \lesssim \|W\|_{\mathcal{H}^s}^2 + \eta^{\frac{2s(2s-d-2)}{2s-d}} \lesssim E_s(W(t)) + \eta^{\frac{2s(2s-d-2)}{2s-d}}$$

because of (1.5). It implies that

$$E_s(W(t)) \lesssim e^{Ct} E_s(W(0)) + \eta^{\frac{2s(2s-d-2)}{2s-d}} (e^{Ct} - 1) \lesssim E_s(W(0)) + \eta^{\frac{2s(2s-d-2)}{2s-d}}$$

for $0 < t < T_*$ and some $C = C(\|\Phi\|_{\mathcal{H}^s}) > 0$. This concludes the proof. □

Remark 4.2 The worst terms in Proposition 4.1 are

$$\left(\overline{u_{\eta,k}} \cdot J^s\overline{w_l}, J^s\nabla \cdot w_3\right)_{L_x^2} \text{ and } \left(\overline{w_k} \cdot J^s\overline{w_l}, J^s\nabla \cdot w_3\right)_{L_x^2}$$

because $\|J^s\nabla \cdot w_3\|_{L_x^2}$ cannot be estimated by $\eta^{-1}\|\Phi\|_{\mathcal{H}^s}$. For example, by Proposition 3.2 and (4.1) with $a = 1$, we have

$$\begin{aligned}\|J^s\nabla \cdot w_3\|_{L_x^2} &\lesssim \|W\|_{\mathcal{H}^{s+1}} \leq \|U_\eta\|_{\mathcal{H}^{s+1}} + \|U_\delta\|_{\mathcal{H}^{s+1}} \\ &\lesssim \|\Phi_\eta\|_{\mathcal{H}^{s+1}} + \|\Phi_\delta\|_{\mathcal{H}^{s+1}} \lesssim (\eta^{-1} + \delta^{-1})\|\Phi\|_{\mathcal{H}^s}.\end{aligned}$$

Since $\delta \leq \eta$, we can only obtain

$$\|J^s\nabla \cdot w_3\|_{L_x^2} \lesssim \delta^{-1}\|\Phi\|_{\mathcal{H}^s},$$

which is weaker estimate than the desired estimate

$$\|J^s\nabla \cdot w_3\|_{L_x^2} \lesssim \eta^{-1}\|\Phi\|_{\mathcal{H}^s}.$$

But we do not have to control $\|J^s\nabla \cdot w_3\|_{L_x^2}$ because the worst terms as above are canceled out by the summation $\mu_1 A_1 + \mu_2 A_2 + \mu_3 A_3$. □

Acknowledgments This work was supported by JSPS KAKENHI Grant numbers JP21K13825, JP 22KJ0446, JP23K03182, and JP23H01079.

References

1. Colin, M., Colin, T.: On a quasilinear Zakharov system describing laser-plasma interactions. Differ. Int. Equ. **17**, 297–330 (2004)
2. Hirayama, H.: Well-posedness and scattering for a system of quadratic derivative nonlinear Schrödinger equations with low regularity initial data. Commun. Pure Appl. Anal. **13**(4), 1563–1591 (2014)
3. Hirayama, H., Kinoshita, S.: Sharp bilinear estimates and its application to a system of quadratic derivative nonlinear Schrödinger equations. Nonlinear Anal. **178**, 205–226 (2019)
4. Hirayama, H., Kinoshita, S., Okamoto, M.: Well-posedness for a system of quadratic derivative nonlinear Schrödinger equations with radial initial data. Ann. Henri Poincaré **21**(8), 2611–2636 (2020)
5. Hirayama, H., Kinoshita, S., Okamoto, M.: Well-posedness for a system of quadratic derivative nonlinear Schrödinger equations in almost critical spaces. J. Math. Anal. Appl. **499**(2), 125028, 29pp. (2021)
6. Hirayama, H., Kinoshita, S., Okamoto, M.: A remark on the well-posedness for a system of quadratic derivative nonlinear Schrödinger equations. Commun. Pure. Appl. Anal. **21**(10), 3309–3334 (2022)
7. Kato, T., Ponce, G.: Commutator estimates and the euler and navier-stokes equations. Commun. Pure Appl. Math. **41**, 891–907 (1988)

Part IV
Harmonic Analysis and Related Topics

Geometric Herz Spaces and Poisson Kernels in the Upper Half-Space

J. Cruz-Lugo, M. Laurel, M. Mitrea, and P. Takemura

Abstract Classical Herz spaces (in the Euclidean setting $\mathbb{R}^{n-1}$) are defined in the spirit of the Littlewood-Paley theory, in which one localizes functions on concentric dyadic annuli, measures their size using L^p norms, and assembles these numerical pieces all together using a "weighted" ℓ^q norm. In this same philosophy, we introduce a new class of Herz space, which we refer to as a *Geometric* Herz space. This new brand of Herz space generalizes its classical counterpart in that now, the functions under consideration are localized on progressively more distant "tubular" regions mimicking the shape of a given fixed closed set in $\mathbb{R}^{n-1}$ of (Hausdorff) dimension $d \in [0, n-1)$. Additionally, we show that these spaces are stable under the real method of interpolation. This permits us to obtain basic boundedness and embedding results which open the door for the treatment of boundary value problems. In particular, we show how Poisson kernels can be effectively employed to ultimately prove well-posedness results for the Dirichlet Problem with boundary data prescribed in such Geometric Herz spaces.

1 Introduction, Preliminaries, and Main Result

Boundary value problems is a venerable topic at the confluence of several areas of mathematics. Here, we focus our efforts in the upper half-space. This geometric setting allows us to use powerful tools to treat boundary value problems, such as the existence of Poisson kernels as established in the pioneering work of Agmon, Douglis, and Nirenberg [1, 2], i.e., functions mirroring the basic properties of the harmonic Poisson kernel, which are used with much success in establishing the solvability of boundary problems involving Legendre-Hadamard strongly elliptic

J. Cruz-Lugo · M. Mitrea · P. Takemura (✉)
Baylor University, Waco, TX, USA
e-mail: jesus_cruz4@baylor.edu; marius_mitrea@baylor.edu; pedro_silva1@baylor.edu

M. Laurel
University of Texas at Arlington, Arlington, TX, USA
e-mail: marcus.laurel@uta.edu

M. R. Ebert et al. (eds.), *New Tools in Mathematical Analysis and Applications*, Research Perspectives, https://doi.org/10.1007/978-3-031-77050-0_14

operators. The potency of the method of Poisson kernels for solving boundary value problems in the upper-half space is also showcased in [7], where the authors consider boundary data from Köthe function spaces, an inclusive functional analytic environment generalizing Lebesgue spaces. Additionally, the method of Poisson kernels has been used in [6] to treat higher-order regularity problems, in which boundary data with an arbitrary amount of smoothness are considered. Here the goal is to use the method of Poisson kernels to treat boundary problems with data belonging to Geometric Herz spaces. The success of this enterprise hinges on an array of harmonic analytic techniques working well in concert. In particular, critical use is made of the method of real interpolation to show that the brand of function spaces we have in mind constitute a friendly environment from the perspective of Poisson kernels.

Let us now present our main result, with the understanding that relevant definitions involved in the statement will be introduced shortly after.

Theorem 1 *Fix some $d \in [0, n-1)$, and a d-Ahlfors regular set $E \subseteq \mathbb{R}^{n-1}$. Select indices $p \in (1, \infty)$, $q \in (0, \infty]$, and $\alpha \in \mathbb{R}$ satisfying*

$$\frac{-(n-1-d)}{p} < \alpha < \frac{n-1-d}{p'}, \tag{1}$$

where $p' \in (1, \infty)$ denotes the Hölder conjugate exponent of p. Let L be a second-order, homogeneous, constant (complex) coefficient, $M \times M$ system in $\mathbb{R}^n$. Finally, fix an aperture parameter $\kappa \in (0, \infty)$, and formulate the following Dirichlet boundary value problem for L in $\mathbb{R}^n_+$ with boundary data in the homogeneous Geometric Herz space $[\dot{\mathrm{K}}^{\alpha}_{p,q}(\mathbb{R}^{n-1}; E)]^M$:

$$\begin{cases} u \in [\mathcal{C}^\infty(\mathbb{R}^n_+)]^M, \quad Lu = 0 \ \text{ in } \mathbb{R}^n_+, \\ \mathcal{N}_\kappa u \in \dot{\mathrm{K}}^{\alpha}_{p,q}(\mathbb{R}^{n-1}; E), \\ u\big|^{\kappa-\mathrm{n.t.}}_{\partial\mathbb{R}^n_+} = f \in [\dot{\mathrm{K}}^{\alpha}_{p,q}(\mathbb{R}^{n-1}; E)]^M. \end{cases} \tag{2}$$

Then the BVP (2) *is well-posed in each of the following scenarios:*

(i) L is Legendre-Hadamard elliptic;
(ii) L is weakly elliptic, $\mathfrak{A}^{\mathrm{dis}}_L \neq \varnothing$, and $\mathfrak{A}^{\mathrm{dis}}_{L^\top} \neq \varnothing$.

More specifically, in each of these cases L has a (unique) Poisson kernel P^L, the (unique) solution $u : \mathbb{R}^n_+ \to \mathbb{C}^M$ for (2) *is given by*

$$u(x', t) := (P^L_t * f)(x') \quad \text{for each } (x', t) \in \mathbb{R}^n_+, \tag{3}$$

and satisfies

$$\|\mathcal{N}_\kappa u\|_{\dot{\mathrm{K}}^{\alpha}_{p,q}(\mathbb{R}^{n-1}; E)} \approx \|f\|_{[\dot{\mathrm{K}}^{\alpha}_{p,q}(\mathbb{R}^{n-1}; E)]^M} \tag{4}$$

with implicit proportionality constants independent of f. Similar conclusions hold for inhomogeneous Geometric Herz spaces $\mathrm{K}^{\alpha}_{p,q}(\mathbb{R}^{n-1}; E)$.

Here and elsewhere we adopt the identification $\partial\mathbb{R}^n_+ \equiv \mathbb{R}^{n-1}$. Pick an aperture parameter $\kappa \in (0, \infty)$ and define the *κ-nontangential approach region* with vertex at $x' \in \mathbb{R}^{n-1}$ by $\Gamma_\kappa(x') := \{y = (y', t) \in \mathbb{R}^n_+ : |y' - x'| < \kappa t\}$. Use this to define the *κ-nontangential maximal operator* $\mathcal{N}_\kappa$ acting on any given $\mathcal{L}^n$-measurable function $u : \mathbb{R}^n_+ \to \mathbb{C}^M$ according to

$$(\mathcal{N}_\kappa u)(x') := \|u\|_{L^\infty(\Gamma_\kappa(x'),\mathcal{L}^n)} \quad \forall\, x' \in \mathbb{R}^{n-1}. \tag{5}$$

Moreover, the *κ-nontangential boundary trace* of u at $x' \in \mathbb{R}^{n-1}$ is given by

$$\Big(u\Big|^{\kappa-\mathrm{n.t.}}_{\partial\mathbb{R}^n_+}\Big)(x') := \lim_{\Gamma_\kappa(x')\ni y\to(x',0)} u(y), \tag{6}$$

provided this (essential) limit exists.

We now briefly elaborate on elliptic systems and Poisson kernels. Fix some $n \in \mathbb{N}$ with $n \geq 2$, along with any $M \in \mathbb{N}$, and consider a second-order, homogeneous, constant complex coefficient, $M \times M$ system in $\mathbb{R}^n$ of the format

$$L := \Big(a^{\beta\gamma}_{jk}\,\partial_j\partial_k\Big)_{1\leq\beta,\gamma\leq M} \tag{7}$$

for some complex numbers $a^{\beta\gamma}_{jk}$ (here and elsewhere, we adopt the summation convention over repeated indices). The "block" $A := \big(a^{\beta\gamma}_{jk}\big)_{\substack{1\leq j,k\leq n\\ 1\leq\beta,\gamma\leq M}}$ is called a *coefficient tensor* for L, and let $\mathfrak{A}_L$ be the collection of all such coefficient tensors. Denote by $L^\top := \Big(a^{\gamma\beta}_{kj}\,\partial_j\partial_k\Big)_{1\leq\beta,\gamma\leq M}$ the (formal real) transpose of L, and define the transpose of A as $A^\top := \big(a^{\gamma\beta}_{kj}\big)_{\substack{1\leq j,k\leq n\\ 1\leq\beta,\gamma\leq M}}$. The action of L on arbitrary vector-valued functions $u = (u_\gamma)_{1\leq\gamma\leq M}$ is described as $Lu := \Big(a^{\beta\gamma}_{jk}\,\partial_j\partial_k u_\gamma\Big)_{1\leq\beta\leq M}$. Define the characteristic matrix of L as $L(\xi) := \Big(a^{\beta\gamma}_{jk}\,\xi_j\xi_k\Big)_{1\leq\beta,\gamma\leq M}$. The system L is called *Legendre-Hadamard elliptic* if $\mathrm{Re}\big\langle L(\xi)\zeta, \overline{\zeta}\big\rangle \geq c|\xi|^2|\zeta|^2$ for all $\xi \in \mathbb{R}^n$ and $\zeta \in \mathbb{C}^M$ for some $c \in (0, \infty)$. Refer to L as *weakly elliptic* if $L(\xi)$ is invertible for all $\xi \in \mathbb{R}^n\setminus\{0\}$. Note that Legendre-Hadamard ellipticity implies weakly ellipticity.

A result from [8] ensures that every weakly elliptic system L has a (matrix-valued) canonical fundamental solution E_L. For each coefficient tensor $A \in \mathfrak{A}_L$ and $\xi \in \mathbb{R}^n$, introduce the *directional derivative operator* along ξ associated with A as $\partial^A_\xi := (\xi_j a^{\beta\gamma}_{jk}\,\partial_k)_{1\leq\beta,\gamma\leq M}$ and call A a *distinguished coefficient tensor* for L provided $\partial^{A^\top}_\xi E_{L^\top}(x) = 0$ for all $x, \xi \in \mathbb{R}^n\setminus\{0\}$ with $x\cdot\xi = 0$. The collection of all such distinguished coefficient tensors for L is denoted by $\mathfrak{A}^{\mathrm{dis}}_L$. To give an example, the Laplacian $L := \Delta$ in $\mathbb{R}^n$ possesses a unique distinguished coefficient tensor, and in fact $\mathfrak{A}^{\mathrm{dis}}_\Delta = \{I_{n\times n}\}$. For more examples, see Sect. 4 and [5, §3.3].

A *Poisson kernel* for a weakly elliptic system L in $\mathbb{R}^n_+$ is a matrix-valued function $P^L : \mathbb{R}^{n-1} \to \mathbb{C}^{M\times M}$ satisfying the following three properties:

1. *Size Condition:* There exists a constant $C \in (0, \infty)$ such that
$$|P^L(x')| \le \frac{C}{(1+|x'|^2)^{n/2}} \quad \text{for each } x' \in \mathbb{R}^{n-1}. \tag{8}$$
2. *Normalization Condition:* The function P^L is Lebesgue measurable and
$$\int_{\mathbb{R}^{n-1}} P^L(x')dx' = I_{M\times M}, \tag{9}$$
where $I_{M\times M}$ denotes the $M \times M$ identity matrix.
3. *PDE Condition:* If $K^L(x', t) := P^L_t(x') := t^{1-n}P^L(x'/t)$ for each $x' \in \mathbb{R}^{n-1}$ and $t > 0$, it follows that the matrix-valued function $K^L : \mathbb{R}^n_+ \to \mathbb{C}^{M\times M}$ satisfies
$$LK^L = 0 \cdot I_{M\times M} \quad \text{in} \quad [\mathcal{D}'(\mathbb{R}^n_+)]^{M\times M}. \tag{10}$$

Remark 1 By elliptic regularity (cf. [10, Theorem 6.5.7, pp. 552–553]), one may conclude that if such a Poisson kernel for a weakly elliptic system L exists, then necessarily $K^L \in \left[\mathcal{C}^\infty(\mathbb{R}^n_+)\right]^{M\times M}$. In particular, this ensures (10) is satisfied in a pointwise fashion in $\mathbb{R}^n_+$, and that $P^L \in \left[\mathcal{C}^\infty(\mathbb{R}^{n-1})\right]^{M\times M}$, since $P^L(\cdot) = K^L(\cdot, 1)$.

Work done in [1, 7, 13] ensures the existence of Poisson kernels. Here is a precise statement:

Theorem 2 *Let L be a homogeneous, second-order, constant complex coefficient, $M \times M$ system in $\mathbb{R}^n$. Then, if either of the following two conditions hold:*

(i) L is Legendre-Hadamard elliptic,
(ii) L is weakly elliptic, $\mathfrak{A}^{\text{dis}}_L \neq \varnothing$, and $\mathfrak{A}^{\text{dis}}_{L^\top} \neq \varnothing$,

it follows that L has a unique Poisson kernel.

Next, consider a d-Ahlfors regular set $E \subseteq \mathbb{R}^{n-1}$ for some $d \in [0, n-1)$, i.e., E is a closed subset of $\mathbb{R}^{n-1}$, and has the property that
$$\mathcal{H}^d\big(B(x', R) \cap E\big) \approx R^d \quad \text{uniformly for } x' \in E \text{ and } R \in \big(0, 2\text{diam}(E)\big). \tag{11}$$

Here, $\mathcal{H}^d$ denotes the d-dimensional Hausdorff measure in $\mathbb{R}^{n-1}$, and $B(x', R)$ denotes the $(n-1)$-dimensional ball in $\mathbb{R}^{n-1}$ with center x' and radius R. Define
$$C_k := \big\{x' \in \mathbb{R}^{n-1} : 2^k < \text{dist}(x', E) \le 2^{k+1}\big\} \quad \text{for each } k \in \mathbb{Z},$$
$$\text{and } \widetilde{C}_0 := \big\{x' \in \mathbb{R}^{n-1} : \text{dist}(x', E) \le 2\big\} \quad \text{and} \quad \widetilde{C}_k := C_k \quad \text{for each } k \in \mathbb{N}. \tag{12}$$

Since $d < n-1$, notice that $\mathcal{L}^{n-1}\big(\mathbb{R}^{n-1} \setminus \cup_{k\in\mathbb{Z}} C_k\big) = \mathcal{L}^{n-1}(E) = 0$, i.e., $\{C_k\}_{k\in\mathbb{Z}}$ covers $\mathbb{R}^{n-1}$ up to a set of Lebesgue measure zero. Pick two integrability exponents $p, q \in (0, \infty]$, along with an index $\alpha \in \mathbb{R}$. For each $\mathcal{L}^{n-1}$-measurable function $f : \mathbb{R}^{n-1} \to \mathbb{C}$ define

$$\|f\|_{\dot{\mathrm{K}}^{\alpha}_{p,q}(\mathbb{R}^{n-1};E)} := \begin{cases} \Big[\sum\limits_{k=-\infty}^{\infty} \big(2^{k\alpha} \|f \cdot \mathbf{1}_{C_k}\|_{L^p(\mathbb{R}^{n-1})}\big)^q \Big]^{1/q} & \text{if } q < \infty, \\ \sup\limits_{k\in\mathbb{Z}} \big\{2^{k\alpha} \|f \cdot \mathbf{1}_{C_k}\|_{L^p(\mathbb{R}^{n-1})}\big\} & \text{if } q = \infty, \end{cases} \tag{13}$$

and

$$\|f\|_{\mathrm{K}^{\alpha}_{p,q}(\mathbb{R}^{n-1};E)} := \begin{cases} \Big[\sum\limits_{k=0}^{\infty} \big(2^{k\alpha} \|f \cdot \mathbf{1}_{\widetilde{C}_k}\|_{L^p(\mathbb{R}^{n-1})}\big)^q \Big]^{1/q} & \text{if } q < \infty, \\ \sup\limits_{k\in\mathbb{N}_0} \big\{2^{k\alpha} \|f \cdot \mathbf{1}_{\widetilde{C}_k}\|_{L^p(\mathbb{R}^{n-1})}\big\} & \text{if } q = \infty. \end{cases} \tag{14}$$

With $\mathcal{M}(\mathbb{R}^{n-1})$ denoting the collection of all $\mathcal{L}^{n-1}$-measurable complex-valued functions in $\mathbb{R}^{n-1}$, define the *homogeneous Geometric Herz space* on $\mathbb{R}^{n-1}$ associated with E as

$$\dot{\mathrm{K}}^{\alpha}_{p,q}(\mathbb{R}^{n-1}; E) := \big\{f \in \mathcal{M}(\mathbb{R}^{n-1}) : \|f\|_{\dot{\mathrm{K}}^{\alpha}_{p,q}(\mathbb{R}^{n-1};E)} < \infty\big\}, \tag{15}$$

and the *inhomogeneous Geometric Herz space* on $\mathbb{R}^{n-1}$ associated with E as

$$\mathrm{K}^{\alpha}_{p,q}(\mathbb{R}^{n-1}; E) := \big\{f \in \mathcal{M}(\mathbb{R}^{n-1}) : \|f\|_{\mathrm{K}^{\alpha}_{p,q}(\mathbb{R}^{n-1};E)} < \infty\big\}. \tag{16}$$

A direct computation shows that, in the diagonal case $p = q \in (0, \infty)$, Geometric Herz spaces coincide (as quasi-normed spaces) with weighted Lebesgue spaces:

$$\begin{aligned} \dot{\mathrm{K}}^{\alpha}_{p,p}(\mathbb{R}^{n-1}; E) &= L^p\big(\mathbb{R}^{n-1}, \operatorname{dist}(x', E)^{\alpha p} dx'\big) \\ \text{and } \mathrm{K}^{\alpha}_{p,p}(\mathbb{R}^{n-1}; E) &= L^p\Big(\mathbb{R}^{n-1}, \big(1 + \operatorname{dist}(x', E)\big)^{\alpha p} dx'\Big). \end{aligned} \tag{17}$$

Some possible choices for the set E include:

1. Taking $E := \{0\} \subseteq \mathbb{R}^{n-1}$, in which case $d = 0$, leads to classical Herz spaces $\dot{\mathrm{K}}^{\alpha}_{p,q}(\mathbb{R}^{n-1})$.
2. Any linear subspace $E \subseteq \mathbb{R}^{n-1}$ of dimension $d < n-1$. This choice yields a new class of Herz-type spaces, which we refer to as *tubular Herz spaces*.

3. Choose $E :=$ graph of a Lipschitz function $\phi : \mathbb{R}^d \to \mathbb{R}$, for some $d < n - 1$. This gives rise to yet another new family of Herz-type spaces, which we shall refer to as *rough-tubular Herz spaces*.

2 Real Interpolation of Geometric Herz Spaces

In this section we identify the intermediate spaces, obtained by the real method of interpolation, between Geometric Herz spaces as yet another Geometric Herz space; succinctly put, this scale of spaces is stable under real interpolation. Stating said result requires some definitions. To set the stage, consider two quasi-normed spaces denoted by $X_0 = (X_0, \|\cdot\|_{X_0})$ and $X_1 = (X_1, \|\cdot\|_{X_1})$ that are compatible, i.e., X_0 and X_1 embed continuously into a Hausdorff topological vector space $\mathcal{X}$. In this scenario, the algebraic sum $X_0 + X_1$ is a quasi-normed space when equipped with

$$\|x\|_{X_0+X_1} := \inf_{\substack{x=x_0+x_1\\ x_0\in X_0,\, x_1\in X_1}} \big(\|x_0\|_{X_0} + \|x_1\|_{X_1}\big) \quad \text{for all } x \in X_0 + X_1. \tag{18}$$

Fix a parameter $\theta \in (0, 1)$ along with some $q \in (0, \infty]$. For each $t \in (0, \infty)$ and $x \in X_0 + X_1$ introduce the K-functional

$$K(t, x, X_0, X_1) := \inf_{\substack{x=x_0+x_1\\ x_0\in X_0,\, x_1\in X_1}} \big(\|x_0\|_{X_0} + t\|x_1\|_{X_1}\big). \tag{19}$$

For each $x \in X_0 + X_1$, bring in the quasi-norm

$$\|x\|_{(X_0,X_1)_{\theta,q}} := \begin{cases} \Big(\displaystyle\int_0^\infty \big[t^{-\theta} K(t, x, X_0, X_1)\big]^q \frac{dt}{t}\Big)^{1/q} & \text{if } 0 < q < \infty, \\ \displaystyle\sup_{t>0} \big[t^{-\theta} K(t, x, X_0, X_1)\big] & \text{if } q = \infty, \end{cases} \tag{20}$$

and define the intermediate space (for the real method of interpolation) between X_0 and X_1 as

$$(X_0, X_1)_{\theta,q} := \big\{x \in X_0 + X_1 : \|x\|_{(X_0,X_1)_{\theta,q}} < \infty\big\}. \tag{21}$$

It is of relevance to observe that (see [11, §1.3])

$$(X_0, X_1)_{\theta,q} = (X_1, X_0)_{1-\theta,q} \quad \text{and} \quad (X_0, X_1)_{\theta,q} \hookrightarrow X_0 + X_1 \quad \text{continuously}. \tag{22}$$

We are now prepared to state and prove our main result in this section.

Theorem 3 *Consider a d-Ahlfors regular set* $E \subseteq \mathbb{R}^{n-1}$ *for some* $d \in [0, n-1)$. *Select exponents* $p \in (0, \infty]$ *and* $q_0, q_1, q \in (0, \infty]$ *along with indices* $\alpha_0, \alpha_1 \in \mathbb{R}$ *such that* $\alpha_0 \neq \alpha_1$. *Then whenever* $\theta \in (0, 1)$ *and* $\alpha := (1-\theta)\alpha_0 + \theta\alpha_1$ *it follows that*

$$\Big(\dot{\mathrm{K}}^{\alpha_0}_{p,q_0}(\mathbb{R}^{n-1}; E), \dot{\mathrm{K}}^{\alpha_1}_{p,q_1}(\mathbb{R}^{n-1}; E)\Big)_{\theta,q} = \dot{\mathrm{K}}^{\alpha}_{p,q}(\mathbb{R}^{n-1}; E). \tag{23}$$

Moreover, a similar result holds for inhomogeneous Geometric Herz spaces.

Proof To begin, we wish to highlight that work done in [14] ensures that Geometric Herz spaces are continuously embedded into the space of $\mathcal{L}^{n-1}$-measurable functions equipped with the convergence in measure (itself a Hausdorff topological vector space). As such the intermediate space in the left-hand side of (23) is meaningfully defined.

Thanks to (22) it suffices to show (23) under the assumption that $\alpha_0 < \alpha_1$. The crux of the matter in proving (23) relies on the following identification:

$$\big(\dot{\mathrm{K}}^{\alpha_0}_{p,1}(\mathbb{R}^{n-1}; E), \dot{\mathrm{K}}^{\alpha_1}_{p,1}(\mathbb{R}^{n-1}; E)\big)_{\theta,q} = \dot{\mathrm{K}}^{\alpha}_{p,q}(\mathbb{R}^{n-1}; E). \tag{24}$$

Indeed, assume momentarily that (24) holds. Select two indices $\beta_1, \beta_2 \in \mathbb{R}$ such that $-\infty < \beta_0 < \alpha_0 < \alpha_1 < \beta_1 < \infty$, and choose $\theta_0, \theta_1 \in (0, 1)$ satisfying $\alpha_0 = (1-\theta_0)\beta_0 + \theta_0\beta_1$ and $\alpha_1 = (1-\theta_1)\beta_0 + \theta_1\beta_1$. According to (24) we have

$$\begin{aligned}
\big(\dot{\mathrm{K}}^{\beta_0}_{p,1}(\mathbb{R}^{n-1}; E), \dot{\mathrm{K}}^{\beta_1}_{p,1}(\mathbb{R}^{n-1}; E)\big)_{\theta_0,q_0} &= \dot{\mathrm{K}}^{\alpha_0}_{p,q_0}(\mathbb{R}^{n-1}; E),\\
\big(\dot{\mathrm{K}}^{\beta_0}_{p,1}(\mathbb{R}^{n-1}; E), \dot{\mathrm{K}}^{\beta_1}_{p,1}(\mathbb{R}^{n-1}; E)\big)_{\theta_1,q_1} &= \dot{\mathrm{K}}^{\alpha_1}_{p,q_1}(\mathbb{R}^{n-1}; E).
\end{aligned} \tag{25}$$

Denote $\widetilde{\theta} := (1-\theta)\theta_0 + \theta\theta_1 \in (0, 1)$ and observe that $\alpha = (1-\widetilde{\theta})\beta_0 + \widetilde{\theta}\beta_1$. Granted these, (24), and (25), we may invoke the Reiteration Theorem [11, Theorem 1.3.9, p. 36] to obtain

$$\begin{aligned}
&\Big(\dot{\mathrm{K}}^{\alpha_0}_{p,q_0}(\mathbb{R}^{n-1}; E), \dot{\mathrm{K}}^{\alpha_1}_{p,q_1}(\mathbb{R}^{n-1}; E)\Big)_{\theta,q}\\
&= \Big(\big(\dot{\mathrm{K}}^{\beta_0}_{p,1}(\mathbb{R}^{n-1}; E), \dot{\mathrm{K}}^{\beta_1}_{p,1}(\mathbb{R}^{n-1}; E)\big)_{\theta_0,q_0},\\
&\qquad \big(\dot{\mathrm{K}}^{\beta_0}_{p,1}(\mathbb{R}^{n-1}; E), \dot{\mathrm{K}}^{\beta_1}_{p,1}(\mathbb{R}^{n-1}; E)\big)_{\theta_1,q_1}\Big)_{\theta,q}\\
&= \Big(\dot{\mathrm{K}}^{\beta_0}_{p,1}(\mathbb{R}^{n-1}; E), \dot{\mathrm{K}}^{\beta_1}_{p,1}(\mathbb{R}^{n-1}; E)\Big)_{(1-\theta)\theta_0+\theta\theta_1,q}\\
&= \dot{\mathrm{K}}^{(1-\widetilde{\theta})\beta_0+\widetilde{\theta}\beta_1}_{p,q}(\mathbb{R}^{n-1}; E) = \dot{\mathrm{K}}^{\alpha}_{p,q}(\mathbb{R}^{n-1}; E).
\end{aligned} \tag{26}$$

This justifies (23).

We are now left with verifying (24). Fix an arbitrary function f belonging to $\dot{\mathrm{K}}^{\alpha}_{p,q}(\mathbb{R}^{n-1};E)$, and make the following observation. Given $j \in \mathbb{Z}$, decompose f as

$$f = \sum_{k=j+1}^{\infty} f \cdot \mathbf{1}_{C_k} + \sum_{k=-\infty}^{j} f \cdot \mathbf{1}_{C_k} =: f_0^{(j)} + f_1^{(j)} \quad \mathcal{L}^{n-1}\text{-a.e. on } \mathbb{R}^{n-1}. \tag{27}$$

We claim that

$$f_0^{(j)} \in \dot{\mathrm{K}}^{\alpha_0}_{p,1}(\mathbb{R}^{n-1};E) \quad \text{and} \quad f_1^{(j)} \in \dot{\mathrm{K}}^{\alpha_1}_{p,1}(\mathbb{R}^{n-1};E). \tag{28}$$

To deduce the first membership in (28), recall that $\alpha_0 - \alpha_1 < 0$, and use Hölder's inequality to estimate

$$\begin{aligned}
\|f_0^{(j)}\|_{\dot{\mathrm{K}}^{\alpha_0}_{p,1}(\mathbb{R}^{n-1};E)} &= \Big\| \sum_{k=j+1}^{\infty} f \cdot \mathbf{1}_{C_k} \Big\|_{\dot{\mathrm{K}}^{\alpha_0}_{p,1}(\mathbb{R}^{n-1};E)} = \sum_{\ell=j+1}^{\infty} 2^{\ell\alpha_0} \|f \cdot \mathbf{1}_{C_\ell}\|_{L^p(\mathbb{R}^{n-1})} \\
&= \sum_{\ell=j+1}^{\infty} \big(2^{\ell\alpha}\|f \cdot \mathbf{1}_{C_\ell}\|_{L^p(\mathbb{R}^{n-1})}\big) 2^{\ell\theta(\alpha_0-\alpha_1)} \\
&\leq \begin{cases} \Big(\displaystyle\sum_{\ell=-\infty}^{\infty} \big(2^{\ell\alpha}\|f \cdot \mathbf{1}_{C_\ell}\|_{L^p(\mathbb{R}^{n-1})}\big)^q\Big)^{1/q} \Big(\displaystyle\sum_{\ell=j+1}^{\infty} 2^{\ell\theta(\alpha_0-\alpha_1)q'}\Big)^{1/q'} & \text{if } q > 1 \\ \Big(\displaystyle\sum_{\ell=-\infty}^{\infty} \big(2^{\ell\alpha}\|f \cdot \mathbf{1}_{C_\ell}\|_{L^p(\mathbb{R}^{n-1})}\big)^q\Big)^{1/q} \Big(\displaystyle\sup_{\ell\geq j+1} 2^{\ell\theta(\alpha_0-\alpha_1)}\Big) & \text{if } q \leq 1 \end{cases} \\
&\leq C 2^{j\theta(\alpha_0-\alpha_1)} \|f\|_{\dot{\mathrm{K}}^{\alpha}_{p,q}(\mathbb{R}^{n-1};E)} < \infty,
\end{aligned} \tag{29}$$

for some constant $C = C(q, \alpha_0, \alpha_1, \theta) \in (0, \infty)$. Regarding the second membership in (28), observe that since $\alpha_1 - \alpha_0 > 0$ we have

$$\begin{aligned}
\|f_1^{(j)}\|_{\dot{\mathrm{K}}^{\alpha_1}_{p,1}(\mathbb{R}^{n-1};E)} &= \Big\| \sum_{k=-\infty}^{j} f \cdot \mathbf{1}_{C_k} \Big\|_{\dot{\mathrm{K}}^{\alpha_1}_{p,1}(\mathbb{R}^{n-1};E)} = \sum_{\ell=-\infty}^{j} 2^{\ell\alpha_1} \|f \cdot \mathbf{1}_{C_\ell}\|_{L^p(\mathbb{R}^{n-1})} \\
&= \sum_{\ell=\infty}^{j} \big(2^{\ell\alpha}\|f \cdot \mathbf{1}_{C_\ell}\|_{L^p(\mathbb{R}^{n-1})}\big) 2^{\ell(1-\theta)(\alpha_1-\alpha_0)} \\
&\leq \begin{cases} \Big(\displaystyle\sum_{\ell=-\infty}^{\infty} \big(2^{\ell\alpha}\|f \cdot \mathbf{1}_{C_\ell}\|_{L^p(\mathbb{R}^{n-1})}\big)^q\Big)^{1/q} \Big(\displaystyle\sum_{\ell=-\infty}^{j} 2^{\ell(1-\theta)(\alpha_1-\alpha_0)q'}\Big)^{1/q'} & \text{if } q > 1 \\ \Big(\displaystyle\sum_{\ell=-\infty}^{\infty} \big(2^{\ell\alpha}\|f \cdot \mathbf{1}_{C_\ell}\|_{L^p(\mathbb{R}^{n-1})}\big)^q\Big)^{1/q} \Big(\displaystyle\sup_{\ell\leq j} 2^{\ell(1-\theta)(\alpha_1-\alpha_0)}\Big) & \text{if } q \leq 1 \end{cases}
\end{aligned}$$

$$\leq C2^{j(1-\theta)(\alpha_1-\alpha_0)}\|f\|_{\dot{\mathrm{K}}^{\alpha}_{p,q}(\mathbb{R}^{n-1};E)} < \infty, \tag{30}$$

for some constant $C = C(q, \alpha_0, \alpha_1, \theta) \in (0, \infty)$. This concludes the proof of (28).

Pressing forward, for simplicity, at each $t \in (0, \infty)$ abbreviate

$$\begin{aligned} K(t, f) &:= K\big(t, f, \dot{\mathrm{K}}^{\alpha_0}_{p,1}(\mathbb{R}^{n-1}; E), \dot{\mathrm{K}}^{\alpha_1}_{p,1}(\mathbb{R}^{n-1}; E)\big) \\ &= \inf_{\substack{f=f_0+f_1 \\ f_0\in\dot{\mathrm{K}}^{\alpha_0}_{p,1}(\mathbb{R}^{n-1};E), \\ f_1\in\dot{\mathrm{K}}^{\alpha_1}_{p,1}(\mathbb{R}^{n-1};E)}} \Big\{\|f_0\|_{\dot{\mathrm{K}}^{\alpha_0}_{p,1}(\mathbb{R}^{n-1};E)} + t\|f_1\|_{\dot{\mathrm{K}}^{\alpha_1}_{p,1}(\mathbb{R}^{n-1};E)}\Big\}, \end{aligned} \tag{31}$$

and set

$$\tau := \frac{-(\alpha-\alpha_0)}{\theta} = \frac{\alpha-\alpha_1}{1-\theta} \in (-\infty, 0). \tag{32}$$

We will prove the present theorem for $q \in (0, \infty)$, as the case $q = \infty$ follows via a similar reasoning. Keeping in mind (27), (28), (32), and that the function $K(\cdot, f)$ defined in (31) is non-decreasing over $(0, \infty)$, estimate

$$\begin{aligned}
\Big(\int_0^\infty [t^{-\theta}K(t,f)]^q \frac{dt}{t}\Big)^{1/q} &= \Big(\sum_{j=-\infty}^{\infty} \int_{2^{(j+1)\tau}}^{2^{j\tau}} [t^{-\theta}K(t,f)]^q \frac{dt}{t}\Big)^{1/q} \\
&\leq C\Big(\sum_{j=\infty}^{\infty} 2^{-j\theta\tau q} K(2^{j\tau}, f)^q \int_{2^{(j+1)\tau}}^{2^{j\tau}} \frac{dt}{t}\Big)^{1/q} \\
&= C\Big(\sum_{j=-\infty}^{\infty} 2^{-j\theta\tau q} K(2^{j\tau}, f)^q\Big)^{1/q} \\
&\leq C\Big\{\sum_{j=-\infty}^{\infty} 2^{-j\theta\tau q}\big(\|f_0^{(j)}\|_{\dot{\mathrm{K}}^{\alpha_0}_{p,1}(\mathbb{R}^{n-1};E)} + 2^{j\tau}\|f_1^{(j)}\|_{\dot{\mathrm{K}}^{\alpha_1}_{p,1}(\mathbb{R}^{n-1};E)}\big)^q\Big\}^{1/q} \\
&\leq C\Big\{\sum_{j=-\infty}^{\infty} 2^{-j\theta\tau q}\Big\|\sum_{k=j+1}^{\infty} f\cdot\mathbf{1}_{C_k}\Big\|^q_{\dot{\mathrm{K}}^{\alpha_0}_{p,1}(\mathbb{R}^{n-1};E)}\Big\}^{1/q} \\
&\quad + C\Big\{\sum_{j=-\infty}^{\infty} 2^{j\tau q(1-\theta)}\Big\|\sum_{k=-\infty}^{j} f\cdot\mathbf{1}_{C_k}\Big\|^q_{\dot{\mathrm{K}}^{\alpha_1}_{p,1}(\mathbb{R}^{n-1};E)}\Big\}^{1/q} \\
&= C\Big\{\sum_{j=-\infty}^{\infty}\Big(\sum_{\ell=j+1}^{\infty} 2^{(j-\ell)(\alpha-\alpha_0)}2^{\ell\alpha}\|f\cdot\mathbf{1}_{C_\ell}\|_{L^p(\mathbb{R}^{n-1})}\Big)^q\Big\}^{1/q}
\end{aligned}$$

$$+C\Big\{\sum_{j=-\infty}^{\infty}\Big(\sum_{\ell=-\infty}^{j}2^{(\ell-j)(\alpha_1-\alpha)}2^{\ell\alpha}\|f\cdot\mathbf{1}_{C_\ell}\|_{L^p(\mathbb{R}^{n-1})}\Big)^q\Big\}^{1/q} =: \mathrm{I}+\mathrm{II}. \tag{33}$$

Estimating I In the case $0 < q \leq 1$, use Fubini's theorem for series and the fact that $\alpha - \alpha_0 > 0$ to further obtain

$$\begin{aligned}\mathrm{I} &\leq C\Big\{\sum_{j=-\infty}^{\infty}\sum_{\ell=j+1}^{\infty}2^{(j-\ell)(\alpha-\alpha_0)q}2^{\ell\alpha q}\|f\cdot\mathbf{1}_{C_\ell}\|^q_{L^p(\mathbb{R}^{n-1})}\Big\}^{1/q}\\ &= C\Big\{\sum_{\ell=-\infty}^{\infty}2^{\ell\alpha q}\|f\cdot\mathbf{1}_{C_\ell}\|^q_{L^p(\mathbb{R}^{n-1})}\sum_{j=-\infty}^{\ell-1}2^{(j-\ell)(\alpha-\alpha_0)q}\Big\}^{1/q}\\ &\leq C\Big\{\sum_{\ell=-\infty}^{\infty}2^{\ell\alpha q}\|f\cdot\mathbf{1}_{C_\ell}\|^q_{L^p(\mathbb{R}^{n-1})}\Big\}^{1/q} = C\|f\|_{\dot{\mathrm{K}}^{\alpha}_{p,q}(\mathbb{R}^{n-1};E)},\end{aligned} \tag{34}$$

for some constant $C = C(\alpha, \alpha_0, q) \in (0, \infty)$. When $1 < q < \infty$, employ Hölder's inequality, Fubini's theorem for series, along with $\alpha - \alpha_0 > 0$ to estimate I as

$$\begin{aligned}\mathrm{I} &\leq \Big\{\sum_{j=-\infty}^{\infty}\Big(\sum_{\ell=j+1}^{\infty}2^{(j-\ell)(\alpha-\alpha_0)q/2}2^{\ell\alpha q}\|f\cdot\mathbf{1}_{C_\ell}\|^q_{L^p(\mathbb{R}^{n-1})}\Big)\\ &\qquad\times\Big(\sum_{\ell=j+1}^{\infty}2^{(j-\ell)(\alpha-\alpha_0)q'/2}\Big)^{q/q'}\Big\}^{1/q}\\ &\leq C\Big\{\sum_{j=-\infty}^{\infty}\sum_{\ell=j+1}^{\infty}2^{(j-\ell)(\alpha-\alpha_0)q/2}2^{\ell\alpha q}\|f\cdot\mathbf{1}_{C_\ell}\|^q_{L^p(\mathbb{R}^{n-1})}\Big\}^{1/q}\\ &= C\Big\{\sum_{\ell=-\infty}^{\infty}2^{\ell\alpha q}\|f\cdot\mathbf{1}_{C_\ell}\|^q_{L^p(\mathbb{R}^{n-1})}\sum_{j=-\infty}^{\ell-1}2^{(j-\ell)(\alpha-\alpha_0)q/2}\Big\}^{1/q}\\ &\leq C\Big\{\sum_{\ell=-\infty}^{\infty}2^{\ell\alpha q}\|f\cdot\mathbf{1}_{C_\ell}\|^q_{L^p(\mathbb{R}^{n-1})}\Big\}^{1/q} = C\|f\|_{\dot{\mathrm{K}}^{\alpha}_{p,q}(\mathbb{R}^{n-1};E)},\end{aligned} \tag{35}$$

for some constant $C = C(\alpha, \alpha_0, q) \in (0, \infty)$.

Estimating II Consider first $0 < q \leq 1$. Then by Fubini's theorem for series, and using that $\alpha_1 - \alpha > 0$, we have

$$\begin{aligned}
\mathrm{II} &\leq C\Big\{\sum_{j=-\infty}^{\infty}\sum_{\ell=-\infty}^{j} 2^{(\ell-j)(\alpha_1-\alpha)q}2^{\ell\alpha q}\|f\cdot\mathbf{1}_{C_\ell}\|^q_{L^p(\mathbb{R}^{n-1})}\Big\}^{1/q}\\
&= C\Big\{\sum_{\ell=-\infty}^{\infty} 2^{\ell\alpha q}\|f\cdot\mathbf{1}_{C_\ell}\|^q_{L^p(\mathbb{R}^{n-1})}\sum_{j=\ell}^{\infty}2^{(\ell-j)(\alpha_1-\alpha)q}\Big\}^{1/q}\\
&\leq C\Big\{\sum_{\ell=-\infty}^{\infty} 2^{\ell\alpha q}\|f\cdot\mathbf{1}_{C_\ell}\|^q_{L^p(\mathbb{R}^{n-1})}\Big\}^{1/q} = C\|f\|_{\dot{\mathrm{K}}^{\alpha}_{p,q}(\mathbb{R}^{n-1};E)}.
\end{aligned} \tag{36}$$

Corresponding to $1 < q < \infty$, use Hölder's inequality, Fubini's theorem for series, and that $\alpha_1 - \alpha > 0$ to get

$$\begin{aligned}
\mathrm{II} &\leq C\Big\{\sum_{j=-\infty}^{\infty}\Big(\sum_{\ell=-\infty}^{j} 2^{(\ell-j)(\alpha_1-\alpha)q/2}2^{\ell\alpha q}\|f\cdot\mathbf{1}_{C_\ell}\|^q_{L^p(\mathbb{R}^{n-1})}\Big)\\
&\qquad\times\Big(\sum_{\ell=-\infty}^{j}2^{(\ell-j)(\alpha_1-\alpha)q'/2}\Big)^{q/q'}\Big\}^{1/q}\\
&\leq C\Big\{\sum_{j=-\infty}^{\infty}\sum_{\ell=-\infty}^{j} 2^{(\ell-j)(\alpha_1-\alpha)q/2}2^{\ell\alpha q}\|f\cdot\mathbf{1}_{C_\ell}\|^q_{L^p(\mathbb{R}^{n-1})}\Big\}^{1/q}\\
&= C\Big\{\sum_{\ell=-\infty}^{\infty} 2^{\ell\alpha q}\|f\cdot\mathbf{1}_{C_\ell}\|^q_{L^p(\mathbb{R}^{n-1})}\sum_{j=\ell}^{\infty}2^{(\ell-j)(\alpha_1-\alpha)q/2}\Big\}^{1/q}\\
&\leq C\Big\{\sum_{\ell=-\infty}^{\infty} 2^{\ell\alpha q}\|f\cdot\mathbf{1}_{C_\ell}\|^q_{L^p(\mathbb{R}^{n-1})}\Big\}^{1/q} = C\|f\|_{\dot{\mathrm{K}}^{\alpha}_{p,q}(\mathbb{R}^{n-1};E)}.
\end{aligned} \tag{37}$$

for some $C = C(\alpha, \alpha_1, q) \in (0, \infty)$. All together, (20) and (33)–(37) ultimately prove that $\dot{\mathrm{K}}^{\alpha}_{p,q}(\mathbb{R}^{n-1}; E) \hookrightarrow \big(\dot{\mathrm{K}}^{\alpha_0}_{p,1}(\mathbb{R}^{n-1}; E), \dot{\mathrm{K}}^{\alpha_1}_{p,1}(\mathbb{R}^{n-1}; E)\big)_{\theta,q}$ continuously.

Let us now deal with the reverse inclusion. To this end, fix an arbitrary function $f \in \big(\dot{\mathrm{K}}^{\alpha_0}_{p,1}(\mathbb{R}^{n-1}; E), \dot{\mathrm{K}}^{\alpha_1}_{p,1}(\mathbb{R}^{n-1}; E)\big)_{\theta,q}$ and write f as

$$f = f_0 + f_1, \quad \text{for some } f_0 \in \dot{\mathrm{K}}^{\alpha_0}_{p,1}(\mathbb{R}^{n-1}; E) \text{ and } f_1 \in \dot{\mathrm{K}}^{\alpha_1}_{p,1}(\mathbb{R}^{n-1}; E). \tag{38}$$

Mindful of this, for each $j \in \mathbb{Z}$ estimate

$$
\begin{aligned}
& 2^{j\alpha}\|f \cdot \mathbf{1}_{C_j}\|_{L^p(\mathbb{R}^{n-1})} \\
& \leq 2^{j\alpha}\|f_0 \cdot \mathbf{1}_{C_j}\|_{L^p(\mathbb{R}^{n-1})} + 2^{j\alpha}\|f_1 \cdot \mathbf{1}_{C_j}\|_{L^p(\mathbb{R}^{n-1})} \\
& = 2^{j(\alpha-\alpha_0)}\Big(2^{j\alpha_0}\|f_0 \cdot \mathbf{1}_{C_j}\|_{L^p(\mathbb{R}^{n-1})} + 2^{j(\alpha_0-\alpha_1)}2^{j\alpha_1}\|f_1 \cdot \mathbf{1}_{C_j}\|_{L^p(\mathbb{R}^{n-1})}\Big) \\
& \leq C2^{j(\alpha-\alpha_0)}\Big(\|f_0\|_{\dot{\mathrm{K}}^{\alpha_0}_{p,1}(\mathbb{R}^{n-1};E)} + 2^{(j+1)(\alpha_0-\alpha_1)}\|f_1\|_{\dot{\mathrm{K}}^{\alpha_1}_{p,1}(\mathbb{R}^{n-1};E)}\Big),
\end{aligned} \tag{39}
$$

for some constant $C = C(\alpha_0, \alpha_1) \in (0, \infty)$. Taking the infimum on the right-hand side over all possible writings of f as in (38), and recalling (31), we arrive at

$$
2^{j\alpha}\|f \cdot \mathbf{1}_{C_j}\|_{L^p(\mathbb{R}^{n-1})} \leq C2^{j(\alpha-\alpha_0)} \cdot K\big(2^{(j+1)(\alpha_0-\alpha_1)}, f\big) \text{ for all } j \in \mathbb{Z}. \tag{40}
$$

Moreover, since $(\alpha_0-\alpha_1)/\tau > 0$, for every $j \in \mathbb{Z}$ and $t \in \big(2^{(j+1)\tau}, 2^{j\tau}\big)$ it is visible that $t \approx 2^{j\tau}$ uniformly on j and $2^{(j+1)(\alpha_0-\alpha_1)} < t^{(\alpha_0-\alpha_1)/\tau} < 2^{j(\alpha_0-\alpha_1)}$. This and the monotonicity of $K(\cdot, f)$ then yield

$$
\int_{2^{(j+1)\tau}}^{2^{j\tau}} \big[t^{(\alpha-\alpha_0)/\tau} K\big(t^{(\alpha_0-\alpha_1)/\tau}, f\big)\big]^q \frac{dt}{t} \geq C\big[2^{j(\alpha-\alpha_0)} K\big(2^{(j+1)(\alpha_0-\alpha_1)}, f\big)\big]^q \tag{41}
$$

for some constant $C = C(\alpha_0, \alpha_1, \theta) \in (0, \infty)$. In turn, use (40) and (41) to estimate

$$
\begin{aligned}
\|f\|_{\dot{\mathrm{K}}^{\alpha}_{p,q}(\mathbb{R}^{n-1};E)} &= \Big\{\sum_{j=-\infty}^{\infty} \big(2^{j\alpha}\|f \cdot \mathbf{1}_{C_k}\|_{L^p(\mathbb{R}^{n-1})}\big)^q\Big\}^{1/q} \\
&\leq C\Big\{\sum_{j=-\infty}^{\infty} \big[2^{j(\alpha-\alpha_0)} K\big(2^{(j+1)(\alpha_0-\alpha_1)}, f\big)\big]^q\Big\}^{1/q} \\
&\leq C\Big\{\sum_{j=-\infty}^{\infty} \int_{2^{(j+1)\tau}}^{2^{j\tau}} \big[t^{(\alpha-\alpha_0)/\tau} K\big(t^{(\alpha_0-\alpha_1)/\tau}, f\big)\big]^q \frac{dt}{t}\Big\}^{1/q} \\
&= C\Big\{\int_0^{\infty} \big[s^{-\theta} K\big(s, f\big)\big]^q \frac{ds}{s}\Big\}^{1/q} = C\|f\|_{\big(\dot{\mathrm{K}}^{\alpha_0}_{p,1}(\mathbb{R}^{n-1};E),\, \dot{\mathrm{K}}^{\alpha_1}_{p,1}(\mathbb{R}^{n-1};E)\big)_{\theta,q}},
\end{aligned} \tag{42}
$$

where in the penultimate equality we made the change of variable $s := t^{(\alpha_0-\alpha_1)/\tau}$ (which implies $ds/s \approx dt/t$), and used the fact that $-\theta = (\alpha - \alpha_0)/(\alpha_0 - \alpha_1)$. Hence, (42) shows that the intermediate space $\big(\dot{\mathrm{K}}^{\alpha_0}_{p,1}(\mathbb{R}^{n-1}; E), \dot{\mathrm{K}}^{\alpha_1}_{p,1}(\mathbb{R}^{n-1}; E)\big)_{\theta,q}$ embeds continuously into $\dot{\mathrm{K}}^{\alpha}_{p,q}(\mathbb{R}^{n-1}; E)$. Altogether, we obtain that (24) holds and the proof of Theorem 3 is thus finished. □

As a consequence of this interpolation result, we can show that Geometric Herz spaces are intermediate spaces of weighted Lebesgue spaces. Specifically, we have the following corollary.

Corollary 1 *Consider a d-Ahlfors regular set $E \subseteq \mathbb{R}^{n-1}$, for some $d \in [0, n-1)$. Select exponents $p \in (0, \infty)$, $q \in (0, \infty]$, and an index $\alpha \in \mathbb{R}$. Then whenever $a, b \in \mathbb{R}$ satisfy $-\infty < a < \alpha p < b < \infty$, there exists $\theta \in (0, 1)$ for which*

$$\dot{\mathrm{K}}^{\alpha}_{p,q}(\mathbb{R}^{n-1}; E) = \Big(L^p\big(\mathbb{R}^{n-1}, \mathrm{dist}(x', E)^a dx'\big), L^p\big(\mathbb{R}^{n-1}, \mathrm{dist}(x', E)^b dx'\big)\Big)_{\theta,q}. \tag{43}$$

In particular, the following continuous embedding holds

$$\dot{\mathrm{K}}^{\alpha}_{p,q}(\mathbb{R}^{n-1}; E) \hookrightarrow L^p\big(\mathbb{R}^{n-1}, \mathrm{dist}(x', E)^a dx'\big) + L^p\big(\mathbb{R}^{n-1}, \mathrm{dist}(x', E)^b dx'\big). \tag{44}$$

Proof To verify (43), choose $q_0 = q_1 := p$, $\alpha_0 := a/p$, and $\alpha_1 := b/p$. Also, by the manner in which a and b were selected, we can find $\theta \in (0, 1)$ with the property that $\alpha = (1-\theta)(a/p) + \theta(b/p)$. With these choices in hand, employ (23) together with (17) to get

$$\begin{aligned} \dot{\mathrm{K}}^{\alpha}_{p,q}(\mathbb{R}^{n-1}; E) &= \Big(\dot{\mathrm{K}}^{a/p}_{p,p}(\mathbb{R}^{n-1}; E), \dot{\mathrm{K}}^{b/p}_{p,p}(\mathbb{R}^{n-1}; E)\Big)_{\theta,q} \\ &= \Big(L^p\big(\mathbb{R}^{n-1}, \mathrm{dist}(x', E)^a dx'\big), L^p\big(\mathbb{R}^{n-1}, \mathrm{dist}(x', E)^b dx'\big)\Big)_{\theta,q}. \end{aligned} \tag{45}$$

This then justifies (43). The continuous embedding claimed in (44) is a direct consequence of (43) and (22). □

Another useful embedding is further implied by (44).

Corollary 2 *Let $E \subseteq \mathbb{R}^{n-1}$ be a d-Ahlfors regular set for some $d \in [0, n-1)$. Having fixed exponents $p \in (1, \infty)$, and $q \in (0, \infty]$, select an index $\alpha \in \mathbb{R}$ such that $\frac{-(n-1-d)}{p} < \alpha < \frac{(n-1-d)}{p'}$ where $p' \in (1, \infty)$ is the Hölder's conjugate exponent of p. Then*

$$\dot{\mathrm{K}}^{\alpha}_{p,q}(\mathbb{R}^{n-1}; E) \hookrightarrow L^1\big(\mathbb{R}^{n-1}, \tfrac{dx'}{1+|x'|^{n-1}}\big) \text{ continuously.} \tag{46}$$

Proof To begin, notice that the range of α specified in the statement allows us to take $a, b \in \mathbb{R}$ satisfying

$$-(n-1-d) < a < \alpha p < b < (p-1)(n-1-d). \tag{47}$$

As such, work done in [5] and [10] ensure that the functions

$$\mathrm{dist}(\cdot, E)^a \text{ and } \mathrm{dist}(\cdot, E)^b \text{ belong to the Muckenhoupt class } A_p(\mathbb{R}^{n-1}), \tag{48}$$

and that the Muckenhoupt weighted Lebesgue spaces $L^p\big(\mathbb{R}^{n-1}, \mathrm{dist}(x', E)^a dx'\big)$ and $L^p\big(\mathbb{R}^{n-1}, \mathrm{dist}(x', E)^b dx'\big)$ continuously embed into $L^1\big(\mathbb{R}^{n-1}, \frac{dx'}{1+|x'|^{n-1}}\big)$. Subsequently, these embeddings and (44) readily imply (46). □

The relevance of identifying Geometric Herz spaces as intermediate spaces between weighted Lebesgue spaces lies in the fact that various operators from Harmonic Analysis and PDE are bounded on Muckenhoupt weighted Lebesgue spaces, thus leading to boundedness results on Geometric Herz spaces. In particular, upon recalling that the Hardy-Littlewood maximal operator acts on any given $\mathcal{L}^{n-1}$-measurable function f in $\mathbb{R}^{n-1}$ as

$$\mathcal{M}f(x') := \sup_{R>0} \frac{1}{|B(x', R)|} \int_{B(x',R)} |f(y')|\, dy', \quad \text{for each } x' \in \mathbb{R}^{n-1}, \tag{49}$$

we have the following boundedness result.

Corollary 3 *Assume $E \subseteq \mathbb{R}^{n-1}$ is a d-Ahlfors regular set, for some $d \in [0, n-1)$. Select exponents $p \in (1, \infty)$, $q \in (0, \infty]$, and pick an index $\alpha \in \mathbb{R}$ such that*

$$\tfrac{-(n-1-d)}{p} < \alpha < \tfrac{(n-1-d)}{p'}, \tag{50}$$

where $p' \in (1, \infty)$ denotes the Hölder conjugate exponent of p. Then the Hardy-Littlewood maximal operator $\mathcal{M}$ is bounded on $\dot{\mathrm{K}}^{\alpha}_{p,q}(\mathbb{R}^{n-1}; E)$.

Proof First, observe that (50) allows us to choose $a, b \in \mathbb{R}$ satisfying (47). Thanks to (48), we have that $\mathcal{M}$ is bounded on $L^p\big(\mathbb{R}^{n-1}, \mathrm{dist}(x', E)^a dx'\big)$ and $L^p\big(\mathbb{R}^{n-1}, \mathrm{dist}(x', E)^b dx'\big)$. In light of this, (43) and [11, Proposition 1.3.7, pp. 33–34] imply the desired boundedness result. □

3 Well-Posedness of the Dirichlet Problem

To treat the Dirichlet Problem formulated in (2), we recall a key result from [7] and [10] regarding nontangential estimates and boundary traces for functions that mimic the behavior of Poisson kernels.

Lemma 1 *Let $P = (P_{\beta\gamma})_{1\leq\beta,\gamma\leq M} : \mathbb{R}^{n-1} \to \mathbb{C}^{M\times M}$ be a Lebesgue measurable function satisfying*

$$|P(x')| \leq \frac{c}{(1+|x'|^2)^{n/2}} \text{ for each } x' \in \mathbb{R}^{n-1} \tag{51}$$

for some $c \in (0,\infty)$, and recall that $P_t(x') := t^{1-n}P(x'/t)$ for each $x' \in \mathbb{R}^{n-1}$ and $t \in (0,\infty)$. Then for each aperture parameter $\kappa \in (0,\infty)$ there exists some constant $C_\kappa \in (0,\infty)$ with the property that for every $f \in \big[L^1\big(\mathbb{R}^{n-1}, \frac{dx'}{1+|x'|^n}\big)\big]^M$ one has

$$\sup_{|x'-y'|<\kappa t} \big|(P_t * f)(y')\big| \leq C_\kappa \mathcal{M} f(x') \ \textit{for each}\ x' \in \mathbb{R}^{n-1}, \tag{52}$$

and at every Lebesgue point $x_0' \in \mathbb{R}^{n-1}$ of f there holds

$$\lim_{\substack{\mathbb{R}^n_+ \ni (x',t) \to (x_0',0) \\ |x'-x_0'|<\kappa t}} \big(P_t * f\big)(x') = \Big(\int_{\mathbb{R}^{n-1}} P(x')\,dx'\Big) f(x_0'). \tag{53}$$

We are now in a position to prove our main result.

Proof of Theorem 1 To begin, let P^L be the (unique) Poisson kernel for L in $\mathbb{R}^n_+$, whose existence is ensured by either of the assumptions (i) or (ii), and Theorem 2. That the function u defined in (3) satisfies the first line in (2) follows from (10), Remark 1, and the definition of u in (3). Moreover, Corollary 2, Corollary 3, (5), (8), and (52) in Lemma 1 give $\|\mathcal{N}_\kappa u\|_{\dot{\mathrm{K}}^\alpha_{p,q}(\mathbb{R}^{n-1};E)} \leq C\|f\|_{[\dot{\mathrm{K}}^\alpha_{p,q}(\mathbb{R}^{n-1};E)]^M} < \infty$, thus showing $\mathcal{N}_\kappa u \in \dot{\mathrm{K}}^\alpha_{p,q}(\mathbb{R}^{n-1}; E)$. Regarding the boundary condition in (2), we employ (6), (9), (53), and Corollary 2 to obtain $u\big|^{\kappa-\mathrm{n.t.}}_{\partial\mathbb{R}^n_+} = f$ at $\mathcal{L}^{n-1}$-a.e. point in $\mathbb{R}^{n-1}$. Also, the size condition in (2) together with the fact that $\Big|u\big|^{\kappa-\mathrm{n.t.}}_{\partial\mathbb{R}^n_+}\Big| \leq \mathcal{N}_\kappa u$ everywhere in $\mathbb{R}^{n-1}$, readily implies (4). This proves that u (as in (3)) solves (2).

Regarding the issue of uniqueness, assume there exists a function u satisfying the homogeneous version of the boundary problem (2), i.e., corresponding to the boundary datum $f \equiv 0$. Hence, in light of (44) such a function u satisfies

$$\begin{cases} u \in [\mathcal{C}^\infty(\mathbb{R}^n_+)]^M, \quad Lu = 0 \ \text{ in } \mathbb{R}^n_+, \\ \mathcal{N}_\kappa u \in L^p\big(\mathbb{R}^{n-1}, \mathrm{dist}(x',E)^a dx'\big) + L^p\big(\mathbb{R}^{n-1}, \mathrm{dist}(x',E)^b dx'\big), \\ u\big|^{\kappa-\mathrm{n.t.}}_{\partial\mathbb{R}^n_+} = 0 \ \text{ at } \mathcal{L}^{n-1} - \text{a.e. in } \mathbb{R}^{n-1}, \end{cases} \tag{54}$$

where, thanks to (1), the parameters $a, b \in \mathbb{R}$ can be chosen so that (47) holds. In this scenario, as highlighted in (48), the functions $\mathrm{dist}(\cdot,E)^a$ and $\mathrm{dist}(\cdot,E)^b$ belong to the Muckenhoupt class $A_p(\mathbb{R}^{n-1})$. Therefore, in the case when L satisfies (ii), i.e., L is weakly elliptic, $\mathfrak{A}^{\mathrm{dis}}_L \neq \varnothing$, and $\mathfrak{A}^{\mathrm{dis}}_{L^\top} \neq \varnothing$, work done in [5] concerning the well-posedness of the Dirichlet Problem with boundary data taken on sums of Muckenhoupt weighted Lebesgue spaces then ensures that $u \equiv 0$ in $\mathbb{R}^n_+$. In the scenario when the system L is Legendre-Hadamard elliptic, the Poisson integral representation formula from [12, Theorem 4.4.1, pp. 799–801] specialized to the setting of $\mathbb{R}^n_+$ gives that any function u satisfying, for an aperture parameter $\kappa > 0$,

$$\begin{cases} u \in \left[\mathcal{C}^\infty(\mathbb{R}^n_+)\right]^M, \quad Lu = 0 \text{ in } \mathbb{R}^n_+, \\ u\Big|^{\kappa-\text{n.t.}}_{\partial\mathbb{R}^n_+} \text{ exists at } \mathcal{L}^{n-1}\text{-a.e. point in } \mathbb{R}^{n-1}, \\ \displaystyle\int_{\mathbb{R}^{n-1}} (\mathcal{N}_\kappa u)(x') \frac{dx'}{1+|x'|^{n-1}} < \infty, \end{cases} \tag{55}$$

may be represented as

$$u(x) = \int_{\mathbb{R}^{n-1}} P^L_t(x'-y')\Big(u\Big|^{\kappa-\text{n.t.}}_{\partial\mathbb{R}^n_+}\Big)(y')\,dy \quad \forall\, x=(x',t) \in \mathbb{R}^n_+, \tag{56}$$

where P^L is the Agmon-Douglis-Nirenberg Poisson kernel for the system L in $\mathbb{R}^n_+$. Notice that the null-solution u in consideration satisfies the third line in (55), thanks to the membership in the second line of (54) and the fact that the sum of Muckenhoupt weighted Lebesgue spaces embeds into $L^1\big(\mathbb{R}^{n-1}, \frac{dx'}{1+|x'|^{n-1}}\big)$ (as pointed out in the proof of Corollary 2). In particular, since the trace $u\big|^{\kappa-\text{n.t.}}_{\partial\mathbb{R}^n_+}$ is zero at $\mathcal{L}^{n-1}$-a.e. point in $\mathbb{R}^{n-1}$ (see (54)), the integral representation (56) yields that u vanishes identically in $\mathbb{R}^n_+$, as desired. This completes the treatment of uniqueness, which ultimately concludes the proof Theorem 1. □

4 Some Concrete Examples

We now present several applications of Theorem 1 to various systems that possess a Poisson kernel. First, it is implicit in the statement of Theorem 1 that BVP (2) is well posed whenever L is Legendre-Hadamard elliptic. Second, we turn our attention to the Laplacian, $\Delta := \partial_1^2 + \cdots + \partial_n^2$. It is worth noting that given the functional analytic setting under consideration, the following well-posedness result is new and nontrivial.

Corollary 4 *Retain the context of Theorem 1. Then the Dirichlet Problem* (2) *with* $L := \Delta$ *is uniquely solvable with naturally accompanying estimates as in* (4)*. Moreover, the solution* $u : \mathbb{R}^n \to \mathbb{R}$ *of* (2) *at any point* $(x', t) \in \mathbb{R}^n_+$ *is given by*

$$u(x',t) := \frac{2}{\omega_{n-1}} \int_{\mathbb{R}^{n-1}} \frac{t}{\big(|x'-y'|^2+t^2\big)^{\frac{n}{2}}} f(y')\,dy'. \tag{57}$$

Proof The Laplacian satisfies both *(i)* and *(ii)* of Theorem 1 and since the (unique) Poisson kernel for Δ is the function

$$P^{\Delta}(x') := \frac{2}{\omega_{n-1}} \frac{1}{(1+|x'|^2)^{\frac{n}{2}}} \quad \text{for all } x' \in \mathbb{R}^{n-1}, \tag{58}$$

the conclusion readily follows from Theorem 1. □

Our next application is a well-posedness result for a complexified version of the Lamé system of elasticity in the upper half-space with boundary data in Geometric Herz spaces.

Corollary 5 *Retain the context of Theorem 1. Introduce two parameters $\lambda, \mu \in \mathbb{C}$ satisfying $\mu \neq 0$, $2\mu+\lambda \neq 0$, and $3\mu+\lambda \neq 0$. Recall the Lamé system of elasticity given by $L_{\mu,\lambda} := \mu\Delta + (\mu+\lambda)\nabla\mathrm{div}$. Then the Dirichlet Problem* (2) *corresponding to $L := L_{\mu,\lambda}$ is uniquely solvable with naturally accompanying estimates as detailed in* (4). *Additionally, the solution $u = (u_\beta)_{1\leq\beta\leq n} : \mathbb{R}^n_+ \to \mathbb{C}^n$ of* (2) *associated with the given boundary datum $f = (f_\beta)_{1\leq\beta\leq n} \in [\mathrm{K}^{\alpha}_{p,q}(\mathbb{R}^{n-1};E)]^n$ is defined at any point $(x',t) \in \mathbb{R}^n_+$ as*

$$\begin{aligned} u_\beta(x',t) :=\ & \frac{4\mu}{3\mu+\lambda}\frac{1}{\omega_{n-1}} \int_{\mathbb{R}^{n-1}} \frac{t}{(|x'-y'|^2+t^2)^{\frac{n}{2}}} f_\beta(y')\,dy' \\ & + \frac{\mu+\lambda}{3\mu+\lambda}\frac{2n}{\omega_{n-1}} \int_{\mathbb{R}^{n-1}} \frac{t(x'-y',t)_\beta(x'-y',t)_\gamma}{(|x'-y'|^2+t^2)^{\frac{n+2}{2}}} f_\gamma(y')\,dy'. \end{aligned} \tag{59}$$

Proof Having $\mu \neq 0$, $2\mu+\lambda \neq 0$ ensure the weak ellipticity of $L_{\mu,\lambda}$ (cf. [8, Proposition 10.14, p. 366]) while $3\mu+\lambda \neq 0$ ensures $\mathfrak{A}^{\mathrm{dis}}_{L_{\mu,\lambda}} = \mathfrak{A}^{\mathrm{dis}}_{L^\top_{\mu,\lambda}} \neq \varnothing$ (cf. [5, Example 3.4, p. 208]), and hence item *(ii)* of Theorem 1 is satisfied. The entries of the (unique) Poisson kernel $P^{L_{\mu,\lambda}} : \mathbb{R}^{n-1} \to \mathbb{C}^{n\times n}$ for the system $L_{\mu,\lambda}$ are defined at every point $x' = (x_1,\ldots,x_{n-1}) \in \mathbb{R}^{n-1}$ according to

$$P^{L_{\mu,\lambda}}_{\beta\gamma}(x') := \frac{4\mu}{3\mu+\lambda}\frac{\delta_{\beta\gamma}}{\omega_{n-1}}\frac{1}{(1+|x'|^2)^{\frac{n}{2}}} + \frac{\mu+\lambda}{3\mu+\lambda}\frac{2n}{\omega_{n-1}}\frac{x_\beta x_\gamma}{(1+|x'|^2)^{\frac{n+2}{2}}} \tag{60}$$

for each $\beta, \gamma \in \{1,\ldots,n\}$ with the understanding that $x_n := 1$. Thus, the desired conclusion follows from Theorem 1. □

Our last application presents a well-posedness result in the upper half-plane with boundary data in Geometric Herz spaces for a specific type of constant coefficient 2×2 system, whose relevance lies in the fact that many 2×2 systems in the plane may be put in this canonical form which depends only on two parameters (as opposed to twelve), cf. [4, Chapter 1, sections 3 & 4].

Corollary 6 *Retain the context of Theorem 1 with $n = M = 2$. Introduce two parameters $k \in \mathbb{C}$ with $\operatorname{Re} k > 0$ and $\rho \in \mathbb{C}\setminus\{0, -k\}$ to define the system*

$$L_{k,\rho} := \begin{pmatrix} 1 & 0 \\ 0 & \frac{\rho}{k^2} \end{pmatrix} \partial_x^2 + \begin{pmatrix} 0 & \frac{\rho-k^2}{k} \\ \frac{\rho-1}{k} & 0 \end{pmatrix} \partial_x \partial_y + \begin{pmatrix} \rho & 0 \\ 0 & 1 \end{pmatrix} \partial_y^2. \tag{61}$$

Then the Dirichlet Problem (2) *corresponding to $L := L_{k,\rho}$ is well-posed with naturally accompanying estimates as in* (4)*. Moreover, the components of the solution $u = (u_1, u_2) : \mathbb{R}^2_+ \to \mathbb{C}^2$ of* (2) *formulated for $L := L_{k,\rho}$ with $n = M = 2$ associated with the boundary datum $f = (f_1, f_2) \in [\dot{\mathrm{K}}^\alpha_{p,q}(\mathbb{R}; E)]^2$ are defined at every point $(x, t) \in \mathbb{R}^2_+$ as*

$$\begin{aligned} u_1(x,t) &:= \frac{k+1}{\pi(\rho+k)} \int_{\mathbb{R}} \frac{[k^2(x-y)+\rho t^2]t\, f_1(y) + (\rho-k^2)(x-y)t^2 f_2(y)}{[(x-y)^2+t^2][k^2(x-y)^2+t^2]}\, dy, \\ u_2(x,t) &:= \frac{k+1}{\pi(\rho+k)} \int_{\mathbb{R}} \frac{k(\rho-1)(x-y)t^2 f_1(y) + k[\rho(x-y)^2+t^2]t\, f_2(y)}{[(x-y)^2+t^2][k^2(x-y)^2+t^2]}\, dy. \end{aligned} \tag{62}$$

Proof The conditions $\rho \neq 0$ and $\operatorname{Re} k > 0$ guarantee that $L_{k,\rho}$ is weakly elliptic. In the particular case when $k \in (0, 1]$ and $\rho \in (-k, k]\setminus\{0, k^2\}$, it is established in [3] that both $L_{k,\rho}$ and $L^\top_{k,\rho}$ have unique distinguished coefficient tensors, which are themselves transpose to one another. Moreover, it is shown that the entries of the (unique) Poisson kernel $P^{L_{k,\rho}} : \mathbb{R} \to \mathbb{R}^{2\times 2}$ for $L_{k,\rho}$ at each $x \in \mathbb{R}$ are defined via

$$\begin{aligned} P^{L_{k,\rho}}_{11}(x) &:= C_{k,\rho} \cdot \varphi(x)(\rho + k^2x^2), & P^{L_{k,\rho}}_{12}(x) &:= C_{k,\rho} \cdot \varphi(x)(\rho - k^2)x, \\ P^{L_{k,\rho}}_{21}(x) &:= C_{k,\rho} \cdot \varphi(x)(\rho - 1)kx, & P^{L_{k,\rho}}_{22}(x) &:= C_{k,\rho} \cdot \varphi(x)(1 + \rho x^2)k, \end{aligned} \tag{63}$$

where $C_{k,\rho} := \dfrac{k+1}{\pi(\rho+k)}$ and $\varphi(x) := \dfrac{1}{(1+x^2)(1+k^2x^2)}$ for every $x \in \mathbb{R}$ (in determining the specific values of the normalization constants, the version of the Residue Theorem from [9] is useful). In such a scenario, the conclusion follows from Theorem 1. The general case when $k \in \mathbb{C}$ with $\operatorname{Re} k > 0$ and $\rho \in \mathbb{C}\setminus\{0, -k\}$ follows from this and unique continuation for holomorphic functions. □

References

1. Agmon, S., Douglis, A., Nirenberg, L.: Estimates near the boundary for solutions of elliptic partial differential equations satisfying general boundary conditions I. Commun. Pure Appl. Math. **12**(4), 623–727 (1959)
2. Agmon, S., Douglis, A., Nirenberg, L.: Estimates near the boundary for solutions of elliptic partial differential equations satisfying general boundary conditions II. Commun. Pure Appl. Math. **17**(1), 35–92 (1964)
3. Cruz-Lugo, J., Mitrea, D., Mitrea, M.: Harmonic and complex analytic techniques for elliptic boundary value problems in the plane. Doctoral thesis, Baylor University (2024)
4. Loo Keng, H., Wei, L., Ci-Quian, W.: Second-Order Systems of Partial Differential Equations in the Plane. Research Notes in Mathemtics, vol. 128, Pitman, Boston, MA, (1985)
5. Marn, J.J., Martell, J.M, Mitrea, D., Mitrea, I, Mitrea, M.: Singular Integral Operators, Quantitative Flatness, and Boundary Problems. Progress in Mathematics, vol. 344. Birkhäuser, Basel (2022)
6. Martell, J.M., Mitrea, D., Mitrea, I., Mitrea, M.: The higher order regularity Dirichlet problem for elliptic systems in the upper-half space. In: Harmonic Analysis and Partial Differential Equations. Contemporary Mathematics, vol. 612, pp. 123–141. American Mathematical Society, Providence (2014)
7. Martell, J.M., Mitrea, D., Mitrea, I., Mitrea, M.: The Dirichlet problem for elliptic systems with data in Köthe function spaces. Rev. Mat. Iber. **32**(3), 913–970 (2016)
8. Mitrea, D.: Distributions, Partial Differential Equations, and Harmonic Analysis. Universitext, 2nd edn. Springer, Cham (2018)
9. Mitrea, D., Mitrea, I., Mitrea, M.: A sharp divergence theorem with nontangential traces. Not. Am. Math. Soc. **67**(9), 1295–1305
10. Mitrea, D., Mitrea, I., Mitrea, M.: Geometric Harmonic Analysis - Volume I: A Sharp Divergence Theorem with Nontangential Pointwise Traces. Developmental Mathematics, vol. 72. Springer, Cham (2022)
11. Mitrea, D., Mitrea, I., Mitrea, M.: Geometric Harmonic Analysis - Volume II: Function Spaces Measuring Size and Smoothness on Rough Sets. Developmental Mathematics, vol. 73. Springer, Cham (2022)
12. Mitrea, D., Mitrea, I., Mitrea, M.: Geometric Harmonic Analysis - Volume III: Integral Representation, Calderón-Zygmund Theory, Fatou Theorems, and Applications to Scattering. Developmental Mathematics, vol. 74. Springer, Cham (2023)
13. Mitrea, D., Mitrea, I., Mitrea, M.: Geometric Harmonic Analysis - Volume V: Fredholm Theorem and Finer Estimates for Integral Operators, with Applications to Boundary Problems. Developmental Mathematics, vol. 76. Springer, Cham (2023)
14. Mitrea, M., Takemura, P.: Singular integrals, herz-type function spaces, and boundary problems. Doctoral thesis, Baylor (2024)

Part V
Partial Differential Equations on Curved Spacetime

Optimal L^∞-Decay Rate of Solutions to a Dissipative Nonlinear Schrödinger Equation System

Naoyasu Kita, Yoshihisa Nakamura, and Yuji Sagawa

Abstract This paper treats an initial value problem of a system of nonlinear Schrödinger equations with critical dissipative power nonlinearities. Our aim is to show that optimal L^∞-decay rate of the solutions is $t^{-1/2}(\ln t)^{-1/2}$. In other words, if a solution $\mathbf{u}(t)$ satisfies $\|\mathbf{u}(t)\|_{L^\infty} = o(t^{-1/2}(\ln t)^{-1/2})$ as $t \to \infty$, then $\mathbf{u}(t, x) \equiv 0$.

1 Introduction

We consider a Cauchy problem for a system of nonlinear Schrödinger equations (NLS):

$$\begin{cases} i\partial_t \mathbf{u} + \frac{1}{2}\partial_x^2 \mathbf{u} = \mathbf{f}(\mathbf{u}) \\ \mathbf{u}(0, x) = \mathbf{u}_0(x), \end{cases} \tag{1}$$

where $(t, x) \in [0, \infty) \times \mathbb{R}$ and $\mathbf{u} = \mathbf{u}(t, x) = (u_1(t, x), u_2(t, x), \cdots, u_n(t, x))^t \in \mathbb{C}^n$ with $n \geq 1$. The nonlinearity is assumed to be a $\mathbb{C}^n$-valued gauge-invariant cubic polynomial of u_j and $\overline{u}_j$ $(j = 1, 2, \cdots, n)$, i.e.,

$$\mathbf{f}(\mathbf{u}) = (f_1(\mathbf{u}), f_2(\mathbf{u}), \cdots, f_n(\mathbf{u}))^t \text{ with } f_j(\mathbf{u}) \in \mathbb{C} \ (j = 1, 2, \cdots, n), \tag{2}$$

$$\mathbf{f}(e^{i\theta}\mathbf{u}) = e^{i\theta}\mathbf{f}(\mathbf{u}) \text{ for any } \theta \in \mathbb{R} \text{ and } \mathbf{u} \in \mathbb{C}^n, \tag{3}$$

$$\mathbf{f}(\rho\mathbf{u}) = \rho^3\mathbf{f}(\mathbf{u}) \text{ for any } \rho > 0 \text{ and } \mathbf{u} \in \mathbb{C}^n. \tag{4}$$

N. Kita (✉) · Y. Nakamura
Faculty of Advanced Science and Technology, Kumamoto University, Kumamoto, Japan
e-mail: nkita@kumamoto-u.ac.jp

Y. Sagawa
Department of Mathematics, Osaka Metropolitan University, Osaka, Japan

M. R. Ebert et al. (eds.), *New Tools in Mathematical Analysis and Applications*, Research Perspectives, https://doi.org/10.1007/978-3-031-77050-0_15

The aim of this manuscript is to clarify the optimal L^∞-decay rate of the solution without size restriction on the initial data $\mathbf{u}_0$. To this end, a structural assumption indicating strong nonlinear dissipation is required, i.e., we assume that there exists some $\rho_1 > 0$ such that, for any $\mathbf{p} = (p_1, p_2, \cdots, p_n)^t \in \mathbb{C}^n$,

$$\mathrm{Im}\left\{\overline{\mathbf{p}}^t\left(\frac{\partial \mathbf{f}}{\partial \mathbf{u}}(\mathbf{u})\mathbf{p} \pm \frac{\partial \mathbf{f}}{\partial \overline{\mathbf{u}}}(\mathbf{u})\overline{\mathbf{p}}\right)\right\} \le -\rho_1 \sum_{j=1}^{n} |u_j|^2 |p_j|^2, \tag{5}$$

where $\overline{\mathbf{p}} = (\overline{p_1}, \overline{p_2}, \cdots, \overline{p_n})^t$, $\partial\mathbf{f}/\partial\mathbf{u}$ (resp. $\partial\mathbf{f}/\partial\overline{\mathbf{u}}$) is a matrix, the jk-entry of which is given by $\partial f_j/\partial u_k$ (resp. $\partial f_j/\partial\overline{u_k}$). We call (5) *the strong dissipative condition* on the nonlinearity. There are some examples of $\mathbf{f}(\mathbf{u})$ satisfying (2)–(5). In the case of $n = 1$, i.e., in the case of single equation, we have $\mathbf{f}(u) = \lambda|u|^2 u$ with $\lambda \in \mathbb{C}$, $\mathrm{Im}\lambda < 0$ and $|\mathrm{Re}\lambda| < \sqrt{3}|\mathrm{Im}\lambda|$ (or equivalently $2|\mathrm{Im}\lambda| - |\lambda| > 0$). In fact, we see that, for $p \in \mathbb{C}$,

$$\begin{aligned}\mathrm{Im}\left\{\overline{p}\cdot\left(\frac{\partial \mathbf{f}}{\partial u}(u)p \pm \frac{\partial \mathbf{f}}{\partial \overline{u}}(u)\overline{p}\right)\right\} &= 2\mathrm{Im}\lambda|u|^2|p|^2 \pm \mathrm{Im}(\lambda u^2\overline{p}^2)\\ &\le -(2|\mathrm{Im}\lambda| - |\lambda|)|u|^2|p|^2,\end{aligned}$$

and we may take $\rho_1 = 2|\mathrm{Im}\lambda| - |\lambda| > 0$. In the case of $n = 2$, we have

$$\mathbf{f}(\mathbf{u}) = \begin{pmatrix}\lambda_{11}|u_1|^2u_1 + \lambda_{12}|u_2|^2u_1\\ \lambda_{21}|u_1|^2u_2 + \lambda_{22}|u_2|^2u_2\end{pmatrix}$$

with $\mathbf{u} = (u_1, u_2)^t$, $\lambda_{jk} \in \mathbb{C}$ $(j, k = 1, 2)$ and

$$\begin{aligned}&\mathrm{Im}\lambda_{jk} < 0,\\ &2|\mathrm{Im}\lambda_{11}| > |\lambda_{11}| + 2|\lambda_{12}| + |\lambda_{21}|,\\ &2|\mathrm{Im}\lambda_{22}| > |\lambda_{22}| + |\lambda_{12}| + 2|\lambda_{21}|.\end{aligned} \tag{6}$$

In fact, we see that

$$\frac{\partial \mathbf{f}(\mathbf{u})}{\partial \mathbf{u}} = \begin{pmatrix}\frac{\partial \mathbf{f}(\mathbf{u})}{\partial u_1} & \frac{\partial \mathbf{f}(\mathbf{u})}{\partial u_2}\end{pmatrix} = \begin{pmatrix}2\lambda_{11}|u_1|^2 + \lambda_{12}|u_2|^2 & \lambda_{12}\overline{u_2}u_1\\ \lambda_{21}\overline{u_1}u_2 & \lambda_{21}|u_1|^2 + 2\lambda_{22}|u_2|^2\end{pmatrix},$$

$$\frac{\partial \mathbf{f}(\mathbf{u})}{\partial \overline{\mathbf{u}}} = \begin{pmatrix}\frac{\partial \mathbf{f}(\mathbf{u})}{\partial \overline{u_1}} & \frac{\partial \mathbf{f}(\mathbf{u})}{\partial \overline{u_2}}\end{pmatrix} = \begin{pmatrix}\lambda_{11}u_1^2 & \lambda_{12}u_2u_1\\ \lambda_{21}u_1u_2 & \lambda_{22}u_2^2\end{pmatrix},$$

In fact, we see that

$$\frac{\partial \mathbf{f}(\mathbf{u})}{\partial \mathbf{u}} = \begin{pmatrix} \frac{\partial \mathbf{f}(\mathbf{u})}{\partial u_1} & \frac{\partial \mathbf{f}(\mathbf{u})}{\partial u_2} \end{pmatrix} = \begin{pmatrix} 2\lambda_{11}|u_1|^2 + \lambda_{12}|u_2|^2 & \lambda_{12}\overline{u_2}u_1 \\ \lambda_{21}\overline{u_1}u_2 & \lambda_{21}|u_1|^2 + 2\lambda_{22}|u_2|^2 \end{pmatrix},$$

$$\frac{\partial \mathbf{f}(\mathbf{u})}{\partial \overline{\mathbf{u}}} = \begin{pmatrix} \frac{\partial \mathbf{f}(\mathbf{u})}{\partial \overline{u_1}} & \frac{\partial \mathbf{f}(\mathbf{u})}{\partial \overline{u_2}} \end{pmatrix} = \begin{pmatrix} \lambda_{11}u_1^2 & \lambda_{12}u_2u_1 \\ \lambda_{21}u_1u_2 & \lambda_{22}u_2^2 \end{pmatrix},$$

and so, by Young's inequality $2|u_1||u_2||p_1||p_2| \le |u_1|^2|p_1|^2 + |u_2|^2|p_2|^2$ and (6), we have

$$\begin{aligned} &\mathrm{Im}\left\{\overline{\mathbf{p}}^t \cdot \left(\frac{\partial \mathbf{f}}{\partial \mathbf{u}}(\mathbf{u})\mathbf{p} \pm \frac{\partial \mathbf{f}}{\partial \overline{\mathbf{u}}}(\mathbf{u})\overline{\mathbf{p}}\right)\right\} \\ &\le -(2|\mathrm{Im}\lambda_{11}| - |\lambda_{11}| - 2|\lambda_{12}| - |\lambda_{21}|)|u_1|^2|p_1|^2 \\ &\quad - (2|\mathrm{Im}\lambda_{22}| - |\lambda_{22}| - |\lambda_{12}| - 2|\lambda_{21}|)|u_2|^2|p_2|^2. \end{aligned}$$

Therefore, by taking $\rho_1 = \min\{2|\mathrm{Im}\lambda_{11}| - |\lambda_{11}| - 2|\lambda_{12}| - |\lambda_{21}|, 2|\mathrm{Im}\lambda_{22}| - |\lambda_{22}| - |\lambda_{12}| - 2|\lambda_{21}|\} > 0$, we see that the nonlinearity satisfies (5).

We here introduce the known results on the dissipative NLS. For the single case, i.e.,

$$\begin{cases} i\partial_t u + \frac{1}{2}\partial_x^2 u = \lambda|u|^2 u \\ u(0,x) = u_0(x), \end{cases} \tag{7}$$

there are some works on the asymptotic dynamics of the solutions. If $\mathrm{Im}\lambda = 0$, Hayashi and Naumkin [4] proved that, under the smallness assumption in some weighted Sobolev norm on the initial data, the solution to (7) decays like $O(t^{-1/2})$ as $t \to \infty$ in L^∞. However the asymptotic leading term of $u(t)$ is given by the phase correction added to the free profile—it suggests that the nonlinear effect is visible in phase. If $\mathrm{Im}\lambda < 0$, Shimomura [24] proved that the small-data-solutions decay like $O(t^{-1/2}(\log t)^{-1/2})$ in L^∞—the nonlinear dissipative effect is now visible in the additional logarithmic decay-rate as well as in the asymptotic leading term. Under some smallness assumptions on the data, the effect caused by nonlinear dissipation has been studied not only for (7) but also for NLS of super-critical and sub-critical power of nonlinearity (see [5, 9, 14]). For higher space-dimensional case, see [1–3]. For derivative type of nonlinearities, see [6].

There are also some works in which the size-restriction on the initial data is removed. Kita and Shimomura [15] assumed the strong dissipative condition on the nonlinearity, i.e.,

$$\mathrm{Im}\lambda < 0, \quad |\mathrm{Re}\lambda| \le \sqrt{3}|\mathrm{Im}\lambda|,$$

and obtained $\|u(t)\|_{L^\infty} = O(t^{-1/2}(\log t)^{-1/2})$ as well as an asymptotic leading term of $u(t)$. The key estimate to remove the smallness assumption is $\|Ju(t)\|_{L^2} \leq \|xu_0\|_{L^2}$, where J is an operator described by $J = x + it\partial_x$. The upper bound of $\|Ju(t)\|_{L^2}$ is derived by the strong dissipative condition ($n = 1$ version of (5)). Jin et al. [11] and Hayashi et al. [7] applied another kind of strong dissipative condition to the sub-critical nonlinearity, and proved the decay estimate and modified asymptotic leading term of the solutions without size-restriction. For the generalization of the initial data, see also [8, 10].

Let us next focus on the works concerning systems of NLS. In early researches, the dissipative structure was not considered. Therefore the decay estimate and asymptotic behavior of the solutions have been considered only for small data solution. This is because the estimate of $\|J\mathbf{u}(t)\|_{L^2}$, which is the key to control the error terms, indicated that $\|J\mathbf{u}(t)\|_{L^2}$ grows and its growth-rate depended on the size of initial data. Therefore, to guarantee the rapid decay of the error terms, one can not help restricting the size of data. Anyway, under the smallness assumption of the data, Katayama and Sakoda [12] considered the case where the nonlinearity contains derivatives of unknown variables, and proved the asymptotic behavior of solutions by imposing a certain structural condition on the nonlinearity—it has something to do with the method of Lyapunov function. In most cases of the systems, it becomes more difficult than in the case of single equation, to detect a concrete form of the asymptotic leading term of the solution, since the well-known technique of gauge-transform of the unknown variables does not work so well. However, in [12], the asymptotic leading term is described by a function satisfying the ordinary differential equation associated with the system of NLS.

Masaki [21] completely classified types of the 2-component systems of the ordinary differential equations associated with NLS systems of gauge-invariant cubic nonlinearities. This work is instructive in the point that we know which nonlinearities are essential for the asymptotic dynamics of nonlinear dispersive equations. For some nonlinearities included in the classification by Masaki [21], Masaki et al. [22] and Kita et al. [16] considered coupled nonlinear Schrödinger equations, for which the solutions decay less rapidly than those of free Schrödinger equations.

On the works where some dissipative structures were taken into account for systems, Kim [13] considered the case of distinct masses (or coefficients of ∂_x^2), and proved that the L^∞-norm of the solutions decays like $O(t^{-1/2}(\log t)^{-1/2})$. Li et al. [20] treated another system, and proved the existence of solutions indicating that one component rapidly decays but the other is asymptotically free. The works introduced above suggest that solutions to the systems of NLS exhibit complicated asymptotic profiles which are not simply expected by the single NLS. For the case of derivative nonlinearities, see [18, 19, 23].

As we have seen above, all of the works on systems imposed the size restriction on the data. However, under the strong dissipative condition (5), Kita et al. [17] proved that, if $\mathbf{u}_0 \in H^1$ and $x\mathbf{u}_0 \in L^2$ without size restriction, there exists a unique global solution to (1) such that

$$\mathbf{u} \in C([0,\infty); H^1) \cap C^1([0,\infty); H^{-1}), \quad x\mathbf{u} \in C([0,\infty); L^2). \tag{8}$$

Furthermore we have an upper bound of L^∞-norm such as

$$\|\mathbf{u}(t,\cdot)\|_{L^\infty} \le Ct^{-1/2}(\ln t)^{-1/2}. \tag{9}$$

The aim of this manuscript is to show that the decay rate described in the right hand side of (9) is optimal. In other words, if $\|\mathbf{u}(t,\cdot)\|_{L^\infty} = o(t^{-1/2}(\ln t)^{-1/2})$ as $t \to \infty$, then the solution must be the trivial one.

Theorem 1.1 (The Optimality of L^∞-Decay Rate) *Let the nonlinearty* $\mathbf{f}(\mathbf{u})$ *satisfy (2)–(5) and* $\mathbf{u}_0 \in H^1$, $x\mathbf{u}_0 \in L^2$. *Assume that the solution* $\mathbf{u}(t,x)$ *satisfying (8) admits*

$$\|\mathbf{u}(t,\cdot)\|_{L^\infty} = o(t^{-1/2}(\ln t)^{-1/2}) \tag{10}$$

as $t \to \infty$ *in* L^∞. *Then* $\mathbf{u}(t,x) = \mathbf{0}$ *for* $(t,x) \in [0,\infty) \times \mathbb{R}$. □

Before closing this section, let us introduce some notation. The function space L^q $(1 \le q \le \infty)$ for $\mathbb{C}^n$-valued functions is defined by

$$L^q = \{\mathbf{u}(x) = (u_1(x), u_2(x), \cdots, u_n(x))^t \ ; \ \|\mathbf{u}\|_{L^q} < \infty\},$$

where $\|\mathbf{u}\|_{L^q} = \left(\sum_{j=1}^n \int_{\mathbb{R}} |u_j(x)|^q\, dx\right)^{1/q}$ if $1 \le q < \infty$, and $\|\mathbf{u}\|_{L^\infty} = \max_{1\le j\le n}\{\text{ess.sup}_{x\in\mathbb{R}}|u_j(x)|\}$. The Sobolev space H^1 is endowed with

$$\|\mathbf{u}\|_{H^1} = \left(\|\mathbf{u}\|_{L^2}^2 + \|\partial_x \mathbf{u}\|_{L^2}^2\right)^{1/2}.$$

The negatively indexed Sobolev space H^{-1} is the dual of H^1. The infinitesimal generator of the Galilei transformation is given by

$$J(t) = U(t)xU(-t) = x + it\partial_x.$$

It commutes with the linear differential operator $i\partial_t + \frac{1}{2}\partial_x^2$.

2 Preliminaries

The next lemma plays an important role in our proof.

Lemma 2.1 *Let* $\mathbf{u}_0 \in H^1$ *and* $x\mathbf{u}_0 \in L^2$. *We assume (2)–(5) for the nonlinearity. Then the solution to (1) satisfies*

$$\|\mathbf{u}(t)\|_{L^2} \le \|\mathbf{u}_0\|_{L^2}, \tag{11}$$

$$\|J\mathbf{u}(t)\|_{L^2} \le \|x\mathbf{u}_0\|_{L^2}, \tag{12}$$

$$\|\partial_x \mathbf{u}(t)\|_{L^2} \le \|\partial_x \mathbf{u}_0\|_{L^2}. \tag{13}$$

Proof of Lemma 2.1 By (3), we see that

$$e^{i\theta}\mathbf{f}(\mathbf{u}) = \mathbf{f}(e^{i\theta}\mathbf{u}).$$

Taking the derivative with respect to θ and substituting $\theta = 0$, we have

$$\mathbf{f}(\mathbf{u}) = \frac{\partial \mathbf{f}(\mathbf{u})}{\partial \mathbf{u}}\mathbf{u} + \frac{\partial \mathbf{f}(\mathbf{u})}{\partial \overline{\mathbf{u}}}\overline{\mathbf{u}}.$$

Then (5) leads to

$$\frac{d\|\mathbf{u}(t)\|_{L^2}^2}{dt} = \mathrm{Im}\int \left\{ \overline{\mathbf{u}}^t \left(\frac{\partial \mathbf{f}(\mathbf{u})}{\partial \mathbf{u}}\mathbf{u} + \frac{\partial \mathbf{f}(\mathbf{u})}{\partial \overline{\mathbf{u}}}\overline{\mathbf{u}} \right) \right\} dx \le 0. \tag{14}$$

Hence (11) follows. We next prove (12). Note that

$$\begin{aligned} J\mathbf{f}(\mathbf{u}) &= \sum_{k=1}^{n} \frac{\partial \mathbf{f}}{\partial u_k}(\mathbf{u}) J u_k - \sum_{k=1}^{n} \frac{\partial \mathbf{f}}{\partial \bar{u}_k}(\mathbf{u}) \overline{J u_k} \\ &= \frac{\partial \mathbf{f}}{\partial \mathbf{u}}(\mathbf{u}) J\mathbf{u} - \frac{\partial \mathbf{f}}{\partial \overline{\mathbf{u}}}(\mathbf{u}) \overline{J\mathbf{u}}. \end{aligned}$$

The strong dissipative condition (5) plays an important role to estimate $\mathrm{Im}\left\{ (\overline{J\mathbf{u}})^t \cdot J\mathbf{f}(\mathbf{u}) \right\}$. In fact, by (5) with $\mathbf{p} = J\mathbf{u}$, the solution $\mathbf{u}$ satisfies

$$\begin{aligned} \frac{d\|J\mathbf{u}(t)\|_{L^2}^2}{dt} &= \int \mathrm{Im}\left\{ \overline{J\mathbf{u}} \cdot \left(\frac{\partial \mathbf{f}}{\partial \mathbf{u}}(\mathbf{u}) J\mathbf{u} - \frac{\partial \mathbf{f}}{\partial \overline{\mathbf{u}}}(\mathbf{u}) \overline{J\mathbf{u}} \right) \right\} dx \\ &\le -\rho_1 \int \sum_{j=1}^{n} |u_j|^2 |J u_j|^2 \, dx \\ &\le 0. \end{aligned}$$

Therefore, for $t \ge 0$, we have (12). By the analogy in deriving (12), we have (13). □

To determine the L^2-decay rate of the solution, the next scale-invariant weighted estimate will be required.

Lemma 2.2 (Scale-Invariant Weighted Estimate) *Let* $\mathbf{u} \in L^2$ *and* $x\mathbf{u} \in L^2$*. Then there exists some positive contant* C *such that*

$$\|\mathbf{u}\|_{L^1} \leq C\|x\mathbf{u}\|_{L^2}^{1/2}\|\mathbf{u}\|_{L^2}^{1/2}. \tag{15}$$

Proof By Cauchy-Schwatz' inequality, it is easy to see that

$$\begin{aligned}\|\mathbf{u}\|_{L^1} &\leq C\|(1+|x|^2)^{1/2}\mathbf{u}\|_{L^2} \\ &= C\left(\|\mathbf{u}\|_{L^2}^2 + \|x\mathbf{u}\|_{L^2}^2\right)^{1/2}.\end{aligned} \tag{16}$$

We here take $\mathbf{u}_\eta(x) = \eta\mathbf{u}(\eta x)$ for $\eta > 0$. Note that $\|\mathbf{u}_\eta\|_{L^1} = \|\mathbf{u}\|_{L^1}$. Then (16) leads to

$$\|\mathbf{u}\|_{L^1} \leq C\left(\eta^{1/2}\|\mathbf{u}\|_{L^2}^2 + \eta^{-1/2}\|x\mathbf{u}\|_{L^2}^2\right)^{1/2}. \tag{17}$$

By taking $\eta = \|x\mathbf{u}\|_{L^2}^2/\|\mathbf{u}\|_{L^2}^2$, the right hand side of (17) is minimized, and we have

$$\|\mathbf{u}\|_{L^1} \leq C\left(2\|\mathbf{u}\|_{L^2}\|x\mathbf{u}\|_{L^2}\right)^{1/2}. \tag{18}$$

It completes the proof. □

By Lemma 2.1, we have a decay estimate of $\|\mathbf{u}(t,\cdot)\|_{L^2}$. The idea of proof is based on S. Masaki.

Proposition 2.3 *Under the assumption of Theorem 1.1, the solution* $\mathbf{u}(t,x)$ *satisfies*

$$\|\mathbf{u}(t)\|_{L^2} \leq C(\ln t)^{-1/3} \tag{19}$$

for $t \geq 2$.

Proof of Proposition 2.3 By Hölder's inequality, we have

$$\|\mathbf{u}(t)\|_{L^2} \leq \|\mathbf{u}(t)\|_{L^\infty}^{1/2}\|\mathbf{u}(t)\|_{L^1}^{1/2}. \tag{20}$$

Lemma 2.2 yields

$$\|\mathbf{u}(t)\|_{L^1} \leq C\|x\mathbf{u}(t)\|_{L^2}^{1/2}\|\mathbf{u}(t)\|_{L^2}^{1/2}.$$

Applying it to the right hand side of (20), we see that

$$\|\mathbf{u}(t)\|_{L^2} \le C\|\mathbf{u}(t)\|_{L^\infty}^{1/2}\|x\mathbf{u}(t)\|_{L^2}^{1/4}\|\mathbf{u}(t)\|_{L^2}^{1/4}.$$

It is simplified as

$$\|\mathbf{u}(t)\|_{L^2} \le C\|\mathbf{u}(t)\|_{L^\infty}^{2/3}\|x\mathbf{u}(t)\|_{L^2}^{1/3}. \tag{21}$$

We are going to consider the estimate of $\|x\mathbf{u}(t,\cdot)\|_{L^2}$ for a while. Since $x = J - it\partial_x$, we have

$$\|x\mathbf{u}(t)\|_{L^2} \le \|J\mathbf{u}(t)\|_{L^2} + t\|\partial_x\mathbf{u}(t)\|_{L^2}.$$

By Lemma 2.1, we see that

$$\begin{aligned}\|x\mathbf{u}(t)\|_{L^2} &\le \|x\mathbf{u}_0\|_{L^2} + t\|\partial_x\mathbf{u}_0\|_{L^2}\\ &\le Ct.\end{aligned} \tag{22}$$

Combining (22) and the assumption (10), (21) leads to

$$\begin{aligned}\|\mathbf{u}(t)\|_{L^2} &\le C(t^{-1/2}(\ln t)^{-1/2})^{2/3}\cdot t^{1/3}\\ &= C(\ln t)^{-1/3}.\end{aligned}$$

It completes the proof of Proposition 2.3. □

3 Proof of Theorem 1.1

Now we are in the position to prove Theorem 1.1. We apply the identity (14), i.e.,

$$\frac{d\|\mathbf{u}(t)\|_{L^2}^2}{dt} = \mathrm{Im}\int\left\{\overline{\mathbf{u}}^t\left(\frac{\partial\mathbf{f}(\mathbf{u})}{\partial\mathbf{u}}\mathbf{u} + \frac{\partial\mathbf{f}(\mathbf{u})}{\partial\overline{\mathbf{u}}}\overline{\mathbf{u}}\right)\right\}dx.$$

Since the integrand of the right hand side is the quartic polynomial of u_j and $\overline{u_j}$ $(j = 1, 2, \cdots, n)$, there exists some positive constant C such that

$$\begin{aligned}\frac{d\|\mathbf{u}(t)\|_{L^2}^2}{dt} &\ge -C\int|\mathbf{u}(t,x)|^4\,dx\\ &\ge -C\|\mathbf{u}(t)\|_{L^\infty}^2\|\mathbf{u}(t)\|_{L^2}^2.\end{aligned}$$

By the assumption (10), we see that, for a small $\varepsilon > 0$, there exists some sufficiently large $T > 0$ such that

$$\frac{d\|\mathbf{u}(t)\|_{L^2}^2}{dt} \geq -C\varepsilon t^{-1}(\ln t)^{-1}\|\mathbf{u}(t)\|_{L^2}^2 \tag{23}$$

holds for any $t \in [T, \infty)$. From (23), it follows that

$$\frac{d}{dt}\left((\ln t)^{C\varepsilon}\|\mathbf{u}(t)\|_{L^2}^2\right) \geq 0.$$

Integrating from T to t, we see that, for $t \in [T, \infty)$,

$$(\ln T)^{C\varepsilon}\|\mathbf{u}(T)\|_{L^2}^2 \leq (\ln t)^{C\varepsilon}\|\mathbf{u}(t)\|_{L^2}^2.$$

Apply Proposition 2.3 to the right hand side. Then we have

$$(\ln T)^{C\varepsilon}\|\mathbf{u}(T)\|_{L^2}^2 \leq (\ln t)^{C\varepsilon} \cdot (\ln t)^{-2/3}.$$

Since $\varepsilon > 0$ is sufficiently small, taking $t \to \infty$ implies that

$$\|\mathbf{u}(T)\|_{L^2}^2 = 0.$$

Since, unlike the heat equation, the nonlinear Schrödinger equation is solved forward and backward in t, the extinction of $\mathbf{u}$ at $t = T$ implies that the solution is extinct for all $t \in [0, \infty)$. Hence $\mathbf{u}(t, x) \equiv 0$. □

Acknowledgments The first author thanks the organizers of the 14th ISAAC Congress at University of São Paulo, Brazil for giving him an opportunity to have a talk. This work was partly supported by Osaka Central Advanced Mathematical Institute (MEXT Joint Usage/Research Center on Mathematics and Theoretical Physics JPMXP0619217849), Osaka Metropolitan University. The first author N.K. is supported by JSPS KAKENHI Grant Number 23K03168.

References

1. Cazenave, T., Han, Z.: Asymptotic behavior for a Schrödinger equation with nonlinear subcritical dissipation. Discrete Contin. Dyn. Syst. **40**, 4801–4819 (2020)
2. Cazenave, T., Naumkin, I.: Modified scattering for the critical nonlinear Schrödinger equation. J. Funct. Anal. **274**, 402–432 (2018)
3. Cazenave, T., Han, Z., Naumkin, I.: Asymptotic behavior for a dissipative nonlinear Schrödinger equation. Nonlinear Anal. **205**, 112243, 37pp. (2021)
4. Hayashi, N., Naumkin, P.I.: Asymptotics for large time of solutions to the nonlinear Schrödinger and Hartree equations. Amer. J. Math. **120**, 369–389 (1998)
5. Hayashi, N., Kaikina, E.I., Naumkin, P.I.: Large time behavior of solutions to the generalized derivative nonlinear Schrödinger equation. Discrete Contin. Dynam. Syst. **5**, 93–106 (1999)

6. Hayashi, N., Naumkin, P.I., Sunagawa, H.: On the Schrödinger equation with dissipative nonlinearities of derivative type. SIAM J. Math. Anal. **40**(1), 278–291 (2008)
7. Hayashi, N., Li, C., Naumkin, P.I.: Time decay for nonlinear dissipative Schrödinger equations in optical fields. Adv. Math. Phys. **2016**, 3702738 (2016)
8. Hoshino, G.: Asymptotic behavior for solutions to the dissipative nonlinear Schrödinger equations with the fractional Sobolev space. J. Math. Phys. **60**, 111504 (2019)
9. Hoshino, G.: Scattering for solutions of a dissipative nonlinear Schrödinger equation. J. Differ. Equ. **266**, 4997–5011 (2019)
10. Hoshino, G.: Dissipative nonlinear Schrödinger equations for large data in one space dimension. Commun. Pure Appl. Anal. **19**, 967–981 (2020)
11. Jin, G., Jin, Y., Li, C.: The initial value problem for nonlinear Schrödinger equations with a dissipative nonlinearity in one space dimension. J. Evol. Equ. **16**, 983–995 (2016)
12. Katayama, S., Sakoda, D.: Asymptotic behavior for a class of derivative nonlinear Schrödinger systems. Partial Differ. Equ. Appl. **1**(3), 12 (2020)
13. Kim, D.: A note on decay rates of solutions to a system of cubic nonlinear Schrödinger equations in one space dimension. Asympt. Anal. **98**, 79–90 (2016)
14. Kita, N., Shimomura, A.: Asymptotic behavior of solutions to Schrödinger equations with a subcritical dissipative nonlinearity. J. Differ. Equ. **242**, 192–210 (2007)
15. Kita, N., Shimomura, A.: Large time behavior of solutions to Schrödinger equations with a dissipative nonlinearity for arbitrarily large initial data. J. Math. Soc. Japan **61**, 39–64 (2009)
16. Kita, N., Masaki, S., Segata, J., Uriya, K.: Polynomial deceleration for a system of cubic nonlinear Schrödinger equations in one space dimension. Nonlinear Anal. TMA **230**, 113216, 22pp. (2023)
17. Kita, N., Nakamura, Y., Sagawa, Y.: Decay and asymptotic behavior of solutions to a system of cubic dissipative nonlinear Schrödinger equations (preprint)
18. Li, C., Sunagawa, H.: On Schrödinger systems with cubic dissipative nonlinearities of derivative type. Nonlinearity **29**, 1537–1563 (2016)
19. Li, C., Sunagawa, H.: Remarks on Derivative Nonlinear Schrödinger Systems with Multiple Masses. Advanced Studies in Pure Mathematics, vol. 81, pp. 173–195. Mathematical Society of Japan, Tokyo (2019)
20. Li, C., Nishii, Y., Sagawa, Y., Sunagawa, H.: Large time asymptotics for a cubic nonlinear Schrödinger system in one space dimension. Funkcialaj Ekvacioj **64**, 361–377 (2021)
21. Masaki, S.: Classification of a class of systems of cubic ordinary differential equations. J. Differ. Equ. **344**, 471–508 (2023)
22. Masaki, S., Segata, J., Uriya, K.: Asymptotic behavior in time of solution to system of cubic nonlinear Schrödinger equations in one space dimension. In: Proceedings of the Workshop Mathematical Physics and its Interactions Proceedings in Mathematics & Statistics Series by Springer. Nonlinear Anal. **230**, Paper No. 113216, 22 pp (2023)
23. Sakoda, D., Sunagawa, H.: Small data global existence for a class of quadratic derivative nonlinear Schrödinger systems in two space dimensions. J. Differ. Equ. **268**, 1722–1749 (2020)
24. Shimomura, A.: Asymptotic behavior of solutions to Schrödinger equations with dissipative nonlinearities. Commun. Partial Differ. Equ. **31**, 1407–1423 (2006)

Global Solutions of the Klein-Gordon Equation Under the Quartic Potential in the de Sitter Spacetime

Makoto Nakamura

Dedicated to Professor Takayoshi Ogawa's 60th birthday

Abstract Global solutions of the Klein-Gordon equation under the quartic potential are considered in the de Sitter spacetime. The existence of global solutions for small rough initial data is considered based on the mechanism of the spontaneous symmetry breaking for the small positive Hubble constant. Some effects of the spatial expansion on the problem on particles with purely imaginary mass are considered based on the results by the author (Nakamura, J. Math. Phys. **62**, 121509 (2021)).

1 Introduction

In this paper, we review some results by the author [9] with several remarks given in the talk at the 14th International ISAAC Congress, which describe characteristic effects of the spatial expansion on the semilinear Klein-Gordon equation. We consider global solutions of the Klein-Gordon equation under the quartic potential in the de Sitter spacetime. The de Sitter spacetime is the solution of the Einstein equations with the cosmological constant in the vacuum under the cosmological principle. For the spatial dimension $n \geq 1$, Greek letters $\alpha, \beta, \gamma, \dots$ run from 0 to n, and the Latin letters $j, k, \ell, \dots$ run from 1 to n. We use the Einstein rule for the sum of indices. The sum is taken for the same upper and lower repeated indices, for example, $T^\alpha{}_\alpha := \sum_{\alpha=0}^n T^\alpha{}_\alpha$. Let $H \in \mathbb{R}$ be the Hubble constant, $c > 0$ be the speed of the light. The de Sitter spacetime that we consider in this paper is the

M. Nakamura (✉)
Osaka University, Graduate School of Information Science and Technology, Osaka, Japan
e-mail: makoto.nakamura.ist@osaka-u.ac.jp

M. R. Ebert et al. (eds.), *New Tools in Mathematical Analysis and Applications*, Research Perspectives, https://doi.org/10.1007/978-3-031-77050-0_16

spacetime with the metric $\{g_{\alpha\beta}\}$ given by

$$-c^2(d\tau)^2 = g_{\alpha\beta}dx^\alpha dx^\beta := -c^2(dx^0)^2 + e^{2Hx^0}\sum_{j=1}^{n}(dx^j)^2, \tag{1}$$

where the spatial curvature is taken as 0, the variable τ denotes the proper time, $x^0 = t$ is the time-variable. The spacetime with (1) reduces to the Minkowski spacetime when $H = 0$.

For the imaginary number i with $i^2 = -1$, let $m_* \in \mathbb{R} \cup i\mathbb{R}$ denote the real mass when $m_* \in \mathbb{R}$, or the purely imaginary mass when $m_* \in i\mathbb{R}$. Let $\hbar > 0$ denote the reduced Planck constant. We consider the Cauchy problem of the semilinear Klein-Gordon equation given by

$$\left(\frac{1}{c^2}\partial_t^2\phi + \frac{nH}{c^2}\partial_t\phi - e^{-2Ht}\Delta\phi + \left(\frac{m_*c}{\hbar}\right)^2\phi + \lambda|\phi|^{p-1}\phi\right)(t,x) = 0 \tag{2}$$

for $(t,x) \in [0,T) \times \mathbb{R}^n$, which is obtained as the Euler-Lagrange equation for the Lagrangian density L of the Klein-Gordon field with a potential V given by

$$L(\phi) := -\frac{1}{2}\overline{\partial^\alpha\phi}\partial_\alpha\phi - V(\phi), \quad V(\phi) := \frac{1}{2}\left(\frac{m_*c}{\hbar}\right)^2|\phi|^2 + \frac{\lambda}{p+1}|\phi|^{p+1}, \tag{3}$$

where $0 < T \le \infty$, $1 < p < \infty$, $\lambda \in \mathbb{R}$, Δ denotes the Laplacian defined by $\Delta := \sum_{j=1}^n(\partial/\partial x^j)^2$, $\partial^\alpha := g^{\alpha\beta}\partial_\beta$, and $g^{\alpha\beta}$ denotes the inverse metric of $g_{\alpha\beta}$. By the change of ϕ to $u := e^{nHt/2}\phi$, the Eq. (2) is rewritten as

$$\frac{1}{c^2}\partial_t^2u - e^{-2Ht}\Delta u + \left\{\left(\frac{m_*c}{\hbar}\right)^2 - \left(\frac{nH}{2c}\right)^2\right\}u + \lambda e^{-n(p-1)Ht/2}|u|^{p-1}u = 0. \tag{4}$$

We note that the term $(m_*c/\hbar)^2 - (nH/2c)^2$ shows new mass, which becomes negative for large $|H|$ even for real mass m_*. Let us consider the real mass $m_* \in \mathbb{R}$ on the Cauchy problem of (2) for data $\phi_0(\cdot) := \phi(0,\cdot)$ and $\phi_1(\cdot) := \partial_t\phi(0,\cdot)$. Yagdjian [15] has shown small global solutions for (2), provided by that the norm of initial data $\|\phi_0\|_{H^s(\mathbb{R}^n)} + \|\phi_1\|_{H^s(\mathbb{R}^n)}$ is sufficiently small for $s > n/2 \ge 1$, where $H^s(\mathbb{R}^n)$ denotes the Sobolev space of order s. Galstian and Yagdjian [2], and Yagdjian [16] have extended this result to the case of the Riemann metric space for each time slices. In [5], the energy solutions for $\phi_0 \in H^1(\mathbb{R}^n)$ and $\phi_1 \in L^2(\mathbb{R}^n)$ have been shown, and they were extended to the case of general Friedmann-Lemaitre-Robertson-Walker spacetime in [1]. We refer to [13] for the numerical simulation for the semilinear Klein-Gordon equation, and [6–8] on the Cauchy problem for the semilinear Schrödinger equation and the semilinear Proca equations in the de Sitter spacetime.

On blowing-up solutions in finite time for the imaginary mass $m_* \in i\mathbb{R}$, we introduce the following result from [9]. The results for $H > 0$ with $m_* \neq 0$ is due to Yagdjian [14, Theorem 1.1] since his method is also applicable to the imaginary mass. We remark that Kato [4] has shown the blowing-up of $\int_{\mathbb{R}^n} u(t,x)dx$ in finite time for the excluded case $H = 0$ with $m_* = 0$.

Proposition 1 *(See [9, Proposition 1]). Let $n \geq 1$, $1 < p < \infty$, $m_* \in i\mathbb{R}$, $H \in \mathbb{R}$. Let $m_* \neq 0$ when $H = 0$. Let $u_0, u_1 \in L^1(\mathbb{R}^n)$ be the functions which have compact supports and satisfy $w_0 := \int_{\mathbb{R}^n} u_0(x)dx > 0$, $w_1 := \int_{\mathbb{R}^n} u_1(x)dx > 0$, and*

$$w_1 \geq c w_0 \left\{ -\left(\frac{m_* c}{\hbar}\right)^2 + \left(\frac{nH}{2c}\right)^2 \right\}^{1/2}. \tag{5}$$

Then the solution with permanently bounded support of the Cauchy problem

$$\begin{cases} \frac{1}{c^2}\partial_t^2 u - e^{-2Ht}\Delta u + \left\{ \left(\frac{m_* c}{\hbar}\right)^2 - \left(\frac{nH}{2c}\right)^2 \right\} u - e^{-n(p-1)Ht/2}|u|^p = 0, \\ u(0,\cdot) = u_0(\cdot), \quad \partial_t u(0,\cdot) = u_1(\cdot) \end{cases} \tag{6}$$

blows up in finite time in the space $L^1(\mathbb{R}^n)$. Namely, $\int_{\mathbb{R}^n} u(t,x)dx$ blows up as $t \to T$ for some positive number $0 < T < \infty$. Moreover, let ω_n denote the volume of the unit ball in $\mathbb{R}^n$, and let r_0 be any number by which the supports of u_0 and u_1 are in the ball of radius r_0 with center at the origin. If H, w_0 and w_1 satisfy $-c/r_0 < H < \infty$ and

$$w_1 \geq \left[(cw_0)^2 \left\{ -\left(\frac{m_* c}{\hbar}\right)^2 + \left(\frac{nH}{2c}\right)^2 \right\} + \frac{2c^2 w_0^{p+1}}{(p+1)(\omega_n r_0^n)^{p-1}} \right]^{1/2} \tag{7}$$

instead of (5)*, then T is estimated by $T = O(w_0^{-(p-1)/2})$ as w_0 tends to infinity. Namely, the blowing-up time T tends to zero as w_0 tends to infinity.*

We remark that if the rate of the spatial expansion, H, is sufficiently large, the conditions (5) and (7) become restrictive to hold, by which we expect the spatial expansion is preferable for global solutions. If r_0 tends to zero, then the condition $-c/r_0 < H < \infty$ becomes weak while the condition (7) becomes restrictive.

Since we have blowing-up solutions for the gauge variant semilinear term with the negative sign in (6), our interest moves to the gauge invariant case with the positive sign. We introduce global solutions for small data when $H > 0$ is small. Put $m_* = im$ in (3) with $m \in \mathbb{R}$, $p = 3$ and $\lambda > 0$. The potential V is rewritten as

$$V(\phi) = -\frac{1}{2}\left(\frac{mc}{\hbar}\right)^2 |\phi|^2 + \frac{\lambda}{4}|\phi|^4.$$

This potential is known as the double well potential or the Mexican hat potential, and it has the minimum when

$$|\phi| = \zeta := \frac{|m|c}{\sqrt{\lambda}\hbar}, \tag{8}$$

while $\phi = 0$ gives the local maximum when $m \neq 0$. Since the equation is invariant for rotational transform for spatial variables, we note that $\zeta \in \mathbb{C}$ with $|\zeta| = |m|c/\sqrt{\lambda}\hbar$ is also treated similarly. We adopt (8) for simplicity. It is expected that the solution around $\phi = 0$ is unstable, and it is stable around $|\phi| = \zeta$, which causes the spontaneous symmetry breaking from $\phi = 0$ to $|\phi| = \zeta$. In this paper, we characterize this breaking from the viewpoint of the Cauchy problem of (2) which is rewritten as

$$\frac{1}{c^2}\partial_t^2\phi + \frac{nH}{c^2}\partial_t\phi - e^{-2Ht}\Delta\phi - \left(\frac{mc}{\hbar}\right)^2\phi + \lambda|\phi|^2\phi = 0. \tag{9}$$

This Eq. (9) is transformed into

$$\frac{1}{c^2}\partial_t^2\phi + \frac{nH}{c^2}\partial_t\phi - e^{-2Ht}\Delta\phi + \lambda\left\{|\phi|^2\phi + \zeta(2\phi\operatorname{Re}\phi + |\phi|^2) + 2\zeta^2\operatorname{Re}\phi\right\} = 0$$

by the shift $\phi \to \phi + \zeta$, which is rewritten as

$$\frac{1}{c^2}\partial_t^2 u - e^{-2Ht}\Delta u - \left(\frac{nH}{2c}\right)^2 u + 2\left(\frac{mc}{\hbar}\right)^2\operatorname{Re} u + h(u) = 0 \tag{10}$$

by the change of the function $u = e^{nHt/2}\phi$, where we have put

$$h(u) := \lambda|u|^2 u e^{-nHt} + \lambda\zeta(2u\operatorname{Re} u + |u|^2)e^{-nHt/2}. \tag{11}$$

We consider the Cauchy problem of (10) for the initial data given by $u(0,\cdot) = u_0(\cdot)$ and $\partial_t u(0,\cdot) = u_1(\cdot)$, and we show that the problem is well-posed. We observe how the purely imaginary mass and the Hubble constant are involved in the solvability of the Cauchy problem.

To sate our results, we use the following notation. We denote the Lebesgue space by $L^q(I)$ for an interval $I \subset \mathbb{R}$ and $1 \le q \le \infty$ with the norm

$$\|u\|_{L^q(I)} := \begin{cases} \left\{\int_I |u(t)|^q dt\right\}^{1/q} & \text{if } 1 \le q < \infty, \\ \operatorname{ess.\,sup}_{t\in I}|u(t)| & \text{if } q = \infty. \end{cases}$$

We use the Sobolev space $H^\mu(\mathbb{R}^n)$, the homogeneous Sobolev space $\dot{H}^\mu(\mathbb{R}^n)$, and the homogeneous Besov space $\dot{B}^\mu_{r,s}(\mathbb{R}^n)$ of order $\mu \ge 0$ for $1 \le r, s \le \infty$. Put $\nabla := (\partial/\partial x^1, \ldots, \partial/\partial x^n)$.

We consider the case $H \geq 0$. For $\mu \geq 0$, $0 < T \leq \infty$ and $Q \geq 0$, define the function space X^μ given by

$$\|u\|_{\dot{X}^\mu(T)} := c^{-1}\|\partial_t u\|_{L^\infty((0,T),\dot{H}^\mu(\mathbb{R}^n))} + \|e^{-Ht}\nabla u\|_{L^\infty((0,T),\dot{H}^\mu(\mathbb{R}^n))}$$
$$+\sqrt{Q}\|u\|_{L^\infty((0,T),\dot{H}^\mu(\mathbb{R}^n))} + \sqrt{H}\|e^{-Ht}\nabla u\|_{L^2((0,T),\dot{H}^\mu(\mathbb{R}^n))},$$
$$X^\mu(T) := \{u;\ \|u\|_{\dot{X}^\nu} < \infty \ \text{ for } \nu = 0, \mu_0, \mu\},$$
$$\dot{D}^\mu := c^{-1}\|u_1\|_{\dot{H}^\mu(\mathbb{R}^n)} + \|\nabla u_0\|_{\dot{H}^\mu(\mathbb{R}^n)} + \sqrt{Q}\|u_0\|_{\dot{H}^\mu(\mathbb{R}^n)}.$$

We define the metric in $X^\mu(T)$ by $d(u, v) := \|u - v\|_{\dot{X}^0(T)}$ for $u, v \in X^\mu(T)$. We introduce the following results from [9] on the existence of local and global solutions, and the asymptotic behaviors of global solutions. The result (5) is originally due to Ginibre and Velo [3, Proposition 3.2] when $H = 0$ and $\mu_0 = \mu = 0$. We put it for the completeness including the case $\mu > 0$.

Theorem 1 (See [9, Theorem 3]) *Let $n \geq 1$, $m \in \mathbb{R}$, $m \neq 0$, $0 \leq H < 2\sqrt{2}|m|c^2/n\hbar$. Put*

$$Q := 2\left(\frac{mc}{\hbar}\right)^2 - \left(\frac{nH}{2c}\right)^2. \tag{12}$$

Let $\lambda > 0$, and let ζ be defined by (8). *Let μ_0 and μ satisfy*

$$\max\left\{0, \frac{n-3}{2}\right\} \leq \mu_0 < \frac{n}{2}, \quad \mu_0 \leq \mu < \infty. \tag{13}$$

Then the following results hold.

(i) For any real-valued functions $u_0 \in H^{\mu+1}(\mathbb{R}^n)$ and $u_1 \in H^\mu(\mathbb{R}^n)$, there exist $0 < T < \infty$ and a unique solution $u \in C([0, T), H^{\mu+1}(\mathbb{R}^n)) \cap C^1([0, T), H^\mu(\mathbb{R}^n)) \cap X^\mu(T)$ of (10).

(ii) The solution u obtained in (i) is continuously dependent on the data. Namely, $d(u, v) \to 0$ as $v_0 \to u_0$ in $H^{\mu_0+1}(\mathbb{R}^n)$ and $v_1 \to u_1$ in $H^{\mu_0}(\mathbb{R}^n)$ for the solution v obtained in (i) for the data $v_0 := v(0, \cdot)$ and $v_1 := \partial_t v(0, \cdot)$.

(iii) Assume $H > 0$, $\mu_0 > 0$, $n \geq 1$. Then the solution u obtained in (i) is the global solution if $\dot{D}^{\mu_0}$ is sufficiently small. Namely, $T = \infty$.

(iv) For the global solution u obtained in (iii), put

$$u_{+0} := u_0 + c^2 \int_0^\infty K_1(s)h(u)(s)ds,$$
$$u_{+1} := u_1 - c^2 \int_0^\infty K_0(s)h(u)(s)ds, \tag{14}$$
$$u_+ := K_0(t)u_{+0} + K_1(t)u_{+1},$$

where K_0 and K_1 are the operators defined by (23)*, below. Then $u_{+0} \in H^{\mu}(\mathbb{R}^n)$, $u_{+1} \in H^{\mu-1}(\mathbb{R}^n)$, and u has the asymptotic behaviors given by*

$$\|u(t) - u_+(t)\|_{H^{\mu-1}(\mathbb{R}^n)} \to 0, \quad \|\partial_t (u(t) - u_+(t))\|_{H^{\mu-1}(\mathbb{R}^n)} \to 0.$$

(v) *The solution u obtained in (i) is the global solution when $H = 0$ and $\mu_0 = 0$. Namely, $T = \infty$.*

The case $H \geq 2\sqrt{2}|m|c^2/n\hbar$ is open for global solutions in Theorem 1.

The purely imaginary mass is imaginarily considered literally. However, it appears naturally as curved mass when the space expands or contracts, typically in the case of the de Sitter spacetime due to the Hubble constant which is known to be small. From the results in Theorem 1, we observe that the purely imaginary mass under the quartic potential yields global solutions for small data by the spontaneous symmetry breaking of the potential when the Hubble constant is sufficiently small.

We remark that the line element $-c^2(d\tau)^2 = -c^2(dt)^2 + a^2(dx)^2$ for a particle yields $d\tau/dt = \sqrt{1 - |av/c|^2}$, and

$$E := mc^2 \frac{dt}{d\tau} = \frac{mc^2}{\sqrt{1 - \left|\frac{av}{c}\right|^2}} \doteqdot mc^2 + \frac{m|av|^2}{2} \tag{15}$$

by the Taylor expansion, where a denotes a general scale-function, and $v := dx/dt$ denotes the speed of the particle in local coordinates. Since the last term is the kinetic energy which is conserved without external forces, we regard E as a conserved quantity. When $E > 0$, and v moves from zero to infinity, m changes from a real number to a purely imaginary number at $v = c/a$ by (15). An example of particles with purely imaginary mass is called tachyon whose speed exceeds the speed of light, where the speed of light c is modified to c/a in the above argument.

We have considered the case of de Sitter spacetime in Theorem 1. The case of Friedmann-Lemaitre-Robertson-Walker spacetimes is a natural question to consider, which would be interesting since the curved mass is dependent on the time variable, and it vanishes in finite time in some of special spacetimes ([12] by the author and Yoshizumi for such spacetimes).

2 Estimates for Linear Solutions

To prove Theorem 1, we prepare several estimates for linear Klein-Gordon equations. Let $H \geq 0$, $\lambda \in \mathbb{R}$ and $\zeta \in \mathbb{R}$. Consider the Cauchy problem

$$\begin{cases} \frac{1}{c^2}\partial_t^2 u - e^{-2Ht}\Delta u - \left(\frac{nH}{2c}\right)^2 u + 2\lambda\zeta^2 \operatorname{Re} u + h = 0, \\ u(0,\cdot) = u_0(\cdot), \quad \partial_t u(0,\cdot) = u_1(\cdot) \end{cases} \tag{16}$$

for any given function h which decays rapidly at spatial infinity. Then the following three results hold from standard argument of energy estimates. When the space is expanding, i.e., $H > 0$, energy estimates are effective to give global solutions.

(i) $\partial_\alpha e^\alpha + e^{n+1} + e^{n+2} = 0$, where

$$e^0 := \frac{1}{2c^2}|\partial_t u|^2 - \frac{1}{2}\left(\frac{nH}{2c}\right)^2 |u|^2 + \lambda\zeta^2(\mathrm{Re}\, u)^2 + \frac{e^{-2Ht}}{2}|\nabla u|^2,$$
$$e^j := -e^{-2Ht}\,\mathrm{Re}\left(\overline{\partial_t u}\partial_j u\right),$$
$$e^{n+1} := He^{-2Ht}|\nabla u|^2, \quad e^{n+2} := \mathrm{Re}\left(\overline{\partial_t u} h\right).$$

(ii) If u is real-valued and $Q := 2\lambda\zeta^2 - (nH/2c)^2 \geq 0$, then the following estimate holds;

$$\frac{1}{c}\|\partial_t u\|_{L^\infty((0,\infty),L^2(\mathbb{R}^n))} + \sqrt{Q}\|u\|_{L^\infty((0,\infty),L^2(\mathbb{R}^n))} + \|e^{-Ht}\nabla u\|_{L^\infty((0,\infty),L^2(\mathbb{R}^n))}$$
$$+\sqrt{H}\|e^{-Ht}\nabla u\|_{L^2((0,\infty)\times\mathbb{R}^n)}$$
$$\lesssim \frac{1}{c}\|u_1\|_{L^2(\mathbb{R}^n)} + \sqrt{Q}\|u_0\|_{L^2(\mathbb{R}^n)} + \|\nabla u_0\|_{L^2(\mathbb{R}^n)} + c\|h\|_{L^1((0,\infty),L^2(\mathbb{R}^n))}.$$

(iii) If $h = h(u)$ is given by (11), then $\partial_0\tilde{e}^0 + \partial_j e^j + \tilde{e}^{n+1} = 0$ holds, where

$$\tilde{e}^0 := e^0 + \frac{\lambda}{4}|u|^4 e^{-nHt} + \lambda\zeta|u|^2\,\mathrm{Re}\, u\, e^{-nHt/2}$$
$$\tilde{e}^{n+1} := e^{n+1} + \frac{\lambda nH}{4}|u|^4 e^{-nHt} + \frac{\lambda\zeta nH}{2}|u|^2\,\mathrm{Re}\, u\, e^{-nHt/2},$$

and e^α is defined in (i) for $0 \leq \alpha \leq n+1$.

We note that $H > 0$ is preferable for global solutions, and it is characterized as the dissipative term $\sqrt{H}\|e^{-Ht}\nabla u\|_{L^2((0,\infty)\times\mathbb{R}^n)}$ in (ii). We use it effectively to show global solutions.

We need to express the solution as a fixed point of some operator by the contraction argument. To do it, we recall the following fundamental results for ordinary differential equations (see, e.g., [10]). Put $D_t := d/dt$. For $H \in \mathbb{R}$ and $Q > 0$, put

$$\widetilde{a} = \widetilde{a}(t,\xi) := \frac{c^2}{e^{2Ht}} \cdot \sum_{j=1}^{n}(\xi^j)^2 + c^2 Q \tag{17}$$

for $\xi = (\xi^1, \ldots, \xi^n)$. Let ρ_0 and ρ_1 be the solutions of the Cauchy problem

$$\begin{cases} \left(D_t^2 + \widetilde{a}(t)\right) \rho_j(t) = 0 \ \text{ for } \ t \in [0, T), \\ \rho_j(0) = \delta_{0j}, \ \ D_t\rho_j(0) = \delta_{1j} \end{cases} \tag{18}$$

for $j = 0, 1$, where $\delta_{00} = \delta_{11} = 1$ and $\delta_{01} = \delta_{10} = 0$. If $D_t\widetilde{a} \le 0$, then we have

$$|\rho_0(\cdot)| \le \sqrt{\frac{\widetilde{a}(0)}{\widetilde{a}(\cdot)}}, \ \ |D_t\rho_0(\cdot)| \le \sqrt{\widetilde{a}(0)}, \ \ |\rho_1(\cdot)| \le \frac{1}{\sqrt{\widetilde{a}(\cdot)}}, \ \ |D_t\rho_1(\cdot)| \le 1. \tag{19}$$

If $D_t\widetilde{a} \ge 0$, then we have

$$|\rho_0(\cdot)| \le 1, \ \ |D_t\rho_0(\cdot)| \le \sqrt{\widetilde{a}(\cdot)}, \ \ |\rho_1(\cdot)| \le \frac{1}{\sqrt{\widetilde{a}(0)}}, \ \ |D_t\rho_1(\cdot)| \le \sqrt{\frac{\widetilde{a}(\cdot)}{\widetilde{a}(0)}}. \tag{20}$$

Let $b \in C([0, T))$, and let ρ be the solution of the equation

$$(D_t^2 + \widetilde{a}(t))\rho(t) = b(t) \tag{21}$$

for $t \in [0, T)$. Then ρ is rewritten as

$$\rho(t) = \rho_0(t)\rho(0) + \rho_1(t)D_t\rho(0) + \int_0^t \rho_2(t, s)b(s)ds, \tag{22}$$

where we have put $\rho_2(t, s) := -\rho_0(t)\rho_1(s) + \rho_1(t)\rho_0(s)$.

Put

$$\begin{aligned} &K_0(t) := F^{-1}\rho_0(t, \cdot)F, \ \ K_1(t) := F^{-1}\rho_1(t, \cdot)F, \\ &K(t, s) := c^2\left\{-K_0(t)K_1(s) + K_1(t)K_0(s)\right\} \end{aligned} \tag{23}$$

for $t, s \in \mathbb{R}$, where F and F^{-1} denote the Fourier transform and its inverse for $(x^1, \ldots, x^n)$. Then the Cauchy problem

$$\frac{1}{c^2}\partial_t^2 u - e^{-2Ht}\Delta u + Qu + h = 0, \ \ u(0) = u_0, \ \ \partial_t u(0) = u_1 \tag{24}$$

for a given function h on $\mathbb{R}^{1+n}$ can be regarded as the solution of the integral equation

$$u(t) = K_0(t)u_0 + K_1(t)u_1 - \int_0^t K(t, s)h(s)ds \tag{25}$$

by (22).

We obtain the following results by the estimates (19), (20) and the Plancherel theorem for the Fourier transform. We refer to [8, Lemma 4.4]. Let $H \geq 0$, $\mu \in \mathbb{R}$, and let $h \in L^1((0,\infty), H^\mu(\mathbb{R}^n))$. Let K_0 and K_1 be the operators defined by (23). Put

$$u_{+0} := u_0 + c^2 \int_0^\infty K_1(s)h(s)ds,$$

$$u_{+1} := u_1 - c^2 \int_0^\infty K_0(s)h(s)ds,$$

$$u_+(t) := K_0(t)u_{+0} + K_1(t)u_{+1}$$

for $t \geq 0$. Let u be the solution of (25). Then the following estimates hold.

$$(i)\ \ \|u_{+0}\|_{H^\mu(\mathbb{R}^n)} \lesssim \|u_0\|_{H^\mu(\mathbb{R}^n)} + c \int_0^\infty \|h(s)\|_{H^\mu(\mathbb{R}^n)} ds.$$

$$(ii)\ \ \|u_{+1}\|_{H^{\mu-1}(\mathbb{R}^n)} \lesssim \|u_1\|_{H^{\mu-1}(\mathbb{R}^n)} + c^2 \int_0^\infty \|h(s)\|_{H^\mu(\mathbb{R}^n)} ds.$$

$$(iii)\ \ \|u_+\|_{H^{\mu-1}(\mathbb{R}^n)} \lesssim \|u_0\|_{H^\mu(\mathbb{R}^n)} + \frac{1}{c}\|u_1\|_{H^{\mu-1}(\mathbb{R}^n)} + c \int_0^\infty \|h(s)\|_{H^\mu(\mathbb{R}^n)} ds.$$

$$(iv)\ \ \|\partial_t u_+\|_{H^{\mu-1}(\mathbb{R}^n)} \lesssim c\|u_0\|_{H^\mu(\mathbb{R}^n)} + \|u_1\|_{H^{\mu-1}(\mathbb{R}^n)} + c^2 \int_0^\infty \|h(s)\|_{H^\mu(\mathbb{R}^n)} ds.$$

$$(v)\ \ \|u(t) - u_+(t)\|_{H^{\mu-1}(\mathbb{R}^n)} \lesssim \frac{1}{c} \int_t^\infty \|h(s)\|_{H^\mu(\mathbb{R}^n)} ds.$$

$$(vi)\ \ \|\partial_t (u(t) - u_+(t))\|_{H^{\mu-1}(\mathbb{R}^n)} \lesssim c^2 \int_t^\infty \|h(s)\|_{H^\mu(\mathbb{R}^n)} ds.$$

We remark that there appears the derivative loss in the left hand sides in the above estimates since the solution u will be obtained in $H^{\mu+1}(\mathbb{R}^n)$ for each $t \geq 0$ in the proof of Theorem 1.

3 Outline of the Proof of Theorem 1

We only give the outline of proofs of (i) and (iii) in Theorem 1 in this section, in which we show the mechanism for the existence of global solutions by the symmetry breaking. The proofs of the other results can be found in [9]. Let ρ_0, ρ_1, K_0, K_1 and K be the functions and the operators defined by (18) and (23) for the function $\tilde{a}$

in (17). Then the solution of (10) is regarded as the fixed point of the operator Φ defined by

$$\Phi(u)(t) := K_0(t)u_0 + K_1(t)u_1 - \int_0^t K(t,s)h(u)(s)ds, \tag{26}$$

where $h(u)$ is defined by (11). For constants $T > 0$, $R_\nu > 0$ for $\nu = 0, \mu_0, \mu$, we define the closed ball by

$$X^\mu(T, R_0, R_{\mu_0}, R_\mu) := \left\{u;\ \|u\|_{\dot{X}^\nu} \le R_\nu \ \text{ for }\ \nu = 0, \mu_0, \mu\right\}, \tag{27}$$

and we show Φ is a contraction mapping on this space for the suitable constants. The solution is obtained as the fixed point of Φ. Let μ_0 and μ satisfy (13). Let $q = \infty$ when $H = 0$, and let $0 \le 1/q \le \min\{1/2, 1/(n-2\mu_0)\}$ when $H > 0$. We define θ, r_*, r_{**}, q_* by

$$\theta := \frac{n-2\mu_0}{3}, \quad \frac{1}{r_*} := \frac{1}{6} - \frac{\mu_0}{3n}, \quad \frac{1}{r_{**}} := \frac{1}{6} + \frac{2\mu_0}{3n}, \quad \frac{1}{q_*} := 1 - \frac{n-2\mu_0}{q}. \tag{28}$$

We note

$$0 \le \theta \le 1, \quad \frac{1}{2} = \frac{2}{r_*} + \frac{1}{r_{**}}, \quad 0 < \frac{1}{r_*} \le \frac{1}{2}, \quad 0 < \frac{1}{r_{**}} \le \frac{1}{2} \tag{29}$$

hold by $(n-3)/2 \le \mu_0 < n/2$. We have

$$\||u|^2u\|_{\dot{H}^\nu} \lesssim \|u\|^2_{L^{r_*}\cap \dot{B}^0_{r_*,2}} \|u\|_{\dot{B}^\nu_{r_{**},2}} \tag{30}$$

for $\nu \ge 0$ by the nonlinear estimate in the Besov spaces (see [11, Lemma 3.1]) and (29). Since we have the embeddings $\dot{H}^{\mu_0+\theta} \hookrightarrow \dot{B}^{\mu_0}_{r_{**},2} \hookrightarrow L^{r_*} \cap \dot{B}^0_{r_*,2}$ and $\dot{H}^{\nu+\theta} \hookrightarrow \dot{B}^\nu_{r_{**},2}$ by $1/r_* = 1/r_{**} - \mu_0/n$ and $1/r_{**} = 1/2 - \theta/n$, we have

$$\begin{aligned}\||u|^2u\|_{\dot{H}^\nu} &\lesssim \|u\|^2_{\dot{H}^{\mu_0+\theta}} \|u\|_{\dot{H}^{\nu+\theta}} \\ &\lesssim \|u\|^{2(1-\theta)}_{\dot{H}^{\mu_0}} \|u\|^{2\theta}_{\dot{H}^{\mu_0+1}} \|u\|^{1-\theta}_{\dot{H}^\nu} \|u\|^\theta_{\dot{H}^{\nu+1}},\end{aligned} \tag{31}$$

where we have used the interpolation inequalities at the last line. Thus, we obtain

$$\begin{aligned}\||u|^2ue^{-nHt}\|_{L^1\dot{H}^\nu} &\lesssim \|A\|_{q_*} \|u\|^{2(1-\theta)}_{L^\infty \dot{H}^{\mu_0}} \|e^{-Ht}u\|^{2\theta}_{L^q\dot{H}^{\mu_0+1}} \|u\|^{1-\theta}_{L^\infty\dot{H}^\nu} \|e^{-Ht}u\|^\theta_{L^q\dot{H}^{\nu+1}} \\ &\lesssim \|A\|_{q_*} B\|u\|^2_{\dot{X}^{\mu_0}} \|u\|_{\dot{X}^\nu}\end{aligned} \tag{32}$$

for $\nu = 0, \mu_0, \mu$ by the Hölder inequality and $1 = 1/q_* + 3\theta/q$, where we have put

$$A = A(t) := e^{-nHt+3\theta Ht} \ \text{ and } \ B := Q^{-3(1-\theta)/2}H^{-3\theta/q}. \tag{33}$$

We note

$$\|A\|_{L^{q_*}((0,T))} = \begin{cases} T^{1/q_*} & \text{if } 1 \le q_* \le \infty, \ \ H\mu_0 = 0, \\ 1 & \text{if } q_* = \infty, \ \ H\mu_0 > 0, \\ \left\{ \frac{1-e^{-2\mu_0 H T q_*}}{2\mu_0 H q_*} \right\}^{1/q_*} & \text{if } q_* < \infty, \ \ H\mu_0 > 0 \end{cases} \tag{34}$$

by a direct calculation.

Similarly to the above estimate, let $\tilde{q} = \infty$ when $H = 0$, and let $0 \le 1/\tilde{q} \le \min\{1/2, 2/(n-2\mu_0)\}$ when $H > 0$. We define $\tilde{\theta}, \tilde{r}_*, \tilde{r}_{**}, \tilde{q}_*$ by

$$\tilde{\theta} := \frac{n-2\mu_0}{4}, \ \ \frac{1}{\tilde{r}_*} := \frac{1}{4} - \frac{\mu_0}{2n}, \ \ \frac{1}{\tilde{r}_{**}} := \frac{1}{4} + \frac{\mu_0}{2n}, \ \ \frac{1}{\tilde{q}_*} := 1 - \frac{n-2\mu_0}{2\tilde{q}}. \tag{35}$$

We note

$$0 \le \tilde{\theta} \le 1, \ \ \frac{1}{2} = \frac{1}{\tilde{r}_*} + \frac{1}{\tilde{r}_{**}}, \ \ 0 < \frac{1}{\tilde{r}_*} \le \frac{1}{2}, \ \ 0 < \frac{1}{\tilde{r}_{**}} \le \frac{1}{2} \tag{36}$$

hold by $(n-4)/2 \le \mu_0 < n/2$. We have

$$\|2u\,\mathrm{Re}\,u + |u|^2\|_{\dot{H}^\nu} \lesssim \|u\|_{L^{\tilde{r}_*} \cap \dot{B}^0_{\tilde{r}_*,2}} \|u\|_{\dot{B}^\nu_{\tilde{r}_{**},2}} \tag{37}$$

by the nonlinear estimate and (36). Since we have the embeddings $\dot{H}^{\mu_0+\tilde{\theta}} \hookrightarrow \dot{B}^{\mu_0}_{\tilde{r}_{**},2} \hookrightarrow L^{\tilde{r}_*} \cap \dot{B}^0_{\tilde{r}_*,2}$ and $\dot{H}^{\nu+\tilde{\theta}} \hookrightarrow \dot{B}^\nu_{\tilde{r}_{**},2}$ by $1/\tilde{r}_* = 1/\tilde{r}_{**} - \mu_0/n$ and $1/\tilde{r}_{**} = 1/2 - \tilde{\theta}/n$, we have

$$\begin{aligned} \|2u\,\mathrm{Re}\,u + |u|^2\|_{\dot{H}^\nu} &\lesssim \|u\|_{\dot{H}^{\mu_0+\tilde{\theta}}} \|u\|_{\dot{H}^{\nu+\tilde{\theta}}} \\ &\lesssim \|u\|_{\dot{H}^{\mu_0}}^{1-\tilde{\theta}} \|u\|_{\dot{H}^{\mu_0+1}}^{\tilde{\theta}} \|u\|_{\dot{H}^\nu}^{1-\tilde{\theta}} \|u\|_{\dot{H}^{\nu+1}}^{\tilde{\theta}}, \end{aligned} \tag{38}$$

where we have used the interpolation inequalities at the last line. Thus, we obtain

$$\begin{aligned} &\|(2u\,\mathrm{Re}\,u + |u|^2)e^{-nHt/2}\|_{L^1\dot{H}^\nu} \\ &\lesssim \|\tilde{A}\|_{\tilde{q}_*} \|u\|_{L^\infty\dot{H}^{\mu_0}}^{1-\tilde{\theta}} \|e^{-Ht}u\|_{L^{\tilde{q}}\dot{H}^{\mu_0+1}}^{\tilde{\theta}} \|u\|_{L^\infty\dot{H}^\nu}^{1-\tilde{\theta}} \|e^{-Ht}u\|_{L^{\tilde{q}}\dot{H}^{\nu+1}}^{\tilde{\theta}} \\ &\lesssim \|\tilde{A}\|_{\tilde{q}_*} \tilde{B} \|u\|_{\dot{X}^{\mu_0}} \|u\|_{\dot{X}^\nu} \end{aligned} \tag{39}$$

for $\mu = 0, \mu_0, \mu$ by the Hölder inequality and $1 = 1/\tilde{q}_* + 2\tilde{\theta}/\tilde{q}$, where we have put $\tilde{A} = \tilde{A}(t) := e^{-nHt/2+2\tilde{\theta}Ht}$ and $\tilde{B} = Q^{-(1-\tilde{\theta})}H^{-2\tilde{\theta}/\tilde{q}}$. We note

$$\|\tilde{A}\|_{L^{\tilde{q}_*}((0,T))} = \begin{cases} T^{1/\tilde{q}_*} & \text{if } 1 \le \tilde{q}_* \le \infty, \quad H\mu_0 = 0, \\ 1 & \text{if } \tilde{q}_* = \infty, \quad H\mu_0 > 0, \\ \left\{ \frac{1-e^{-\mu_0 HT\tilde{q}_*}}{\mu_0 H \tilde{q}_*} \right\}^{1/\tilde{q}_*} & \text{if } \tilde{q}_* < \infty, \quad H\mu_0 > 0 \end{cases} \tag{40}$$

by a direct calculation.

By (32) and (39), we have

$$\|h(u)\|_{L^1\dot{H}^\nu} \lesssim \lambda \|A\|_{q_*} B R_{\mu_0}^2 R_\nu + \lambda r_* \|\tilde{A}\|_{\tilde{q}_*} \tilde{B} R_{\mu_0} R_\nu \tag{41}$$

for $\nu = 0, \mu_0, \mu$, and any $u \in X^\mu(T, R_0, R_{\mu_0}, R_\mu)$. Since we have

$$\|\Phi(u)\|_{\dot{X}^\nu} \lesssim \frac{1}{c}\|u_1\|_{\dot{H}^\nu} + \|\nabla u_0\|_{\dot{H}^\nu} + \sqrt{Q}\|u_0\|_{\dot{H}^\nu} + c\|h(u)\|_{L^1\dot{H}^\nu}$$

by the linear estimates, we have

$$\|\Phi(u)\|_{\dot{X}^\nu} \le C_0 \dot{D}^\nu + Cc\lambda \left(\|A\|_{q_*} B R_{\mu_0} + \zeta \|\tilde{A}\|_{\tilde{q}_*} \tilde{B} \right) R_{\mu_0} R_\nu \le R_\nu \tag{42}$$

for $\nu = 0, \mu_0, \mu$ for some constants $C_0 > 0$, $C > 0$, and any $u \in X^\mu(T, R_0, R_{\mu_0}, R_\mu)$ by (41) if R_0, R_{μ_0} and R_μ satisfy

$$R_\nu \ge 2C_0 \dot{D}^\nu, \quad Cc\lambda \left(\|A\|_{q_*} B R_{\mu_0} + \zeta \|\tilde{A}\|_{\tilde{q}_*} \tilde{B} \right) R_{\mu_0} \le \frac{1}{2} \tag{43}$$

for $\nu = 0, \mu_0, \mu$. We also have

$$\begin{aligned} &\|h(u) - h(v)\|_{L^1L^2} \\ &\lesssim \lambda \|A\|_{q_*} B \max_{w=u,v} \|w\|_{\dot{X}^{\mu_0}}^2 \|u - v\|_{\dot{X}^0} + \lambda\zeta \|\tilde{A}\|_{\tilde{q}_*} \tilde{B} \max_{w=u,v} \|w\|_{\dot{X}^{\mu_0}} \|u - v\|_{\dot{X}^0} \\ &\lesssim \lambda \|A\|_{q_*} B R_{\mu_0}^2 d(u, v) + \lambda\zeta \|\tilde{A}\|_{\tilde{q}_*} \tilde{B} R_{\mu_0} d(u, v) \end{aligned}$$

for any $u, v \in X^\mu(T, R_0, R_{\mu_0}, R_\mu)$ similarly to (41). Therefore, we obtain

$$d(\Phi(u), \Phi(v)) \lesssim c\|h(u) - h(v)\|_{L^1L^2}$$

and

$$d(\Phi(u), \Phi(v)) \leq C c \lambda \left(\|A\|_{q_*} B R_{\mu_0} + \zeta \|\tilde{A}\|_{\tilde{q}_*} \tilde{B} \right) R_{\mu_0} d(u, v) \leq \frac{1}{2} d(u, v) \tag{44}$$

by the linear estimates similarly to (42) under the second condition in (43).

We take $q = \tilde{q} = \infty$, thus, $q_* = \tilde{q}_* = 1$, and $R_\nu = 2C_0 \dot{D}^\nu$ for $\nu = 0, \mu_0, \mu$. We note that $\|A\|_{q_*}$ and $\|\tilde{A}\|_{\tilde{q}_*}$ tend to zero as T tends to zero by (34) and (40). Then the second condition in (43) is satisfied for sufficiently small $T > 0$, and T depends on the size of $\dot{D}^{\mu_0}$. So that, Φ is a contraction mapping, and we obtain the local in time solutions in (i) in Theorem 1.

We prove the result in (iii) in Theorem 1. We take $q_* < \infty$, $\tilde{q}_* < \infty$. Then the second condition in (43) is satisfied by

$$C c \lambda \left((2\mu_0 H q_*)^{-1/q_*} B R_{\mu_0} + \zeta (\mu_0 H \tilde{q}_*)^{-1/\tilde{q}_*} \tilde{B} \right) R_{\mu_0} \leq \frac{1}{2}. \tag{45}$$

Since (45) holds for $T = \infty$ and sufficiently small $\dot{D}^{\mu_0} > 0$, we obtain the global solutions.

We remark that the semilinear terms from the quartic potential are well-controlled by the dissipative term in energy estimates in the case of $H > 0$. We obtain global solutions for any data if H is sufficiently large since the condition (45) holds for large H.

Acknowledgments The author is thankful to two anonymous referees for several comments to revise the paper. This work was supported by JSPS KAKENHI Grant Numbers 22K18671, 23K03183, 24K06855.

References

1. Galstian, A., Yagdjian, K.: Global solutions for semilinear Klein-Gordon equations in FLRW spacetimes. Nonlinear Anal. Theory Methods Appl. **113**, 339–356 (2015)
2. Galstian, A., Yagdjian, K.: Global in time existence of self-interacting scalar field in de Sitter spacetimes. Nonlinear Anal. Real World Appl. **34**, 110–139 (2017)
3. Ginibre, J., Velo, G.: The global Cauchy problem for the nonlinear Klein-Gordon equation. II. Ann. Inst. H. Poincaré Anal. Non Linéaire **6**(1), 15–35 (1989)
4. Kato, T.: Blow-up of solutions of some nonlinear hyperbolic equations. Commun. Pure Appl. Math. **33**(4), 501–505 (1980)
5. Nakamura, M.: The Cauchy problem for semi-linear Klein-Gordon equations in de Sitter spacetime. J. Math. Anal. Appl. **410**(1), 445–454 (2014)
6. Nakamura, M.: On nonlinear Schrödinger equations derived from the nonrelativistic limit of nonlinear Klein-Gordon equations in de Sitter spacetime. J. Differ. Equ. **259**, 3366–3388 (2015)
7. Nakamura, M.: On the nonrelativistic limit of a semilinear field equation in a uniform and isotropic space. Kyoto J. Math. **60**(4), 1333–1359 (2020)
8. Nakamura, M.: On the Cauchy problem for the semilinear Proca equations in the de Sitter spacetime. J. Differ. Equ. **270**, 1218–1257 (2021)

9. Nakamura, M.: The Cauchy problem for the Klein-Gordon equation under the quartic interaction in the de Sitter spacetime. J. Math. Phys. **62**, 121509 (2021)
10. Nakamura, M.: The Cauchy problem for a semilinear ordinary differential equation in homogeneous and isotropic spaces. Tsukuba J. Math. **47**(2), 153–189 (2023)
11. Nakamura, M., Ozawa, T.: Small data scattering for nonlinear Schrödinger, wave and Klein-Gordon equations. Ann. della Scuola Normale Superiore di Pisa. Classe di Scienze. Serie V **1**, 435–460 (2002)
12. Nakamura, M., Yoshizumi, T.: The Cauchy problem for semi-linear Klein-Gordon equations in Friedmann-Lemaitre-Robertson-Walker spacetimes (preprint; arXiv:2411.02872)
13. Tsuchiya, T., Nakamura, M.: On the numerical analysis of the Cauchy problem for semi-linear Klein-Gordon equations in the de Sitter spacetime. J. Comput. Appl. Math. **361**, 396–412 (2019)
14. Yagdjian, K.: The semilinear Klein-Gordon equation in de Sitter spacetime. Discrete Contin. Dyn. Syst. Ser. S **2**(3), 679–696 (2009)
15. Yagdjian, K.: Global existence of the scalar field in de Sitter spacetime. J. Math. Anal. Appl. **396**(1), 323–344 (2012)
16. Yagdjian, K.: Global existence of the self-interacting scalar field in the de Sitter universe. J. Math. Phys. **60**(5), 051503, 29pp. (2019)

About Critical Exponents in Semi-Linear de Sitter Models

M. R. Ebert, J. Marques, and M. Reissig

Dedicated to Prof. Karen Yagdjian (Edinburg, Texas) on the occasion of his 70th birthday

Abstract In this paper we consider the Cauchy problem for semi-linear de Sitter models. The model of interest is

$$\phi_{tt} - e^{-2t}\Delta\phi + n\phi_t + m^2\phi = F_p(\phi), \quad (\phi(0,x), \phi_t(0,x)) = (f(x), g(x)),$$

where n is the space dimension, m^2 is a non-negative constant and p is a positive parameter. Our main goal is to verify that, in general, one cannot observe a critical exponent $p_{crit} = p_{crit}(n)$ in the family of non-linearities $\{F_p(\phi)\}_{p>1} = \{|\phi|^p\}_{p>1}$.

MSC 2020: 35L05, 35B71

M. R. Ebert
Department of Computer Science and Mathematics, University of São Paulo, Ribeirão Preto, Brazil
e-mail: ebert@ffclrp.usp.br

J. Marques
Centre for Business and Economics Research (CeBER), Faculty of Economics, University of Coimbra, Coimbra, Portugal
e-mail: jmarques@fe.uc.pt

M. Reissig (✉)
Faculty for Mathematics and Computer Science, Technical University Bergakademie Freiberg, Freiberg, Germany
e-mail: michael.reissig@extern.tu-freiberg.de

M. R. Ebert et al. (eds.), *New Tools in Mathematical Analysis and Applications*, Research Perspectives, https://doi.org/10.1007/978-3-031-77050-0_17

1 Introduction

In this paper we prove the global (in time) existence of small data Sobolev solutions for the Cauchy problem

$$\begin{cases} \phi_{tt} - e^{-2t}\Delta\phi + n\phi_t + m^2\phi = F_p(\phi), & (t,x) \in (0,\infty) \times \mathbb{R}^n, \\ (\phi(0,x), \phi_t(0,x)) = (f(x), g(x)), & x \in \mathbb{R}^n. \end{cases} \tag{1}$$

Here n is the space dimension, m^2 is a non-negative constant and $F_p(\phi) = |\phi|^p$, $p > 1$. Our main goal is to prove global (in time) existence of Sobolev solutions to the Cauchy problem (1) for some range of parameters $p \in (1, p_1(n))$.

In [12], under the assumption that the right-hand side F is Lipschitz continuous in the Sobolev space $H^s(\mathbb{R}^n)$, $s > \frac{n}{2}$, global existence (in time) of small data solutions to the model (1) is proved for $m \in (0, \frac{\sqrt{n^2-1}}{2}) \cup [\frac{n}{2}, \infty)$. Some generalization of these results are obtained in [5] including the case $m = \frac{\sqrt{n^2-1}}{2}$. In some cases, for instance, if the Cauchy problem has a vanishing first initial data, the range $(\frac{\sqrt{n^2-1}}{2}, \frac{n}{2})$ for m is allowed. In [11] the previous results were extended to the case of a complex $m \in \mathbb{C}$ that includes, in particular, the interval $(\frac{\sqrt{n^2-1}}{2}, \frac{n}{2})$ for m.

Recently, in [10] blow-up results have been proved for $m = 0$ in (1) , that is, it was shown that local (in time) classical solutions blow-up in finite time for all $p > 1$.

To study the model (1), in [3] the authors divided it into three groups according to the size of m: de Sitter models with dominant dissipation ($m \in (0, \frac{n}{2})$), de Sitter models with dominant mass ($m > \frac{n}{2}$) and de Sitter models with balanced mass and dissipation ($m = \frac{n}{2}$). It was proved global (in time) existence and uniqueness of energy solutions to the Cauchy problem (1) for admissible $p > \bar{p}(n,m) > 1$ in space dimension $n \geq 2$.

In the recent paper [4] the authors proved, under the assumption of compactly supported data, global (in time) existence (no uniqueness) of small data Sobolev solutions to (1) for $n = 1$, $m = \frac{1}{2}$ and for all $p > 1$.

In order to explain the classification introduced in [3] we apply the change of variable $\phi(t,x) = e^{r\,t}u(t,x)$ with some constant $r < 0$, to the Cauchy problem (1) to get (see [6, 7] in the case of constant speed of propagation)

$$\begin{cases} u_{tt} - e^{-2t}\Delta u + (2r+n)u_t + (r^2 + rn + m^2)u = e^{r(p-1)t}|u|^p, \\ \qquad (t,x) \in (0,\infty) \times \mathbb{R}^n, \\ u(0,x) = f(x) =: u_0(x), \quad u_t(0,x) = g(x) - rf(x) =: u_1(x),\ x \in \mathbb{R}^n. \end{cases} \tag{2}$$

Then we split our analysis into three cases.

1.1 De Sitter Model with Dominant Dissipation

If we choose in (2) the parameter

$$r := -\frac{n-\mu}{2}, \qquad \mu := \sqrt{n^2 - 4m^2},$$

then $r^2 + rn + m^2 = 0$ and we obtain the following model with dominant dissipation:

$$\begin{cases} u_{tt} - e^{-2t}\Delta u + \mu\, u_t = e^{r(p-1)t}|u|^p, \quad (t,x) \in (0,\infty) \times \mathbb{R}^n, \\ u(0,x) = f(x) =: u_0(x), \quad u_t(0,x) \\ \qquad = g(x) + \frac{n-\sqrt{n^2-4m^2}}{2} f(x) =: u_1(x), x \in \mathbb{R}^n. \end{cases} \tag{3}$$

We expect some improving influence of the dissipation term. For this reason we assume $m < \frac{n}{2}$. In the paper [2] the authors introduced some classification of damping terms for the Cauchy problem

$$v_{tt} - a(t)^2 \Delta v + b(t) v_t = 0, \qquad v(0,x) = v_0(x), \qquad v_t(0,x) = v_1(x).$$

Due to this classification it turns out that the dissipative term in (3) is non-effective if $\mu \in (0,1)$, that is, $m \in (\frac{\sqrt{n^2-1}}{2}, \frac{n}{2})$. In the case of non-effective damping term the treatment of semi-linear wave models with power non-linearity $|u|^p$ is a challenging problem. For some special models the blow-up part is established in [8] and [9]. Some special cases are treated in [1].

1.2 De Sitter Model with Dominant Mass

If we choose in (2) the parameter $r = -\frac{n}{2}$, then we obtain the following model with dominant mass:

$$\begin{cases} u_{tt} - e^{-2t}\Delta u + \left(m^2 - \frac{n^2}{4}\right) u = e^{-\frac{n}{2}(p-1)t}|u|^p, \quad (t,x) \in (0,\infty) \times \mathbb{R}^n, \\ u(0,x) = f(x) =: u_0(x), \quad u_t(0,x) = g(x) + \frac{n}{2} f(x) =: u_1(x), \quad x \in \mathbb{R}^n. \end{cases} \tag{4}$$

We hope to have some improving influence of the mass term, too. For this reason we assume $m > \frac{n}{2}$.

1.3 De Sitter Model with Balanced Dissipation and Mass

If we choose in (2) the parameters $r = -\frac{n}{2}$ and $m = \frac{n}{2}$, then we obtain the following model with a balance between mass and dissipation:

$$\begin{cases} u_{tt} - e^{-2t}\Delta u = e^{-\frac{n}{2}(p-1)t}|u|^p, & (t,x) \in (0,\infty) \times \mathbb{R}^n, \\ u(0,x) = f(x) =: u_0(x), \quad u_t(0,x) = g(x) + \frac{n}{2}f(x) =: u_1(x), & x \in \mathbb{R}^n. \end{cases} \tag{5}$$

In this paper, we use the following notation.

Notation 1 Let $f, g : [0,\infty) \to (0,\infty)$ be two positive functions. We use the notation $f \lesssim g$ if there exists a positive constant C such that $f(t) \leq Cg(t)$, for all $t \geq 0$. □

The present paper is organized as follows. In Sects. 2 to 4 we consider the cases $m \in (0, \frac{n}{2})$ (model (3)), $m > \frac{n}{2}$ (model (4)) and $m = \frac{n}{2}$ (model (5)), respectively. Conclusions in Sect. 5 complete the paper.

2 De Sitter Model with Dominant Dissipation: Case $m \in (0, \frac{n}{2})$

In this section we consider with $m \in (0, \frac{n}{2})$ the Cauchy problem (3), that is,

$$\begin{cases} u_{tt} - e^{-2t}\Delta u + \mu\, u_t = e^{r(p-1)t}|u|^p, & (t,x) \in (0,\infty) \times \mathbb{R}^n, \\ u(0,x) = u_0(x), \quad u_t(0,x) = u_1(x), & x \in \mathbb{R}^n, \end{cases} \tag{6}$$

where $\mu = \sqrt{n^2 - 4m^2} \in (0,n)$ and $r := -\frac{n-\mu}{2} < 0$. According to Duhamel's principle, we may interpret weak solutions of (6) as solutions of the non-linear integral equation (in this paper we understand solutions only in this sense)

$$u(t,x) = K_0(t,0,x) *_{(x)} u_0(x) + K_1(t,0,x) *_{(x)} u_1(x) + Gu(t,x),$$

where

$$Gu(t,x) := \int_0^t e^{r(p-1)s} K_1(t,s,x) *_{(x)} |u(s,x)|^p\, ds.$$

Here $K_j = K_j(t,s,x)$, $j = 0, 1$, $0 \le s \le t$, are the fundamental solutions to the corresponding linear parameter-dependent Cauchy problem with vanishing right-hand side

$$\begin{cases} K_{tt} - e^{-2t}\Delta K + \mu K_t = 0, & (t,x) \in (s,\infty) \times \mathbb{R}^n, \\ K(s,s,x) = \delta_{0j}, \quad K_t(s,s,x) = \delta_{1j}, & x \in \mathbb{R}^n, \end{cases}$$

with $\delta_{kj} = 1$ for $k = j$, and zero otherwise. So, Duhamel's principle explains that we have to take account of solutions to a family of parameter-dependent Cauchy problems.

2.1 Estimates of Solutions to Parameter-Dependent Linear Models

Let us consider for $\mu > 0$ the parameter-dependent Cauchy problem for the damped wave equation

$$\begin{cases} u_{tt} - e^{-2t}\Delta u + \mu u_t = 0, & (t,x) \in (s,\infty) \times \mathbb{R}^n, \\ u(s,x) = \varphi(s,x), \quad u_t(s,x) = \psi(s,x), & x \in \mathbb{R}^n. \end{cases} \tag{7}$$

Using the representation of Sobolev solutions given in Section 2.1 in [3], we are able to verify the following result.

Proposition 1 *Let us consider* (7) *with* $\mu \in (0, n)$. *If* $\varphi(0,\cdot) \in H^{\gamma+\left[\frac{1-\mu}{2}\right]_+}(\mathbb{R}^n)$ *with* $\gamma \ge 0$, *then for all* $t \in [0,\infty)$ *we have*

$$\left\| K_0(t,0,\cdot) *_{(x)} \varphi(0,\cdot) \right\|_{\dot{H}^\gamma} \lesssim \|\varphi(0,\cdot)\|_{H^{\gamma+\left[\frac{1-\mu}{2}\right]_+}}. \tag{8}$$

If $\psi(s,\cdot) \in L^2(\mathbb{R}^n)$, *then for all* $t \in [s,\infty)$ *and* $0 \le \gamma < \min\{1, \frac{1+\mu}{2}\}$ *it holds*

$$\left\| K_1(t,s,\cdot) *_{(x)} \psi(s,\cdot) \right\|_{\dot{H}^\gamma} \lesssim e^{\gamma s} \|\psi(s,\cdot)\|_{L^2}. \tag{9}$$

Proof Following ideas from Section 2.1 in [3], we split the extended phase space into three zones

$$Z_1(s) := \left\{\xi : |\xi| e^{-s} \le N\right\}, \quad Z_2(t,s) := \left\{Ne^s \le |\xi| \le Ne^t\right\},$$
$$Z_3(t) := \left\{|\xi| e^{-t} \ge N\right\}.$$

In $Z_1(s)$ we have

$$|\xi|^\gamma|\widehat{u}(t,\xi)| \lesssim |\xi|^\gamma|\widehat{\varphi}(s,\xi)|+|\xi|^{\gamma_1}e^{\gamma_2 s}|\hat{\psi}(s,\xi)| \text{ for } \gamma,\gamma_1,\gamma_2 \geq 0 \text{ and } \gamma=\gamma_1+\gamma_2.$$

In $Z_3(t)$ we have

$$|\xi|^\gamma|\widehat{u}(t,\xi)| \lesssim |\xi|^\gamma\Big(|\widehat{\varphi}(s,\xi)|(|\xi|e^{-s})^{\frac{1-\mu}{2}}+|\widehat{\psi}(s,\xi)|(|\xi|e^{-s})^{-\frac{1+\mu}{2}}\Big)(|\xi|e^{-t})^{\frac{\mu-1}{2}}.$$

If $\psi\equiv 0$ and $s=0$, then

$$|\xi|^\gamma|\widehat{u}(t,\xi)| \lesssim |\xi|^{\gamma+\left[\frac{1-\mu}{2}\right]_+}|\widehat{\varphi}(0,\xi)|.$$

If $\varphi\equiv 0$, taking into consideration that for $\gamma<\min\{1,\frac{1+\mu}{2}\}$ we have $|\xi|^{\gamma-1}\lesssim e^{(\gamma-1)t}$ and $e^{\left(\gamma-\frac{1+\mu}{2}\right)t}\lesssim e^{\left(\gamma-\frac{1+\mu}{2}\right)s}$ in $Z_3(t)$, it follows immediately

$$\begin{aligned}|\xi|^\gamma|\widehat{u}(t,\xi)| &\lesssim |\xi|^{\gamma-1}e^{\frac{1-\mu}{2}t}e^{\frac{1+\mu}{2}s}|\hat{\psi}(s,\xi)| \lesssim e^{\left(\gamma-\frac{1+\mu}{2}\right)t}e^{\frac{1+\mu}{2}s}|\hat{\psi}(s,\xi)|\\ &\lesssim e^{\gamma s}|\hat{\psi}(s,\xi)|.\end{aligned}$$

In $Z_2(t,s)$ we have

$$\begin{aligned}|\xi|^\gamma|\widehat{u}(t,\xi)| \lesssim |\xi|^\gamma&\Big(|\widehat{\varphi}(s,\xi)|(|\xi|e^{-s})^{\frac{1-\mu}{2}}+|\widehat{\psi}(s,\xi)|(|\xi|e^{-s})^{-\frac{1+\mu}{2}}\Big)\\ &\times\left(1+(|\xi|e^{-t})^\mu\right).\end{aligned}$$

If $\psi\equiv 0$ and $s=0$, then we have the same estimate as in $Z_3(t)$. If $\varphi\equiv 0$, then by using the definition of $Z_2(t,s)$ we may conclude that $1+(|\xi|e^{-t})^\mu$ is bounded and due to $|\xi|^{\gamma-\frac{1+\mu}{2}}\lesssim e^{\left(\gamma-\frac{1+\mu}{2}\right)s}$ we may verify for $\gamma<\frac{1+\mu}{2}$ again

$$|\xi|^\gamma|\widehat{u}(t,\xi)| \lesssim |\xi|^{\gamma-\frac{1+\mu}{2}}|\hat{\psi}(s,\xi)|e^{\frac{1+\mu}{2}s} \lesssim e^{\gamma s}|\hat{\psi}(s,\xi)|.$$

Summarizing, the desired estimate (8) and (9) are concluded by Plancherel's theorem. □

Corollary 1 *Consider the linear Cauchy problem* (1) *with vanishing right-hand side. Assume that* $f\in H^{\gamma+\left[\frac{1-\mu}{2}\right]_+}(\mathbb{R}^n)$ *and* $g\in L^2(\mathbb{R}^n)$ *with* $0\leq\gamma<\min\{1,\frac{1+\mu}{2}\}$. *Then, the solution* ϕ *satisfies the following a-priori estimate with the parameter* $\mu=\sqrt{n^2-4m^2}\in(0,n)$*:*

$$\|\phi(t,\cdot)\|_{\dot{H}^\gamma} \lesssim e^{-\frac{n-\mu}{2}t}\Big(\|f\|_{H^{\gamma+\left[\frac{1-\mu}{2}\right]_+}}+\|g\|_{L^2}\Big).$$

2.2 Global (in Time) Existence of Small Data Sobolev Solutions

In the next result let us choose $m \in \left(\frac{\sqrt{n^2-4\gamma^2}}{2}, \frac{n}{2}\right)$ in (1). Under this choice there exists $\gamma < \min\{1, \frac{1+\mu}{2}\}$, such that $\mu = \sqrt{n^2-4m^2} \in (0, 2\gamma)$:

Theorem 1 *Consider for $n \geq 1$ the Cauchy problem (1) with $m \in \left(\frac{\sqrt{n^2-4\gamma^2}}{2}, \frac{n}{2}\right)$ for some $\frac{\mu}{2} < \gamma < \min\{1, \frac{1+\mu}{2}\}$. Moreover, let $f \in H^{\gamma+\frac{[1-\mu]_+}{2}}(\mathbb{R}^n)$ and $g \in L^2(\mathbb{R}^n)$. If $p > p_{n,\mu,\gamma} := 1 + \frac{2\gamma}{n-\mu}$ and $p \leq \frac{n}{[n-2\gamma]_+}$, then there exists a constant $\varepsilon_0 > 0$ such that, for every small data satisfying*

$$\|f\|_{H^{\gamma+\frac{[1-\mu]_+}{2}}} + \|g\|_{L^2} \leq \varepsilon \text{ for } \varepsilon \leq \varepsilon_0,$$

there exists a uniquely determined global (in time) Sobolev solution

$$\phi \in C\big([0,\infty), H^\gamma(\mathbb{R}^n)\big).$$

Moreover, the solution satisfies the decay estimate

$$\|\phi(t,\cdot)\|_{H^\gamma} \lesssim e^{-\frac{(n-\mu)t}{2}}\big(\|f\|_{H^{\gamma+\frac{[1-\mu]_+}{2}}} + \|g\|_{L^2}\big).$$

Proof It is enough to prove the global (in time) existence of small data solutions to (6). We define the space

$$X(t) := \Big\{u \in C\big([0,t], H^\gamma(\mathbb{R}^n)\big) : \|u\|_{X(t)} := \sup_{\tau\in[0,t]} \{\|u(\tau,\cdot)\|_{H^\gamma}\} < \infty\Big\}$$

with the usual norm in $H^\gamma(\mathbb{R}^n)$. Using Proposition 1 for $s = 0$ we have

$$\|K_0(t,0,x) *_{(x)} u_0(x) + K_1(t,0,x) *_{(x)} u_1(x)\|_{X(t)} \lesssim \|u_0\|_{H^{\gamma+\frac{[1-\mu]_+}{2}}} + \|u_1\|_{L^2}. \tag{10}$$

Now fractional Gagliardo-Nirenberg inequality comes into play. We may estimate

$$\||u(s,\cdot)|^p\|_{L^2} = \|u(s,\cdot)\|^p_{L^{2p}} \lesssim \|u(s,\cdot)\|^{p(1-\theta)}_{L^2}\|u(s,\cdot)\|^{p\theta}_{\dot{H}^\gamma} \lesssim \|u\|^p_{X(s)},$$

where

$$\theta = \frac{n}{\gamma}\Big(\frac{1}{2} - \frac{1}{2p}\Big) \in [0,1] \text{ iff } p \in \Big[1, \frac{n}{[n-2\gamma]_+}\Big].$$

Using again Proposition 1 we obtain for $\kappa \in [0, \gamma]$ the estimate

$$\|Gu(t,\cdot)\|_{\dot{H}^\kappa} \lesssim \int_0^t e^{(p-1)rs+\kappa s} \| |u(s,x)|^p \|_{L^2} ds \lesssim \|u\|^p_{X(t)},$$

thanks to

$$(p-1)r + \gamma < 0 \text{ iff } (p-1)(n-\mu) > 2\gamma \text{ iff } p > p_{n,\mu,\gamma}.$$

Therefore, by (10) it follows that the operator

$$P : u = u(t,x) \to Pu(t,x) := K_0(t,0,x) *_{(x)} u_0(x) \\ + K_1(t,0,x) *_{(x)} u_1(x) + Gu(t,x),$$

satisfies

$$\|Pu\|_{X(t)} \lesssim \|u_0\|_{H^{\gamma+\frac{[1-\mu]_+}{2}}} + \|u_1\|_{L^2} + \|u\|^p_{X(t)}. \tag{11}$$

Using Hölder's inequality and Gagliardo-Nirenberg inequality we obtain

$$\|Pu - Pv\|_{\dot{H}^\gamma} \lesssim \|u-v\|_{X(t)} \big(\|u\|^{p-1}_{X(t)} + \|v\|^{p-1}_{X(t)} \big)$$

for the admissible range of p. Summarizing all the estimates we have

$$\|Pu - Pv\|_{X(t)} \lesssim \|u-v\|_{X(t)} \big(\|u\|^{p-1}_{X(t)} + \|v\|^{p-1}_{X(t)} \big) \tag{12}$$

for any $u, v \in X(t)$. Using (11) and (12) the proof is completed. □

Remark 1 If $\mu \in (0,1]$, i.e., $m \in [\frac{\sqrt{n^2-1}}{2}, \frac{n}{2})$, then we may choose $\gamma \in (\frac{\mu}{2}, \frac{1+\mu}{2})$ for $n \geq 1$.

If $\mu \in (1,2)$, i.e., $m \in [\frac{\sqrt{[n^2-4]_+}}{2}, \frac{\sqrt{n^2-1}}{2})$, then one can choose a suitable $\gamma \in (\frac{1}{2}, 1)$ such that $\frac{\sqrt{n^2-4\gamma^2}}{2} < m < \frac{\sqrt{n^2-1}}{2}$. Hence, taking $\gamma \to 1$, we can get a global (in time) existence result for $\frac{\sqrt{[n^2-4]_+}}{2} < m < \frac{\sqrt{n^2-1}}{2}$. In this way we get a global (in time) well-posedness result for $m \in (0, \frac{n}{2})$ in dimensions $n = 1$ and $n = 2$.

Remark 2 The requirement $p_{n,\mu,\gamma} < p \leq \frac{n}{n-2\gamma}$ implies $\mu < 2\gamma$. We have to choose γ such that

$$\frac{n(p-1)}{2p} \leq \gamma < \min\Big\{1, \frac{1+\mu}{2}, (n-\mu)(p-1)/2\Big\}.$$

This implies for $n \geq 1$ the condition $\frac{n}{n-\mu} < p < \frac{n}{[n-1]_+}$.

At a first glance we could use $p_{n,\mu,\gamma} \to 1$ as $\gamma \to 0$ in Theorem 1. But due to the condition $\mu = \sqrt{n^2 - 4m^2} < 2\gamma$, for a fixed $m < \frac{n}{2}$ we cannot take γ arbitrarily small. But in the case $m \geq \frac{n}{2}$ it will be allowed to use (see Theorems 2 and 3). On the other hand, the statements of Theorem 1 verify that in the moment we are not able to get a critical exponent $p_{crit} = p_{crit}(n) > 1$ for the family of de Sitter models (1) with dominant dissipation.

3 De Sitter Model with Dominant Mass: Case $m \in (\frac{n}{2}, \infty)$

In this section we consider the Cauchy problem (4), that is,

$$\begin{cases} u_{tt} - e^{-2t}\Delta u + \mu^2 u = e^{-\frac{n}{2}(p-1)t}|u|^p, & (t,x) \in (0,\infty) \times \mathbb{R}^n, \\ u(0,x) = u_0(x), \quad u_t(0,x) = u_1(x), & x \in \mathbb{R}^n, \end{cases} \tag{13}$$

where $\mu^2 = m^2 - \frac{n^2}{4} > 0$, so $m \in (\frac{n}{2}, \infty)$. According to Duhamel's principle, we may interpret weak solutions of (13) as solutions of the non-linear integral equation (in this paper we understand solutions only in this sense)

$$u(t,x) = K_0(t,0,x) *_{(x)} u_0(x) + K_1(t,0,x) *_{(x)} u_1(x) + Gu(t,x),$$

where

$$Gu(t,x) := \int_0^t e^{-\frac{n}{2}(p-1)s} K_1(t,s,x) *_{(x)} |u(s,x)|^p \, ds$$

and $K_j = K_j(t,s,x)$, $j = 0, 1$, $0 \leq s \leq t$, are the fundamental solutions to the corresponding linear parameter-dependent Cauchy problem with vanishing right-hand side

$$\begin{cases} K_{tt} - e^{-2t}\Delta K + \mu^2 K = 0, & (t,x) \in (s,\infty) \times \mathbb{R}^n, \\ K(s,s,x) = \delta_{0j}, \quad K_t(s,s,x) = \delta_{1j}, & x \in \mathbb{R}^n, \end{cases}$$

with $\delta_{kj} = 1$ for $k = j$, and zero otherwise.

3.1 Estimates of Solutions to Parameter-Dependent Linear Models

We consider for $\mu^2 > 0$ and $s \geq 0$ the family of parameter-dependent Cauchy problems

$$\begin{cases} u_{tt} - e^{-2t}\Delta u + \mu^2 u = 0, & (t,x) \in (s,\infty)\times\mathbb{R}^n, \\ u(s,x) = \varphi(s,x), \quad u_t(s,x) = \psi(s,x), & x \in \mathbb{R}^n. \end{cases} \tag{14}$$

Using the representation of Sobolev solutions given in Section 3.1 in [3], we are able to verify the following result:

Proposition 2 *Let us consider* (14) *with* $\mu > 0$. *If* $\varphi(0,\cdot) \in H^{\gamma+\frac{1}{2}}(\mathbb{R}^n)$, *with* $\gamma \geq 0$, *then for all* $t \in [0,\infty)$ *we have*

$$\|K_0(t,0,\cdot) *_{(x)} \varphi(0,\cdot)\|_{\dot{H}^\gamma} \lesssim \|\varphi(0,\cdot)\|_{H^{\gamma+\frac{1}{2}}}. \tag{15}$$

If $\psi(s,\cdot) \in L^2(\mathbb{R}^n)$, *then for all* $t \in [s,\infty)$ *and* $\gamma \in [0,\frac{1}{2}]$ *it holds*

$$\left\|K_1(t,s,\cdot) *_{(x)} \psi(s,\cdot)\right\|_{\dot{H}^\gamma} \lesssim e^{\gamma s}\|\psi(s,\cdot)\|_{L^2}. \tag{16}$$

Proof We split the extended phase space into three zones

$$Z_1(s) := \{\xi : |\xi|e^{-s} \leq N\}, \quad Z_2(t,s) := \{Ne^s \leq |\xi| \leq Ne^t\},$$
$$Z_3(t) := \{|\xi|e^{-t} \geq N\}.$$

In $Z_1(s)$ we have

$$|\xi|^\gamma|\widehat{u}(t,\xi)| \lesssim |\xi|^\gamma\left(|\widehat{\varphi}(s,\xi)| + |\widehat{\psi}(s,\xi)|\right) \quad \text{for } \gamma \geq 0.$$

In $Z_3(t)$ we have

$$|\xi|^\gamma|\widehat{u}(t,\xi)| \lesssim |\xi|^\gamma\left(|\widehat{\varphi}(s,\xi)|(|\xi|e^{-s})^{\frac{1}{2}} + |\widehat{\psi}(s,\xi)|(|\xi|e^{-s})^{-\frac{1}{2}}\right)(|\xi|e^{-t})^{-\frac{1}{2}}.$$

If $s = 0$ and $\psi(0,\cdot) \equiv 0$, then

$$|\xi|^\gamma|\widehat{u}(t,\xi)| \lesssim |\xi|^{\gamma+\frac{1}{2}}|\widehat{\varphi}(0,\xi)|.$$

If $\varphi(s,\cdot) \equiv 0$ and $\gamma \in [0,\frac{1}{2}]$, then

$$|\xi|^\gamma|\widehat{u}(t,\xi)| \lesssim |\xi|^{\gamma-1}e^{\frac{s+t}{2}}|\widehat{\psi}(s,\xi)| \lesssim e^{(\gamma-\frac{1}{2})t}e^{\frac{s}{2}}|\widehat{\psi}(s,\xi)| \lesssim e^{\gamma s}|\widehat{\psi}(s,\xi)|.$$

In $Z_2(t,s)$ we have

$$|\xi|^\gamma |\widehat{u}(t,\xi)| \lesssim |\xi|^\gamma \left(|\widehat{\varphi}(s,\xi)| (|\xi| e^{-s})^{\frac{1}{2}} + |\widehat{\psi}(s,\xi)| (|\xi| e^{-s})^{-\frac{1}{2}} \right).$$

If $s = 0$ and $\psi(0,\cdot) \equiv 0$, we have the same estimate as in $Z_3(t)$ and (15) follows.
If $\varphi(s,\cdot) \equiv 0$, using the definition of $Z_2(t,s)$ we may conclude

$$|\xi|^\gamma |\widehat{u}(t,\xi)| \lesssim |\xi|^{\gamma-\frac{1}{2}} e^{\frac{1}{2}s} |\widehat{\psi}(s,\xi)| \lesssim e^{\gamma s} |\widehat{\psi}(s,\xi)|.$$

Summarizing, the desired estimate (16) is proved. □

Corollary 2 *Consider the linear Cauchy problem* (1) *with vanishing right-hand side and* $m > \frac{n}{2}$. *Assume that* $f \in H^{\gamma+\frac{1}{2}}$ *and* $g \in L^2(\mathbb{R}^n)$ *with* $\gamma \in [0, \frac{1}{2}]$. *Then, the solution* ϕ *satisfies the following a-priori estimate:*

$$\|\phi(t,\cdot)\|_{\dot{H}^\gamma} \lesssim e^{-\frac{n}{2}t} \Big(\|f\|_{H^{\gamma+\frac{1}{2}}} + \|g\|_{L^2} \Big).$$

3.2 Global (in Time) Existence of Small Data Sobolev Solutions

We are interested in the global (in time) existence of Sobolev solutions to the Cauchy problem (1) for all $p > 1$, where $m > \frac{n}{2}$. Differently from Theorem 1, in the following result we may take γ arbitrarily small and (16) is the key estimate to verify it.

Theorem 2 *Let us consider for* $n \geq 1$ *the Cauchy problem* (1) *for* $p > 1$, $m \in (\frac{n}{2}, \infty)$, *and with data* $f \in H^{\gamma+\frac{1}{2}}(\mathbb{R}^n)$, $\gamma \in (0, \frac{1}{2})$, *and* $g \in L^2(\mathbb{R}^n)$. *Then, there exists a constant* $\varepsilon_0 > 0$ *such that, for every small data satisfying*

$$\|f\|_{H^{\gamma+\frac{1}{2}}} + \|g\|_{L^2} \leq \varepsilon \ \textit{for}\ \varepsilon \leq \varepsilon_0,$$

there exists a uniquely determined global (in time) Sobolev solution

$$\phi \in C\big([0,\infty), H^\gamma(\mathbb{R}^n)\big) \quad \textit{with} \quad \frac{n(p-1)}{2p} \leq \gamma < \frac{1}{2}\min\{1, n(p-1)\}.$$

Moreover, the solution satisfies the decay estimate

$$\|\phi(t,\cdot)\|_{H^\gamma} \lesssim e^{-\frac{n}{2}t} \big(\|f\|_{H^{\gamma+\frac{1}{2}}} + \|g\|_{L^2} \big).$$

Proof It is enough to prove the global (in time) existence of small data solutions to (6). We define the space

$$X(t) := \Big\{u \in \mathcal{C}\big([0,t], H^\gamma(\mathbb{R}^n)\big) \,:\, \|u\|_{X(t)} := \sup_{\tau\in[0,t]} \{\|u(\tau,\cdot)\|_{H^\gamma}\} < \infty\Big\}$$

with the usual norm in $H^\gamma(\mathbb{R}^n)$. Using (15) and (16) for $s = 0$ we have

$$\|K_0(t,0,x)*_{(x)}u_0(x)+K_1(t,0,x)*_{(x)}u_1(x)\|_{X(t)} \lesssim \|u_0\|_{H^{\gamma+\frac{1}{2}}}+\|u_1\|_{L^2}. \tag{17}$$

Now fractional Gagliardo-Nirenberg inequality comes into play. We may estimate

$$\||u(s,\cdot)|^p\|_{L^2} = \|u(s,\cdot)\|^p_{L^{2p}} \lesssim \|u(s,\cdot)\|^{p(1-\theta)}_{L^2}\|u(s,\cdot)\|^{p\theta}_{\dot H^\gamma} \lesssim \|u\|^p_{X(s)},$$

where

$$\theta = \frac{n}{\gamma}\Big(\frac{1}{2}-\frac{1}{2p}\Big) \in [0,1] \text{ iff } p \in \Big[1,\frac{n}{n-2\gamma}\Big].$$

Hence,

$$\begin{aligned}\|Gu(t,\cdot)\|_{L^2} &\lesssim \int_0^t e^{-\frac{n}{2}(p-1)s}\big\|K_1(t,s,x)*_{(x)}|u(s,x)|^p\big\|_{L^2}ds\\ &\lesssim \int_0^t e^{-\frac{n}{2}(p-1)s}\||u(s,x)|^p\|_{L^2}ds \lesssim \|u\|^p_{X(t)}\end{aligned}$$

thanks to the assumption $p > 1$. For $0 < \gamma < \frac{1}{2}$, using Proposition 2 we may estimate

$$\|Gu(t,\cdot)\|_{\dot H^\gamma} \lesssim \int_0^t e^{-\frac{n}{2}(p-1)s+\gamma s}\||u(s,x)|^p\|_{L^2}ds \lesssim \|u\|^p_{X(t)},$$

since

$$-\frac{n(p-1)}{2}+\gamma < 0 \text{ iff } n(p-1) > 2\gamma \text{ iff } p > 1+\frac{2\gamma}{n}.$$

Therefore, by (17) it follows that the operator

$$\begin{aligned}P:\ u = u(t,x) \to\ Pu(t,x) &:= K_0(t,0,x)*_{(x)}u_0(x)\\ &+ K_1(t,0,x)*_{(x)}u_1(x) + Gu(t,x),\end{aligned}$$

satisfies

$$\|Pu\|_{X(t)} \lesssim \|u_0\|_{H^{\gamma+\frac{1}{2}}} + \|u_1\|_{L^2} + \|u\|^p_{X(t)}. \tag{18}$$

Using that

$$||u|^p - |v|^p| \le C|u-v|(|u|+|v|)^{p-1},$$

Hölder's inequality implies

$$\||u(t,\cdot)|^p - |v(t,\cdot)|^p\|_{L^2} \le C\|u(t,\cdot)-v(t,\cdot)\|_{L^{2p}}\big(\|u\|_{L^{2p}}^{p-1} + \|v\|_{L^{2p}}^{p-1}\big).$$

Finally, by Gagliardo-Nirenberg inequality we obtain

$$\|Pu - Pv\|_{\dot{H}^\gamma} \lesssim \|u-v\|_{X(t)}\big(\|u\|_{X(t)}^{p-1} + \|v\|_{X(t)}^{p-1}\big),$$

for the admissible range of p. Summarizing all the estimates we have

$$\|Pu - Pv\|_{X(t)} \lesssim \|u-v\|_{X(t)}\big(\|u\|_{X(t)}^{p-1} + \|v\|_{X(t)}^{p-1}\big) \tag{19}$$

for any $u, v \in X(t)$. Using (18) and (19) the proof is completed. □

Remark 3 The requirement $\gamma < \frac{1}{2}\min\{1, n(p-1)\}$ implies $p > \frac{n+2\gamma}{n}$. The requirement $\frac{n(p-1)}{2p} \le \gamma$ implies $p \le \frac{n}{n-2\gamma}$. We may choose $\gamma \in (0, \frac{1}{2})$. This implies for $n \ge 1$ the condition $1 < p < \frac{n}{[n-1]_+}$.

4 De Sitter Model with Balanced Dissipation and Mass: Case $m = \frac{n}{2}$

In this section we consider with $m = \frac{n}{2}$ in (1) the Cauchy problem (5), that is,

$$\begin{cases} u_{tt} - e^{-2t}\Delta u = e^{-\frac{n}{2}(p-1)t}|u|^p, & (t,x) \in (0,\infty)\times\mathbb{R}^n, \\ u(0,x) = u_0(x), \quad u_t(0,x) = u_1(x), & x \in \mathbb{R}^n. \end{cases} \tag{20}$$

According to Duhamel's principle, we may interpret weak solutions of (20) as solutions of the non-linear integral equation (in this paper we understand solutions only in this sense)

$$\begin{aligned} u(t,x) = {} & K_0(t,0,x) *_{(x)} u_0(x) + K_1(t,0,x) *_{(x)} u_1(x) + \\ & + \int_0^t e^{-\frac{n}{2}(p-1)s} K_1(t,s,x) *_{(x)} |u(s,x)|^p \, ds. \end{aligned}$$

Here $K_j = K_j(t, s, x)$, $j = 0, 1$, $0 \le s \le t$, are the fundamental solutions to the corresponding linear parameter-dependent Cauchy problem with vanishing right-hand side

$$\begin{cases} K_{tt} - e^{-2t}\Delta K = 0, & (t, x) \in (s, \infty) \times \mathbb{R}^n, \\ K(s, s, x) = \delta_{0j}, \quad K_t(s, s, x) = \delta_{1j}, & x \in \mathbb{R}^n, \end{cases}$$

with $\delta_{kj} = 1$ for $k = j$, and zero otherwise.

4.1 Estimates of Solutions to Parameter-Dependent Linear Models

Let us consider, with $s \ge 0$, the family of parameter-dependent Cauchy problems

$$\begin{cases} u_{tt} - e^{-2t}\Delta u = 0, & (t, x) \in (s, \infty) \times \mathbb{R}^n, \\ u(s, x) = \varphi(s, x), \quad u_t(s, x) = \psi(s, x), & x \in \mathbb{R}^n. \end{cases} \tag{21}$$

Using the representation of solutions given in Section 4.1 in [3] we are able to verify the following result:

Proposition 3 *Let us consider* (21). *If* $\varphi(0, \cdot) \in H^{\gamma+\frac{1}{2}}(\mathbb{R}^n)$ *with* $\gamma \ge 0$, *then for all* $t \in [0, \infty)$ *we have*

$$\|K_0(t, 0, \cdot) *_{(x)} \varphi(0, \cdot)\|_{\dot{H}^\gamma} \lesssim (1 + t)\|\varphi(0, \cdot)\|_{H^{\gamma+\frac{1}{2}}}. \tag{22}$$

If $\psi(s, \cdot) \in L^2(\mathbb{R}^n)$, *then for any* $t \in [s, \infty)$ *we have*

$$\|K_1(t, s, \cdot) *_{(x)} \psi(s, \cdot)\|_{\dot{H}^\gamma} \lesssim (1 + t)e^{\gamma s}\|\psi(s, \cdot)\|_{L^2} \text{ with } \gamma \in \Big[0, \frac{1}{2}\Big]. \tag{23}$$

Proof We split the extended phase space into three zones

$$Z_1(s) := \{\xi : |\xi|e^{-s} \le N\}, \quad Z_2(t, s) := \{Ne^s \le |\xi| \le Ne^t\},$$
$$Z_3(t) := \{|\xi|e^{-t} \ge N\}.$$

First let us put $s = 0$ and $\psi(0, \cdot) \equiv 0$. Then in $Z_1(s)$ we have

$$|\xi|^\gamma|\widehat{u}(t, \xi)| \lesssim (1 + t)|\xi|^\gamma|\widehat{\varphi}(0, \xi)|.$$

In $Z_3(t)$ we have

$$|\xi|^\gamma|\widehat{u}(t,\xi)| \lesssim e^{\frac{t}{2}}|\xi|^\gamma|\widehat{\varphi}(s,\xi)| \lesssim |\xi|^{\gamma+\frac{1}{2}}|\widehat{\varphi}(s,\xi)|.$$

In $Z_2(t,s)$ we have

$$|\xi|^\gamma|\widehat{u}(t,\xi)| \lesssim |\xi|^{\frac{1}{2}}(1+t+\log(|\xi|))|\xi|^\gamma|\widehat{\varphi}(0,\xi)| \lesssim (1+t)|\xi|^{\gamma+\frac{1}{2}}|\widehat{\varphi}(0,\xi)|.$$

Summing up we have proved (22). Now let $\varphi(s,\cdot) \equiv 0$. Then in $Z_1(s)$ we have

$$|\xi|^\gamma|\widehat{u}(t,\xi)| \lesssim (1+t)|\xi|^\gamma|\widehat{\psi}(s,\xi)| \lesssim (1+t)e^{\gamma s}|\widehat{\psi}(s,\xi)|.$$

In $Z_3(t)$ we have for $\gamma \in [0,\frac{1}{2}]$ the estimate

$$|\xi|^\gamma|\widehat{u}(t,\xi)| \lesssim |\xi|^{\gamma-1}e^{\frac{s+t}{2}}|\widehat{\psi}(s,\xi)| \lesssim e^{(\gamma-\frac{1}{2})t}e^{\frac{t}{2}}|\widehat{\psi}(s,\xi)| \lesssim e^{\gamma s}|\widehat{\psi}(s,\xi)|.$$

Using the definition of $Z_2(t,s)$ and again that $\gamma \in [0,\frac{1}{2}]$ we may conclude

$$|\xi|^\gamma|\widehat{u}(t,\xi)| \lesssim (1+t)|\xi|^{\gamma-\frac{1}{2}}e^{\frac{s}{2}}|\widehat{\psi}(s,\xi)| \lesssim (1+t)e^{\gamma s}|\widehat{\psi}(s,\xi)|.$$

In this way (23) is also proved. □

Corollary 3 *Consider the linear Cauchy problem* (1) *with vanishing right-hand side and* $m = \frac{n}{2}$. *Assume that* $f \in H^{\gamma+\frac{1}{2}}$ *and* $g \in L^2(\mathbb{R}^n)$ *with* $\gamma \in [0,\frac{1}{2}]$. *Then, the solution* ϕ *satisfies the following a-priori estimate:*

$$\|\phi(t,\cdot)\|_{\dot{H}^\gamma} \lesssim (1+t)e^{-\frac{n}{2}t}\Big(\|f\|_{H^{\gamma+\frac{1}{2}}} + \|g\|_{L^2}\Big).$$

4.2 Global (in Time) Existence of Small Data Sobolev Solutions

If we compare the derived estimates for the solutions to the linear Cauchy problems from Sects. 3.1 and 4.1, then they differ only by the factor $(1+t)$ (compare (15), (16) with (22), (23)). This factor will not have any influence on the admissible set of powers p for the global (in time) existence of small data Sobolev solutions to the Cauchy problem (1). For this reason we mainly restrict ourselves to formulate the next result.

Theorem 3 *Consider for $n \geq 1$ the Cauchy problem* (1) *with $p > 1$ and $m = \frac{n}{2}$, and with data $f \in H^{\gamma+\frac{1}{2}}(\mathbb{R}^n)$, $\gamma \in (0, \frac{1}{2})$, and $g \in L^2(\mathbb{R}^n)$. Then, there exists a constant $\varepsilon_0 > 0$ such that, for every small data satisfying*

$$\|f\|_{H^{\gamma+\frac{1}{2}}} + \|g\|_{L^2} \leq \varepsilon \ \ for \ \ \varepsilon \leq \varepsilon_0,$$

there exists a uniquely determined global (in time) Sobolev solution

$$\phi \in C\big([0,\infty), H^\gamma(\mathbb{R}^n)\big) \quad with \quad \frac{n(p-1)}{2p} \leq \gamma < \frac{1}{2}\min\{1, n(p-1)\}.$$

Moreover, the solution satisfies the decay estimate

$$\|\phi(t,\cdot)\|_{H^\gamma} \lesssim (1+t)e^{-\frac{n}{2}t}\big(\|f\|_{H^{\gamma+\frac{1}{2}}} + \|g\|_{L^2}\big).$$

Remark 4 The conclusions of Remark 3 are still valid for the parameters in Theorem 3.

5 Conclusions

- The mass term has an essential influence on the question for the critical exponent.
- Let us consider

$$\begin{cases} \phi_{tt} - e^{-2t}\Delta\phi + a\phi_t + m^2\phi = |\phi|^p, & (t,x) \in (0,\infty) \times \mathbb{R}^n, \\ (\phi(0,x), \phi_t(0,x)) = (f(x), g(x)), & x \in \mathbb{R}^n. \end{cases} \tag{24}$$

  If $a > 0$, then for a family of mass terms $m^2\phi$ we cannot expect a critical exponent $p_{crit} > 1$. If we replace in (1) the damping term $n\phi_t$ by $a\phi_t$, then in the proof to Theorem 2 we get the condition

$$\frac{n(p-1)}{2p} \leq \gamma < \frac{1}{2}\min\{1, a(p-1)\}.$$

  This condition is not empty iff $p > \frac{n}{a}$, that is, p increases as $a \to 0$ and $p > 1$ for $a \geq n$. The coefficient m^2 of the mass term satisfies $m > \frac{a}{2}$.
- If $m = 0$, then we have blow-up results for solutions to (24) for all $p > 1$.
- Let us choose $a = 0$. Could we expect global (in time) small data well-posedness for solutions to (24) for all $p > 1$? This would verify the improving influence of the mass term in comparison with damping term.
- The last two observations propose a lot of interesting questions for models with mass and dissipation and $L^1(\mathbb{R}^+)$ propagation speed.

Acknowledgments The first author is partially supported by "Conselho Nacional de Desenvolvimento Cientifico e Tecnológico (CNPq)" Grant 304408/2020-4. The second author is partially supported by " FCTFundação para a Ciência e a Tecnologia, I.P." Grant/Award Number: UIDB/05037/2020.

References

1. D'Abbicco, M., Lucente, S., Reissig, M.: A shift in the Strauss exponent for semilinear wave equations with a not effective damping. J. Differ. Equ. **259**, 5040–5073 (2015)
2. Ebert, M.R., Reissig, M.: Theory of damped wave models with integrable and decaying in time speed of propagation. J. Hyper. Differ. Equ. **13**(2), 417–439 (2016)
3. Ebert, M.R., Reissig, M.: Regularity theory and global existence of small data solutions to semi-linear de Sitter models with power non-linearity. Nonlinear Anal. Real World Appl. **40**, 14–54 (2018). http://dx.doi.org/10.1016/j.nonrwa.2017.08.009
4. Ebert, M.R., Reissig, M.: A note to semilinear de Sitter models in 1d with balanced mass and dissipation. Nonlinear Anal. Real World Appl. **71**, 103835 (2023)
5. Galstian, A., Yagdjian, K.: Global in time existence of self-interacting scalar field in de Sitter spacetimes. Nonlinear Anal. Real World Appl. **34**, 110–139 (2017)
6. Nunes do Nascimento, W.: Klein-Gordon models with non-effective potential, Ph.D. Thesis, Universidade Federal de São Carlos/Technical University Bergakademie Freiberg, 183pp. (2016)
7. Nunes do Nascimento, W., Palmieri, A., Reissig, M.: Semi-linear wave models with power non-linearity and scale-invariant time-dependent mass and dissipation. Math. Nachr. **290**, 1779–1805 (2017)
8. Palmieri, A., Reissig, M.: A competition between Fujita and Strauss type exponents for blow-up of semi-linear wave equations with scale-invariant damping and mass. J. Differ. Equ. **266**, 1176–1220 (2019)
9. Palmieri, A., Takamura, H.: A note on blow-up results for semilinear wave equations in de Sitter and anti-de Sitter spacetimes. J. Math. Anal. Appl. **514**(1), 126266 (2022)
10. Tsutaya, K., Wakasugi, Y.: Blow up of solutions of semilinear wave equations related to nonlinear waves in de Sitter spacetime. Partial Differ. Equ. Appl. **3**, 6 (2022)
11. Yagdjian, K.: Global existence of the self-interacting scalar field in the de Sitter universe. J. Math. Phys. **60**, 051503 (2019). https://doi.org/10.1063/1.5082653
12. Yagdjian, K.: Global existence of the scalar field in de Sitter spacetime. J. Math. Anal. Appl. **396**, 323–344 (2012)

Part VI
PDEs in Fluid Mechanics

A Family of Systems Including the Herschel-Bulkley Fluid Equations

Nikolai V. Chemetov and Marcelo M. Santos

Abstract We analyze the equations of motion for incompressible fluids with the "viscous stress tensor" $\mathbb{S}$ in a family which includes the Bingham model for viscoplastic fluids (more generally, the Herschel-Bulkley model). $\mathbb{S}$ is the subgradient of a convex potential, which can depend on the space-time variables. The potential has its one-sided directional derivatives uniformly bounded from below and above by a p-power function. For $p \geqslant 2.2$ we solve an initial boundary value problem for those fluid systems, in a bounded region in $\mathbb{R}^3$. We take a nonlinear boundary condition, which encompasses the Navier friction/slip boundary condition.

1 Introduction

The system

$$\mathbf{v}_t + (\mathbf{v} \cdot \nabla)\,\mathbf{v} = \operatorname{div} \mathbb{T}, \quad \operatorname{div} \mathbf{v} = 0, \tag{1}$$

models an incompressible fluid with particle velocity $\mathbf{v} = \mathbf{v}(\mathbf{x}, t)$ at a position $\mathbf{x}$ in a time t. It is well established that $\mathbb{T} = -\mathrm{p}\,\mathrm{I} + \mathbb{S}$, where p is pressure and $\mathbb{S}$ is the "viscous part" of the Cauchy stress tensor (see e.g. [16]). We shall call $\mathbb{S}$ the *viscous stress tensor* and throughout the paper, to shorten the notation, we set $\mathbb{B} = \nabla \mathbf{v}$.

N. V. Chemetov (✉)
Department of Computing and Mathematics, University of São Paulo, Ribeirão Preto, Brazil
e-mail: nvchemetov@gmail.com

M. M. Santos
Universidade Estadual de Campinas, IMECC-UNICAMP, Campinas, Brazil
e-mail: mmsantos@unicamp.br

M. R. Ebert et al. (eds.), *New Tools in Mathematical Analysis and Applications*, Research Perspectives, https://doi.org/10.1007/978-3-031-77050-0_18

For viscoplastic fluids with rotating particles (*micropolar viscoplastic fluid*), the viscous stress tensor $\mathbb{S}$ depends on both the symmetric and the antisymmetric parts of $\mathbb{B}$

$$\mathbb{B}_s = \left(\mathbb{B} + \mathbb{B}^T\right)/2 \quad \text{and} \quad \mathbb{B}_a = \left(\mathbb{B} - \mathbb{B}^T\right)/2$$

(see e.g. [13, 15, 20, 21]). The *Cosserat-Bingham model*, proposed by Shelukhin and Ružička[20], has the viscous stress tensor $\mathbb{S}$, in the part of the fluid domain where the stress surpasses the yield stress, given as a function of $\mathbb{B}_s$ (the symmetric part of $\mathbb{B}$) and $\mathrm{R} = \mathbb{B}_a - \Omega$, where $\Omega = \Omega(\mathbf{x}, t)$ is the *micro-rotational velocity tensor*, being an antisymmetric matrix for each $(\mathbf{x}, t)$. In this paper we are interested mostly in the system (1) and we consider Ω as a known function. For given (constant) viscosity coefficients $\mu_1 > 0$, $\mu_2 \geqslant 0$ and (constant) *plug parameters* $\nu \geqslant 0$, $\tau_* > 0$, a particular model for $\mathbb{S}$ in [20] can be written as follows:

$$\begin{cases} \mathbb{S} = \mathbb{B}_{\mu,p} + \tau_* \mathbb{P}_{\nu,p}, & \text{if } \mathbb{B}_\nu = \mathbb{B}_s + \nu \mathrm{R} \neq 0; \\ |\mathbb{S}| \leqslant \tau_*, & \text{if } \mathbb{B}_\nu = 0; \end{cases} \tag{2}$$

where

$$\mathbb{B}_{\mu,p} = \mu_1 |\mathbb{B}_s|^{p-2} \mathbb{B}_s + \mu_2 |\mathrm{R}|^{p-2} \mathrm{R} \qquad \text{with } \mu = (\mu_1, \mu_2), \tag{3}$$

and the *plastic operator* is defined as

$$\mathbb{P}_{\nu,p} = \frac{\mathbb{B}_{\nu,p}}{|\mathbb{B}_{\nu,p}|} \qquad \text{with } \mathbb{B}_{\nu,p} = |\mathbb{B}_s|^{p-2} \mathbb{B}_s + \nu |\mathrm{R}|^{p-2} \mathrm{R}. \tag{4}$$

Motivated by the model (2)–(4), we consider the system (1) with the following model for the viscous stress tensor $\mathbb{S}$:

$$\begin{cases} \mathbb{S} = \mathbb{B}_{\mu,p} + \widehat{\tau}_* \dfrac{\mathbb{B}_{\nu,q}}{|\widehat{\mathbb{B}}_{\nu,q}|^{\frac{2(q-1)}{q}}}, & \text{if } \mathbb{B}_\nu \neq 0; \\ |\mathbb{S}| \leqslant \tau_*, \quad \text{if } \mathbb{B}_\nu = 0; \end{cases} \tag{5}$$

with given constant exponents $p, q \geqslant 2$, $\widehat{\tau}_* = \frac{\tau_*}{\max(1, \sqrt[q]{\nu})}$ and

$$\mathbb{B}_{\nu,q} = |\mathbb{B}_s|^{q-2} \mathbb{B}_s + \nu |\mathrm{R}|^{q-2} \mathrm{R}, \qquad \widehat{\mathbb{B}}_{\nu,q} = |\mathbb{B}_s|^{\frac{q-2}{2}} \mathbb{B}_s + \sqrt{\nu} |\mathrm{R}|^{\frac{q-2}{2}} \mathrm{R}. \tag{6}$$

This model reduces to the power law model known as Herschel-Bulkley model (see e.g. [20, (3.1)]) if $\nu = \mu_2 = 0$ (viscoplastic fluid without particles) and $q = p$ (as well as, the model (2), if $\nu = \mu_2 = 0$), becoming Bingham model when $p = 2$. In addition, it coincides with the model (2) in the case $\nu = 0$ or $\nu = 1$ and the plastic operator (4) has exponent $p = 2$ (see [11] for more details).

An important feature of (5) is that these tensors can be achieved by the condition

$$\mathbb{S}(\mathbf{x},t) \in \partial V(\mathbb{B}(\mathbf{x},t)), \qquad \text{when } \nu > 0 \text{ or } \nu = \mu_2 = 0,$$

for the potential[1]

$$\begin{aligned} V(\mathbf{X}) &= V_{(\mathbf{x},t)}(\mathbf{X}) = U(\mathbf{X}) + \widehat{\tau}_* W(\mathbf{X}) \qquad \text{with} \\ U(\mathbf{X}) &= \frac{\mu_1}{p}|\mathbf{X}_s|^p + \frac{\mu_2}{p}|\mathrm{R}|^p, \quad W(\mathbf{X}) = \sqrt[q]{|\mathbf{X}_s|^q + \nu|\mathrm{R}|^q}. \end{aligned} \tag{7}$$

For $\mathbb{S}$ given by (5), we consider the following three dimensional initial boundary value problem (ibvp) and show the existence of a solution of the following system in a bounded domain $\mathcal{O} \subset \mathbb{R}^3$ with a C^2-smooth boundary Γ, for any fixed time $T > 0$:

$$\begin{cases} \mathbf{v}_t + (\mathbf{v}\cdot\nabla)\,\mathbf{v} = \operatorname{div}\mathbb{T}, \quad \operatorname{div}\mathbf{v} = 0 & \text{in} \quad \mathcal{O}_T = \mathcal{O}\times(0,T), \\ \mathbf{v}\cdot\mathbf{n} = 0, \quad (\mathbb{T}\,\mathbf{n} + \nabla g(\mathbf{v}))\cdot\boldsymbol{\tau} = 0 & \text{on} \quad \Gamma_T = \Gamma\times(0,T), \\ \mathbf{v} = \mathbf{v}_0 & \text{on} \quad \mathcal{O}\times\{t=0\}, \end{cases} \tag{8}$$

where $\mathbf{n} = \mathbf{n}(\mathbf{x})$ is the exterior normal to Γ and $\boldsymbol{\tau} = \boldsymbol{\tau}(\mathbf{x})$ denotes any vector tangent to Γ at the point $\mathbf{x} \in \Gamma$.

As for the boundary function, $g = g_{(\mathbf{x},t)}(\mathbf{v})$ is a Caratheodory function $g : \Gamma_T \times \mathbb{R}^3 \to \mathbb{R}$, being differentiable and convex with respect to $\mathbf{v}$, satisfying the non-negativity and boundedness conditions:

$$0 \leqslant \nabla g_{(\mathbf{x},t)}(\mathbf{v})\cdot\mathbf{v}, \quad |\nabla g_{(\mathbf{x},t)}(\mathbf{v})| \leqslant \mathrm{c}|\mathbf{v}|, \quad \forall \mathbf{v} \in \mathbb{R}^3,\ \forall (\mathbf{x},t) \in \Gamma_T, \tag{9}$$

where the gradient operator ∇ acts only on the variable $\mathbf{v}$ and c is a positive constant, independent of $\mathbf{v}$ and $(\mathbf{x},t)$. The particular case $g = \frac{1}{2}\alpha|\mathbf{v}|^2$ with $\alpha : \Gamma_T \to \mathbb{R}$ being a non-negative function in $L^\infty(\Gamma_T)$, satisfies (9) and yields the well-known Navier friction/slip boundary condition $(\mathbb{T}\,\mathbf{n} + \alpha\mathbf{v})\cdot\boldsymbol{\tau} = 0$ (see [3–9, 12]).

The initial data $\mathbf{v}_0$ is a given function in the space H defined below.

For the majority of this paper, we omit for simplicity the explicit dependence of the potentials (as already mentioned in a previous footnote) and of the "boundary function" g on the variables $\mathbf{x}$ and t . That is, we will use the notation V or $V(\mathbb{B})$ and $g(\mathbf{v})$ instead of $V_{(\mathbf{x},t)}(\mathbb{B})$ and $g_{(\mathbf{x},t)}(\mathbf{v})$.

We end this introduction providing a brief description of the organization of this paper. Due to the limitation of the number of pages for publication, we confine ourselves here to the above concise introduction, present only the statements of the

[1] We use the notation $V_{(\mathbf{x},t)}$ to indicate the dependence of V on the variable $(\mathbf{x},t)$, i.e., instead of write $V(\mathbf{x},t)$ or $V(\mathbf{x},t,\mathbf{X})$ we prefer to write $V_{(\mathbf{x},t)}$ or $V_{(\mathbf{x},t)}(\mathbf{X})$. Besides, most of the time we will omit the explicit dependence of V on $(\mathbf{x},t)$. Notice that V depends on $(\mathbf{x},t)$ as well due the given matrix function $\Omega = \Omega(\mathbf{x},t)$.

main results and give an overview of the proof of Theorem 1 below. However, more details and proofs can be found in the more comprehensive version of this paper available in [11] (we also refer to the results [10, 18, 19], which are related with a "linear" Cosserat-Bingham model, i.e. when $p = 2$ and $\nu = 0$).

In Sect. 2 we state our main results and in Sect. 3 we collect some important results we use in their proof. In Sect. 4 we give the main ideas to prove Theorem 1.

2 Main Results

We use the function spaces

$$\begin{aligned} H &= \{\mathbf{v} \in L^2(\mathcal{O}) : \text{div}\, \mathbf{v} = 0 \quad \text{in } \mathcal{D}'(\mathcal{O}), \quad \mathbf{v} \cdot \mathbf{n} = 0 \quad \text{in } H^{-1/2}(\Gamma)\}, \\ \mathbb{V} &= \{\mathbf{v} \in H^1(\mathcal{O}) : \text{div}\, \mathbf{v} = 0 \quad \text{a.e. in } \mathcal{O}, \quad \mathbf{v} \cdot \mathbf{n} = 0 \quad \text{in } H^{1/2}(\Gamma)\}, \\ V_p &= \{\mathbf{v} \in \mathbb{V} : \quad |\nabla \mathbf{v}| \in L^p(\mathcal{O})\}, \end{aligned} \tag{10}$$

where we have used standard notations of Lebesgue and Sobolev spaces and of the space of regular distributions $\mathcal{D}'(\mathcal{O})$. The space V_p is endowed with the norm $\|\mathbf{v}\|_{V_p} = \|\mathbf{v}\|_{L^2(\mathcal{O})} + \|\nabla \mathbf{v}\|_{L^p(\mathcal{O})}$. We solve the ibvp (8) in the sense of the following theorem.

Theorem 1 *Let $d = 3$. Let V be the potential given in (7), being $\Omega = \Omega(\mathbf{x}, t)$ an antisymmetric matrix function in $L^p(\mathcal{O}_T; \mathbb{R}^{3\times 3})$, and g satisfying (9). Then for any*

$$\mathbf{v}_0 \in H, \quad p \geqslant 2.2 \quad \textit{and} \quad q \geqslant 2,$$

there exists a pair $(\mathbf{v}, \mathbb{S})$ of functions with $\mathbb{S}(\mathbf{x}, t) \in \partial V_{(\mathbf{x},t)}(\mathbb{B}(\mathbf{x}, t))$ for almost all $(\mathbf{x}, t) \in \mathcal{O}_T$ (in particular,[2] $\mathbb{S}$ satisfies (5) for almost all $(\mathbf{x}, t) \in \mathcal{O}_T$), such that

$$\mathbf{v} \in L^\infty(0, T; H) \cap L^p(0, T; V_p), \ \mathbf{v}_t \in L^{p'}(0, T; V_p^*) \textit{ and } \mathbb{S} \in L^{p'}(\mathcal{O}_T; \mathbb{R}^3 \times \mathbb{R}^3),$$

where $p' = \frac{p}{p-1}$ and V_p^* is the dual space of V_p. The pair $(\mathbf{v}, \mathbb{S})$ satisfies the integral equality

$$\int_{\mathcal{O}_T} [\mathbf{v} \cdot \partial_t \boldsymbol{\varphi} + (\mathbf{v} \otimes \mathbf{v} - \mathbb{S}) : \nabla \boldsymbol{\varphi}]\, d\mathbf{x}dt + \int_{\mathcal{O}} \mathbf{v}_0 \cdot \boldsymbol{\varphi}(\mathbf{x}, 0)\, d\mathbf{x} = \int_{\Gamma_T} \nabla g(\mathbf{v}) \cdot \boldsymbol{\varphi}\, d\Gamma dt \tag{11}$$

for any function $\boldsymbol{\varphi} \in C^1(\overline{\mathcal{O}_T})$ such that $\boldsymbol{\varphi}(\cdot, T) = 0$, $(\boldsymbol{\varphi} \cdot \mathbf{n})|\Gamma_T = 0$ and $\text{div}\, \boldsymbol{\varphi}(\cdot, t) = 0$, for each $t \in [0, T]$.

---

[2] See Corollary 2.

The next theorem characterizes quite completely the subdifferential of the potential V defined in (7). Notice that V is differentiable at matrices $\mathbf{X}$ such that $\mathbf{X}_\nu \neq 0$ and that $\mathbf{X}_\nu = 0$ means that $\mathbf{X} = \Omega$ if $\nu \neq 0$ and $\mathbf{X}_s = 0$ but $\mathbf{X}_a$ is arbitrary, if $\nu = 0$.

Theorem 2 *Let $p, q \geqslant 2$. Then, the potential V defined in* (7) *is convex, differentiable at any $\mathbf{X} \in \mathbb{R}^{d\times d}$ such that $\mathbf{X}_\nu \neq 0$, with*

$$\nabla V(\mathbf{X}) = \mathbf{X}_{\mu,p} + \widehat{\tau}_* |\widehat{\mathbf{X}}_{\nu,q}|^{-\frac{2(q-1)}{q}} \mathbf{X}_{\nu,q}, \tag{12}$$

where $\mathbf{X}_\nu = \mathbf{X}_s + \nu R$, $R = \mathbf{X}_a - \Omega$, and $\mathbf{X}_{\mu,p}$, $\mathbf{X}_{\nu,q}$ $\widehat{\mathbf{X}}_{\nu,q}$ are defined by (3), (6), *replacing $\mathbb{B}$ by $\mathbf{X}$. Moreover:*

(a) $B_{r_*}(0) \subset \partial V(\Omega) \subset B_{\tau_*}(0)$, where $B_r(0)$ is the closed ball of radius r and center $\mathbf{X}$ in $\mathbb{R}^{d\times d} = \mathbb{R}^{d^2}$, and

$$r_* = \widehat{\tau}_* r_q, \quad r_q = \begin{cases} \sqrt[q]{\nu}/(1+\nu^{\frac{1}{q-2}})^{\frac{q-2}{2q}}, & \text{if } q > 2; \\ \min(1, \sqrt{\nu}), & \text{if } q = 2; \end{cases}$$

where $\widehat{\tau}_*$ is introduced in (5).

(b) The subdifferential $\partial V(\Omega)$ has an "ellipsoidal form", i.e.,

$$\partial V(\Omega) = \{\mathbf{X}^* \in \mathbb{R}^{d\times d} : |\mathbf{X}_s^*|^{q'} + \nu^{1-q'} |\mathbf{X}_a^*|^{q'} \leqslant (\widehat{\tau}_*)^{q'}\}, \quad q' = q/(q-1). \tag{13}$$

(c) If $\nu = 0$, the subdifferential $\partial V(\mathbf{X})$ at a matrix $\mathbf{X}$, such that $\mathbf{X}_\nu = 0$ (i.e., $\mathbf{X}_s = 0$ and $\mathbf{X}_a$ arbitrary) is given by

$$\partial V(\mathbf{X}) = \mu_2 |\mathrm{R}|^{p-2}\mathrm{R} + \{\mathbf{X}^* \in \mathbb{R}^{d\times d} : \mathbf{X}_a^* = 0 \text{ and } |\mathbf{X}_s^*| \leqslant \tau_*\}. \tag{14}$$

As consequences of the above theorem we have the following corollaries.

Corollary 1 *For any matrix $\mathbf{X} \in \mathbb{R}^{d\times d}$, such that $\mathbf{X}_\nu \neq 0$, one has the estimate*

$$\frac{|\mathbf{X}_{\nu,q}|}{|\widehat{\mathbf{X}}_{\nu,q}|^{\frac{2(q-1)}{q}}} \leqslant \max(1, \sqrt[q]{\nu}). \tag{15}$$

We have also the following lemma regarding the potential (7).

Lemma 1 *Let Ω be a fixed $d \times d$ antisymmetric real matrix. For given $p, q \geqslant 2$, let us consider the convex potential* (7) *and the approximations (where U is defined in* (7)*)*

$$V^n(\mathbf{X}) = U(\mathbf{X}) + \widehat{\tau}_* W^n(\mathbf{X}), \quad W^n(\mathbf{X}) = \sqrt[q]{|\mathbf{X}_s|^q + \nu |\mathrm{R}|^q + n^{-1}},$$

$$\mathbb{S}^n = \mathbf{X}_{\mu,p} + \widehat{\tau}_* \frac{\mathbf{X}_{\nu,q}}{\sqrt[q]{(|\widehat{\mathbf{X}}_{\nu,q}|^2 + n^{-1})^{q-1}}}, \qquad \mathbf{X} \in \mathbb{R}^{d\times d}, \qquad n \in \mathbb{N}. \tag{16}$$

Here, $\mathrm{R} = \mathbf{X}_a - \Omega$ *and* $\mathbf{X}_{\mu,p}$, $\mathbf{X}_{\nu,q}$, $\widehat{\mathbf{X}}_{\nu,q}$ *are defined by* (3), (6), *replacing* $\mathbb{B}$ *by* $\mathbf{X}$. *Then the following statements are true:*

(a) *Both potentials* V *and* V^n *satisfy the estimates*

$$\mu_1|\mathbf{X}_s|^p - \mu_2 2^{p-2}|\Omega|^p - \tau_*|\Omega| \leqslant V'(\mathbf{X};\mathbf{X}) \leqslant c_1|\mathbf{X}|^p + c_2|\Omega|^p + \tau_*|\mathbf{X}| \tag{17}$$

for any matrix $\mathbf{X} \in \mathbb{R}^d \times \mathbb{R}^d$, $\forall n \in \mathbb{N}$ *(we can replace* V *by* V^n *in* (17)*), where*

$$c_1 = \mu_1 + 2^{p-2}\mu_2(1 + \frac{1}{p}) \quad \text{and} \quad c_2 = 2^{p-2}\mu_2(1 - \frac{1}{p}).$$

(b) *For any given* $\mathbf{X} \in \mathbb{R}^{d\times d}$, *we have that* $\mathbb{S}^n = \nabla V^n(\mathbf{X})$. *Since* V^n *is differentiable and convex, this statement is equivalent to the variational inequality*

$$V^n(\mathbf{Y}) - V^n(\mathbf{X}) \geqslant \mathbb{S}^n : (\mathbf{Y} - \mathbf{X}), \quad \forall\, \mathbf{Y} \in \mathbb{R}^{d\times d}.$$

Moreover, $\mathbb{S}^n$ *satisfies the estimate*

$$|\mathbb{S}^n| \leqslant \mu_1|\mathbf{X}_s|^{p-1} + \mu_2|R|^{p-1} + \tau_*, \qquad \forall \mathbf{X} \in \mathbb{R}^{d\times d},\ \forall n \in \mathbb{N}. \tag{18}$$

The proof of this lemma is obtained by straightforward computations, Theorem 2 and Corollary 1.

Corollary 2 *Let* $\mathbb{S}$ *be a tensor in* $\mathbb{R}^{d\times d}$ *such that* $\mathbb{S} \in \partial V(\mathbb{B})$ *(more precisely,* $\mathbb{S}(\mathbf{x}, t) \in \partial V_{(\mathbf{x},t)}(\mathbb{B}(\mathbf{x}, t))$*), i.e., satisfies the variational inequality*

$$V_{(\mathbf{x},t)}(\mathbf{X}) - V_{(\mathbf{x},t)}(\mathbb{B}(\mathbf{x}, t)) \geqslant \mathbb{S}(\mathbf{x}, t) : (\mathbf{X} - \mathbb{B}(\mathbf{x}, t)), \quad \forall\, \mathbf{X} \in \mathbb{R}^{d\times d},$$

for almost all $(\mathbf{x}, t) \in \mathcal{O}_T$. *Then* $\mathbb{S}$ *accomplishes the relation* (5) *for almost all* $(\mathbf{x}, t) \in \mathcal{O}_T$, *when* $\nu > 0$ *or* $\nu = \mu_2 = 0$.

Corollary 3 *The "modified plastic operator"*

$$\mathcal{P}_{\nu,q}(\mathbb{B}_s, \mathrm{R}) = \frac{\mathbb{B}_{\nu,q}}{|\widehat{\mathbb{B}}_{\nu,q}|^{\frac{2(q-1)}{q}}} = \frac{|\mathbb{B}_s|^{q-2}\mathbb{B}_s + \nu|\mathrm{R}|^{q-2}\mathrm{R}}{(\sqrt[q]{|\mathbb{B}_s|^q + \nu|\mathrm{R}|^q})^{q-1}}, \qquad \forall q \geqslant 2, \tag{19}$$

is monotone, when $\nu > 0$ *or* $\nu = \mu_2 = 0$.

3 Auxiliary Results

In this section, we present some useful facts for our paper. We begin with basic facts on convex analysis.

Let $f : \mathbb{R}^m \to \mathbb{R}$ be a convex function. We recall that for each $\mathbf{y} \in \mathbb{R}^m$ there exists the one-sided directional derivative $f'(\mathbf{x}; \mathbf{y})$ of f at $\mathbf{x}$ with respect to the vector $\mathbf{y}$:

$$f'(\mathbf{x}; \mathbf{y}) = \lim_{\lambda \to 0+} \lambda^{-1}[f(\mathbf{x} + \lambda \mathbf{y}) - f(\mathbf{x})] = \inf_{\lambda > 0} \lambda^{-1}[f(\mathbf{x} + \lambda \mathbf{y}) - f(\mathbf{x})].$$

A vector $\mathbf{x}^*$ is said to be a *subgradient* of f at $\mathbf{x}$ if

$$f(\mathbf{y}) - f(\mathbf{x}) \geqslant \mathbf{x}^* \cdot (\mathbf{y} - \mathbf{x}), \quad \forall \mathbf{y} \in \mathbb{R}^m,$$

where the dot "$\cdot$" denotes the usual inner product in $\mathbb{R}^m$. The set of all subgradients of f at $\mathbf{x}$ is called the *subdifferential* of f at $\mathbf{x}$ and is denoted by $\partial f(\mathbf{x})$. A vector $\mathbf{x}^*$ is a subgradient of f at $\mathbf{x}$ if and only if $f'(\mathbf{x}; \mathbf{y}) \geqslant \mathbf{x}^* \cdot \mathbf{y}$, for all $\mathbf{y} \in \mathbb{R}^m$. (See [17] or [1].)

An elementary but important remark is that for any function $f : \mathbb{R}^m \to \mathbb{R}$ positively homogeneous of order 1 we have that $f'(0; \mathbf{y}) = f(\mathbf{y})$, for all $\mathbf{y} \in \mathbb{R}^m$.

It is also useful to note that the l^p norm in $\mathbb{R}^m$, i.e., $\|\mathbf{x}\|_{l^p(\mathbb{R}^m)} = \sqrt[p]{|x_1|^p + \ldots + |x_m|^p}$, $\mathbf{x} = (x_1, \cdots, x_m) \in \mathbb{R}^m$, is decreasing with respect to $p \in [1, \infty]$.

An auxiliary result that is often used is the equivalence of the norms

$$\|\mathbf{v}\|_{V_p} = \|\mathbf{v}\|_{L^2(\mathcal{O})} + \|\mathbf{v}\|_{L_p(\mathcal{O})}, \quad \|\mathbf{v}\|_{W^{1,p}(\mathcal{O})} = \|\mathbf{v}\|_{L^p(\mathcal{O})} + \|\mathbf{v}\|_{L^p(\mathcal{O})},$$

for $p \geqslant 2$, where $W^{1,p}(\mathcal{O})$ denotes the usual Sobolev space. Since $p \geqslant 2$, we have that $\|\mathbf{v}\|_{L^2(\mathcal{O})} \leqslant |\mathcal{O}|^{(p-2)/2p}\|\mathbf{v}\|_{L^p(\mathcal{O})}$ (where $|\mathcal{O}|$ is the Lebesgue measure of $\mathcal{O}$). Thus, the above equivalence is a direct consequence of the following lemma.

Lemma 2 *For any $p \geqslant 2$ there exists a constant $C = C(\Omega, p)$ such that*

$$\|\mathbf{v}\|_{L^p(\mathcal{O})} \leqslant C\|\mathbf{v}\|_{V_p}, \quad \forall\, \mathbf{v} \in W^{1,p}(\mathcal{O}). \tag{20}$$

This lemma is a consequence of Ehrling's lemma (also known as Lions' lemma - see [2, Exercise 6.12, p. 173]) by observing the continuous embeddings $W^{1,p}(\mathcal{O}) \subset L^p(\mathcal{O}) \subset L^2(\mathcal{O})$ being compact the first one. In the same way we can slightly improve the Korn's inequality

$$\|\mathbb{B}\|_{L^p(\mathcal{O})} \leqslant C\left(\|\mathbf{v}\|_{L^p(\mathcal{O})} + \|\mathbb{B}_s\|_{L^p(\mathcal{O})}\right), \quad \forall p \in (1, \infty),$$

(see [23, (1.70), p. 20]) to obtain the following result.

Proposition 1 *For any $p \in [2, \infty)$ there exists a constant C such that*

$$\|\mathbb{B}\|_{L^p(\mathcal{O})} \leqslant C\left(\|\mathbf{v}\|_{L^2(\mathcal{O})} + \|\mathbb{B}_s\|_{L^p(\mathcal{O})}\right), \quad \forall\, \mathbf{v} \in W^{1,p}(\mathcal{O}).$$

Other important results are the following. We use the continuity of the trace map from $V_p = W^{1,p}(\mathcal{O})$ to $L^p(\Gamma)$ to obtain the convergences (24) below in $L^p(\Gamma_T)$. The Aubin-Lions-Simon compactness result (see [22, 24]) is used to prove the strong convergence in (24). The lower semicontinuity of functionals of the form "$\mathcal{F}(u) = \int F(\mathbf{x}, u(\mathbf{x}))d\mathbf{x}$" in the weak topology [14, Theorem 4.3, p. 123] is essential to conclude the proof of Theorem 1, in order to show that our potential (5) and the boundary condition are achieved (accordingly to the integral equality (11)). Finally, we would like to mention the interpolation inequality in Lebesgue spaces

$$\|\mathbf{v}\|_{L^r(\mathcal{O})} \leqslant \|\mathbf{v}\|^{\theta_1}_{L^{s_1}(\mathcal{O})}\|\mathbf{v}\|^{\theta_2}_{L^{s_2}(\mathcal{O})}, \quad \forall\, \mathbf{v} \in L^{s_1}(\mathcal{O}) \cap L^{s_2}(\mathcal{O}),$$

$$\text{for } s_1, s_2, r \in [1, +\infty], \quad \theta_1, \theta_2 \in [0, 1], \quad \theta_1 + \theta_2 = 1, \quad \frac{\theta_1}{s_1} + \frac{\theta_2}{s_2} = \frac{1}{r},$$

used in the proof of Lemma 3, and the Sobolev imbedding $L^r(\mathcal{O}) \subset V_p$ with $r = dp/(d-p)$, if $1 \leqslant p < d$, any $r < \infty$, if $p = d$, and $r = \infty$, if $p > d$.

Some of the results/facts stated in this section does not appears explicitly where they are used in this abridged version of our work, but the reader is kindly invited to visit our more comprehensive version available in [11] and see where all those results/facts are used.

4 Sketch of the Proof of Theorem 1

In this section we provide the main ideas to prove Theorem 1.

We consider the following approximation to the ibvp (8), where $\mathbb{T}^n = -p^n\,\mathrm{I} + \mathbb{S}^n$ and $\mathbb{S}^n$ is the approximation for our viscous stress tensor (5), which was given in Lemma 1 (see (16)):

$$\begin{cases} \mathbf{v}^n_t + (\mathbf{v}^n \cdot \nabla)\,\mathbf{v}^n = \operatorname{div} \mathbb{T}^n, \qquad \operatorname{div} \mathbf{v}^n = 0, \qquad \text{in } \ \mathcal{O}_T, \\ \mathbf{v}^n \cdot \mathbf{n} = 0, \quad (\mathbb{T}^n\,\mathbf{n} + \nabla g(\mathbf{v}^n)) \cdot \boldsymbol{\tau} = 0 \qquad \text{on } \Gamma_T, \\ \mathbf{v}^n|_{t=0} = \mathbf{v}^n_0 \qquad \text{in } \ \mathcal{O}. \end{cases} \tag{21}$$

The approximate solutions $(\mathbf{v}_n)$ satisfy the energy equality

$$\frac{1}{2}\int_{\mathcal{O}} |\mathbf{v}^n(t)|^2 - |\mathbf{v}^n_0|^2\, d\mathbf{x} = -\int_{\mathcal{O}_t} \mathbb{S}^n : \mathbb{B}^n\ d\mathbf{x}\,ds - \int_{\Gamma_t} \nabla g(\mathbf{v}^n) \cdot \mathbf{v}^n\, d\Gamma\, ds, \tag{22}$$

for any $t \in (O, T)$. Then the non-negativity condition in (9) for the boundary function g, the coercivity condition (estimate from below) in (17) for the approximate potential V^n and an appropriate version of Korn's inequatility (see Proposition 1) yield the estimates

$$\begin{aligned} \|\mathbf{v}^n\|_{L^\infty(0,T;H)} + \|\mathbf{v}^n\|_{L^p(0,T;V_p)} \leqslant C, \\ \|\partial_t \mathbf{v}^n\|_{L^{p'}(0,T;\, W^*)} \leqslant C, \quad \|\mathbb{S}^n\|_{L^{p'}(\mathcal{O}_T)} \leqslant C, \end{aligned} \tag{23}$$

where W^* is dual space of $W = W^{3,2}(\mathcal{O}) \cap V$ and the positive constants $C = C(p, q, \mathbf{v}_0, \mu_1, \alpha, \Omega)$ do not depend on n. Then, we have the following convergences (up to subsequences), with the help of the continuity of the trace map from $V_p = W^{1,p}(\mathcal{O})$ to $L^p(\Gamma)$, the Aubin-Lions-Simon compactness result [22, 24] and the boundedness condition in (9) for the gradient of the boundary function g:

$$\begin{array}{llll} \mathbf{v}^n & \rightharpoonup \mathbf{v} & \text{weakly in} & L^\infty(0,T;H) \cap L^p(0,T;V_p) \cap L^p(\Gamma_T), \\ \mathbf{v}^n & \to \mathbf{v} & \text{strongly in} & L^2(\mathcal{O}_T), \\ \partial_t \mathbf{v}^n & \rightharpoonup \partial_t \mathbf{v} & \text{weakly in} & L^{p'}(0,T;W^*), \\ \mathbb{S}^n & \rightharpoonup \mathbb{S} & \text{weakly in} & L^{p'}(\mathcal{O}_T), \\ \nabla g(\mathbf{v}^n) & \rightharpoonup \mathbf{u} & \text{weakly in} & L^p(\Gamma_T), \end{array} \tag{24}$$

for some $\mathbb{S} \in L^{p'}(\mathcal{O}_T) = L^{p'}(\mathcal{O}_T; \mathbb{R}^{3\times 3})$, $\mathbf{u} \in L^p(\Gamma_T) = L^p(\Gamma_T; \mathbb{R}^3)$. Thus, we can take $n \to \infty$ in the weak formulation of (21) and obtain (11) with $\mathbf{u}$ in place of $\nabla g(\mathbf{v})$. At this point, it will remain to show only that

$$\mathbb{S} \in \partial V(\mathbb{B}) \text{ (the relation (5))} \quad \text{and} \quad \mathbf{u} = \nabla g(\mathbf{v}). \tag{25}$$

We can show this by the argument which we explain in the sequel for the convenience of the reader. Although the argument is standard, we would like to draw attention to the fact that here enters the condition $p \geqslant 2.2 = 11/5$. More precisely, in this part of the proof of Theorem 1 we utilize the following lemma.

Lemma 3 *Under the hypothesis stated in Theorem 1 - in particular we stress the condition $p \geqslant 2.2$ -, the function $\mathbf{v} \in L^\infty(0,T;H) \cap L^p(0,T;V_p)$ fulfilling the integral equality* (11) *belongs to the space $L^{p'}(0,T;\ V_p^*)$.*

The main ingredients to show that $\mathbb{S} \in \partial V(\mathbb{B})$ and $\mathbf{u} = \nabla g(\mathbf{v})$ appears to be the lower semicontinuity of functionals of the form "$\mathcal{F}(u) = \int F(\mathbf{x}, u(\mathbf{x}))d\mathbf{x}$" in the weak topology [14, Theorem 4.3, p. 123], the condition $p \geqslant 2.2$ and the energy equality (22). Indeed, from [14, Theorem 4.3, p. 123] and the convexity of the functions V, g, we have for any $\mathbb{X} \in L^2(\mathcal{O}_T)$, $\mathbf{w} \in L^2(\Gamma_T)$ and $t \in (0, T)$:

$$\begin{aligned} &\int_{\mathcal{O}_t} V(\mathbb{X}) - V(\mathbb{B})\, d\mathbf{x}ds + \int_{\Gamma_t} g(\mathbf{w}) - g(\mathbf{v})\, d\Gamma ds \\ &\qquad \geqslant \liminf_{n\to\infty} \Big[\int_{\mathcal{O}_t} V^n(\mathbb{X}) - V^n(\mathbb{B}^n)\, d\mathbf{x}ds + \int_{\Gamma_t} g(\mathbf{w}) - g(\mathbf{v}^n)\, d\Gamma ds \Big] \end{aligned}$$

$$\geqslant \liminf_{n\to\infty}[\int_{\mathcal{O}_t} \mathbb{S}^n : (\mathbb{X} - \mathbb{B}^n)\, d\mathbf{x}ds + \int_{\Gamma_t} \nabla g(\mathbf{v}^n)\cdot(\mathbf{w}-\mathbf{v}^n)\, d\Gamma ds]$$

$$= \int_{\mathcal{O}_t} \mathbb{S} : \mathbb{X}\, d\mathbf{x}ds + \int_{\Gamma_t} \mathbf{u}\cdot \mathbf{w} d\Gamma dt$$

$$+ \liminf_{n\to\infty}[-\int_{\mathcal{O}_t} \mathbb{S}^n : \mathbb{B}^n\, d\mathbf{x}ds - \int_{\Gamma_t} \nabla g(\mathbf{v}^n)\cdot \mathbf{v}^n\, d\Gamma ds].$$

Substituting (22) in the above obtained inequality, we obtain

$$\int_{\mathcal{O}_t} V(\mathbb{X}) - V(\mathbb{B})\, d\mathbf{x}ds + \int_{\Gamma_t} g(\mathbf{w}) - g(\mathbf{v})\, d\Gamma ds$$

$$\geqslant \int_{\mathcal{O}_t} \mathbb{S} : \mathbb{X}\, d\mathbf{x}ds + \int_{\Gamma_t} \mathbf{u}\cdot \mathbf{w} d\Gamma dt + \liminf_{n\to\infty}[\tfrac{1}{2}\int_{\mathcal{O}} |\mathbf{v}^n(t)|^2 - |\mathbf{v}_0^n|^2\, d\mathbf{x}]$$

then, using the strong convergence $\mathbf{v}^n \to \mathbf{v}$ in $L^2(\mathcal{O}_T)$ by (24) and $\mathbf{v}_0^n \to \mathbf{v}_0$ also strongly in $L^2(\mathcal{O})$, it follows that

$$\int_{\mathcal{O}_t} V(\mathbb{X}) - V(\mathbb{B})\, d\mathbf{x}ds + \int_{\Gamma_t} g(\mathbf{w}) - g(\mathbf{v})\, d\Gamma ds$$

$$\geqslant \int_{\mathcal{O}_t} \mathbb{S} : \mathbb{X}\, d\mathbf{x}ds + \int_{\Gamma_t} \mathbf{u}\cdot \mathbf{w} d\Gamma dt + \frac{1}{2}\int_{\mathcal{O}} |\mathbf{v}(t)|^2 - |\mathbf{v}_0|^2\, d\mathbf{x} \tag{26}$$

for a.a. $t \in (0, T)$. We now apply the counterpart formula (22) for $\mathbf{v}$ in the form

$$\frac{1}{2}\int_{\mathcal{O}} |\mathbf{v}(t)|^2 - |\mathbf{v}_0|^2\, d\mathbf{x} = -\int_{\mathcal{O}_t} \mathbb{S} : \mathbb{B}\, d\mathbf{x}ds - \int_{\Gamma_t} \mathbf{u}\cdot \mathbf{v}\, d\Gamma\, ds. \tag{27}$$

It can be obtained from the weak formulation (11) with $\mathbf{g}$ in place of $\nabla g(\mathbf{v})$, by choosing test functions approaching $\mathbf{v}$. In fact, using Lemma 3 we can put $\boldsymbol{\varphi} = \zeta_\varepsilon(t)\mathbf{v}$ in the equation

$$\int_{\mathcal{O}_T} [\mathbf{v}\partial_t\boldsymbol{\varphi} + (\mathbf{v}\otimes\mathbf{v} - \mathbb{S}) : \nabla\boldsymbol{\varphi}]d\mathbf{x}dt + \int_{\mathcal{O}} \mathbf{v}_0\boldsymbol{\varphi}(0)d\mathbf{x} = \int_{\Gamma_T} \mathbf{u}\cdot\boldsymbol{\varphi} d\Gamma dt$$

with a smooth sequence ζ_ε converging pointwisely to the characteristic function $\chi_{(0,t)}$ of the time interval $(0, t)$. As $\varepsilon \to 0$, we get

$$-\int_{\mathcal{O}} |\mathbf{v}(t)|^2 d\mathbf{x} + \int_0^t \langle \mathbf{v}_t, \mathbf{v}\rangle_{V_p^*, V_p} ds - \int_{\mathcal{O}_t} \mathbb{S} : \nabla\mathbf{v} d\mathbf{x}ds + \int_{\mathcal{O}} |\mathbf{v}_0|^2 d\mathbf{x} = \int_{\Gamma_t} \mathbf{u}\cdot\mathbf{v} d\Gamma ds$$

for a.a. $t \in (0, T)$. But $\int\limits_{\mathcal{O}} |\mathbf{v}(t)|^2 - |\mathbf{v}_0|^2 \, d\mathbf{x} = \int_0^t 2\langle \mathbf{v}_t, \mathbf{v}\rangle_{V_p^*, V_p} ds$, so we obtain (27). From (26) and (27) we have that

$$\int\limits_{\mathcal{O}_t} V(\mathbb{X}) - V(\mathbb{B})\, d\mathbf{x} ds + \int\limits_{\Gamma_t} g(\mathbf{w}) - g(\mathbf{v})\, d\Gamma ds$$
$$\geqslant \int\limits_{\mathcal{O}_t} \mathbb{S} : (\mathbb{X} - \mathbb{B})\, d\mathbf{x} ds + \int\limits_{\Gamma_t} \mathbf{u} \cdot (\mathbf{w} - \mathbf{v})\, d\Gamma\, ds. \tag{28}$$

Since the matrix function $\mathbb{X} \in L^2(\mathcal{O}_T)$ and the vector function $\mathbf{w} \in L^2(\Gamma_T)$ are arbitrary, from this inequality it is possible to prove that $V'(\mathbb{B}; \mathbb{Z}) \geqslant \mathbb{S} : \mathbb{Z}$ and $g'(\mathbf{v}; \mathbf{z}) \geqslant \mathbf{u} \cdot \mathbf{z}$ for any matrix function $\mathbb{Z} \in L^2(\mathcal{O}_T)$ and vector function $\mathbf{z} \in L^2(\Gamma_T)$, and thus, we get (25).

Acknowledgments The authors would like to thank Prof. Anderson L. A. de Araujo (UFV, Brazil) for useful discussions under the article subject.

The work of N.V. Chemetov was supported by FAPESP (Fundação de Amparo à Pesquisa do Estado de São Paulo), project 2021/03758-8, "Mathematical problems in fluid dynamics" and the Visiting Professor Projects 2023/11369-7, "Fluid-rigid body" interaction problem", 2023/11593-4, "Non-newtonian fluid - Rigid body" interaction problem".

References

1. Bauschke, H.H., Combettes, L.: Convex Analysis and Monotone Operator Theory in Hilbert Spaces. Springer-Verlag, New York (2011)
2. Brezis, H.: Functional Analysis, Sobolev Spaces and Partial Differential Equations. Springer-Verlag, New York (2011)
3. Chemetov, N.V., Antontsev, S.N.: Euler equations with non-homogeneous Navier slip boundary condition. Phys. D: Nonlinear Phenomena **237**, 92–10 (2008)
4. Chemetov, N.V., Cipriano, F.: Problema de camada limite: equações de Navier-Stokes e de Euler. Boletim da Sociedade Portuguesa de Matematica **Special**, 31–35 (2012)
5. Chemetov, N.V., Cipriano, F.: Boundary layer problem: Navier-Stokes equations and Euler equations. Nonlinear Anal. Real World Appl. **14**(6), 2091–2104 (2013)
6. Chemetov, N.V., Cipriano, F.: The inviscid limit for slip boundary conditions. Hyperbolic Prob. Theory Numer. Appl. **8**, 431–439 (2014)
7. Chemetov, N.V., Cipriano, F.: Inviscid limit for Navier-Stokes equations in domains with permeable boundaries. Appl. Math. Letters **33**, 6–11 (2014)
8. Chemetov, N.V., Cipriano, F.: Injection-suction control for two-dimensional Navier–Stokes equations with slippage. SIAM J. Control Optim. **56**, 1253–1281 (2018)
9. Chemetov, N.V., Cipriano, F.: A boundary control problem for stochastic 2D-Navier-Stokes equations. J. Optim. Theory Appl. 1–33 (2024)
10. Chemetov, N.V., de Araujo, A.L.A.: Well-posedness of the Cosserat-Bingham fluid equations. Nonlinear Differ. Equ. Appl. **29**(31) (2022)
11. Chemetov, N.V., Santos M.M.: A family of systems including the Herschel-Bulkley fluid equations (2024). https://arxiv.org/abs/2401.13830
12. Chemetov, N.V., Cipriano, F., Gavrilyuk S.: Shallow water model for the lake with friction and penetration. Math. Methods Appl. Sci. **29**(6), 687–703 (2010)
13. Eringen, A.C.: Microcontinuum Field Theories. Springer-Verlag, New York (1999)

14. Giusti, E.: Direct Methods in the Calculus of Variations. World Scientific Publishing Co. Pte. Ltd., Singapore (2003)
15. Lukaszewicz, G.: Microploar Fluids, Theory and Applications. Birkhäuser Inc., Boston (1999)
16. Malek, J., Nečas, J., Rokyta, M., Růžička, M.: Weak and Measure-valued Solutions to Evolutionary PDEs. Chapman&Hall, London (1996)
17. Rockafellar, R.T.: Convex Analysis. Princeton University Press, Princeton (1970)
18. Ružička, M., Shelukhin, V.V., Santos, M.M.: Steady flows of Cosserat-Bingham fluids. Math. Meth. Appl. Sci. **40**(7), 2746–2761 (2017)
19. Shelukhin, V.V., Chemetov, N.V.: Global solvability of the 1-D Cosserat Bingham fluid equations. J. Math. Fluid Mech. **17**(3), 495–511 (2015)
20. Shelukhin, V.V., Ružička, M.: On Cosserat–Bingham fluids. Z. Angew. Math. Mech., **93**(1), 57–72 (2013)
21. Teisseyre, R., Teisseyre-Jerenska, M.: Asymmetric Continuum: Extreme Processes in Solids and Fluids. Springer-Heldelberg, London (2014)
22. Simon, J.: Compact sets in the space $L^p(0, T; B)$. Ann. Mat. Pura Appl. **146**, 65–96 (1987)
23. Temam, R.: Mathematical Problems in Plasticity. Gauthier-Villars, Jura (1985)
24. Temam, R.: Navier-Stokes Equations, Theory and Numerical Analysis. AMS Chelsea Publishing, Providence (2001)

On the Well-posedness for the Inviscid Boussinesq Equations in Besov-Herz Spaces

Daniel Ferreira Machado

Abstract This paper is dedicated to the study of the incompressible inviscid Boussinesq equations in the whole space $\mathbb{R}^n$ ($n \geq 2$). We obtain the local well-posedness and blow-up criterion in a new framework, namely Besov-Herz spaces $BK^{\alpha,s}_{p,q,r}(\mathbb{R}^n)$, covering critical and supercritical cases of regularity ($s \geq n/p + 1$). To obtain such results, we need to obtain appropriate linear estimates for a linearized inviscid Boussinesq system in our setting.

1 Introduction

This paper is concerned with the following inviscid Boussinesq equations

$$\begin{cases} \partial_t u + u \cdot \nabla u + \nabla \pi = \theta e_n, \\ \partial_t \theta + u \cdot \nabla \theta = 0, \\ \operatorname{div} u = 0, \\ (u, \theta)_{|_{t=0}} = (u_0, \theta_0), \end{cases} \qquad (x,t) \in \mathbb{R}^n \times \mathbb{R}^+, \tag{1}$$

where $n \geq 2$, θ is the temperature, $u = (u_1, \ldots, u_n)$ is the velocity field of the fluid and π denotes the scalar pressure. Moreover, e_n is standard to denote the vector $(0, \ldots, 0, 1)$ and the pressure term $\nabla\pi$ can be expressed as

$$\nabla \pi = (-\Delta)^{-1} \nabla \operatorname{div}(u \cdot \nabla u) - (-\Delta)^{-1} \nabla \operatorname{div}(\theta e_n).$$

Boussinesq systems are widely used to describe the dynamics of the oceans or the atmosphere, where rotation and stratification are important, as well as other physical problems. For more details on other physical problems, see [1, 5]. Furthermore, the inviscid 2D Boussinesq equations can be used to model the 3D axisymmetric Euler equations with swirl from the symmetric axis $r = 0$ (see [11]) and also

D. F. Machado (✉)
Universidade Estadual de Campinas, IMECC - Departamento de Matemática, Campinas - SP, Brasil

M. R. Ebert et al. (eds.), *New Tools in Mathematical Analysis and Applications*, Research Perspectives, https://doi.org/10.1007/978-3-031-77050-0_19

describe the movement of fluids under the action of gravitational forces. Due to the open problems involving the Boussinesq equations (for example, viscous, partially viscous, inviscid Boussinesq equations, among others), the system (1) has been extensively studied, over the years, by several authors and in different functional spaces (see, e.g., [6, 7, 9, 11, 12, 15, 19, 24, 26, 28, 29]). In particular, Chae and Nam [6] proved a local well-posedness (existence and uniqueness) result and also a blow-up criterion for the system (1) in Sobolev spaces $W^{s,2}(\mathbb{R}^2)$ with $s > 2$. In the context of Besov spaces, Liu et al.[24] presented a result by proving the local well-posedness of solutions and a blow-up criterion in critical space $B^{2/p+1}_{p,1}(\mathbb{R}^2)$ with $1 < p < \infty$, and also, Yuan [29] proved a result of local existence and uniqueness of solutions in $B^s_{p,r}(\mathbb{R}^2)$ with $s \geq 2/p + 1$, $1 < p < \infty$ and $1 \leq r \leq \infty$, where $r = 1$ if $s = 2/p + 1$, as well as a blow-up criterion. After, Cui et al. [9] proved that (1) has a unique solution in the Hölder spaces $C^s(\mathbb{R}^n)$ with $s > 1$ and $n \geq 2$, and also presented a blow-up criterion. Xiang and Yan [26] and Bie et al. [3] showed similar results in Triebel-Lizorkin-Lorentz spaces $F^{s,r}_{p,q}(\mathbb{R}^n)$ with $n \geq 2$, $s > n/p + 1$, $1 < r \leq p < \infty$ and $1 < q < \infty$, and Besov-Morrey spaces $N^s_{p,q,r}(\mathbb{R}^n)$ with $n \geq 2$, $s \geq n/p+1$, $1 < q \leq p < \infty$ and $1 \leq r \leq \infty$, where $r = 1$ if $s = n/p+1$, respectively. Moreover, among the 2D results, we highlight Xu [28], who presented local well-posedness of solutions for isentropic compressible Boussinesq equations, as well as a blow-up criterion.

In this scenario, adapting the methods used by Ferreira and Machado in [13], and Ferreira and Pérez-López in [14], in this work our purpose is to establish the local well-posedness and also a blow-up criterion for the system (1) in a new framework of Besov-type spaces, more precisely, nonhomogeneous Besov-Herz spaces $BK^{\alpha,s}_{p,q,r}$, which are modeled on nonhomogeneous Besov spaces based on Herz spaces $K^{\alpha}_{p,q}$, and have been introduced in [27]. Our results cover critical ($s = n/p + 1$) and supercritical ($s > n/p + 1$) cases of regularity.

Below we present our main results for the well-posedness of solutions and for the blow-up criterion.

Theorem 1 *Let* $1 < p < \infty$, $1 \leq q, r \leq \infty$, $0 \leq \alpha < n(1 - 1/p)$ *and also* $s \geq n/p + 1$ *with* $r = 1$ *if* $s = n/p + 1$. *Assume that the initial data* $(u_0, \theta_0) \in BK^{\alpha,s}_{p,q,r} \times BK^{\alpha,s}_{p,q,r}$ *with* $\operatorname{div} u_0 = 0$. *Then, there exists a time* $T \in (0, \infty)$ *such that the inviscid Boussinesq system* (1) *admits a unique solution* (u, θ) *satisfying*

$$(u, \theta) \in C([0, T]; BK^{\alpha,s}_{p,q,r}) \times C([0, T]; BK^{\alpha,s}_{p,q,r}). \tag{2}$$

Based on the existence of solutions, we establish a blow-up criterion in Besov-Herz spaces.

Remark 1 (Blow-up Criterion) Consider $0 < T^* < \infty$ and u_0, θ_0 as in Theorem 1. Then, the aforementioned local solution (u, θ) blows up at time $T^* > T$ in space $BK^{\alpha,s}_{p,q,r}$, namely

$$\limsup_{t \nearrow T^*} \|u(t)\|_{BK^{\alpha,s}_{p,q,r}} = \infty \quad \text{or} \quad \limsup_{t \nearrow T^*} \|\theta(t)\|_{BK^{\alpha,s}_{p,q,r}} = \infty,$$

if and only if

$$\int_0^{T^*} \|\nabla \times u(t)\|_{\dot{B}^0_{\infty,\infty}} dt = \infty \quad \left(\text{resp.} \int_0^{T^*} \|\nabla \times u(t)\|_{\dot{B}^0_{\infty,1}} dt = \infty\right),$$

for $s > n/p + 1$ and $1 \leq r \leq \infty$ (resp. $s = n/p + 1$ with $r = 1$). The respective blow-up criterion above can be proved following essentially the same reasoning used in the proof of Proposition 1 combined with the logarithm inequality,

$$\|u\|_{L^\infty} \lesssim 1 + \|u\|_{\dot{B}^0_{\infty,\infty}} \left(\log^+ \|u\|_{BK^{\alpha,s'}_{p,q,r}} + 1\right) \qquad \text{for some } s' > n/p,$$

or the estimate $\|\nabla u\|_{L^\infty} \lesssim \|\nabla u\|_{\dot{B}^0_{\infty,1}}$, according to the respective case. Moreover, if we denote by $\mathcal{P}$ a singular integral operator and by M a constant matrix, we need the following decomposition $\nabla u = \mathcal{P}w + \mathrm{M}w$, where $w = \nabla \times u$, to complete the proof of blow-up criterion.

Remark 2 It is worth highlighting that the nonhomogeneous Besov-Herz space $BK^{\alpha,s}_{p,q,r}$ is larger than the nonhomogeneous Besov space $B^s_{p,r}$ for $\alpha = 0$, $q = \infty$ and $p < \infty$ and that the Herz space $K^\alpha_{p,q}$ is not contained in the Morrey space $M_{q,\lambda}$ for $\alpha = 0$ and $q = \infty$ (see Remark 5). Then, since spaces $BK^{0,s}_{p,\infty,r}$ contain Besov spaces $B^s_{p,r}$, Theorem 1 recovers and generalizes previous results in classical Besov spaces, such as [24] and [29], as well as providing a new class of solutions for (1).

This paper is organized as described below. In Sect. 2, we introduce some notations, recall the definition of Herz and Besov-Herz spaces, as well as some suitable basic estimates, in particular, for volume-preserving maps in the framework of Herz spaces. Section 3 is devoted to product, commutator and divergent-type estimates. Now, in the Sect. 4 we derive estimates for a linearized system associated with (1). Finally, the subject of Sect. 5 is the proof of Theorem 1 through of an approximate linear system, uniform estimates for the approximate scheme, and also contraction estimates, which is divided into three subsections.

2 Preliminaries

In this section we will review some definitions and also properties about our functional spaces and some operators, as well as the Littlewood-Paley theory, the Bony paraproduct, and related useful facts.

Here, we use the notation $A \lesssim B$ when there is a generic constant $C > 0$, which can be different in each case, such that $A \leq CB$, and we use $A \approx B$ when $A \lesssim B$ and $B \lesssim A$. Moreover, the notation $[F_1, F_2] = F_1F_2 - F_2F_1$ denotes the commutator between operators F_1 and F_2. On the other hand, $C_0^\infty(\mathbb{R}^n)$ represents the space of compactly supported smooth functions, $\mathcal{S} = \mathcal{S}(\mathbb{R}^n)$ the space of

Schwartz functions and $\mathcal{S}' = \mathcal{S}'(\mathbb{R}^n)$ stands for the space of tempered distributions. And also, $\mathbb{N}_0 := \mathbb{N} \cup \{0\}$ denotes the set of non-negative integers.

Due to the purpose of this paper, our framework are nonhomogeneous Besov-Herz spaces and, for the convenience of the reader, we introduce the definition of Herz spaces, as well as some of their properties. See, e.g., [13, 14, 16, 18, 20, 23, 27] for more details.

Definition 1 For $\alpha \in \mathbb{R}$, and $1 \leq p, q \leq \infty$, we define the nonhomogeneous Herz space $K^{\alpha}_{p,q} = K^{\alpha}_{p,q}(\mathbb{R}^n)$ as the set of functions $u \in L^p_{\mathrm{loc}}(\mathbb{R}^n)$ such that

$$\|u\|_{K^{\alpha}_{p,q}} := \begin{cases} \left(\sum_{k \geq -1} \left(2^{\alpha k} \|u\|_{L^p(A_k)} \right)^q \right)^{1/q} < \infty, & \text{if } q < \infty, \\ \sup_{k \geq -1} 2^{\alpha k} \|u\|_{L^p(A_k)} < \infty, & \text{if } q = \infty. \end{cases}$$

where $A_{-1} := \{x \in \mathbb{R}^n : |x| < 2^{-1}\}$ and $A_k := \{x \in \mathbb{R}^n : 2^{k-1} \leq |x| < 2^k\}$, for $k \geq 0$.

Remark 3 In the case $p < \infty$, $K^{\alpha}_{p,q}$ is a Banach space with the norm $\|\cdot\|_{K^{\alpha}_{p,q}}$, see, e.g., [17]. Moreover, in the case where $p = 1$, the space $K^{\alpha}_{1,q}$ is considered a space of signed measures and $\|u\|_{L^1(A_k)}$ represents the total variation of u on A_k.

Next, we will recall the Littlewood-Paley decomposition theory and some related facts. For more details on this theory we suggest, e.g., [8, 22]. Initially, consider two radial functions $\psi, \varphi \in C_0^{\infty}(\mathbb{R}^n)$ satisfying $\mathrm{supp}(\psi) \subset \mathcal{B}$ and $\mathrm{supp}(\varphi) \subset \mathcal{C}$, where $\mathcal{B} := \{\xi \in \mathbb{R}^n : |\xi| \leq 4/3\}$ and $\mathcal{C} := \{\xi \in \mathbb{R}^n : 3/4 \leq |\xi| \leq 8/3\}$, such that

$$\sum_{j \in \mathbb{Z}} \varphi(2^{-j}\xi) = 1, \quad \text{for } \xi \in \mathbb{R}^n \setminus \{0\}, \quad \text{and}$$

$$\psi(\xi) + \sum_{j \in \mathbb{N}_0} \varphi(2^{-j}\xi) = 1, \quad \text{for } \xi \in \mathbb{R}^n. \tag{3}$$

Now, if $\varphi_j(\xi) := \varphi(2^{-j}\xi)$ and $u \in \mathcal{S}'(\mathbb{R}^n)$, we define the localization operators $\dot{\Delta}_j$, Δ_j and S_j as follows $\dot{\Delta}_j u := \mathcal{F}^{-1}[\varphi_j \mathcal{F} u]$ for all $j \in \mathbb{Z}$,

$$\Delta_j u := \begin{cases} 0, & j \leq -2, \\ \mathcal{F}^{-1}[\psi \mathcal{F} u], & j = -1, \\ \dot{\Delta}_j u, & j \geq 0. \end{cases} \quad \text{and} \quad S_j u := \sum_{l \leq j-1} \Delta_l u,$$

where $\widehat{u} = \mathcal{F}u$ and $\mathcal{F}^{-1}u$ represents the Fourier transform and the inverse Fourier transform of u on $\mathbb{R}^n$, respectively. Thus, if we denote the set of polynomials with n variables by $\mathcal{P} = \mathcal{P}[\mathbb{R}^n]$, we obtain the Littlewood-Paley decompositions

$$u = \sum_{j\in\mathbb{Z}} \dot{\Delta}_j u, \quad \text{for all } u \in \mathcal{S}'(\mathbb{R}^n)/\mathcal{P}[\mathbb{R}^n], \quad \text{and}$$

$$u = \sum_{j\in\mathbb{Z}} \Delta_j u, \quad \text{for all } u \in \mathcal{S}'(\mathbb{R}^n),$$

satisfying the following conditions $\Delta_j(\Delta_k u) \equiv 0$ if $|j-k| \geq 2$, and also $\Delta_j(S_{k-1}u\Delta_k u) \equiv 0$ if $|j-k| \geq 5$.

In what follows, we recall the definitions of nonhomogeneous and homogeneous Besov-Herz spaces, which are type-Besov spaces based on Herz $K^{\alpha}_{p,q}$ spaces (see, e.g., [13, 14, 27] for more details).

Definition 2 For indices $1 \leq p,q,r \leq \infty$, and $\alpha, s \in \mathbb{R}$, we define the nonhomogeneous Besov-Herz space $BK^{\alpha,s}_{p,q,r} = BK^{\alpha,s}_{p,q,r}(\mathbb{R}^n)$ as the set of all $u \in \mathcal{S}'$ such that $\Delta_j u \in K^{\alpha}_{p,q}$ for every $j \geq -1$, and that

$$\|u\|_{BK^{\alpha,s}_{p,q,r}} := \begin{cases} \left(\sum_{j\geq -1} (2^{sj}\|\Delta_j u\|_{K^{\alpha}_{p,q}})^r\right)^{1/r} < \infty, & \text{if } r < \infty, \\ \sup_{j\geq -1} 2^{sj}\|\Delta_j u\|_{K^{\alpha}_{p,q}} < \infty, & \text{if } r = \infty. \end{cases}$$

The definition of the homogeneous Besov-Herz space $\dot{B}K^{\alpha,s}_{p,q,r} = \dot{B}K^{\alpha,s}_{p,q,r}(\mathbb{R}^n)$ is similar, just consider $\dot{\Delta}_j u \in K^{\alpha}_{p,q}$ for every $j \in \mathbb{Z}$ and $u \in \mathcal{S}'/\mathcal{P}$ instead of $\Delta_j u \in K^{\alpha}_{p,q}$ for every $j \geq -1$ and $u \in \mathcal{S}'$, respectively.

Now, we recall some properties about homogeneous and nonhomogeneous Besov-Herz spaces.

Remark 4 In general, consider $1 \leq p,q,r \leq \infty$, and $\alpha, s \in \mathbb{R}$.

1. For $p < \infty$, the pairs $\left(\dot{B}K^{\alpha,s}_{p,q,r}, \|\cdot\|_{\dot{B}K^{\alpha,s}_{p,q,r}}\right)$ and $\left(BK^{\alpha,s}_{p,q,r}, \|\cdot\|_{BK^{\alpha,s}_{p,q,r}}\right)$ are Banach spaces.
2. Moreover, in the case $s > 0$ we have that

$$BK^{\alpha,s}_{p,q,r} = \dot{B}K^{\alpha,s}_{p,q,r} \cap K^{\alpha}_{p,q} \quad \text{with} \quad \|u\|_{BK^{\alpha,s}_{p,q,r}} \approx \|u\|_{\dot{B}K^{\alpha,s}_{p,q,r}} + \|u\|_{K^{\alpha}_{p,q}}. \tag{4}$$

3. On the other hand, for $\alpha \geq 0$ and $s \geq n/p$ with $r = 1$ if $s = n/p$, it follows that

$$BK^{\alpha,s}_{p,q,r} \hookrightarrow L^{\infty} \quad \text{and} \quad \dot{B}K^{\alpha,n/p}_{p,q,1} \hookrightarrow L^{\infty}. \tag{5}$$

4. And for $s_1 > s_2$ and $1 \leq r_1, r_2 \leq \infty$, the inclusion relation $BK^{\alpha,s_1}_{p,q,r_1} \subset BK^{\alpha,s_2}_{p,q,r_2}$ holds.

In what follows, we highlight some relations between Herz $K^{\alpha}_{p,q}$, Lebesgue L^p and Morrey $M_{p,\lambda}$ spaces, as well as the relation between Besov-Herz $BK^{s}_{p,r}$ and Besov $B^{s}_{p,r}$ spaces (see, [14, Lemma 3.3]). For more details about the spaces $B^{s}_{p,r}$ and $M_{p,\lambda}$, we suggest [2, 21], in this order.

Remark 5 If $1 \leq p < \infty$, $1 \leq r \leq \infty$ and $s \in \mathbb{R}$, the strict inclusions $L^p \subset K^{0}_{p,\infty}$ and $B^{s}_{p,r} \subset BK^{0,s}_{p,\infty,r}$ there holds. Moreover, for $1 \leq q < \infty$ and $0 \leq \lambda < n$ with $\lambda \neq n(1 - q/p)$ when $q < p$, we have $K^{0}_{p,\infty} \not\subset M_{q,\lambda}$.

Finally, we recall the decomposition of the Bony paraproduct and, for more details, see [4]. For $u, v \in \mathcal{S}'$, the product of u and v can be written as following $uv = T_u v + T_v u + \mathcal{R}(u, v)$, where

$$T_u v := \sum_{j\in\mathbb{Z}} S_{j-1} u \Delta_j v \quad \text{and} \quad \mathcal{R}(u, v) := \sum_{j\in\mathbb{Z}} \Delta_j u \widetilde{\Delta}_j v, \qquad \text{for}$$

$$\widetilde{\Delta}_j v := \sum_{i=-1}^{1} \Delta_{j-i} v.$$

2.1 Some Basic Estimates in the Framework of Herz Spaces $K^{\alpha}_{p,q}$

First, we recall the Hölder-type inequality in Herz spaces (see, e.g., [13, 14, 25]).

Lemma 1 *Consider* $1 \leq p, p_1, p_2, q, q_1, q_2 \leq \infty$ *and* $\alpha, \alpha_1, \alpha_2 \in \mathbb{R}$ *satisfying* $1/p = 1/p_1 + 1/p_2$, $1/q = 1/q_1 + 1/q_2$ *and* $\alpha = \alpha_1 + \alpha_2$. *Then, for all* $u \in K^{\alpha_1}_{p_1,q_1}$ *and* $v \in K^{\alpha_2}_{p_2,q_2}$, *there is a universal constant* $C > 0$ *such that*

$$\|uv\|_{K^{\alpha}_{p,q}} \leq C \|u\|_{K^{\alpha_1}_{p_1,q_1}} \|v\|_{K^{\alpha_2}_{p_2,q_2}}, \tag{6}$$

In particular, taking $(\alpha_1, p_1, q_1) = (0, \infty, \infty)$ *in (6), if* $u \in L^\infty$ *and* $v \in K^{\alpha}_{p,q}$, *we obtain*

$$\|uv\|_{K^{\alpha}_{p,q}} \leq C \|u\|_{L^\infty} \|v\|_{K^{\alpha}_{p,q}}. \tag{7}$$

Next, we will also need a Young-type inequality in these spaces (see [14, Lemma 2.1]).

Lemma 2 *Let* $\alpha \in \mathbb{R}$ *and* $1 \leq p, q \leq \infty$. *Additionally, consider* $\phi \in L^1$ *satisfying*

$$M_\phi := \begin{cases} \max\left\{\|\phi\|_{L^1}, \||\cdot|^{\beta}\phi\|_{L^1}, \||\cdot|^{2\beta+\alpha p}\phi\|_{L^1}\right\} < \infty, & \text{if } \alpha \geq 0, \\ \max\left\{\|\phi\|_{L^1}, \||\cdot|^{\beta-\alpha p}\phi\|_{L^1}, \||\cdot|^{2\beta}\phi\|_{L^1}\right\} < \infty, & \text{if } \alpha < 0, \end{cases}$$

for some $\beta > 0$. Then, for all $u \in K^{\alpha}_{p,q}$, there is a constant $C > 0$, independent of ϕ, such that

$$\|\phi * u\|_{K^{\alpha}_{p,q}} \leq CM_{\phi}\|u\|_{K^{\alpha}_{p,q}}. \tag{8}$$

The following result compiles some Bernstein-type inequalities in the context of Herz spaces partially found in the literature (see [14, Lemma 2.2]). For constants $0 < R_1 < R_2$, we establish the notations $\mathcal{B}(0, R_1) := \{\xi \in \mathbb{R}^n; |\xi| \leq R_1\}$ and $\mathcal{C}(0, R_1, R_2) := \{\xi \in \mathbb{R}^n; R_1 \leq |\xi| \leq R_2\}$.

Lemma 3 *Let $1 \leq p, q \leq \infty$, $\alpha \in \mathbb{R}$, and also $0 < R_1 < R_2$ and $j \in \mathbb{Z}$.*

1. *For $u \in K^{\alpha}_{p,q}$ such that $\operatorname{supp}\widehat{u} \subset \mathcal{B}(0, R_1 2^j)$, there is a constant $C > 0$, independent of j, such that the following estimate holds*

$$\|\partial^{\beta} u\|_{K^{\alpha}_{p,q}} \leq C2^{j|\beta|}\|u\|_{K^{\alpha}_{p,q}}. \tag{9}$$

2. *For $u \in K^{\alpha}_{p,q}$ such that $\operatorname{supp}\widehat{u} \subset \mathcal{C}(0, R_1 2^j, R_2 2^j)$, there is a constant $C > 0$, independent of j, such that the following estimate holds*

$$C^{-1}2^{j|\beta|}\|u\|_{K^{\alpha}_{p,q}} \leq \|\partial^{\beta} u\|_{K^{\alpha}_{p,q}} \leq C2^{j|\beta|}\|u\|_{K^{\alpha}_{p,q}}. \tag{10}$$

3. *For $\alpha \geq 0$ and $u \in K^{\alpha}_{p,q}$ such that $\operatorname{supp}\widehat{u} \subset \mathcal{B}(0, R_1 2^j)$, there is a constant $C > 0$, independent of j, such that the following estimate holds*

$$\|u\|_{L^{\infty}(\mathbb{R}^n)} \leq C2^{jn/p}\|u\|_{K^{\alpha}_{p,q}}, . \tag{11}$$

As a expected consequence of Lemma 3, we obtain the following remark.

Remark 6 Under the conditions of Lemma 3 and assuming $|\beta| = k$ with $k \in \mathbb{N}_0$, there exists a universal constant $C > 0$ such that

$$\|\partial^{\beta} u\|_{BK^{\alpha,s}_{p,q,r}} \leq C\|u\|_{BK^{\alpha,s+k}_{p,q,r}} \quad \text{and}$$

$$C^{-1}\|u\|_{\dot{B}K^{\alpha,s+k}_{p,q,r}} \leq \|\partial^{\beta} u\|_{\dot{B}K^{\alpha,s}_{p,q,r}} \leq C\|u\|_{\dot{B}K^{\alpha,s+k}_{p,q,r}}.$$

2.2 Volume-Preserving Maps

In this section we recall an estimate for volume-preserving maps X in the context of Herz spaces. For a full proof of the result below, we refer the reader to [14, Lemma 4.1].

Lemma 4 *Let $1 \leq p, q \leq \infty$, $\alpha \geq 0$, and $X : \mathbb{R}^n \to \mathbb{R}^n$ a volume-preserving diffeomorphism satisfying*

$$|X^{\pm 1}(x_0) - x_0| \leq \gamma, \qquad \text{for all } x_0 \in \mathbb{R}^n, \tag{12}$$

and some fixed $\gamma > 0$. Then, for all $u \in K^{\alpha}_{p,q}$, there exists a constant $C > 0$ such that

$$C^{-1}\|u\|_{K^{\alpha}_{p,q}} \leq \|u \circ X\|_{K^{\alpha}_{p,q}} \leq C\|u\|_{K^{\alpha}_{p,q}}. \tag{13}$$

For handling purposes, the following remark will be useful in our context.

Remark 7 Note that, if X is the flow associated with the ODE

$$\begin{cases} \partial_t X(y,t) = v(X(y,t),t) \\ X(y,0) = y \end{cases}, \qquad (y,t) \in \mathbb{R}^n \times \mathbb{R}^+, \tag{14}$$

then X satisfies the estimate $|X^{\pm 1}(y,t) - y| \leq \int_0^t \|v(\tau)\|_{L^\infty} d\tau$. Moreover, if $\operatorname{div} v = 0$, we have that X is a volume-preserving diffeomorphism for each $t \geq 0$ and thus satisfies the estimate (12) under appropriate time-integrability conditions on v.

3 Product, Commutator and Divergent-type Estimates

In this section we present some estimates for the product and commutator operators as well as divergent-type estimates in the framework of nonhomogeneous Besov-Herz spaces. For similar estimates, the reader is referred to [13, 14]. First, we have estimates for the product.

Lemma 5 *Let $1 \leq p, q, r \leq \infty$ and $\alpha \in \mathbb{R}$.*

1. *For $s > 0$ and $u, v \in BK^{\alpha,s}_{p,q,r} \cap L^\infty$, there exists a constant $C > 0$ such that the following estimate holds*

$$\|uv\|_{BK^{\alpha,s}_{p,q,r}} \leq C\left(\|u\|_{L^\infty}\|v\|_{BK^{\alpha,s}_{p,q,r}} + \|v\|_{L^\infty}\|u\|_{BK^{\alpha,s}_{p,q,r}}\right). \tag{15}$$

2. *For $p < \infty$, $\alpha \geq 0$, $s \geq n/p + 1$ (with $r = 1$ if $s = n/p + 1$) and also $u \in BK^{\alpha,s-1}_{p,q,r}$ and $v \in BK^{\alpha,s}_{p,q,r}$, there exists a constant $C > 0$ such that the following estimate holds*

$$\|u \cdot \nabla v\|_{BK^{\alpha,s-1}_{p,q,r}} \leq C\|u\|_{BK^{\alpha,s-1}_{p,q,r}}\|v\|_{BK^{\alpha,s}_{p,q,r}}. \tag{16}$$

Next, we present commutator estimates in Besov-Herz spaces, see [13, Lemma 3.3] for a full proof or [10, 14, 30] for similar estimates. Remember that we consider $\|\cdot\|_{\ell^r}$ with $j \geq -1$.

Lemma 6 *Let $1 \leq p < \infty$, $1 \leq q, r \leq \infty$, $\alpha \geq 0$ and v be a vector field such that* $\operatorname{div} v = 0$.

1. *For $s > 0$, $u, v \in BK^{\alpha,s}_{p,q,r}$ and $\nabla u, \nabla v \in L^\infty$, there exists a constant $C > 0$ such that the estimate holds*

$$\|2^{sj}\|[\Delta_j, v \cdot \nabla]u\|_{K^\alpha_{p,q}}\|_{\ell^r} \leq C\left(\|\nabla u\|_{L^\infty}\|v\|_{BK^{\alpha,s}_{p,q,r}} + \|\nabla v\|_{L^\infty}\|u\|_{BK^{\alpha,s}_{p,q,r}}\right). \tag{17}$$

2. *For $s \geq n/p + 1$ (with $r = 1$ if $s = n/p + 1$) and $u, v \in BK^{\alpha,s}_{p,q,r}$, there exists a constant $C > 0$ such that the estimate holds*

$$\|2^{sj}\|[\Delta_j, v \cdot \nabla]u\|_{K^\alpha_{p,q}}\|_{\ell^r} \leq C\|u\|_{BK^{\alpha,s}_{p,q,r}}\|v\|_{BK^{\alpha,s}_{p,q,r}}. \tag{18}$$

Moreover, if $u \in BK^{\alpha,s-1}_{p,q,r}$ and $v \in BK^{\alpha,s}_{p,q,r}$, the following estimate holds

$$\|2^{(s-1)j}\|[\Delta_j, v \cdot \nabla]u\|_{K^\alpha_{p,q}}\|_{\ell^r} \leq C\|u\|_{BK^{\alpha,s-1}_{p,q,r}}\|v\|_{BK^{\alpha,s}_{p,q,r}}. \tag{19}$$

To treat the pressure term $\nabla\pi$, we essentially need to estimate the term $(-\Delta)^{-1}\nabla\operatorname{div}(u \cdot \nabla v)$ in our context of Besov-Herz spaces. The next lemma gives us the necessary estimates. For a full proof, see, for example, [14, Lemma 4.5].

Lemma 7 *Let $1 \leq p < \infty$, $1 \leq q, r \leq \infty$. Let us assume $0 \leq \alpha < n(1 - 1/p)$ when $p \neq 1$ and $\alpha = 0$ when $p = 1$. Moreover, let u, v be divergence-free vector fields.*

1. *For $1 < p < \infty$, $s > 1$, $u, v \in BK^{\alpha,s}_{p,q,r}$ and $\nabla u, \nabla v \in L^\infty$, there exists a constant $C > 0$ such that the following estimate holds*

$$\begin{aligned}\|(-\Delta)^{-1}\nabla\operatorname{div}(u \cdot \nabla v)\|_{BK^{\alpha,s}_{p,q,r}} \\ \leq C\left(\|\nabla u\|_{L^\infty}\|v\|_{BK^{\alpha,s}_{p,q,r}} + \|\nabla v\|_{L^\infty}\|u\|_{BK^{\alpha,s}_{p,q,r}}\right).\end{aligned} \tag{20}$$

2. *For $s \geq n/p$ (with $r = 1$ if $s = n/p$) and $u, v \in BK^{\alpha,s}_{p,q,r}$, there exists a constant $C > 0$ such that the following estimate holds*

$$\|(-\Delta)^{-1}\nabla\operatorname{div}(u \cdot \nabla v)\|_{BK^{\alpha,s}_{p,q,r}} \leq C\|u\|_{BK^{\alpha,s}_{p,q,r}}\|v\|_{BK^{\alpha,s}_{p,q,r}}. \tag{21}$$

Moreover, if $u \in BK^{\alpha,s-1}_{p,q,r}$ *and* $v \in BK^{\alpha,s}_{p,q,r}$*, the following estimate holds*

$$\|(-\Delta)^{-1}\nabla \mathrm{div}(u \cdot \nabla v)\|_{BK^{\alpha,s-1}_{p,q,r}} \leq C\|u\|_{BK^{\alpha,s-1}_{p,q,r}}\|v\|_{BK^{\alpha,s}_{p,q,r}}. \tag{22}$$

And also, if $u \in BK^{\alpha,s}_{p,q,r}$ *and* $v \in BK^{\alpha,s-1}_{p,q,r}$*, the following estimate holds*

$$\|(-\Delta)^{-1}\nabla \mathrm{div}(u \cdot \nabla v)\|_{BK^{\alpha,s-1}_{p,q,r}} \leq C\|u\|_{BK^{\alpha,s}_{p,q,r}}\|v\|_{BK^{\alpha,s-1}_{p,q,r}}. \tag{23}$$

Remark 8 The main point in the proof Lemma 7 is that Riesz transforms are bounded in $BK^{\alpha,s}_{p,q,r}$, due to their boundedness in Herz spaces $K^{\alpha}_{p,q}$ for $-n/p < \alpha < n(1-1/p)$ (see, e.g., [14, Lemma 2.4] or [23, Corollary 2.1]).

4 Linear Estimates

Due to our purpose, in this section, we develop estimates in our setting for a problem associated with the inviscid Boussinesq system (1), which will play a fundamental role in the proof of Theorem 1.

Proposition 1 *Let* $1 < p < \infty$, $1 \leq q, r \leq \infty$, $0 \leq \alpha < n(1-1/p)$, $s \geq n/p+1$ *(with* $r = 1$ *if* $s = n/p+1$*) and* $T > 0$. *Assume that* $u_0, \theta_0 \in BK^{\alpha,s}_{p,q,r}$ *and* $v \in L^{\infty}_T(BK^{\alpha,s}_{p,q,r})$ *with* $\mathrm{div}\, v = 0$. *Then, if* $(u,\theta) \in L^{\infty}_T(BK^{\alpha,s}_{p,q,r}) \times L^{\infty}_T(BK^{\alpha,s}_{p,q,r})$ *is a solution of the linearized inviscid Boussinesq system*

$$\begin{cases} \partial_t u + v \cdot \nabla u + \nabla \pi = \theta e_n \\ \partial_t \theta + v \cdot \nabla \theta = 0 \\ \mathrm{div}\, u = 0 \\ u(\cdot,0) = u_0, \quad \theta(\cdot,0) = \theta_0 \end{cases}, \qquad (x,t) \in \mathbb{R}^n \times \mathbb{R}^+, \tag{24}$$

there exists a constant $C > 0$ *such that*

$$\|u(t)\|_{BK^{\alpha,s}_{p,q,r}} \leq C\Big[\|u_0\|_{BK^{\alpha,s}_{p,q,r}} + \int_0^t \Big(\|\theta(\tau)\|_{BK^{\alpha,s}_{p,q,r}} + \|u(\tau)\|_{BK^{\alpha,s}_{p,q,r}}\|v(\tau)\|_{BK^{\alpha,s}_{p,q,r}}\Big)\, d\tau\Big], \tag{25}$$

and also

$$\|\theta(t)\|_{BK^{\alpha,s}_{p,q,r}} \leq C\left(\|\theta_0\|_{BK^{\alpha,s}_{p,q,r}} + \int_0^t \|\theta(\tau)\|_{BK^{\alpha,s}_{p,q,r}}\|v(\tau)\|_{BK^{\alpha,s}_{p,q,r}}\, d\tau\right). \tag{26}$$

In particular, the following estimate holds

$$\|u\|_{L^\infty_T(BK^{\alpha,s}_{p,q,r})} + \|\theta\|_{L^\infty_T(BK^{\alpha,s}_{p,q,r})}$$
$$\leq C\exp\left(C\int_0^T \left(1+\|v(\tau)\|_{BK^{\alpha,s}_{p,q,r}}\right)d\tau\right)\left(\|u_0\|_{BK^{\alpha,s}_{p,q,r}} + \|\theta_0\|_{BK^{\alpha,s}_{p,q,r}}\right). \tag{27}$$

Proof Applying the operator localization Δ_j to (24), using the commutator properties, and considering the flow X (according to Remark 7) associated with the vector field v, we have

$$\partial_t \Delta_j u\,(X(y,t),t) = [v\cdot\nabla, \Delta_j]u(X(y,t),t) - \Delta_j\nabla\pi(X(y,t),t)$$
$$+\Delta_j\theta(X(y,t),t)e_n \tag{28}$$

and

$$\partial_t\Delta_j\theta(X(y,t),t) = [v\cdot\nabla,\Delta_j]\theta(X(y,t),t) \tag{29}$$

for $(y,t)\in\mathbb{R}^n\times\mathbb{R}^+$. Next, integrating from 0 to t the Eqs. (28)–(29) and using the initial condition in (14), applying the Herz norm $\|\cdot\|_{K^\alpha_{p,q}}$ together with Lemma 4, leads us to

$$\|\Delta_j u(t)\|_{K^\alpha_{p,q}} \lesssim \|\Delta_j u_0\|_{K^\alpha_{p,q}} + \int_0^t \left(\|\Delta_j\nabla\pi(\tau)\|_{K^\alpha_{p,q}} + \|\Delta_j\theta(\tau)\|_{K^\alpha_{p,q}}\right)d\tau$$
$$+\int_0^t \|[v\cdot\nabla,\Delta_j]u(\tau)\|_{K^\alpha_{p,q}}\,d\tau \tag{30}$$

and also

$$\|\Delta_j\theta(t)\|_{K^\alpha_{p,q}} \lesssim \|\Delta_j\theta_0\|_{K^\alpha_{p,q}} + \int_0^t \|[v\cdot\nabla,\Delta_j]\theta(\tau)\|_{K^\alpha_{p,q}}\,d\tau. \tag{31}$$

Finally, multiplying both sides of estimates (30) and (31) by 2^{sj} and afterwards applying the ℓ^r-norm (with $j\geq -1$), we obtain the following estimates

$$\|u(t)\|_{BK^{\alpha,s}_{p,q,r}} \lesssim \|u_0\|_{BK^{\alpha,s}_{p,q,r}} + \int_0^t \left(\|\nabla\pi(\tau)\|_{BK^{\alpha,s}_{p,q,r}} + \|\theta(\tau)\|_{BK^{\alpha,s}_{p,q,r}}\right)d\tau \tag{32}$$
$$+\int_0^t \|2^{sj}\|[v\cdot\nabla,\Delta_j]u(\tau)\|_{K^\alpha_{p,q}}\|_{\ell^r}\,d\tau$$

and

$$\|\theta(t)\|_{BK^{\alpha,s}_{p,q,r}} \leq C\left(\|\theta_0\|_{BK^{\alpha,s}_{p,q,r}} + \int_0^t \|2^{sj}\|[v\cdot\nabla, \Delta_j]\theta(\tau)\|_{K^{\alpha}_{p,q}}\|_{\ell^r}\, d\tau\right). \tag{33}$$

Observe that the estimate (26) follows directly from (33) along with the estimate (18) in Lemma 6.

On the other hand, we know that the pressure term can be expressed as follows

$$\nabla\pi = (-\Delta)^{-1}\nabla\text{div}(v\cdot\nabla u) - (-\Delta)^{-1}\nabla\text{div}(\theta e_n).$$

Thus, we obtain that

$$\|\nabla\pi\|_{BK^{\alpha,s}_{p,q,r}} \leq \|(-\Delta)^{-1}\nabla\text{div}(v\cdot\nabla u)\|_{BK^{\alpha,s}_{p,q,r}} + \|(-\Delta)^{-1}\nabla\text{div}(\theta e_n)\|_{BK^{\alpha,s}_{p,q,r}}. \tag{34}$$

Consequently, by the estimate (21) in Lemma 7 and boundedness of the Riesz transform in $BK^{\alpha,s}_{p,q,r}$, according with Remark 8, from (34) we arrive at

$$\|\nabla\pi\|_{BK^{\alpha,s}_{p,q,r}} \leq \|u\|_{BK^{\alpha,s}_{p,q,r}}\|v\|_{BK^{\alpha,s}_{p,q,r}} + \|\theta\|_{BK^{\alpha,s}_{p,q,r}}. \tag{35}$$

Therefore, from estimate (32) and (35), along with the estimate (18) in Lemma 6, we conclude (25). Finally, by Grönwall inequality we obtain the estimate (27) directly from (25) and (26).

5 Local-time Existence and Uniqueness of Solution

For clarity, in this section we will present the several steps necessary to prove Theorem 1.

5.1 Construction and Uniform Boundedness of the Approximate Scheme

To prove the Theorem 1 we will use an interaction argument (see, e.g., [3, 13, 14]). To do this, we consider $(u^0, \theta^0) := (u_0, \theta_0)$ and we suppose that the pair (u^m, θ^m) is well defined for $t > 0$. For each $m \in \mathbb{N}_0$ consider (u^{m+1}, θ^{m+1}) solution of the inviscid Boussinesq equations

$$\begin{cases} \partial_t u^{m+1} + u^m \cdot \nabla u^{m+1} + \nabla \pi^{m+1} = \theta^{m+1} e_n, \\ \partial_t \theta^{m+1} + u^m \cdot \nabla \theta^{m+1} = 0, \\ \operatorname{div} u^m = \operatorname{div} u^{m+1} = 0, \\ u^{m+1}(\cdot, 0) := S_{m+1} u_0, \quad \theta^{m+1}(\cdot, 0) := S_{m+1} \theta_0. \end{cases} \tag{36}$$

In the follows, we present the uniform boundedness of the approximate solutions (u^m, θ^m) constructed above in (36) through an interaction process. For each $m \in \mathbb{N}_0$, let X^m the flow associated with the vector field u^m according to Remark 7 and let $T > 0$. From this, for each $t \in [0, T]$, see that

$$\left|(X^m)^{\pm 1}(y,t) - y\right| \le \int_0^t \|u^m(\tau)\|_{L^\infty} \, d\tau \le CT \|u^m\|_{L^\infty_T(BK^{\alpha,s}_{p,q,r})} \le \gamma, \tag{37}$$

provided that $\|u^m\|_{L^\infty_T(BK^{\alpha,s}_{p,q,r})} \le (CT)^{-1}\gamma$, where $\gamma > 0$ is a constant. Assuming (37), we can make use of the estimate (27) in Proposition 1 to obtain a constant $C > 0$ which is independent of m such that

$$\begin{aligned} &\|u^{m+1}\|_{L^\infty_T(BK^{\alpha,s}_{p,q,r})} + \|\theta^{m+1}\|_{L^\infty_T(BK^{\alpha,s}_{p,q,r})} \\ &\quad \le C \exp\left(C \int_0^T \left(1 + \|u^m(\tau)\|_{BK^{\alpha,s}_{p,q,r}}\right) d\tau \right) \left(\|u_0\|_{BK^{\alpha,s}_{p,q,r}} + \|\theta_0\|_{BK^{\alpha,s}_{p,q,r}} \right), \end{aligned} \tag{38}$$

since $\|S_{m+1}u_0\|_{BK^{\alpha,s}_{p,q,r}} \lesssim \|u_0\|_{BK^{\alpha,s}_{p,q,r}}$ and $\|S_{m+1}\theta_0\|_{BK^{\alpha,s}_{p,q,r}} \lesssim \|\theta_0\|_{BK^{\alpha,s}_{p,q,r}}$ holds true. Thus, from the estimate (38), we can apply a standard induction process. First, for $m = 0$, by (37) and (38) since $u^0 = u_0$ we have the following estimates

$$\begin{aligned} &\|u^1\|_{L^\infty_T(BK^{\alpha,s}_{p,q,r})} + \|\theta^1\|_{L^\infty_T(BK^{\alpha,s}_{p,q,r})} \\ &\qquad \le C \exp\left(CT(1 + \|u_0\|_{BK^{\alpha,s}_{p,q,r}}) \right) \left(\|u_0\|_{BK^{\alpha,s}_{p,q,r}} + \|\theta_0\|_{BK^{\alpha,s}_{p,q,r}} \right), \end{aligned} \tag{39}$$

and

$$\left|(X^0)^{\pm 1}(y,t) - y\right| \le CT \|u_0\|_{BK^{\alpha,s}_{p,q,r}} \le \gamma, \tag{40}$$

for some suitably chosen γ. So if we choose $0 < T_1 \le T$ such that it satisfies $CT_1(1 + \|u_0\|_{BK^{\alpha,s}_{p,q,r}}) \le \ln(2)$, we have that $\exp\left(CT_1(1 + \|u_0\|_{BK^{\alpha,s}_{p,q,r}})\right) \le 2$ and estimate (40) holds with $\gamma = \ln(2)$. Consequently, if $C_0 := 2C$, we arrive at

$$\|u^1\|_{L^\infty_{T_1}(BK^{\alpha,s}_{p,q,r})} + \|\theta^1\|_{L^\infty_{T_1}(BK^{\alpha,s}_{p,q,r})} \le C_0 \left(\|u_0\|_{BK^{\alpha,s}_{p,q,r}} + \|\theta_0\|_{BK^{\alpha,s}_{p,q,r}} \right).$$

Then, if we choose $0 < T_2 \leq T_1$ such that

$$CT_2\left(1 + C_0\left(\|u_0\|_{BK^{\alpha,s}_{p,q,r}} + \|\theta_0\|_{BK^{\alpha,s}_{p,q,r}}\right)\right) \leq \ln(2),$$

and proceeding by induction, for all $m \in \mathbb{N}_0$ we obtain the following estimate

$$\|u^{m+1}\|_{L^\infty_{T_2}(BK^{\alpha,s}_{p,q,r})} + \|\theta^{m+1}\|_{L^\infty_{T_2}(BK^{\alpha,s}_{p,q,r})} \leq C_0\left(\|u_0\|_{BK^{\alpha,s}_{p,q,r}} + \|\theta_0\|_{BK^{\alpha,s}_{p,q,r}}\right). \tag{41}$$

Then, it follows from (41) that (u^{m+1}, θ^{m+1}) is uniformly boundedness with respect to m in space $L^\infty_{T_2}(BK^{\alpha,s}_{p,q,r})$.

5.2 *Convergence of the Approximation Scheme*

In this part, we will show that there is a time $0 < T_3 \leq T_2$ such that $\{(u^m, \theta^m)\}_{m\in\mathbb{N}_0}$ is a Cauchy sequence in space $L^\infty_{T_3}(BK^{\alpha,s-1}_{p,q,r}) \times L^\infty_{T_3}(BK^{\alpha,s-1}_{p,q,r})$. To this end, let $\delta u^{m+1} := u^{m+1} - u^m$, $\delta\theta^{m+1} := \theta^{m+1} - \theta^m$ and $\delta\pi^{m+1} := \pi^{m+1} - \pi^m$. From (36), we see that the difference $(\delta u^{m+1}, \delta\theta^{m+1})$ satisfies the following systems

$$\begin{cases} \partial_t \delta u^{m+1} + u^m \cdot \nabla \delta u^{m+1} + \nabla \delta\pi^{m+1} = -\delta u^m \cdot \nabla u^m + \delta\theta^{m+1} e_n, \\ \partial_t \delta\theta^{m+1} + u^m \cdot \nabla \delta\theta^{m+1} = -\delta u^m \cdot \nabla \theta^m, \\ \operatorname{div} \delta u^m = \operatorname{div} \delta u^{m+1} = 0, \\ \delta u^{m+1}(\cdot, 0) = \Delta_{m+1} u_0, \quad \delta\theta^{m+1}(\cdot, 0) = \Delta_{m+1}\theta_0. \end{cases} \tag{42}$$

Consequently, by a process similar to that used to prove the estimates (25) and (26) of Proposition 1, together with the product estimate (16) given in Lemma 5, the commutator estimate (19) of Lemma 6 and the estimates (22)–(23) of Lemma 7, from (42) gives us the following inequalities

$$\begin{aligned} \|\delta u^{m+1}\|_{L^\infty_T(BK^{\alpha,s-1}_{p,q,r})} \leq C \Bigg[& \|\Delta_{m+1} u_0\|_{BK^{\alpha,s-1}_{p,q,r}} + \int_0^T \|\delta\theta^{m+1}(\tau)\|_{BK^{\alpha,s-1}_{p,q,r}} d\tau \\ & + \int_0^T \left(\|\delta u^m(\tau)\|_{BK^{\alpha,s-1}_{p,q,r}} + \|\delta u^{m+1}(\tau)\|_{BK^{\alpha,s-1}_{p,q,r}}\right) \|u^m(\tau)\|_{BK^{\alpha,s}_{p,q,r}} d\tau \Bigg], \end{aligned} \tag{43}$$

and also

$$\begin{aligned}
&\|\delta\theta^{m+1}\|_{L^\infty_T(BK^{\alpha,s-1}_{p,q,r})}\\
&\quad\le C\left[\|\Delta_{m+1}\theta_0\|_{BK^{\alpha,s-1}_{p,q,r}}+\int_0^T\|\delta\theta^{m+1}(\tau)\|_{BK^{\alpha,s-1}_{p,q,r}}\|u^m(\tau)\|_{BK^{\alpha,s}_{p,q,r}}d\tau\right.\\
&\qquad\left.+\int_0^T\|\delta u^m(\tau)\|_{BK^{\alpha,s-1}_{p,q,r}}\|\theta^m(\tau)\|_{BK^{\alpha,s}_{p,q,r}}\,d\tau\right]. \qquad (44)
\end{aligned}$$

Thus, from (43)–(44) and also using the Bernstein inequality, we can found a constant $C_0 > 0$ (independent of m) such that the following estimate holds

$$\begin{aligned}
&\|\delta u^{m+1}\|_{L^\infty_T(BK^{\alpha,s-1}_{p,q,r})}+\|\delta\theta^{m+1}\|_{L^\infty_T(BK^{\alpha,s-1}_{p,q,r})}\\
&\quad\le C_0 2^{-(m+1)}\left(\|u_0\|_{BK^{\alpha,s}_{p,q,r}}+\|\theta_0\|_{BK^{\alpha,s}_{p,q,r}}\right) \qquad (45)\\
&\quad+C_0T\left(\|\delta u^{m+1}\|_{L^\infty_T(BK^{\alpha,s-1}_{p,q,r})}+\|\delta\theta^{m+1}\|_{L^\infty_T(BK^{\alpha,s-1}_{p,q,r})}\right)\left(1+\|u^m\|_{L^\infty_T(BK^{\alpha,s}_{p,q,r})}\right)\\
&\quad+C_0T\|\delta u^m\|_{L^\infty_T(BK^{\alpha,s-1}_{p,q,r})}\left(\|u^m\|_{L^\infty_T(BK^{\alpha,s}_{p,q,r})}+\|\theta^m\|_{L^\infty_T(BK^{\alpha,s}_{p,q,r})}\right).
\end{aligned}$$

And by uniform boundedness of the sequence $\{(u^m,\theta^m)\}_{m\in\mathbb{N}_0}$, there exists a constant $C_1 > 0$ such that $\|u^m\|_{L^\infty_T(BK^{\alpha,s}_{p,q,r})}+\|\theta^m\|_{L^\infty_T(BK^{\alpha,s}_{p,q,r})}\le C_1$, and with this we obtain the estimate

$$\begin{aligned}
&\|\delta u^{m+1}\|_{L^\infty_T(BK^{\alpha,s-1}_{p,q,r})}+\|\delta\theta^{m+1}\|_{L^\infty_T(BK^{\alpha,s-1}_{p,q,r})}\\
&\quad\le C_0 2^{-(m+1)}\left(\|u_0\|_{BK^{\alpha,s}_{p,q,r}}+\|\theta_0\|_{BK^{\alpha,s}_{p,q,r}}\right)+C_0C_1T\|\delta u^m\|_{L^\infty_T(BK^{\alpha,s-1}_{p,q,r})}\\
&\qquad+C_0(1+C_1)T\left(\|\delta u^{m+1}\|_{L^\infty_T(BK^{\alpha,s-1}_{p,q,r})}+\|\delta\theta^{m+1}\|_{L^\infty_T(BK^{\alpha,s-1}_{p,q,r})}\right). \qquad (46)
\end{aligned}$$

So if we choose $0 < T_3 \le T_2$ independent of m such that $C_0(1+2C_1)T_3 \le 1/2$, by (46) we obtain that

$$\begin{aligned}
&\|\delta u^{m+1}\|_{L^\infty_{T_3}(BK^{\alpha,s-1}_{p,q,r})}+\|\delta\theta^{m+1}\|_{L^\infty_{T_3}(BK^{\alpha,s-1}_{p,q,r})}\\
&\quad\le C_0\frac{1}{2^m}\left(\|u_0\|_{BK^{\alpha,s}_{p,q,r}}+\|\theta_0\|_{BK^{\alpha,s}_{p,q,r}}\right)+\frac{1}{2}\|\delta u^m\|_{L^\infty_{T_3}(BK^{\alpha,s-1}_{p,q,r})},
\end{aligned}$$

for all $m \geq 1$. Thus, since $u^0 = u_0$ and proceeding recurrently, we arrive at the estimate

$$\|\delta u^{m+1}\|_{L^\infty_{T_3}(BK^{\alpha,s-1}_{p,q,r})} + \|\delta\theta^{m+1}\|_{L^\infty_{T_3}(BK^{\alpha,s-1}_{p,q,r})} \leq \widetilde{C}\left(\frac{m+1}{2^m}\right),$$

implying that $\{(u^m, \theta^m)\}_{m\in\mathbb{N}_0}$ is a Cauchy sequence in space $L^\infty_{T_3}(BK^{\alpha,s-1}_{p,q,r})$ the limit of which is denoted by (u, θ).

Remark 9 According to Sect. 5.2, if (u, θ) is the limit of the sequence $\{(u^m, \theta^m)\}_{m\in\mathbb{N}_0}$ and $T > 0$ is the existence time, by a duality argument, the uniform boundedness (Sect. 5.1) and by uniqueness of the weak-$*$ limit, we obtain that

$$(u^m, \theta^m) \longrightarrow (u, \theta) \text{ in } C([0, T]; BK^{\alpha,s-1}_{p,q,r}) \times C([0, T]; BK^{\alpha,s-1}_{p,q,r}).$$

This property of strong convergence allows us to pass the limit in our approximate system (36) and obtain that (u, θ) is in fact a solution of system (1). The desired time-continuity for (u, θ) from $[0, T]$ to $BK^{\alpha,s}_{p,q,r}$ follows by a standard regularity argument, since they satisfy (1) in a suitable integral sense, together with the estimates obtained Sect. 5.1, namely

$$(u, \theta) \in C([0, T]; BK^{\alpha,s}_{p,q,r}) \times C([0, T]; BK^{\alpha,s}_{p,q,r}).$$

5.3 Uniqueness of Solution

For the proof of uniqueness, assume that

$$(u^i, \theta^i) \in C([0, T_3]; BK^{\alpha,s}_{p,q,r}) \times C([0, T_3]; BK^{\alpha,s}_{p,q,r})$$

for $i = 1, 2$, are two solutions of our system (1) with the same initial data (u_0, θ_0). Moreover, if we define $(\delta u, \delta\theta, \nabla\delta\pi) := (u^2 - u^1, \theta^2 - \theta^1, \nabla\pi^2 - \nabla\pi^1)$, we obtain that $(\delta u, \delta\theta)$ satisfies the following system

$$\begin{cases} \partial_t \delta u + u^2 \cdot \nabla\delta u + \nabla\delta\pi = -\delta u \cdot \nabla u^1 + \delta\theta e_n, \\ \partial_t \delta\theta + u^2 \cdot \nabla\delta\theta = -\delta u \cdot \nabla\theta^1, \\ \operatorname{div} u^1 = \operatorname{div} u^2 = 0, \\ \delta u(\cdot, 0) = \delta\theta(\cdot, 0) = 0. \end{cases}$$

Thus, proceeding as before (see the arguments to get (45)), for all $0 < T \leq T_3$, we obtain the following estimate

$$\begin{aligned}
&\|\delta u\|_{L^\infty_T(BK^{\alpha,s-1}_{p,q,r})} + \|\delta\theta\|_{L^\infty_T(BK^{\alpha,s-1}_{p,q,r})} \\
&\leq C_2 T\left(\|\delta u\|_{L^\infty_T(BK^{\alpha,s-1}_{p,q,r})} + \|\delta\theta\|_{L^\infty_T(BK^{\alpha,s-1}_{p,q,r})}\right)\left(1+\|u^2\|_{L^\infty_T(BK^{\alpha,s}_{p,q,r})}\right) \\
&\quad + C_2 T\|\delta u\|_{L^\infty_T(BK^{\alpha,s-1}_{p,q,r})}\left(\|u^1\|_{L^\infty_T(BK^{\alpha,s}_{p,q,r})} + \|\theta^1\|_{L^\infty_T(BK^{\alpha,s}_{p,q,r})}\right). \qquad (47)
\end{aligned}$$

From hypothesis, we can consider $T > 0$ sufficiently small such that

$$2C_2 T\left(1+\|u^2\|_{L^\infty_T(BK^{\alpha,s}_{p,q,r})}\right) + C_2 T\left(\|u^1\|_{L^\infty_T(BK^{\alpha,s}_{p,q,r})} + \|\theta^1\|_{L^\infty_T(BK^{\alpha,s}_{p,q,r})}\right) \leq \frac{1}{2}. \qquad (48)$$

Considering the above condition (48) in the estimate (47), we obtain that

$$\|\delta u\|_{L^\infty_T(BK^{\alpha,s-1}_{p,q,r})} + \|\delta\theta\|_{L^\infty_T(BK^{\alpha,s-1}_{p,q,r})} \leq \frac{1}{2}\left(\|\delta u\|_{L^\infty_T(BK^{\alpha,s-1}_{p,q,r})} + \|\delta\theta\|_{L^\infty_T(BK^{\alpha,s-1}_{p,q,r})}\right), \qquad (49)$$

which implies the desired uniqueness property.

From Sects. 5.1, 5.2, 5.3 and Remark 9, we finish the proof of Theorem 1.

References

1. Berestycki, H., Constantin, P., Ryzhik, L.: Non-planar fronts in Boussinesq reactive flows. Ann. Inst. H. Poincaré Anal. Non Linéaire **23**(4), 407–437 (2006)
2. Bergh, J., Löfström, J.: Interpolation spaces. An introduction. Grundlehren der Mathematischen Wissenschaften, vol.223. Springer-Verlag, Berlin-New York (1976)
3. Bie, Q., Wang, Q., Yao, Z.: On the well-posedness of the inviscid Boussinesq equations in the Besov-Morrey spaces. Kinet. Relat. Models **8**(3), 395–411 (2015)
4. Bony, J.-M.: Calcul symbolique et propagation des singularités pour les équations aux dérivées partielles non linéaires. Ann. Sci. École Norm. Sup. (4) **14**(2), 209–246 (1981)
5. Brenier, Y.: Optimal transport, convection, magnetic relaxation and generalized Boussinesq equations. J. Nonlinear Sci. **19**(5). 547–570 (2009)
6. Chae, D., Nam, H.: Local existence and blow-up criterion for the Boussinesq equations. Proc. R. Soc. Edinb. Sect. A **127**(5), 935–946 (1997)
7. Chae, D., Kim, S.-K., Nam, H.-S.: Local existence and blow-up criterion of Hölder continuous solutions of the Boussinesq equations. Nagoya Math. J. **155**, 55–80 (1999)
8. Chemin, J.-Y.: Fluides parfaits incompressibles. Astérisque **230**, 177 pp. (1995)
9. Cui, X., Dou, C., Jiu, Q.: Local well-posedness and blow-up criterion for the inviscid Boussinesq system in Hölder spaces. J. Partial Differ. Equ. **25**(3), 220–238 (2012)
10. Danchin, R.: On the well-posedness of the incompressible density-dependent Euler equations in the L^p framework. J. Differ. Equ. **248**(8), 2130–2170 (2010)

11. E, W., Shu, C.-W.: Small-scale structures on Boussinesq convection. Phys. Fluids **6**(1), 49–58 (1994)
12. Fan, J., Zhou, Y.: A note on regularity criterion for the 3D Boussinesq system with partial viscosity. Appl. Math. Lett. **22**(5), 802–805 (2009)
13. Ferreira, L.C.F., Machado, D.F.: On the well-posedness in Besov-Herz spaces for the inhomogeneous incompressible Euler equations. Dyn. Partial Differ. Equ. **21**(1), 1–29 (2024)
14. Ferreira, L.C.F., Pérez-López, J.E.: On the theory of Besov-Herz spaces and Euler equations. Isr. J. Math. **220**(1), 283–332 (2017)
15. Guo, B.L.: Spectral method for solving two-dimensional Newton-Boussinesq equation. Acta Math. Appl. Sinica **5**(3), 208-218 (1989)
16. Grafakos, L., Li, X., Yang, D.: Bilinear operators on Herz-type Hardy spaces. Trans. Am. Math. Soc. **350**(3), 1249-1275 (1998)
17. Hernández, E., Yang, D.: Interpolation of Herz spaces and applications. Math. Nachr. **205**, 69–87 (1999)
18. Herz, C.S.: Lipschitz spaces and Bernstein's theorem on absolutely convergent Fourier transforms. J. Math. Mech. **18**, 283–323 (1968/1969)
19. Inci, H.: On the well-posedness of the inviscid 2D Boussinesq equation. Z. Angew. Math. Phys. **69**(4), Paper No. 103, 8 pp. (2018)
20. Johnson, R.: Lipschitz spaces, Littlewood-Paley spaces, and convoluteurs. Proc. Lond. Math. Soc. (3) **29**, 127–141 (1974)
21. Kato, T.: Strong solutions of the Navier-Stokes equation in Morrey spaces. Bol. Soc. Brasil. Mat. (N.S.) **22**(2), 127–155 (1992)
22. Lemarié-Rieusset, P.G.: Recent Developments in the Navier-Stokes Problem. Chapman & Hall/CRC Research Notes in Mathematics , vol. 431. Chapman & Hall/CRC, Boca Raton (2002)
23. Li, X., Yang, D.: Boundedness of some sublinear operators on Herz spaces. Ill. J. Math. **40**(3), 484–501 (1996)
24. Liu, X., Wang, M., Zhang, Z.: Local well-posedness and blow-up criterion of the Boussinesq equations in critical Besov spaces. J. Math. Fluid Mech. **12**(2), 280–292 (2010)
25. Tsutsui, Y.: The Navier-Stokes equations and weak Herz spaces. Adv. Differ. Equ. **16**(11–12), 1049–1085 (2011)
26. Xiang, Z., Yan, W.: On the well-posedness of the Boussinesq equation in the Triebel-Lizorkin-Lorentz spaces. Abstr. Appl. Anal. Art. ID 573087, 17 pp. (2012)
27. Xu, J.: Equivalent norms of Herz-type Besov and Triebel-Lizorkin spaces. J. Funct. Spaces Appl. **3**(1), 17–31 (2005)
28. Xu, X.: Local existence and blow-up criterion of the 2-D compressible Boussinesq equations without dissipation terms. Discrete Contin. Dyn. Syst. **25**(4), 1333–1347 (2009)
29. Yuan, B.: Local existence and continuity conditions of solutions to the Boussinesq equations in Besov spaces. Acta Math. Sinica (Chin. Ser.) **53**(3), 455–468 (2010)
30. Zhou, Y.: Local well-posedness and regularity criterion for the density dependent incompressible Euler equations. Nonlinear Anal. **73**(3), 750–766 (2010)

Blowing Up and Dissipation for a Couple of One-dimensional Non-local Conservation Laws

E. Abreu, M. Huacasi-Machacha, J. Pérez, and J. C. Valencia-Guevara

Abstract We are interested in studying solutions to the non-local conservation law $\partial_t u + \chi \partial_x (u \Lambda^{\alpha-1} \mathcal{H} u) = 0$ subject to initial data in $u_0 \in L^1 \cap L^\infty$, where $\chi = \pm 1$. We proved that there is a finite existence time T^* (maximal time) such that the solution u is finite in $[0, T^*)$ and blows up at T^* in the sense of mass concentration ($\chi = -1$). We proved that in the case $\chi = 1$ the solution exhibits a spreading of the initial datum, which is verified via numerical insights. We also improved comprehension of the results of Biler and Karch (J. Evol. Equ. 10:247–262, 2010) and Li and Zhang (Commun. Pure Appl. Anal. 9(6):1591–1606, 2010) to simplify the analysis and to circumvent the requirement of initial concentration such that our results hold for initial data in $L^1 \cap L^\infty$.

1 Introduction

We are concerned with the study of solutions to the nonlocal conservation law

$$\partial_t u + \chi \partial_x (u \Lambda^{\alpha-1} \mathcal{H} u) = 0, \tag{1}$$

with initial data $u_0 \in L^1 \cap L^\infty$, where $\chi = \pm 1$. Concretely, we are interested in the mass concentration ($\chi = -1$) and the mass spread ($\chi = 1$) of the solution of (1). The first important step in this study is to exhibit a rigorous structure of aggregation

E. Abreu (✉)
Universidade Estadual de Campinas, IMECC-UNICAMP-SP, Campinas, Brazil
e-mail: eabreu@unicamp.br

M. Huacasi-Machaca · J. C. Valencia-Guevara
Departamento Académico de Matemáticas, Universidad Nacional de San Agustin Arequipa, Arequipa, Perú
e-mail: mhuacasim@unsa.edu.pe; jvalenciag@unsa.edu.pe

J. Pérez
Instituto Tecnológico Metropolitano, Antioquia, Colombia
e-mail: jhonperez@itm.edu.co

M. R. Ebert et al. (eds.), *New Tools in Mathematical Analysis and Applications*, Research Perspectives, https://doi.org/10.1007/978-3-031-77050-0_20

equation that has the conservation law (1). In fact, by using Fourier transform, we can see that model (1) is an aggregation equation with interaction kernel given by $W(x) = -\frac{b_\alpha}{1-\alpha}|x|^{1-\alpha}$, see model (4) in Sect. 2. So, the corresponding model has an attractive nature when $\chi = -1$ and is repulsive when $\chi = 1$.

Although there is extensive literature concerned with aggregation models and concretely with the study of the blowing up of solutions, we have not been able to find the respective result in our specific case. Let us make a nonexhaustive bibliographic revision of such a matter. In [5] it is studied the well-possedness of the aggregation equation when the Osgood condition is satisfied. On the contrary, when this condition is violated, the finite time blowing up of the solutions is established for a space dimension greater or equal to 2, with initial data compactly supported. In a precursor work of the previous comment, [4], it has been proved finite time blowing up of solutions for potentials with a Lipschitz point at the origin, which does not correspond with our case. A similar result, for diffusive aggregation models, was established in [9] for interaction kernels with bounded gradients. For the one space dimension case, blowing up of solutions for aggregation equations, like model (4) below, has been proved for values of $\alpha < 0$ in the work [12]. As commented in [11], a typical potential producing blow-up of the solution is of the form $W(x) \approx |x|^\gamma$ for $\gamma \geq 1$. Our case shows that the same fact is expected for $0 < \gamma = 1 - \alpha < 1$.

Similar works have been developed for models of type Patlack-Keller-Segel in which, among other properties, the blowing-up in finite time has been established. See for example, [3, 7, 8]. Note that model (1) with $\chi = 1$ corresponds to the Porous medium equation with fractional potential pressure, studied by Caffarelli and Vazquez in [10], which has motivated a lot of works in this matter.

In this work, we combine the ideas of [6] and [13]. It is worth commenting that the technique of [6] does not apply directly, since it makes strong use of the dimension greater than one, which ensures the integrability of some involved expressions. On another hand, the ideas of [13] are technically more involved and require an *a priori* concentration on the initial data. In this work, we use the ideas of [6] to get a rate of decreasing the moment of the solution in the concentration case and a rate of increasing the moment for the spreading case. Moreover, the technique of [6] allows us to circumvent the requirement of initial concentration in [13], and to simplify the arguments. Thus, the result holds for a generic initial data in $L^1 \cap L^\infty$.

In the proof, we assume the local well-posedness and some basic properties of model (1) which are more or less established facts. Indeed, papers like [10, 13], for example, established solutions $u(t) \in L^1 \cap L^\infty$ for all $t > 0$ when initial data $u_0 \in L^1 \cap L^\infty$, nonnegativity of the solution when the initial data is nonnegative and the mass conservation property. All these facts will be assumed in our proofs.

2 Interpretation as Aggregation Models and Some Properties

In this section, we show that models in Eq. (1) admit a structure of aggregation equation of repulsive nature for $\chi = 1$ and attractive nature for $\chi = -1$. From now on, we adopt the notation $L^p = L^p(\mathbb{R})$, for $1 \leq p \leq \infty$. The basic important tool is the following result.

Lemma 1 *Consider the constant*

$$b_\alpha = \frac{1}{2\Gamma(1-\alpha)\cos\frac{\alpha\pi}{2}}, \tag{2}$$

for $0 < \alpha < 1$. *Then, the identity,*

$$\Lambda^{\alpha-1}\mathcal{H}f(x) = b_\alpha \int_{\mathbb{R}} \frac{(x-y)}{|x-y|^{\alpha+1}} f(y)\,\mathrm{d}y, \tag{3}$$

holds true for all $f \in \mathcal{S}(\mathbb{R})$.

Proof We break the proof in some steps.

Step 1 For the singular kernel $k_\alpha(x) = b_\alpha \dfrac{x}{|x|^{\alpha+1}}$, we have that

$$\widehat{k_\alpha}(\xi) = -i(2\pi)^{\alpha-1}\frac{\xi}{|\xi|^{2-\alpha}}.$$

In fact, assume initially $\frac{1}{2} < \alpha < 1$ and note that

$$k_\alpha = k_\alpha 1_{]-1,1[} + k_\alpha 1_{]-1,1[^c} \in L^1 + L^2.$$

Therefore, $\widehat{k_\alpha} \in L^\infty + L^2$ and it exists as a function. Now, since k_α is an odd and homogeneous function of degree $-\alpha$, $\widehat{k_\alpha}$ is also odd and homogeneous of degree $\alpha - 1$. Thus, we can deduce that $\widehat{k_\alpha}(\xi) = \xi|\xi|^{\alpha-2}\widehat{k_\alpha}(1)$. Now, by standard computations we know that

$$\begin{aligned}\widehat{k_\alpha}(1) &= -2ib_\alpha(2\pi)^{\alpha-1}\int_0^\infty \frac{\sin t}{t^\alpha}\,\mathrm{d}t \\ &= -2ib_\alpha(2\pi)^{\alpha-1}\Gamma(1-\alpha)\cos\frac{\alpha\pi}{2} = -i(2\pi)^{\alpha-1},\end{aligned}$$

and we are done in this case. For $0 < \alpha < \frac{1}{2}$, it is known that $\widehat{k_\alpha}$ is defined in the distributional sense (in $\mathcal{S}'(\mathbb{R})$). Also note that $k_\alpha = ib_\alpha(2\pi)^\alpha\widehat{k_{1-\alpha}}$. By using the distributional definition of the Fourier transform, we get

$$\widehat{k_\alpha}(\xi) = -i(2\pi)^\alpha b_\alpha k_{1-\alpha}(\xi) = -i(2\pi)^\alpha b_\alpha b_{1-\alpha}\frac{\xi}{|\xi|^{2-\alpha}}.$$

By the Euler's reflection formula, i.e. $\Gamma(\alpha)\Gamma(1-\alpha) = \dfrac{\pi}{\sin \pi\alpha}$, we found that $b_\alpha b_{1-\alpha} = \frac{1}{2\pi}$. The case $\alpha = \frac{1}{2}$ is proved by an approximation argument, and we omit this.

Step 2 $\widehat{\Lambda^{\alpha-1}\mathcal{H}f} = -i(2\pi)^{\alpha-1}\dfrac{\xi}{|\xi|^{2-\alpha}}\widehat{f}$, for all $f \in \mathcal{S}(\mathbb{R})$. Indeed, by the Fourier representation of the Hilbert transform and the Riesz potential operator, the claim follows. The lemma is a direct consequence of *Step 1* and *Step 2*.

□

Thanks to the previous lemma, we can define the interaction potential $W(x) = -\frac{b_\alpha}{1-\alpha}|x|^{1-\alpha}$. Thus, models in (1) can be rewritten as

$$\partial_t u = \chi \partial_x (u(W' * u)), \tag{4}$$

which is a one-dimensional aggregation model with repulsive potential when $\chi = 1$ and attractive potential when $\chi = -1$.

3 Blowing Up and Dissipation of the Solutions

The main objective of this section is to show that solutions of (1) concentrate mass in finite time into a point when $\chi = -1$ and dissipate mass when $\chi = 1$. For this, Lemma 1 plays a crucial role. We start this section by giving a notion of weak solution we employ.

Definition 1 By a weak solution of (1), we mean a funtion $u(x,t) \in L^1_{loc}([0,T); L^1 \cap L^\infty)$ which satisfies:

$$\begin{aligned} \frac{\mathrm{d}}{\mathrm{d}t}\int_{\mathbb{R}} u(x,t)\psi(x)\,\mathrm{d}x = \frac{\chi b_\alpha}{2}\int\int_{\mathbb{R}^2} \frac{(x-y)}{|x-y|^{\alpha+1}}(\psi'(x) \\ - \psi'(y))u(x,t)u(y,t)\,\mathrm{d}y\,\mathrm{d}x, \end{aligned} \tag{5}$$

for all $\psi \in C_c^\infty(\mathbb{R})$ and all $0 < t < T$ respectively. Here b_α is the constant defined in Eq. (2).

The following result is a technical lemma, adapted from [13], for the one-dimensional case. We give the precise statement since it was posed originally in dimension two.

Lemma 2 (Three Point Inequality) *Let $0 < \alpha < 1$. Then, for any three pairwise different numbers $x, y, z \in \mathbb{R}$, we have*

$$\frac{(x-y)}{|x-y|^{\alpha+1}} \cdot \frac{(x-z)}{|x-z|^{\alpha+1}} + \frac{(y-x)}{|y-x|^{\alpha+1}} \cdot \frac{(y-z)}{|y-z|^{\alpha+1}} + \frac{(z-x)}{|z-x|^{\alpha+1}} \cdot \frac{(z-y)}{|z-y|^{\alpha+1}}$$
$$\geq (2^{1+\alpha} - 2^{2\alpha})(|x-y| + |y-z| + |z-x|)^{-2\alpha}. \tag{6}$$

With this lemma in mind, we adapt the following result from [13] to the one-dimensional case.

Proposition 1 *Let $0 < \beta < 1$ and $R > 0$ fixed. Consider a function $u \in L^1 \cap L^\infty$ such that $u(x) \geq 0$ a.e. in $\mathbb{R}$ and*

$$\int_{|x|\leq R} u(x)\, \mathrm{d}x \geq \beta \|u\|_{L^1}. \tag{7}$$

Then,

$$\int_{\mathbb{R}} u(x) \left(\Lambda^{\alpha-1}\mathcal{H}u(x)\right)^2 \, \mathrm{d}x \geq \frac{(2^{1-\alpha} - 1)}{3^{1+2\alpha}} R^{-2\alpha} \beta^3 \|u\|_{L^1}^3. \tag{8}$$

Proof First, note that the integral in (8) is well defined thanks to Lemma 1, since $\|\Lambda^{\alpha-1}\mathcal{H}u\|_{L^\infty} \lesssim \|u\|_{L^1}^{1-\alpha} \|u\|_{L^\infty}^{\alpha}$. Now, by symmetrizing in the variables x, y and z, we can write

$$\int_{\mathbb{R}} u(x) \left(\Lambda^{\alpha-1}\mathcal{H}u(x)\right)^2 \, \mathrm{d}x = \frac{1}{3} \int_{\mathbb{R}\times\mathbb{R}\times\mathbb{R}} N(x, y, z) u(x) u(y) u(z) \, \mathrm{d}x \, \mathrm{d}y \, \mathrm{d}z,$$

where $N(x, y, z)$ denotes the left hand side of (6). Thus, by using Lemma 2 we can estimate as follows:

$$\begin{aligned}
\int_{\mathbb{R}} u(x) \left(\Lambda^{\alpha-1}\mathcal{H}u(x)\right)^2 \, \mathrm{d}x &\geq \frac{2^{1-\alpha} - 1}{3} \int_{\mathbb{R}^3} \frac{u(x)u(y)u(z)}{(|x| + |y| + |z|)^{2\alpha}} \, \mathrm{d}x \, \mathrm{d}y \, \mathrm{d}z \\
&\geq \frac{2^{1-\alpha} - 1}{3} \int_{[-R,R]^3} (3R)^{-2\alpha} u(x)u(y)u(z) \, \mathrm{d}x \, \mathrm{d}y \, \mathrm{d}z \\
&\geq \frac{(2^{1-\alpha} - 1)}{3^{1+2\alpha}} R^{-2\alpha} \beta^3 \|u\|_{L^1}^3,
\end{aligned}$$

which is the assertion. □

The main tool to show concentration and dissipation for a general initial data is the control of the moment,

$$I_\theta(t) := I_\theta(u(t)) = \int_{\mathbb{R}} \psi_\theta(x) u(x, t) \, \mathrm{d}x, \tag{9}$$

of the weak solution $u(x,t)$ of model (1). Here, $\psi_\theta(x) = (1+x^2)^{\frac{\theta}{2}} - 1$ and $\theta > 0$. First, we ensure the finiteness and differentiability of such a moment.

Proposition 2 *Let u be a nonnegative weak solution of* (1) *in the sense of Definition 1 with nonnegative initial data $u_0 \in L^1 \cap L^\infty$. If $I_\theta(u_0)$ is finite, then $I_\theta(u(t))$ is finite and differentiable for all $t \in [0, T)$.*

Proof For $\epsilon > 0$ consider $\phi_\epsilon(x) = e^{-\epsilon\sqrt{1+x^2}}$ and

$$I_{\theta,\epsilon}(t) = \int_{\mathbb{R}} \psi(x)\phi_\epsilon(x)u(x,t)\,\mathrm{d}x. \tag{10}$$

Now, from Definition 1 we get

$$\begin{aligned}
\frac{\mathrm{d}}{\mathrm{d}t} I_{\theta,\epsilon}(t) &= \chi \int_{\mathbb{R}} \Lambda^{\alpha-1}\mathcal{H}u(x,t)\,\Big(\psi'(x)\phi_\epsilon(x) \\
&\qquad -\epsilon \frac{x}{\sqrt{1+x^2}}\psi(x)\phi_\epsilon(x)\Big)\,u(x,t)\,\mathrm{d}x \\
&\le \int_{\mathbb{R}} |\Lambda^{\alpha-1}\mathcal{H}u(x,t)| \left(\frac{\theta}{2}(\psi(x)+1)\phi_\epsilon(x) + \epsilon\psi(x)\phi_\epsilon(x)\right) u(x,t)\,\mathrm{d}x \\
&\le C\|u_0\|_{L^1}^{1-\alpha}\|u(t)\|_{L^\infty}^{\alpha}\left[\left(\frac{\theta}{2}+\epsilon\right) I_{\theta,\epsilon}(t) + \frac{\theta}{2}\|u_0\|_{L^1}\right].
\end{aligned}$$

Thus, Gronwall's Lemma and Monotone Convergence Theorem imply that $I_\theta(u(t))$ is finite for every $t \in [0, T)$. By integrating in time the first equality and making $\epsilon \to 0$, we get the assertion about the differentiability. □

The first main result of the present work is the following theorem, which contains the dissipation and, essentially, the concentration facts.

Theorem 1 *Let us assume that $u(x,t) \ge 0$ is a weak solution of* (1), *in the sense of Definition 1 with nonnegative initial data $u_0 \in L^1 \cap L^\infty$. Also assume that $I_\theta(u_0)$ is finite, for $1 < \theta < \frac{2}{2-\alpha}$, and denote by $T^* > 0$ the maximal time of existence of $u(x,t)$. Then, there exists a constant $C > 0$ such that:*

1. *(Concentration) for $\chi = -1$, $I_\theta(t)$ is decreasing and*

$$\int_0^{T^*} \|u(\tau)\|_{L^\infty}^{\alpha-1}\,\mathrm{d}\tau \le C\|u_0\|_{L^1}^{\frac{\alpha\theta-2\theta-2}{\theta}} \left(\|u_0\|_{L^1} + I_\theta(0)\right)^{\frac{2}{\theta}}. \tag{11}$$

2. *(Dissipation) for* $\chi = 1$, $I_\theta(t)$ *is increasing and*

$$\left(\|u_0\|_{L^1} + I_\theta(t)\right)^{\frac{2}{\theta}} \geq \left(\|u_0\|_{L^1} + I_\theta(0)\right)^{\frac{2}{\theta}} + C\|u_0\|_{L^1}^{\frac{2-\alpha\theta+2\theta}{\theta}} \int_0^t \|u(\tau)\|_{L^\infty}^{\alpha-1} \, \mathrm{d}\tau, \tag{12}$$

for $0 \leq t < T^*$.

Proof We focus only on the concentration case since the dissipation case is very similar. By appealing to Proposition 2 we have

$$\frac{\mathrm{d}I_\theta}{\mathrm{d}t}(t) = -\, b_\alpha \int_{\mathbb{R}\times\mathbb{R}} \frac{(x-y)}{|x-y|^{\alpha+1}} \left(\psi_\theta'(x) - \psi_\theta'(y)\right) u(x,t)u(y,t) \, \mathrm{d}x \, \mathrm{d}y.$$

Now, from the elementary fact $\psi_\theta''(z) \geq \theta(\theta-1)(1+z^2)^{\frac{\theta}{2}-1}$, we can do the following estimate:

$$\begin{aligned}
(\psi_\theta'(x) - \psi_\theta'(y))(x-y) &\geq \theta(\theta-1)(x-y)^2 \int_0^1 (1+(sx+(1-s)y)^2)^{\frac{\theta}{2}-1} \, \mathrm{d}s \\
&\geq \theta(\theta-1)(x-y)^2 \int_0^1 (1+sx^2+(1-s)y^2)^{\frac{\theta}{2}-1} \, \mathrm{d}s \\
&\geq 3^{1-\theta/2}\theta(\theta-1)\frac{(x-y)^2}{1+|x|^{2-\theta}+|y|^{2-\theta}},
\end{aligned}$$

where we have used that $\frac{\theta}{2} - 1 < 0$. Putting together the last two facts, it follows that

$$\frac{\mathrm{d}I_\theta}{\mathrm{d}t}(t) \leq -3^{1-\theta/2}\theta(\theta-1)b_\alpha \int_{\mathbb{R}\times\mathbb{R}} \frac{|x-y|^{1-\alpha}}{1+|x|^{2-\theta}+|y|^{2-\theta}} u(x,t)u(y,t) \, \mathrm{d}x \, \mathrm{d}y. \tag{13}$$

At this point we can see that $I_\theta(t)$ is decreasing. Now, we are going to estimate the right-hand side of (13). The conservation mass property of the solutions of (1) and Hölder's inequality implies that,

$$\begin{aligned}
\|u_0\|_{L^1}^2 \leq &\left(\int_{\mathbb{R}\times\mathbb{R}} \frac{|x-y|^{p\nu}}{(1+|x|^{2-\theta}+|y|^{2-\theta})^{p\delta}} u(x,t)u(y,t) \, \mathrm{d}x\,\mathrm{d}y\right)^{1/p} \\
&\times \left(\int_{\mathbb{R}\times\mathbb{R}} \frac{(1+|x|^{2-\theta}+|y|^{2-\theta})^{p'\delta}}{|x-y|^{p'\nu}} u(x,t)u(y,t) \, \mathrm{d}x\,\mathrm{d}y\right)^{1/p'},
\end{aligned}$$

for any $\nu, \delta > 0$ and $p, p' \geq 1$ such that $\frac{1}{p} + \frac{1}{p'} = 1$. By imposing $p\nu = 1 - \alpha$ and $p\delta = 1$, we deduce that $p'\delta = \frac{1}{p-1}$ and $p'\nu = \frac{1-\alpha}{p-1}$. Thus,

$$\|u_0\|_{L^1}^2 \leq 3^{\frac{1}{p}} \left(\int_{\mathbb{R}\times\mathbb{R}} \frac{|x-y|^{1-\alpha}}{1+|x|^{2-\theta}+|y|^{2-\theta}} u(x,t)u(y,t)\, \mathrm{d}x\, \mathrm{d}y \right)^{1/p}$$
$$\times \left(\int_{\mathbb{R}\times\mathbb{R}} \frac{(1+|x|^{\frac{2-\theta}{p-1}}+|y|^{\frac{2-\theta}{p-1}})}{|x-y|^{\frac{1-\alpha}{p-1}}} u(x,t)u(y,t)\, \mathrm{d}x\, \mathrm{d}y \right)^{1/p'}.$$

Continuing, we recall the well known estimate

$$\left| \int_{\mathbb{R}} \frac{u(y)}{|x-y|^s}\, \mathrm{d}y \right| \leq C\|u\|_{L^1}^{1-s} \|u\|_{L^\infty}^s,$$

for all $x \in \mathbb{R}$, where $0 < s < 1$ and $C > 0$. Therefore, taking $p = \frac{2}{\theta}$ we can follow the estimating as

$$\|u_0\|_{L^1}^2 \leq C \left(\int_{\mathbb{R}\times\mathbb{R}} \frac{|x-y|^{1-\alpha}}{1+|x|^{2-\theta}+|y|^{2-\theta}} u(x,t)u(y,t)\, \mathrm{d}x\, \mathrm{d}y \right)^{\theta/2}$$
$$\times \left(\int_{\mathbb{R}} (1+2|x|^{\theta}) u(x,t)\, \mathrm{d}x \right)^{(2-\theta)/2} \|u_0\|_{L^1}^{\frac{2-2\theta+\alpha\theta}{2}} \|u(t)\|_{L^\infty}^{\frac{(1-\alpha)\theta}{2}}.$$

Now, note that there exists constants $C_1, C_2 > 0$ such that $|x|^{\theta} \leq C_1\psi_{\theta}(x) + C_2$, thus

$$\|u_0\|_{L^1}^{\frac{2+2\theta-\alpha\theta}{\theta}} \leq C \int_{\mathbb{R}\times\mathbb{R}} \frac{|x-y|^{1-\alpha}}{1+|x|^{2-\theta}+|y|^{2-\theta}} u(x,t)u(y,t)\, \mathrm{d}x\, \mathrm{d}y$$
$$\times \left(\|u_0\|_{L^1} + I_{\theta}(t)\right)^{(2-\theta)/\theta} \|u(t)\|_{L^\infty}^{1-\alpha}, \tag{14}$$

for some positive constant C. Putting inequality (14) into (13) we arrive at

$$\left(\|u_0\|_{L^1} + I_{\theta}(t)\right)^{\frac{2-\theta}{\theta}} \frac{\mathrm{d}I_{\theta}}{\mathrm{d}t}(t) \leq -C\|u_0\|_{L^1}^{\frac{2+2\theta-\alpha\theta}{\theta}} \|u(t)\|_{L^\infty}^{1-\alpha}.$$

By integrating on $t > 0$ the last inequality, it follows the assertion. □

Corollary 1 (Mass Confinement) *Under the assumptions of Theorem 1 and for* $0 < \beta < 1$, *there exists* $R > 0$ *such that, if* $u(x,t)$ *is a weak solution of* (1), *with* $\chi = -1$ *and maximal time of existence* $T^* \leq \infty$, *then*

$$\int_{|x|\geq R} u(x,t)\, \mathrm{d}x \leq (1-\beta)\|u_0\|_{L^1}, \tag{15}$$

for all $0 < t < T^*$.

Proof In fact, for $R > 0$ and recalling that there exist constants $C_1, C_2 > 0$ such that $|x|^\theta \leq C_1\psi_\theta(x) + C_2$, we have

$$\begin{aligned}\int_{|x|\geq R} u(x,t)\, \mathrm{d}x &\leq \frac{1}{R^\theta}\int_{\mathbb{R}} |x|^\theta u(x,t)\, \mathrm{d}x \\ &\leq \frac{C_1 I_\theta(t) + C_2\|u_0\|_{L^1}}{R^\theta} \leq \frac{C_1 I_\theta(0) + C_2\|u_0\|_{L^1}}{R^\theta}.\end{aligned}$$

So, by taking $R^\theta \geq \frac{C_1 I_\theta(0)+C_2\|u_0\|_{L^1}}{(1-\beta)\|u_0\|_{L^1}}$ the result follows. □

Now, we are in a position to prove the concentration in finite time result of the solution of (1) with $\chi = -1$. Here, we follow the proof of [13] even though, with the improvement, we do not need to restrict the initial data.

Theorem 2 *Let* $u(x,t) \geq 0$ *a weak solution of* (1) *in the sense of definition 1, with* $\chi = -1$ *and nonnegative initial data* $u_0 \in L^1 \cap L^\infty$ *and such that* $I_\theta(u_0)$ *is finite for some* $\theta \in (1, \frac{2}{2-\alpha})$. *Then, the maximal time of existence of* u, $T^* > 0$, *is finite, and the mass concentration of the solution occurs in a finite time. Also we have the bound*

$$T^* \leq \frac{b_\alpha 3^{1-2\alpha} R^{2\alpha} \int_{\mathbb{R}\times\mathbb{R}} |x-y|^{1-\alpha} u_0(x)u_0(y)\, \mathrm{d}x\, \mathrm{d}y}{\beta^3(1-\alpha)(2^{2-\alpha}-2)\|u_0\|_{L^1}}. \tag{16}$$

Proof Arguin as in Proposition 2 and using Lemma 1, we get

$$\begin{aligned}&\frac{\mathrm{d}}{\mathrm{d}t}\int_{\mathbb{R}\times\mathbb{R}} |x-y|^{1-\alpha} u(x,t)u(y,t)\, \mathrm{d}x\, \mathrm{d}y \\ &= \int_{\mathbb{R}\times\mathbb{R}} |x-y|^{1-\alpha}\Big(\partial(u\Lambda^{\alpha-1}\mathcal{H}u)(x,t)u(y,t) \\ &\quad + u(x,t)\partial(u\Lambda^{\alpha-1}\mathcal{H}u)(y,t)\Big)\, \mathrm{d}x\, \mathrm{d}y \\ &= -\frac{2(1-\alpha)}{b_\alpha}\int_{\mathbb{R}} u(x,t)((\Lambda^{\alpha-1}\mathcal{H}u)(x,t))^2\, \mathrm{d}x.\end{aligned}$$

Now, let $0 < \beta < 1$ and consider R depending on β given by Corollary 1. Proposition 1 implies that

$$\begin{aligned}&\frac{\mathrm{d}}{\mathrm{d}t}\int_{\mathbb{R}\times\mathbb{R}} |x-y|^{1-\alpha} u(x,t)u(y,t)\ \mathrm{d}x\,\mathrm{d}y\\ &\le -\frac{2(1-\alpha)}{b_\alpha}\frac{(2^{1-\alpha}-1)}{3^{1-2\alpha}} R^{-2\alpha}\beta^3 \|u_0\|^3_{L^1},\end{aligned}$$

by the mass conservation property. Integrating in time, we have

$$\begin{aligned}&\int_{\mathbb{R}\times\mathbb{R}} |x-y|^{1-\alpha} u(x,t)u(y,t)\ \mathrm{d}x\,\mathrm{d}y\\ &\le \int_{\mathbb{R}\times\mathbb{R}} |x-y|^{1-\alpha} u_0(x)u_0(y)\ \mathrm{d}x\,\mathrm{d}y - \frac{(1-\alpha)}{b_\alpha}\frac{(2^{2-\alpha}-2)}{3^{1-2\alpha}R^{2\alpha}}\beta^3\|u_0\|_{L^1}t.\end{aligned}$$

The last inequality implies estimate (16). In fact, on the contrary the expression

$$\inf_{y\in\mathbb{R}}\int_{\mathbb{R}} |x-y|^{1-\alpha} u(x,t)\ \mathrm{d}x,$$

becomes negative, which is a contradiction. □

Remark 1 If we denote by $\bar{x}$ the point which realizes the infimun

$$\inf_{y\in\mathbb{R}}\int_{\mathbb{R}} |x-y|^{1-\alpha} u_0(x)\ \mathrm{d}x = \int_{\mathbb{R}} |x-\bar{x}|^{1-\alpha} u(x,t)\ \mathrm{d}x,$$

the proof of the precedent result suggests that the mass is concentrate at $\bar{x}$.

4 Numerical Insights

Blowing-up illustration: considering the initial data of type

$$u_0(x) = Me^{-x^2}, \tag{17}$$

thus $\|u_0\|_{L^1} = M\sqrt{\pi}$, for $M > 0$. We can see in Figs. 1, 2, and 3 the blowing-up formation and that the time of explosion T^* decreases as the mass increases. This fact can be visualized from the theory by looking at Eq. (16).

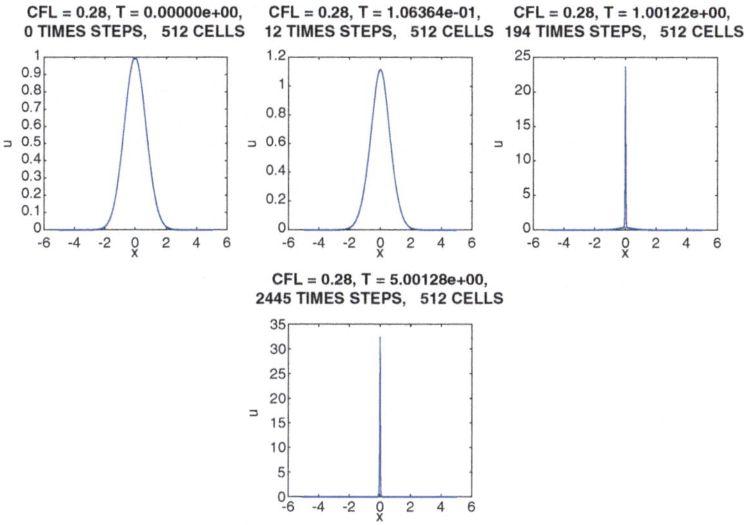

Fig. 1 Mass concentration ($\chi = -1$) with Gaussian initial data (17), and $M = 1$, for $t = 0, 0.1, 1, 5$ (left to right)

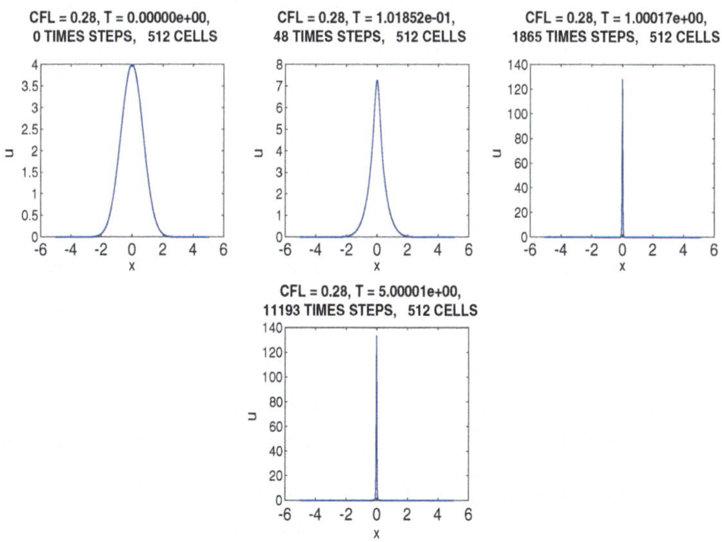

Fig. 2 Mass concentration ($\chi = -1$) with Gaussian initial data (17), and $M = 4$, for $t = 0, 0.1, 1, 5$ (left to right)

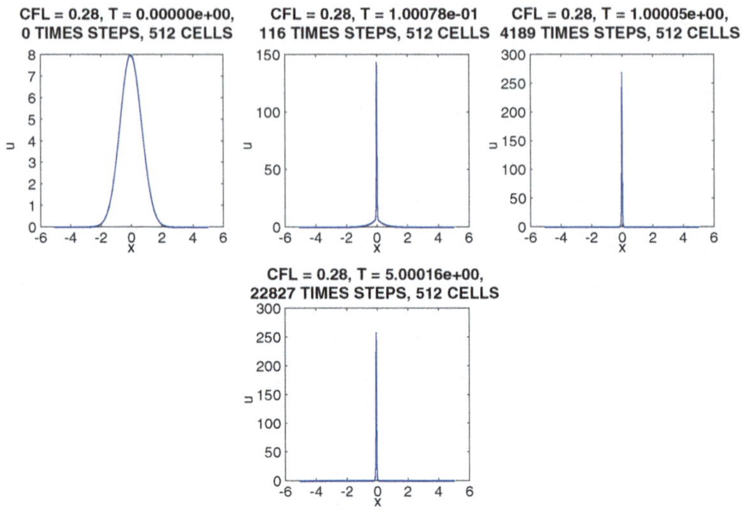

Fig. 3 Mass concentration ($\chi = -1$) with Gaussian initial data (17), and $M = 8$, for $t = 0, 0.1, 1, 5$ (left to right)

Geometrical place of concentration: according to Remark 1, we can suspect that concentration happens at the point $\bar{x}$ that realizes the infimum $\inf_{y \in \mathbb{R}} \int_{\mathbb{R}} |x - y|^{1-\alpha} u_0(x)\ \mathrm{d}x$. We explore this fact by using the initial condition

$$u_0(x) = e^{-x^2} + e^{-(x-8)^2}. \tag{18}$$

In such a case, we guess that $\bar{x} = 4$ is the mass center. Thus, in Figs. 4 and 5, it is illustrated this fact under double refinement on the spatial variable.

Mass spreading: consider the initial data

$$u_0(x) = 10e^{-(x-4)^2/R}, \tag{19}$$

for $R > 0$. According to Figs. 6 and 7, we can see how the mass is spreading along the x axis, and this spreadness is faster as soon as the initial mass increases. This fact can be anticipated from Theorem 1 inequality (12).

Initial positive negative data: finally, consider the initial data

$$u_0(x) = e^{-x^2} - e^{-(x-a)^2}, \tag{20}$$

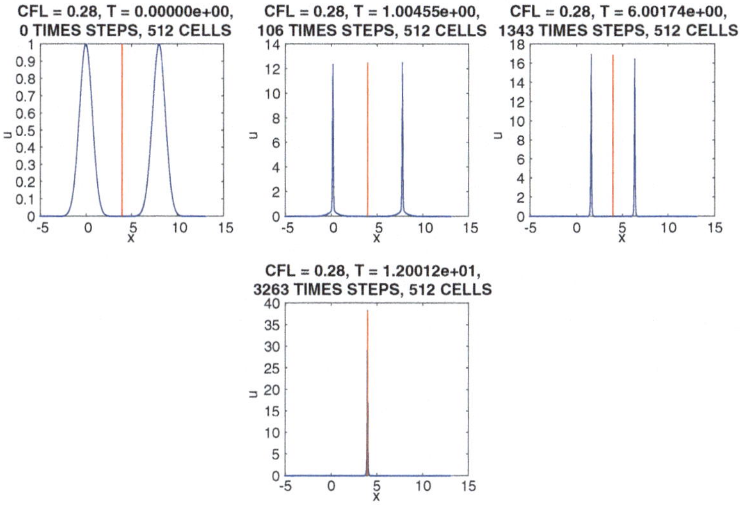

Fig. 4 Geometric location of the mass concentration point ($\chi = -1$), with initial data (18), for $t = 0, 1, 6, 12$ (left to right), 512 cells

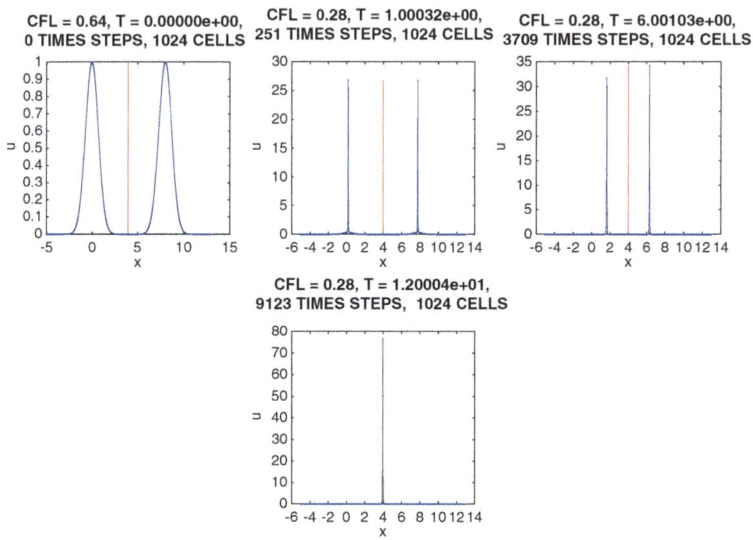

Fig. 5 Geometric location of the mass concentration point ($\chi = -1$), with initial data (18), for $t = 0, 1, 6, 12$ left to right, 1024 cells

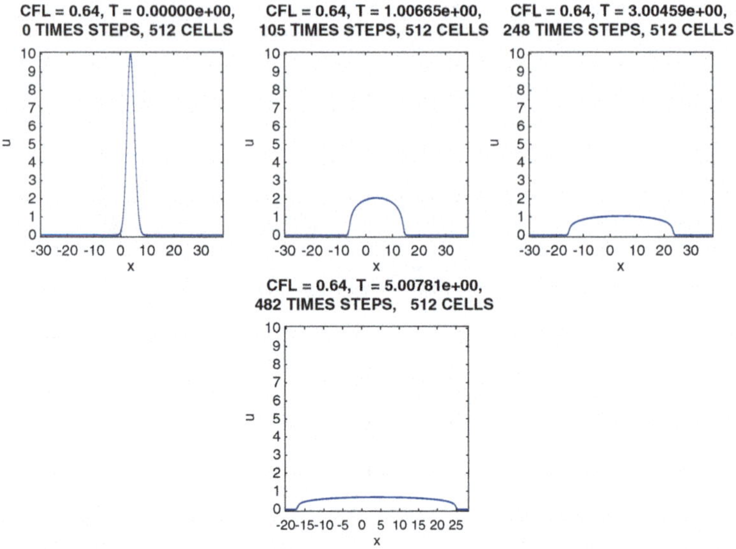

Fig. 6 Mass spreading of solution ($\chi = 1$), for initial data (19), with $R = 2$ for $t = 0, 1, 3, 5$ (left to right)

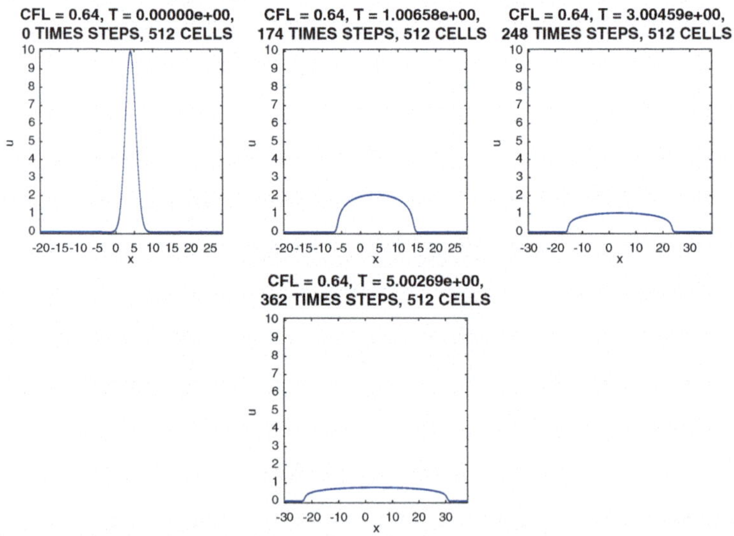

Fig. 7 Mass spreading of solution ($\chi = 1$), for initial data (19), with $R = 4$ for $t = 0, 1, 3, 5$ (left to right)

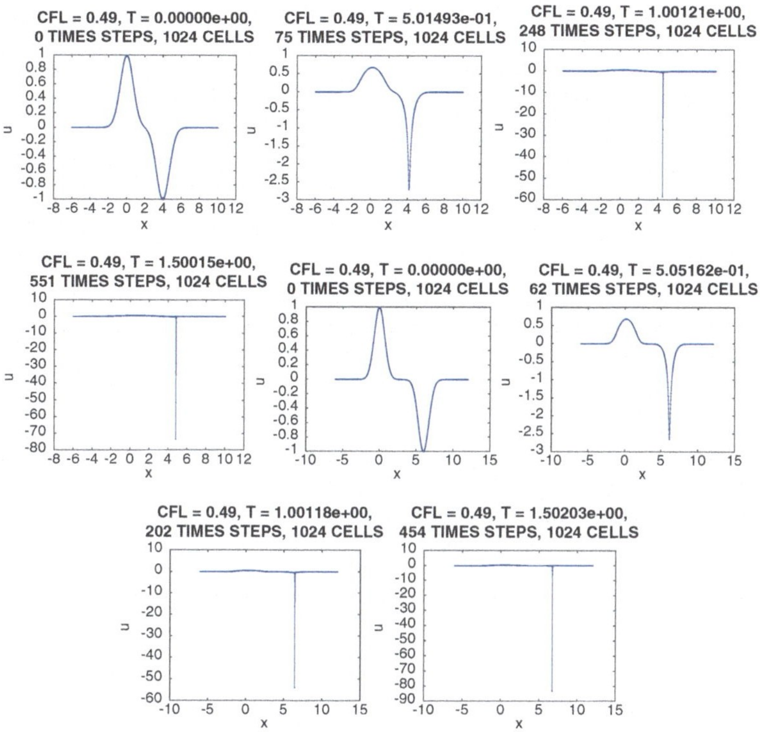

Fig. 8 Solution for initial positive negative data (20), with $a = 4, 6$, top to bottom, and $t = 0, 0.5, 1, 1.5$, left to right

for $a \neq 0$. The most interesting fact that we can see is that the positive part of the solution experiments mass dissipation while the negative part presents concentration (Fig. 8). Furthermore, the dissipation of the mass pushes the concentration point in the direction of the dissipation (Fig. 8). We refer the interested reader to [1, 2], and references therein, for a broader overview of the mathematical study on the computational approach employed in this work.

References

1. Abreu, E., Pérez, J.: A fast, robust, and simple Lagrangian–Eulerian solver for balance laws and applications. Comput. Math. Appl. **77**(9), 2310–2336 (2019)
2. Abreu, E., Ferreira, L.C.F., Galeano, J., Pérez, J.: On a 1D model with nonlocal interactions and mass concentrations: an analytical-numerical approach. Nonlinearity **35**(4), 1734 (2022)
3. Bedrossian, J., Rodriguez, N., Bertozzi, A.L.: Local and global well-posedness for aggregation equations and Patlak–Keller–Segel models with degenerate diffusion. Nonlinearity **24**, 1683 (2011)

4. Bertozzi, A.L., Laurent, T.: Finite-Time Blow-up of Solutions of an Aggregation Equation in $\mathbb{R}^n$. Commun. Math. Phys. **274**, 717–735 (2007)
5. Bertozzi, A.L., Carrillo, J.A., Laurent, T.: Blow-up in multidimensional aggregation equations with mildly singular interaction kernels. Nonlinearity **22**, 683 (2009)
6. Biler, P., Karch, G.: Blowup of solutions to generalized Keller–Segel model. J. Evol. Equ. **10**, 247–262 (2010)
7. Biler, P., Zienkiewicz, J.: Blowing up radial solutions in the minimal Keller–Segel model of chemotaxis. J. Evol. Equ. **19**, 71–90 (2019)
8. Biler. P., Karch, G., Zienkiewicz, J.: Morrey spaces norms and criteria for blowup in chemotaxis models. Netw. Heterog. Media **11**(2), 239–250 (2016)
9. Biler, P., Boritchev, A., Karch, G., Laurençot, P.: Concentration phenomena in a diffusive aggregation model. J. Differ. Equ. **271**, 1092–1108 (2021)
10. Caffarelli, L., Vazquez, J.L.: Nonlinear porous medium flow with fractional potential pressure. Arch Ration. Mech. Anal. **202**, 537–565 (2011)
11. Carrillo, J.A., DiFrancesco, M., Figalli, A., Laurent, T. Slepčev, D.: Global-in-time weak measure solutions and finite-time aggregation for nonlocal interaction equations. Duke Math. J. **156**, 229–271 (2011)
12. Li, H., Toscani, G. Long-time asymptotics of kinetic models of granular flows. Arch. Ration. Mech. Anal. **172**, 407–428 (2004)
13. Li, D., Zhang, X.: Global wellposedness and blowup of solutions to a nonlocal evolution problem with singular kernels. Commun. Pure Appl. Anal. **9**(6), 1591–1606 (2010)

Part VII
Recent Progress in Evolution Equations

$L^p - L^q$ Estimates for Very Strongly Damped Wave Equations

Marcello D'Abbicco and Antonio Lagioia

Abstract In this paper we determine $L^p - L^q$ estimates, $1 \leq p \leq q \leq \infty$, for the solution to the Cauchy problem for the damped wave equation

$$\begin{cases} u_{tt} - \Delta u + (-\Delta)^{\frac{\theta}{2}} u_t = 0, & t \geq 0, x \in \mathbb{R}^n \\ u(0,x) = 0, \; u_t(0,x) = u_1(x); \end{cases}$$

in the case of very strong damping, $\theta > 2$.

1 Introduction

The damped wave equation

$$\begin{cases} u_{tt} - \Delta u + (-\Delta)^{\frac{\theta}{2}} u_t = 0, & t \geq 0, x \in \mathbb{R}^n \\ u(0,x) = 0, \; u_t(0,x) = u_1(x), \end{cases} \tag{1}$$

with the so-called structural damping, that is, $\theta > 0$, has been investigated by several authors in recent years. With respect to the case of the classical wave equation with exterior damping, $\theta = 0$, the structural damping influences both the asymptotic profile of the solution as well as its regularity, in different ways. For $L^p - L^q$ estimates when $\theta = 0$ in (1) we address the reader to [17, 19, 20, 22].

In the case of *effective damping* (see the classification introduced in [5]), $\theta \in (0, 1)$, a phenomenon of *double diffusion* appears. The asymptotic profile of the solution is described by the solution to two anomalous diffusion equations (one or the other being prevalent according to different scenarios), see [3] (see also [18]).

M. D'Abbicco (✉) · A. Lagioia
University of Bari, Bari, Italy
e-mail: marcello.dabbicco@uniba.it; a.lagioia21@studenti.uniba.it

M. R. Ebert et al. (eds.), *New Tools in Mathematical Analysis and Applications*, Research Perspectives, https://doi.org/10.1007/978-3-031-77050-0_21

This phenomenon is analogous to the diffusion phenomenon which appears in the classical damped wave equation, $\theta = 0$, see [15, 19, 22].

In the case of *non effective damping*, $\theta \in (1, 2]$, the presence of the oscillations at low frequencies influences the asymptotic profile of the solution [6, 9, 23, 24]. The decay rate of $L^p - L^q$ estimates depends on the quantity

$$d(p, q) = n\left(\frac{1}{p} - \frac{1}{q}\right) + (n-1)\max\left\{\frac{1}{2} - \frac{1}{p}, \frac{1}{q} - \frac{1}{2}\right\}. \tag{2}$$

In particular, when $d(p, q) \leq 1$, the decay rate is the same as per the wave equation with no damping, and it is independent on θ.

Results which take into account of the presence of both an effective and a noneffective damping terms are provided in [2, 11].

In the limit case $\theta = 1$, the equation is homogeneous, so that the possible presence of oscillations does not modify the decay rate structure of the equation [1].

The regularity of the solution in the case of very strong damping $\theta > 2$ has been investigated by Ghisi et al. [14], in the setting of problems in Hilbert spaces with boundary conditions.

The aim of this paper is to determine $L^p - L^q$ estimates, with $1 \leq p \leq q \leq \infty$, for the solution to the Cauchy problem related to the wave equation with very strong damping. One possible application of these estimates is to determine the Fujita critical exponent for semilinear perturbations of these problems (see, for instance, [4, 7, 8, 12, 13, 21]).

In order to deal with low frequencies, it is sufficient to apply Theorem 1 in [6], so this paper will focus in dealing with high frequencies estimates, since the results in [6] do not apply to wave equations with very strong damping. It is well-known that a phenomenon of regularity-loss type of decay appears, but such phenomenon has not been investigated in the $L^p - L^q$ setting, so far.

We will prove the following.

Theorem 1 *Let $\theta > 2$ and assume that $1 \leq p_0 \leq p_1 \leq q \leq \infty$ are such that*

$$\frac{n}{\theta}\left(\frac{1}{p_1} - \frac{1}{q}\right) \leq 1, \tag{3}$$

with a strict inequality, i.e., $n(1/p_1 - 1/q) < \theta$, if $p_1 = 1$ or $q = \infty$. Let $d = d(p_0, q)$ as in (2)*. Then the solution to* (1) *satisfies*

$$\begin{aligned} \|u(t, \cdot)\|_{L^q} &\leq C\,(1+t)^{1-n\left(\frac{1}{p_0}-\frac{1}{q}\right)+(d-1)_+\left(1-\frac{1}{\theta}\right)}\|u_1\|_{L^{p_0}} \\ &\quad + C\,(1+t)^{-\frac{\theta}{\theta-2}+\frac{n}{\theta-2}\left(\frac{1}{p_1}-\frac{1}{q}\right)}\|u_1\|_{L^{p_1}}, \end{aligned} \tag{4}$$

provided that we replace the first term in the right-hand side of (4) *by*

$$C\,(\log(e+t))^{\frac{1}{2}}\,\|u_1\|_{L^{p_0}},$$

if $n = 2$ *and* $(p_0, q) \in \{(1, 2), (2, \infty)\}$. *If* $p_1 = q$, *we may replace the second term in the right-hand side of* (4) *by*

$$C\,(1+t)^{-\frac{\theta+a}{\theta-2}}\,\|u_1\|_{W^{a,q}},$$

for any $a \geq 0$, *where* $W^{a,q}$ *denotes the Bessel potential space* $\{f \in L^q : \mathcal{F}^{-1}(\langle\xi\rangle^a \hat{f}) \in L^q\}$, *where* $\langle\xi\rangle = (1+|\xi|^2)^{\frac{1}{2}}$.

In Theorem 1 and in the following, C and C' denote generic positive constants whose values change in every formula.

What Is New?
The original contribution of this paper consists in the proof of the subsequent Lemma 1, which allows us to describe the influence of the regularity-loss structure of decay in the setting of $L^p - L^q$ estimates. This structure also appears in other evolution models, like the damped plate equation with rotational inertia (see, for instance, [10, 26]).

The proof of Lemma 1 is based on the use of Fourier inversion formula and integration by parts method to show that a multiplier $m(t, \xi)$ is in $\mathcal{F}(L^q)$ for any t and to estimate its norm $\|m(t, \cdot)\|_{M_1^q}$ (see Sect. 2). This approach is a variant of the approach used, for instance, in [25, Theorem 2.3]. However, our approach is also applicable to multipliers that are not in L^1, since the Fourier inversion formula is only applied to appropriate derivatives of the multiplier. Smooth multipliers that are not in L^1 may occur in other models, when a problem is localized at high frequencies.

The case $p = 1$ and $q = \infty$
Dealing with high-frequencies $L^p - L^q$ estimates when $1 < p \leq q < \infty$ for the solution to problem (4) is very easy. Indeed, Hardy-Littlewood-Sobolev theorem and Mikhlin-Hörmander multiplier theorem (see later, Sect. 2) effectively apply, due to the lack of oscillations in the multiplier associated to the fundamental solution to (4). Those theorems, however, do not apply in the case $p = 1$ or $q = \infty$. Also, the lack of the smoothing effect and of the exponential decay when $\theta > 2$ in (4), makes this study slightly more challenging than the case $\theta \in (1, 2]$ treated in [6, 24].

2 Some Tools from Fourier Analysis

As usual, $\mathcal{S}'(\mathbb{R}^n)$ denotes the space of tempered distributions. For any $T \in \mathcal{S}'(\mathbb{R}^n)$, $\mathcal{F}(T) = \hat{T}$ denotes the Fourier transform of T and $\mathcal{F}^{-1}(T)$ its inverse Fourier transform. When $f(t,x)$ is a slowly increasing function, $\hat{f}(t,\xi)$ denotes its Fourier transform with respect to the space variable. In particular, if $f(t,\cdot) \in L^1$ then

$$\hat{f}(t,\xi) = \int_{\mathbb{R}^n} e^{-ix\cdot\xi} f(t,x)\,dx.$$

To deal with Fourier multipliers, we use the notation introduced by Hörmander [16].

Notation 1 Let $m \in \mathcal{S}'(\mathbb{R}^n)$, the space of tempered distributions. We say that m is a Fourier multiplier of type (p,q) and we write $m \in M_p^q$, where $1 \leq p \leq q \leq \infty$, if $\mathcal{F}^{-1}(m\hat{f}) \in L^q$ for any $f \in \mathcal{S}(\mathbb{R}^n)$, and the quantity

$$\|m\|_{M_p^q} = \sup\left\{\|\mathcal{F}^{-1}(m\hat{f})\|_{L^q} : f \in \mathcal{S},\ \|f\|_{L^p} = 1\right\}$$

is finite. The quantity $\|\cdot\|_{M_p^q}$ is a norm on M_p^q. The operator $f \mapsto \mathcal{F}^{-1}(m\hat{f})$ is then extended by density to a bounded linear operator from L^p to L^q, with norm $\|m\|_{M_p^q}$. □

We mention that $M_p^q = M_{q'}^{p'}$, where $p' = p/(p-1)$ denotes the Hölder conjugate of p, for any $p \in [1,\infty]$ (here $p' = \infty$ if $p = 1$, and viceversa). For $p = 1, \infty$, $M_1^1 = M_\infty^\infty$ is the space of the Fourier transforms of bounded measures [16, Theorem 1.4]. In particular, $\mathcal{F}(L^1) \hookrightarrow M_1^1$ and

$$\|m\|_{M_1^1} \leq \|\mathcal{F}^{-1}(m)\|_{L^1}.$$

For $p \in (1,\infty]$, M_1^p is the space of the Fourier transforms of functions in L^p, and $\mathcal{F}(L^p) = M_1^p$, with

$$\|m\|_{M_1^p} = \|\mathcal{F}^{-1}(m)\|_{L^p}.$$

If $1 \leq p \leq q \leq 2$, then $M_p^q \subset L^2_{\mathrm{loc}}$. On the other hand, the space M_p^q contains distributions that are not L^1_{loc} functions if $p < 2 < q$ (though in this paper, we will not deal with multipliers that are not functions). In the both cases, due to Young convolution theorem, $\mathcal{F}(L^r) \hookrightarrow M_p^q$, with

$$\|m\|_{M_p^q} \leq \|\mathcal{F}^{-1}(m)\|_{L^r}, \quad \frac{1}{r} = 1 - \frac{1}{p} + \frac{1}{q}.$$

Notation 2 For any multi-index $\beta \in \mathbb{N}^n$, we use the notation

$$\partial_\xi^\beta g(\xi) = \partial_{\xi_1}^{\beta_1} \cdots \partial_{\xi_n}^{\beta_n} g(\xi); \qquad (ix)^\beta = (ix_1)^{\beta_1} \cdots (ix_n)^{\beta_n}.$$

where $\beta = (\beta_1, \ldots, \beta_n)$. We say that

$$|\beta| = \beta_1 + \ldots + \beta_n$$

is the length of β. □

We recall that

$$\mathcal{F}(\partial_x^\beta T) = (i\xi)^\beta \hat{T}, \quad \mathcal{F}^{-1}(\partial_\xi^\beta T) = (-ix)^\beta \mathcal{F}^{-1} T,$$

for any $T \in \mathcal{S}'(\mathbb{R}^n)$ and $\beta \in \mathbb{N}^n$.

In this paper, we make use of the Hardy-Littlewood-Sobolev inequality and of Mikhlin-Hörmander multiplier theorem; in particular,

$$|\xi|^{-n\left(\frac{1}{p}-\frac{1}{q}\right)} \in M_p^q, \qquad 1 < p \leq q < \infty,$$

and if $m \in C^k(\mathbb{R}^n \setminus \{0\})$ with $k \geq (n+1)/2$, then

$$|\partial_\xi^\beta m(\xi)| \leq C\,|\xi|^{-|\beta|}, \quad 1 < p < \infty \Rightarrow m \in M_p^p. \tag{5}$$

Combining the two results, one finds that

$$|\partial_\xi^\beta m(\xi)| \leq C\,|\xi|^{-n\left(\frac{1}{p}-\frac{1}{q}\right)-|\beta|}, \quad 1 < p \leq q < \infty \Rightarrow m \in M_p^q. \tag{6}$$

3 Proof of Theorem 1

Applying the Fourier transform $\mathcal{F}$ with respect to the space variable, we obtain

$$\begin{cases} \widehat{u}_{tt} + |\xi|^2 \widehat{u} + |\xi|^\theta \widehat{u}_t = 0, & t \geq 0,\, x \in \mathbb{R}^n \\ \widehat{u}(0,\xi) = 0,\ \widehat{u}_t(0,\xi) = \hat{u}_1(\xi). \end{cases}$$

The characteristic equation associated to the full symbol of the equation in (1) is

$$\lambda^2 + |\xi|^\theta \lambda + |\xi|^2 = 0,$$

from which we may get the characteristic roots

$$\lambda_\pm(\xi) = -\frac{|\xi|^\theta}{2} \pm i|\xi|\sqrt{1-|\xi|^{2(\theta-1)}/4}, \qquad \text{if } |\xi|^{2(\theta-1)} < 4,$$

$$\lambda_\pm(\xi) = \frac{|\xi|^\theta}{2}\left(-1 \pm \sqrt{1-4|\xi|^{-2(\theta-1)}}\right), \qquad \text{if } |\xi|^{2(\theta-1)} > 4.$$

Let $\varphi_1 \in C^\infty$ be such that $\varphi_1(\xi) = 0$ for $|\xi| \leq M$, $0 \leq \varphi_1 \leq 1$, and $\varphi_1(\xi) = 1$ for $|\xi| \geq 2M$, for sufficiently large $M \gg 1$.

Applying Theorem 1 in [6], we find that

$$\|\mathcal{F}^{-1}(\hat{u}(t,\cdot)\,(1-\varphi_1))\|_{L^q} \leq C\,(1+t)^{1-n\left(\frac{1}{p_0}-\frac{1}{q}\right)+(d-1)_+\left(1-\frac{1}{\theta}\right)}\|u_1\|_{L^{p_0}},$$

with the exception that if $n = 2$ and $(p_0, q) \in \{(1,2), (2,\infty)\}$, we replace the previous estimate by

$$\|\mathcal{F}^{-1}(\hat{u}(t,\cdot)\,(1-\varphi_1))\|_{L^q} \leq C\,(\log(e+t))^{\frac{1}{2}}\,\|u_1\|_{L^{p_0}}.$$

Therefore, it remains to consider $L^{p_1}-L^q$ estimates for $\mathcal{F}^{-1}(\hat{u}(t,\cdot)\,\varphi_1)$, the solution localized at high frequencies. Due to

$$\widehat{u}(t,\xi) = \frac{e^{\lambda_+ t} - e^{\lambda_- t}}{\lambda_+ - \lambda_-}\,\widehat{u}_1(\xi)$$

for $|\xi| \gg 1$, we shall estimate the multiplier norm $\|m_\pm(t,\cdot)\|_{M_p^q}$ (see Sect. 2), where

$$m_\pm(t,\xi) = \frac{e^{\lambda_\pm(\xi)t}}{\lambda_+(\xi) - \lambda_-(\xi)}\,\varphi_1(\xi).$$

Following as in the proof of Theorem 4 in [6], it is easy to prove that

$$\|m_-(t,\cdot)\|_{M_{p_1}^q} \leq C\,e^{-ct},$$

provided that (3) holds, with a strict inequality if $p_1 = 1$ or $q = \infty$. Indeed, for $1 < p_1 \leq q < \infty$, by (6), thanks to (3), we immediately obtain

$$\|m_-(t,\xi)\|_{M_{p_1}^q} \leq C\,e^{-ct}.$$

Now let $p_1 = 1$ and $q \in [1,\infty]$ (by duality, $M_1^q = M_{q'}^\infty$), so that $M_1^q = \mathcal{F}(L^q)$ if $q > 1$ and $\mathcal{F}(L^1) \subset M_1^1$ if $q = 1$. Thanks to the strict inequality $n(1-1/q) < \theta$, by using Lemma 5 in [6], we obtain

$$\|\mathcal{F}^{-1}m_-(t,\cdot))\|_{L^q} \leq Ce^{-ct}.$$

It remains to estimate $\|m_+(t,\cdot)\|_{M^q_{p_1}}$. We estimate the derivatives of $m_+(t,\xi)$ with respect to ξ. Using Taylor's formula for $|\xi| \gg 1$, we may write

$$\lambda_+(\xi) = -|\xi|^{-(\theta-2)} - \mu(\xi), \qquad \text{where}$$

$$\mu(\xi) = \frac{2}{|\xi|^{3\theta-4}} g\left(\frac{4}{|\xi|^{2(\theta-1)}}\right), \qquad g(s) = \int_0^1 (1-\sigma s)^{-3/2}(1-\sigma)\,d\sigma.$$

It is easy to prove that the following estimates hold for any $\beta \in \mathbb{N}^n$ and $|\xi| \gg 1$:

$$|\partial_\xi^\beta \lambda_-(\xi)| \leq C\,|\xi|^{\theta-|\beta|},$$

$$|\partial_\xi^\beta \lambda_+(\xi)| \leq C\,|\xi|^{-(\theta-2)-|\beta|},$$

$$|\partial_\xi^\beta\,(\lambda_+(\xi) - \lambda_-(\xi))^{-1}\,| \leq C\,|\xi|^{-\theta-|\beta|}.$$

Therefore,

$$|\partial_\xi^\beta e^{\lambda_+(\xi)t}| \leq C\,|\xi|^{-|\beta|}\,e^{-ct|\xi|^{-(\theta-2)}};$$

hence,

$$|\partial_\xi^\beta m_+(t,\xi)| \leq C\,|\xi|^{-\theta-|\beta|}\,e^{-ct|\xi|^{-(\theta-2)}}. \tag{7}$$

It is well known that the presence of a very strong damping $\theta > 2$ causes the loss of the exponential decay in time, of the solution and its derivatives, in m_+. However, we may produce a polynomial decay in time. Namely, let $t > 1$; for any $k > 0$, we may estimate

$$e^{-ct|\xi|^{2-\theta}} = t^{-k}(t|\xi|^{(2-\theta)})^k e^{-ct|\xi|^{2-\theta}}\,|\xi|^{k(\theta-2)} \leq Ct^{-k}|\xi|^{k(\theta-2)}. \tag{8}$$

Estimates like the one in (8) hints to a so-called regularity-loss structure of decay: assuming more regular data, it is possible to produce arbitrarily large decay rate of polynomial type. Using (8) and recalling that $|\xi| \geq M$, we may improve estimate (7), replacing it by

$$|\partial_\xi^\beta m_+(t,\xi)| \leq C(1+t)^{-k}|\xi|^{-\theta-|\beta|+k(\theta-2)}, \quad \forall k \geq 0. \tag{9}$$

The conclusion of the proof of Theorem 1 follows if we prove the following result (we put $p_1 = p$ for brevity).

Lemma 1 *Let $\theta > 2$ and assume that $1 \leq p \leq q \leq \infty$ are such that*

$$\frac{n}{\theta}\left(\frac{1}{p} - \frac{1}{q}\right) \leq 1, \tag{10}$$

with a strict inequality, i.e., $n(1/p - 1/q) < \theta$, if $p = 1$ or $q = \infty$. Then

$$\|m_+(t,\cdot)\|_{M_p^q} \leq C\,(1+t)^{-\frac{\theta}{\theta-2}+\frac{n}{\theta-2}\left(\frac{1}{p}-\frac{1}{q}\right)}. \tag{11}$$

Moreover, for any $q \in [1,\infty]$ and $a \geq 0$, it holds

$$\|\langle\xi\rangle^{-a} m_+(t,\cdot)\|_{M_q^q} \leq C\,(1+t)^{-\frac{\theta+a}{\theta-2}}. \tag{12}$$

Proof First, let $1 < p \leq q < \infty$ be such that (10) holds. Thanks to (6), using (9) with

$$k = \frac{1}{\theta-2}\left(\theta - n\left(\frac{1}{p}-\frac{1}{q}\right)\right),$$

we get

$$\|m_+(t,\xi)\|_{M_p^q} \leq C\,(1+t)^{-\frac{1}{\theta-2}\left(\theta-n\left(\frac{1}{p}-\frac{1}{q}\right)\right)}.$$

Similarly, for any $q \in (1,\infty)$ and $a \geq 0$, by (5), using (9) with $k = (\theta+a)/(\theta-2)$, we get

$$\|\langle\xi\rangle^{-a}\, m_+(t,\xi)\|_{M_q^q} \leq C(1+t)^{-\frac{\theta+a}{\theta-2}}.$$

Now let $p = 1$ and $q \in [1,\infty]$ (by duality, $M_1^q = M_{q'}^\infty$), so that $M_1^q = \mathcal{F}(L^q)$ if $q > 1$ and $\mathcal{F}(L^1) \subset M_1^1$ if $q = 1$.

Recalling (9), for a given $k \geq 0$, we choose γ such that $|\gamma|+\theta-k(\theta-2) > n$, so that $\partial_\xi^\gamma m_+ \in L^1$. As a consequence, we may use the L^1-Fourier inversion formula in $\mathcal{S}'$ (namely, if $m \in L^1$ then $\mathcal{F}^{-1}(m) \in C_0$) to obtain the pointwise expression:

$$\begin{aligned}(-ix)^\gamma(\mathcal{F}^{-1}m_+)(t,x) &= \mathcal{F}^{-1}(\partial_\xi^\gamma m_+)(t,x)\\ &= (2\pi)^{-n}\int_{|\xi|\geq M} e^{ix\xi}\,\partial_\xi^\gamma m_+(t,\xi)\,d\xi.\end{aligned} \tag{13}$$

For large $|x| \geq M^{-1}$, using (9), we may directly estimate

$$|(-ix)^\gamma(\mathcal{F}^{-1}m_+)(t,x)| \leq C\,(1+t)^{-k}\int_{|\xi|\geq M} |\xi|^{-\theta-|\gamma|+k(\theta-2)}\,d\xi = C'\,(1+t)^{-k},$$

for any $k \geq 0$. For any given integer N, varying $\gamma \in \mathbb{N}^n$ so that $|\gamma| = N$, the previous estimate implies that

$$|(\mathcal{F}^{-1}m_+)(t,x)| \leq C\,(1+t)^{-k}\,|x|^{-N}, \tag{14}$$

for any given $k \geq 0$. Namely, we have an arbitrarily large polynomial decay both in time and space.

We now consider small $|x| \leq M^{-1}$. In this case, we split the integral in (13) into two parts. On the one hand, using (9), we obtain

$$\begin{aligned} |I_1| &= \left| \int_{|\xi| \geq |x|^{-1}} e^{ix\xi} \partial_\xi^\gamma m_+(t, \xi) \, d\xi \right| \\ &\leq C \, (1+t)^{-k} \int_{|\xi| \geq |x|^{-1}} |\xi|^{-\theta - |\gamma| + k(\theta - 2)} \, d\xi \\ &\leq C' \, (1+t)^{-k} |x|^{\theta + |\gamma| - k(\theta-2) - n}. \end{aligned}$$

To deal with

$$I_2 = \int_{M \leq |\xi| \leq |x|^{-1}} e^{ix\xi} \partial_\xi^\gamma m_+ \, d\xi,$$

we perform $|\gamma|$ steps of integration by parts, thanks to the fact that $J = \{\xi : M \leq |\xi| \leq |x|^{-1}\}$ is a compact, so that $m_+ \in L^1(J)$. We stress that no boundary term appears when $|\xi| = M$, since $\varphi_1 \in C^\infty$ is supported in $\{\xi : |\xi| \geq M\}$.

We assume that k is sufficiently large to get

$$\theta - k(\theta - 2) < n, \tag{15}$$

so that, using (9), we obtain

$$\begin{aligned} |I_2| &\leq C |x|^{|\gamma|} \int_{M \leq |\xi| \leq |x|^{-1}} |m_+(t, \xi)| \, d\xi \\ &\quad + C \sum_{\beta \leq \gamma} |x|^{|\beta| - 1} \int_{|\xi| = |x|^{-1}} |\partial_\xi^{\gamma - \beta} m_+(t, \xi)| \, d\xi \\ &\leq C' \, (1+t)^{-k} |x|^{\theta + |\gamma| - k(\theta - 2) - n}. \end{aligned}$$

Varying $\gamma \in \mathbb{N}^n$ as we did for $|x| \geq M^{-1}$, we derive

$$|\mathcal{F}^{-1} m_+(t, \xi)| \leq C \, (1+t)^{-k} \, |x|^{\theta - k(\theta - 2) - n}, \tag{16}$$

for any given $k \geq 0$ such that (15) holds; in view of (14), estimate (16) holds uniformly with respect to $x \in \mathbb{R}^n$.

We distinguish two cases. First, let $\theta < n$. In this case, (15) is verified for any $k \geq 0$.

Recalling that $n(1-1/q)<\theta$, that is, $(\theta-n)q>-n$, we now use (16) with $k=0$ if $|x|\leq(1+t)^{-\frac{1}{\theta-2}}$ and with $k=k_2>\theta/(\theta-2)$ if $|x|\geq(1+t)^{-\frac{1}{\theta-2}}$, so that we obtain

$$\begin{aligned}\|\mathfrak{F}^{-1}m_+(t,\cdot)\|_{L^q} &\leq C\Big(\int_{|x|\leq(1+t)^{-\frac{1}{\theta-2}}}|x|^{(\theta-n)q}\,dx\Big)^{\frac{1}{q}}\\&\quad+C\,(1+t)^{-k_2}\Big(\int_{|x|\geq(1+t)^{-\frac{1}{\theta-2}}}|x|^{-(n+k_2(\theta-2)-\theta)q}\,dx\Big)^{\frac{1}{q}}\\&\leq C'\,(1+t)^{-\frac{\theta}{\theta-2}+\frac{n}{\theta-2}\left(1-\frac{1}{q}\right)}.\end{aligned}$$

We now consider the case $\theta\geq n$. In this case, we cannot take $k=0$, due to (15), so we assume that $k=k_1$ with

$$\frac{\theta-n}{\theta-2}<k_1<\frac{\theta-n+n/q}{\theta-2}$$

if $|x|\leq(1+t)^{-\frac{1}{\theta-2}}$. Within this range, (15) holds and $(\theta-k_1(\theta-2)-n)q>-n$. Therefore, letting $k=k_2>\theta/(\theta-2)$ if $|x|\geq(1+t)^{-\frac{1}{\theta-2}}$ as before, we obtain

$$\begin{aligned}\|\mathfrak{F}^{-1}m_+(t,\cdot)\|_{L^q} &\leq C\,(1+t)^{-k_1}\Big(\int_{|x|\leq(1+t)^{-\frac{1}{\theta-2}}}|x|^{(\theta-k_1(\theta-2)-n)q}\,dx\Big)^{\frac{1}{q}}\\&\quad+C\,(1+t)^{-k_2}\Big(\int_{|x|\geq(1+t)^{-\frac{1}{\theta-2}}}|x|^{-(n+k_2(\theta-2)-\theta)q}\,dx\Big)^{\frac{1}{q}}\\&\leq C'\,(1+t)^{-\frac{\theta}{\theta-2}+\frac{n}{\theta-2}\left(1-\frac{1}{q}\right)}.\end{aligned}$$

In the case $q=1$, for any given $a\geq0$, we also consider $\langle\xi\rangle^{-a}\,m_+(t,\xi)$. We proceed as before, but now

$$\begin{aligned}|I_1| &= \Big|\int_{|\xi|\geq|x|^{-1}} e^{ix\xi}\,\partial_\xi^\gamma\,(\langle\xi\rangle^{-a}\,m_+(t,\xi))\,d\xi\Big|\\&\leq C\,(1+t)^{-k}\int_{|\xi|\geq|x|^{-1}}|\xi|^{-\theta-a-|\gamma|+k(\theta-2)}\,d\xi\\&\leq C'\,(1+t)^{-k}|x|^{\theta+a+|\gamma|-k(\theta-2)-n}.\end{aligned}$$

We replace (15) by

$$\theta+a-k(\theta-2)<n, \tag{17}$$

so that, using (9), we obtain

$$
\begin{aligned}
|I_2| &\leq C|x|^{|\gamma|} \int_{M\leq|\xi|\leq|x|^{-1}} |\langle\xi\rangle^{-a}\, m_+(t,\xi)|\, d\xi \\
&\quad + C \sum_{\beta\leq\gamma} |x|^{|\beta|-1} \int_{|\xi|=|x|^{-1}} |\partial_\xi^{\gamma-\beta}(\langle\xi\rangle^{-a} m_+(t,\xi))|\, d\xi \\
&\leq C'\,(1+t)^{-k}|x|^{\theta+a+|\gamma|-k(\theta-2)-n};
\end{aligned}
$$

hence,

$$
|\mathcal{F}^{-1}(\langle\xi\rangle^{-a}\, m_+(t,\xi))| \leq C\,(1+t)^{-k}\,|x|^{\theta+a-k(\theta-2)-n}, \tag{18}
$$

for any given $k \geq 0$ such that (17) holds; estimate (18) holds uniformly with respect to $x \in \mathbb{R}^n$.

Proceeding as we did in the previous case, we fix $k_1 = 0$ if $\theta + a < n$ and

$$
\frac{\theta+a-n}{\theta-2} < k_1 < \frac{\theta+a}{\theta-2}
$$

if $\theta + a \geq n$, so that (17) is verified for $k = k_1$. We fix $k_2 > (\theta+a)/(\theta-2)$, so that we obtain

$$
\begin{aligned}
\|\mathcal{F}^{-1}(\langle\xi\rangle^{-a}\, m_+(t,\cdot))\|_{L^1} &\leq C\,(1+t)^{-k_1} \int_{|x|\leq(1+t)^{-\frac{1}{\theta-2}}} |x|^{\theta+a-k_1(\theta-2)-n}\, dx \\
&\quad + C\,(1+t)^{-k_2} \int_{|x|\geq(1+t)^{-\frac{1}{\theta-2}}} |x|^{-(n+k_2(\theta-2)-\theta-a)}\, dx \\
&\leq C'\,(1+t)^{-\frac{\theta+a}{\theta-2}}.
\end{aligned}
$$

This concludes the proof of Lemma 1. □

Acknowledgments The first author has been supported by: INdAM-GNAMPA Project, CUP_E55F22000270001 and by PNR-MUR project CUP_H93C22000450007.

References

1. D'Abbicco, M.: A benefit from the L^1 smallness of initial data for the semilinear wave equation with structural damping. In: Mityushev, V., Ruzhansky, M. (eds.) Current Trends in Analysis and its Applications, pp. 209–216 (2015). Proceedings of the 9th ISAAC Congress, Krakow. http://www.springer.com/br/book/9783319125763
2. D'Abbicco. M.: $L^1 - L^1$ estimates for a doubly dissipative semilinear wave equation. Nonlinear Differ. Equ. Appl. **24**, Article 5 (2017). https://doi.org/10.1007/s00030-016-0428-4

3. D'Abbicco, M., Ebert, M.R.: Diffusion phenomena for the wave equation with structural damping in the $L^p - L^q$ framework. J. Differ. Equ. **256**, 2307–2336 (2014). http://dx.doi.org/10.1016/j.jde.2014.01.002
4. D'Abbicco, M., Ebert, M.R.: An application of $L^p - L^q$ decay estimates to the semilinear wave equation with parabolic-like structural damping. Nonlinear Anal. **99**, 16–34 (2014). http://dx.doi.org/10.1016/j.na.2013.12.021
5. D'Abbicco, M., Ebert, M.R.: A classification of structural dissipations for evolution operators. Math. Meth. Appl. Sci. **39**, 2558–2582 (2016). http://dx.doi.org/10.1002/mma.3713
6. D'Abbicco, M., Ebert, M.R.: Sharp $L^p - L^q$ estimates for a class of dissipative wave equations (2024). https://arxiv.org/abs/2311.03173
7. D'Abbicco, M., Girardi, G.: A structurally damped σ-evolution equation with nonlinear memory. Math. Methods Appl. Sci. 1–19 (2020). https://doi.org/10.1002/mma.6633
8. D'Abbicco, M., Girardi, G.: Second order p-evolution equations with critical nonlinearity. Bruno Pini Math. Anal. Seminar **14**(2), 263–280 (2023) https://doi.org/10.6092/issn.2240-2829/18879
9. D'Abbicco, M., Girardi, G., Liang, J.: $L^1 - L^1$ estimates for the strongly damped plate equation. J. Math. Anal. Appl. **478**, 476–498 (2019). https://doi.org/10.1016/j.jmaa.2019.05.039
10. Charão, R.C., da Luz, C.R.: Asymptotic properties for a semilinear plate equation in unbounded domains. J. Hyperbolic Differ. Equ. **6**, 269–294 (2009)
11. Chen, W., D'Abbicco, M., Girardi, G.: Global small data solutions for semilinear waves with two dissipative terms. Annali di Matematica Pura ed Applicata **201**, 529–560 (2022). https://doi.org/10.1007/s10231-021-01128-z
12. Ebert, M.R., Reissig, M.: Methods for partial differential equations, Qualitative Properties of Solutions, Phase Space Analysis. Semilinear Models. Birkhäuser/Springer, Cham (2018)
13. Ebert, M.R., Girardi, G., Reissig, M.: Critical regularity of nonlinearities in semilinear classical damped wave equations. Math. Ann. **378**, 1311–1326 (2020). https://doi.org/10.1007/s00208-019-01921-5
14. Ghisi, G., Gobbino, M., Haraux, A.: Local and global smoothing effects for some linear hyperbolic equations with a strong dissipation. Trans. Am. Math. Soc. **368**, 2039–2079 (2016)
15. Han, Y., Milani, A.: On the diffusion phenomenon of quasilinear hyperbolic waves. Bull. Sci. Math. **124**(5), 415–433 (2000)
16. Hörmander, L.: Estimates for translation invariant operators in L^p spaces. Acta Math. **104**, 93–140 (1960)
17. Ikeda M., Inui, T., Okamoto, M., Wakasugi, Y.: $L^p - L^q$ estimates for the damped wave equation and the critical exponent for the nonlinear problem with slowly decaying data. Commun. Pure. Appl. Anal. **18**, 1967–2008 (2019)
18. Karch, G.: Selfsimilar profiles in large time asymptotics of solutions to damped wave equations. Stud. Math. **143**(2), 175–197 (2000)
19. Marcati, P., Nishihara, K.: The L^p-L^q estimates of solutions to one-dimensional damped wave equations and their application to the compressible flow through porous media. J. Differ. Equ. **191**, 445–469 (2003)
20. Matsumura, A.: On the asymptotic behavior of solutions of semi-linear wave equations. Publ. RIMS. **12**, 169–189 (1976)
21. Narazaki, T.: $L^p - L^q$ estimates for damped wave equations and their applications to semilinear problem. J. Math. Soc. Jpn. **56**, 586–626 (2004)
22. Nishihara, K.: $L^p - L^q$ estimates for solutions to the damped wave equations in 3-dimensional space and their applications. Math. Z. **244**, 631–649 (2003)
23. Ponce, G.: Global existence of small solutions to a class of nonlinear evolution equations. Nonlinear Anal. TMA **9**(5), 399–418 (1985)
24. Shibata, Y.: On the rate of decay of solutions to linear viscoelastic equation. Math. Methods Appl. Sci. **23**, 203–226 (2000)

25. Shibata, Y., Shimizu, S.: A decay property of the fourier transform and its application to the Stokes problem. J. Math. Fluid Mech. **3**, 213–230 (2001).
26. Sugitani, Y., Kawashima, S.: Decay estimates of solutions to a semi-linear dissipative plate equation. J. Hyperbolic Differ. Equ. **7**, 471–501 (2010)

Decay Estimates for Space-time Fractional Equations with Structural Damping and Nonlinear Memory

Nelson Faustino and Jorge Marques

Abstract We study a wide class of space-time fractional equations with structural damping and nonlinear memory:

$$\partial_t^{2\alpha} u(t,x) + \mu(-\Delta)^{\frac{\sigma}{2}} \partial_t^{\alpha} u(t,x) + (-\Delta)^{\sigma} u(t,x) = I_{0^+}^{1-\alpha} f(t,x),$$

$t \geq 0$, $x \in \mathbf{R}^n$, where the parameters satisfying $1/2 < \alpha \leq 1$, $\mu > 0$ and $\sigma > 1$. We take the so-called Caputo-Djrbashian derivative with respect to time t, ∂_t^{κ} of order κ with $\kappa = 2\alpha, \alpha$, the fractional Laplacian in the space variable x, $(-\Delta)^{\delta}$ with $\delta = \frac{\sigma}{2}, \sigma$, and the Riemann-Liouville integral $I_{0^+}^{1-\alpha} f(t,x)$. We first obtain solution representations for Cauchy type problems expressed in terms of Mittag-Leffler's functions $E_{\alpha,\beta}(z)$ ($z \in \mathbb{C}$) by using a Duhamel type formula with the aid of Laplace (in time) and Fourier (in space) transforms. Taking into account that the Fourier multipliers are radially symmetric functions, the convolution kernels are represented through the Hankel transform. Then, our purpose is to establish decay estimates of the derivatives of its solutions for initial data with L^p and Sobolev regularity. We highlight that the strategy to be employed relies on harmonic analysis tools. We also provide applications employing Strichartz's estimates to prove existence results for mild solutions in particular cases, in which $f(t,x) = |u(t,x)|^k$ with $k > 1$ or f belongs to a time weighted Banach space.

MSC2020: 26A33, 33C10, 33E12, 35R11, 42B37, 44A20.

N. Faustino
Center for R&D in Mathematics and Applications (CIDMA), University of Aveiro, Aveiro, Portugal

J. Marques (✉)
Centre for Business and Economics Research (CeBER), Faculty of Economics, University of Coimbra, Coimbra, Portugal
e-mail: jmarques@fe.uc.pt

M. R. Ebert et al. (eds.), *New Tools in Mathematical Analysis and Applications*, Research Perspectives, https://doi.org/10.1007/978-3-031-77050-0_22

1 Introduction

Let us begin with a few facts concerning the theory of *structurally damped equations*. After Russell's fundamental paper [22] on mathematical models of structural mechanics describing elastic beams much more attention has been paid rightly to the strongly damped plate/wave problems and to its further applications on the crossroads of control theory [1], dynamical systems [20, 25] and transmission problems [19]. The current trends on *structurally damped equations*, exhibited on the contribution of Pham et al. [21], and more recently on D'Abbicco and its collaborators papers [2–8], paved the way for considering space-fractional model problems such as the inhomogeneous $\sigma-$evolution equation

$$\partial_t^2 u(t,x) + \mu(-\Delta)^{\frac{\sigma}{2}} \partial_t u(t,x) + (-\Delta)^{\sigma} u(t,x) = f(t,x). \tag{1}$$

We refer the reader to the works of Carvalho et al. [11] and Denk and Schnaubelt [12] just to mention a few papers on analysis of PDEs towards structurally damped plate equations (case of $\sigma = 2$) and damped wave equations (case of $\sigma = 1$). Also, we refer the seminal work of Karch [16] for a general account on the asymptotic behaviour of structurally damped models.

Our aim here is to investigate a space-time-fractional counterpart of (1) that encodes memory effects, ubiquitous in natural phenomena [13, 14, 23]. In our setting it will be replaced the time derivatives ∂_t and ∂_t^2 by the so-called Caputo-Djrbashian derivatives ∂_t^κ of order κ, defined as

$$\partial_t^\kappa u(t,x) := \begin{cases} \displaystyle\int_0^t \frac{(t-\tau)^{m-\kappa-1}}{\Gamma(m-\kappa)} \frac{\partial^m u(\tau,x)}{\partial \tau^m} d\tau, & m-1 < \kappa < m, \\ \displaystyle\frac{\partial^m u(t,x)}{\partial t^m}, & \kappa = m, \end{cases} \tag{2}$$

where $\Gamma(\cdot)$ denotes the Euler's Gamma function and $m = \lfloor \kappa \rfloor + 1$ ($\lfloor \kappa \rfloor$ stands for the integer part of κ). For the time-fractional counterpart of the inhomogeneous term $f(t,x)$, it will be adopted the Riemann-Liouville integral

$$I_{0+}^\zeta f(t,x) := \begin{cases} \displaystyle\int_0^t \frac{(t-\tau)^{\zeta-1}}{\Gamma(\zeta)} f(\tau,x) d\tau, & 0 < \zeta < 1, \\ f(t,x), & \zeta = 0. \end{cases} \tag{3}$$

The fractional differential operator $(-\Delta)^{\frac{\eta}{2}}$ ($\eta \geq 0$), is defined by the spectral formula

$$\mathcal{F}\left((-\Delta)^{\frac{\eta}{2}} \varphi\right)(\xi) = |\xi|^\eta \widehat{\varphi}(\xi), \tag{4}$$

or equivalently, as $(-\Delta)^{\frac{\eta}{2}} = \mathcal{F}^{-1}|\xi|^{\eta}\mathcal{F}$, where $\mathcal{F}$ and $\mathcal{F}^{-1}$ are the partial Fourier transform and its inverse with respect to space x, respectively.

2 Our Model: Fractional PDE with Structural Damping and Nonlinear Memory

We consider the Cauchy type problem defined on $[0, \infty[\times\mathbb{R}^n$:

$$\begin{cases} \partial_t^{2\alpha} u(t,x) + \mu(-\Delta)^{\frac{\sigma}{2}}\partial_t^{\alpha} u(t,x) + (-\Delta)^{\sigma} u(t,x) = I_{0+}^{1-\alpha} f(t,x), \\ u(0,x) = u_0(x), x \in \mathbb{R}^n, \\ \partial_t u(0,x) = u_1(x), x \in \mathbb{R}^n, \end{cases} \tag{5}$$

where the parameters satisfying $\frac{1}{2} < \alpha \leq 1$, $\mu > 0$ and $\sigma > \frac{\alpha}{2\alpha-1}$. From here comes $\sigma > 1$ for all α. It is well-known that $\sigma = 1$ is the limit case, in which we have the wave equation with critical structural damping for $\alpha = 1$. Hence we assume $\sigma > \frac{\alpha}{2\alpha-1}$ to get the desired estimates for derivatives of the solution of (5). We take the so-called Caputo-Djrbashian derivative with respect to time t, ∂_t^{κ} of order κ with $\kappa = 2\alpha, \alpha$, the fractional Laplacian in the space variable x, $(-\Delta)^{\delta}$ with $\delta = \frac{\sigma}{2}, \sigma$, and the Riemann-Liouville integral $I_{0+}^{1-\alpha} f(t,x)$. The fractional PDE includes the following:

- Elastic/Kinetic term: $(-\Delta)^{\sigma} u(t,x)$;
- Space-time fractional damping term: $\mu(-\Delta)^{\frac{\sigma}{2}}\partial_t^{\alpha} u(t,x)$;
- Inhomogeneous term with memory: $I_{0+}^{1-\alpha} f(t,x)$.

Let $u_0 \in \mathcal{S}(\mathbb{R}^n)$, $u_1 \in \mathcal{S}(\mathbb{R}^n)$ and $f \in L^1([0,\infty[;\mathcal{S}(\mathbb{R}^n))$. By using Laplace (in time) and Fourier (in space) transforms, the mild solution of the (5) is expressed in terms of Mittag-Leffler's functions $E_{\alpha,\beta}$ by (see section 6. in [15])

$$u(t,x) = \left(N_{\sigma,n}(t,\cdot|\alpha,\mu) * u_0\right)(x) + (J_{\sigma,n}(t,\cdot|\alpha,\mu) * u_1)(x) + u^{nl}(t,x) \tag{6}$$

with

$$u^{nl}(t,x) := \int_0^t \left[\, M_{\sigma,n}(t-\tau,\cdot|\alpha,\mu) * f(\tau,\cdot)\,\right](x)\, d\tau \tag{7}$$

where

$$N_{\sigma,n}(t,x|\alpha,\mu) =$$

$$= \begin{cases} \mathcal{F}^{-1}\left(\frac{E_{\alpha,1}(-\lambda_+|\xi|^{\sigma}t^{\alpha})}{1-(\lambda_+)^2} + \frac{E_{\alpha,1}(-\lambda_-|\xi|^{\sigma}t^{\alpha})}{1-(\lambda_-)^2}\right)(x), & \mu \neq 2, \\ \mathcal{F}^{-1}\left(E_{\alpha,1}(-|\xi|^{\sigma}t^{\alpha}) + \frac{|\xi|^{\sigma}t^{\alpha}}{\alpha} E_{\alpha,\alpha}(-|\xi|^{\sigma}t^{\alpha})\right)(x), & \mu = 2; \end{cases}$$

$$J_{\sigma,n}(t,x|\alpha,\mu) =$$

$$= \begin{cases} t\mathcal{F}^{-1}\left(\frac{\lambda_+ E_{\alpha,2}(-\lambda_+|\xi|^\sigma t^\alpha) - \lambda_- E_{\alpha,2}(-\lambda_-|\xi|^\sigma t^\alpha)}{\lambda_+ - \lambda_-} \right)(x), & \mu \neq 2, \\ \frac{t}{\alpha}\mathcal{F}^{-1}\left(E_{\alpha,1}(-|\xi|^\sigma t^\alpha) - (\alpha-1)E_{\alpha,2}(-|\xi|^\sigma t^\alpha) \right)(x), & \mu = 2; \end{cases}$$

$$M_{\sigma,n}(t,x|\alpha,\mu) =$$

$$= \begin{cases} t^\alpha\mathcal{F}^{-1}\left(\frac{\lambda_+ E_{\alpha,\alpha+1}(-\lambda_+|\xi|^\sigma t^\alpha) - \lambda_- E_{\alpha,\alpha+1}(-\lambda_-|\xi|^\sigma t^\alpha)}{\lambda_+ - \lambda_-} \right)(x), & \mu \neq 2, \\ \frac{t^\alpha}{\alpha}\mathcal{F}^{-1}\left(E_{\alpha,\alpha}(-|\xi|^\sigma t^\alpha) \right)(x), & \mu = 2 \end{cases}$$

with

$$\lambda_\pm = \begin{cases} \frac{\mu}{2} \pm i\sqrt{1-\frac{\mu^2}{4}}, & 0 < \mu < 2, \\ \frac{\mu}{2} \pm \sqrt{\frac{\mu^2}{4}-1}, & \mu \geq 2. \end{cases} \tag{8}$$

Remark 1 For $\alpha = 1$ we recover the Duhamel's principle to (6) since

$$M_{\sigma,n}(t,\cdot|\alpha,\mu) = J_{\sigma,n}(t,\cdot|\alpha,\mu). \tag{9}$$

In particular, we get the same solution $u(t,x)$ obtained in [21] taking into account that $I_{0^+}^{1-\alpha} f(t,x) = f(t,x)$.

3 Methodology: Decay Estimates of the Fourier Multipliers

We have five differents profiles for Mittag-Leffler's functions: $E_{\alpha,\beta}$ with $\beta = 1, 2, \alpha, \alpha+1$ and $|\xi|^\sigma E_{\alpha,\alpha}$. Hence, from now on we consider

$$K^\eta_{\sigma,n}(t,x|\alpha,\beta,\lambda) := \frac{1}{(2\pi)^n}\int_{\mathbb{R}^n} \phi(t,|\xi|)e^{ix\cdot\xi}d\xi \tag{10}$$

with $\phi(t,|\xi|) = |\xi|^\eta E_{\alpha,\beta}(-\lambda|\xi|^\sigma t^\alpha)$, $\eta \geq 0$, $0 < \alpha \leq 1$, $\sigma, \beta > 0$, $\lambda \in \{\lambda_+, \lambda_-\}$. In our strategy, the convolution kernels are represented through the Hankel Transform, $\widetilde{\mathcal{H}}_{\frac{n}{2}-1}$:

$$K^\eta_{\sigma,n}(t,x|\alpha,\beta,\lambda) = \frac{1}{(2\pi)^{\frac{n}{2}}}\widetilde{\mathcal{H}}_{\frac{n}{2}-1}\phi(t,|x|), \tag{11}$$

since the multipliers are radially symmetric. By using the Hausdorff-Young's inequality for the Hankel transform, we may estimate:
If $1 \leq r' \leq 2$ then

$$\left\| \widetilde{\mathcal{H}}_{\frac{n}{2}-1}\phi(t,|x|) \right\|_{L^r} \lesssim \| \rho^\eta E_{\alpha,\beta}(-\lambda\rho^\sigma t^\alpha) \|_{L^{r'}} \tag{12}$$

where $\rho = |\xi|$ and r' is the conjugate exponent of r (see [10]).

Remark 2 We claim that the condition $r' \leq 2$ can be relaxed by using the Mihlin-Hörmander's theorem and the Hardy-Littlewood-Sobolev theorem combined with asymptotic expansions of $E_{\alpha,\beta}$ ([18]) as we could see in [9] (Remark 2.3). Since the bounds for the partial derivatives of $\phi(t,\rho)$ are uniform with respect to time then we could apply Mihlin's multiplier theorem. If $\mu \geq 2$ then $E_{\alpha,\beta}(-\lambda\rho^\sigma t^\alpha)$ have a property of the complete monotonicity with respect to ρ for $\beta \geq \alpha$, $\lambda > 0$ (see (8)) and $t > 0$. In the other case, $\mu < 2$, we have to control the oscillations of $E_{\alpha,\beta}(z)$ for $z = \lambda\rho^\sigma t^\alpha$, where $\lambda \in \mathbb{C}\backslash(-\infty, 0]$ with $|\lambda| = 1$ (see (8)), by using an analysis localized at low and high frequencies. The desired estimates are consistent with the ones from Proposition 4.4 ([3]) in the case $\alpha = 1$ and $\mu = 1$.

First, we obtain L^r estimates for Mittag-Leffler's functions in polar coordinates for $r < \frac{n}{\max\{n+\gamma-\sigma,0\}}$. This latter condition is necessary to avoid the lack of integrability which appears for large r. We can distinguish two cases in Lemma 1 for these estimates.

Lemma 1 ([15]) *Let $0 < \alpha \leq 1$ and $n > \sigma - \gamma$. If $1 \leq r < \frac{n}{n+\gamma-\sigma}$ for $0 \leq \gamma < \sigma$ or if $\frac{n}{n+\gamma} < r < \frac{n}{n+\gamma-\sigma}$ for $-n < \gamma < 0$ then it holds*

$$\| \rho^\eta E_{\alpha,\beta}\left(-\lambda\rho^\sigma t^\alpha \right) \|_{L^r} \lesssim$$
$$\lesssim \begin{cases} t^{-\frac{\alpha}{\sigma}\left(n\left(1-\frac{1}{r}\right)+\gamma\right)}, & \eta = \gamma, \\ t^{-\alpha-\frac{\alpha}{\sigma}\left(n\left(1-\frac{1}{r}\right)+\gamma\right)}, & \beta = \alpha \wedge \eta = \sigma + \gamma. \end{cases}$$

We now choose $\gamma = \frac{(1-\alpha)\sigma}{\alpha}$ and $\sigma > \frac{\alpha}{2\alpha-1}$ for $\frac{1}{2} < \alpha \leq 1$. Then we apply the Young's inequality in order to obtain $L^p - L^q$ estimates.

Lemma 2 *If $n > \frac{(2\alpha-1)\sigma}{\alpha}$ then for $1 \leq p \leq q$ such that $\dfrac{1}{p} - \dfrac{1}{q} < \dfrac{(2\alpha-1)\sigma}{\alpha n}$ and for all $t > 0$ it holds*

$$\| \left(K^\eta_{\sigma,n}(t,x|\alpha,\beta,\lambda) * \varphi\right)(x) \|_{L^q} \lesssim$$
$$\lesssim \begin{cases} t^{-\frac{\alpha n}{\sigma}\left(\frac{1}{p}-\frac{1}{q}\right)+\alpha-1} \|\varphi\|_{L^p}, & \eta = \frac{(1-\alpha)\sigma}{\alpha}, \\ t^{-\frac{\alpha n}{\sigma}\left(\frac{1}{p}-\frac{1}{q}\right)-1} \|\varphi\|_{L^p}, & \beta = \alpha \wedge \eta = \frac{\sigma}{\alpha}. \end{cases}$$

Remark 3 (Lower Dimension with Respect to (α, σ)) If $n < \frac{(2\alpha-1)\sigma}{\alpha}$ then the previous estimates holds for all $1 \leq p \leq \infty$.
If $n = \frac{(2\alpha-1)\sigma}{\alpha}$ then the previous estimates holds for all $1 \leq p < \infty$.

4 Results

Our focus is to establish decay estimates of the derivatives of its solutions for initial data with L^p and Sobolev regularity. We introduce the homogeneous Bessel potential spaces, which are defined by

$$\dot{H}^{\eta,p}(\mathbb{R}^n) := \left\{ \varphi \in L^p(\mathbb{R}^n) : (-\Delta)^{\frac{\eta}{2}}\varphi \in L^p(\mathbb{R}^n) \right\}. \tag{13}$$

By using Lemma 1 ($\eta = 0$) and Lemma 2 ($\eta = \gamma$) we obtain decay estimates of derivatives of (6) for the linear part, i.e., $f(t,x) = 0$.

Lemma 3 *Let $\eta = \frac{(1-\alpha)\sigma}{\alpha}$. We assume that $u_0 \in \dot{H}^{\eta,p}(\mathbb{R}^n)$, $u_1 \in L^p(\mathbb{R}^n)$ and $f(t,x) = 0$. If $n > \frac{(2\alpha-1)\sigma}{\alpha}$ then for $1 \leq p \leq q$ such that $\frac{1}{p} - \frac{1}{q} < \frac{(2\alpha-1)\sigma}{\alpha n}$ and for all $t > 0$ it holds*

$$\| u(t,x) \|_{\dot{H}^{\eta,q}} \lesssim t^{-\frac{\alpha n}{\sigma}\left(\frac{1}{p}-\frac{1}{q}\right)} \|u_0\|_{\dot{H}^{\eta,p}} + t^{\alpha\left(1-\frac{n}{\sigma}\left(\frac{1}{p}-\frac{1}{q}\right)\right)} \|u_1\|_{L^p}. \tag{14}$$

Remark 4 We point out that we have $0 \leq \eta < \sigma$ and $\sigma > \frac{\alpha}{2\alpha-1}$ for $\frac{1}{2} < \alpha \leq 1$. If $\alpha = 1$ then $\eta = 0$, so the fractional derivatives give rise to sobolev regularity of its solutions.

We will provide two applications employing Strichartz's estimates to prove existence results for derivatives of mild solutions. In the first one, we consider $f(u) = |u|^k$ with $k > 1$ fixed, and in the second one the inhomogeneous term f belongs to a time weighted Banach space (see Definition 1).

Definition 1 For $1 \leq s < \infty$, we define the Banach space $L^s([0,T]; L^p(\mathbb{R}^n))$, $0 < T < \infty$, with norm

$$\|v\|_{L^s([0,T];L^p(\mathbb{R}^n))} = \left(\int_0^T \|v(t,\cdot)\|_{L^p}^s dt \right)^{\frac{1}{s}} \quad (1 \leq p \leq \infty). \tag{15}$$

Moreover, the space $L^\infty([0,T]; L^p(\mathbb{R}^n))$ is defined by the norm

$$\|v\|_{L^\infty([0,T];L^p(\mathbb{R}^n))} = \operatorname*{ess\ sup}_{t\in[0,T]} \|v(t,\cdot)\|_{L^p} \quad (1 \leq p \leq \infty). \tag{16}$$

For $\theta > 0$, we also define the time-weighted Banach space $C_\theta([0, T]; L^p(\mathbb{R}^n))$ as the set of functions $f \in C([0, T]; L^p(\mathbb{R}^n))$ with norm

$$\|f\|_{C_\theta([0,T];L^p(\mathbb{R}^n))} = \sup_{t\in[0,T]} t^{1/\theta}\|f(t,\cdot)\|_{L^p}. \tag{17}$$

Definition 2 It is called that (p, q, s) is an admissible triplet with respect to (α, σ) if

$$\frac{1}{s} = \frac{\alpha n}{\sigma}\left(\frac{1}{p} - \frac{1}{q}\right) \tag{18}$$

where p and q satisfies either:

(i) $1 \le p < \frac{\alpha n}{(2\alpha-1)\sigma}$ and $p \le q < \frac{\alpha n p}{\alpha n-(2\alpha-1)\sigma p}$ or

(ii) $p > \max\left\{1, \frac{\alpha n}{(2\alpha-1)\sigma}\right\}$ and $p \le q \le \infty$ or

(iii) $p = \max\left\{1, \frac{\alpha n}{(2\alpha-1)\sigma}\right\}$ and $p \le q < \infty$.

We remark that $\frac{1}{2\alpha-1} < s \le \infty$ follows from $\frac{1}{p} - \frac{1}{q} < \frac{(2\alpha-1)\sigma}{\alpha n}$.
In order to estimate the inhomogeneous term (nonlinear part) we consider

$$U^{nl}(t,x) := \int_0^t \Big[(-\Delta)^{\frac{\eta}{2}} M_{\sigma,n}(t-\tau,\cdot|\alpha,\mu) * f(\tau,\cdot) \Big](x)\, d\tau. \tag{19}$$

Proposition 1 (Estimate of the Nonlinear Part) *Let (p, q, s) is an admissible triplet with respect to (α, σ). If $f \in L^{s/k}([0, T]; L^{q/k}(\mathbb{R}^n))$ with $1 < k < q$ and $s > k$ then*

$$\left\|U^{nl}(t,x)\right\|_{L^\infty([0,T];L^p(\mathbb{R}^n))} + \|U^{nl}(t,x)\|_{L^s([0,T];L^q(\mathbb{R}^n))} \lesssim T^{2\alpha-\frac{\alpha n(k-1)}{p\sigma}}\|f\|_{L^{s/k}([0,T];L^{q/k}(\mathbb{R}^n))}$$

for $q < kp$ and

$$\left\|U^{nl}(t,x)\right\|_{L^\infty([0,T];L^p(\mathbb{R}^n))} + \|U^{nl}(t,x)\|_{L^s([0,T];L^q(\mathbb{R}^n))} \lesssim T^{2\alpha-\frac{\alpha n(k-1)}{p\sigma}}\||f|^{1/k}\|^{k\theta}_{L^\infty([0,T];L^p(\mathbb{R}^n))}\||f|^{1/k}\|^{k(1-\theta)}_{L^s([0,T];L^q(\mathbb{R}^n))}$$

for $q \ge kp$ with $\theta = \frac{q-kp}{(q-p)k}$.

Proof Let $\eta = \frac{(1-\alpha)\sigma}{\alpha}$ and $\sigma > \frac{\alpha}{2\alpha-1}$ for $\frac{1}{2} < \alpha \leq 1$. Assume that the non-homogeneous term $f \in L^{s/k}([0,T]; L^{q/k}(\mathbb{R}^n))$ with $1 < k < q$ and $s > k$.
We start with the estimate of the $L^{\infty}([0,T]; L^p(\mathbb{R}^n))$ norm:

1. For $k < q < kp$ then by Lemma 2 we have

$$\left\| M_{\sigma,n}(t-\tau,\cdot|\alpha,\mu) * f(\tau,\cdot) \right\|_{\dot{H}^{\eta,p}} \lesssim$$
$$\left\| (-\Delta)^{\frac{\eta}{2}} M_{\sigma,n}(t-\tau,\cdot|\alpha,\mu) \right\|_{L^r} \|f\|_{L^{q/k}} \lesssim (t-\tau)^{2\alpha-1-\frac{\alpha n}{\sigma}\left(\frac{k}{q}-\frac{1}{p}\right)} \|f\|_{L^{q/k}} \tag{20}$$

   where $1 + \frac{1}{p} = \frac{1}{r} + \frac{k}{q}$.
2. By Hölder's inequality with respect to τ we obtain

$$\int_0^t \left\| (-\Delta)^{\frac{\eta}{2}} M_{\sigma,n}(t-\tau,\cdot|\alpha,\mu) * f(\tau,\cdot) \right\|_{L^p} d\tau \lesssim$$
$$\left(\int_0^t (t-\tau)^{\left(2\alpha-1-\frac{\alpha n}{\sigma}\left(\frac{k}{q}-\frac{1}{p}\right)\right)\phi} d\tau \right)^{1/\phi} \left(\int_0^t \|f\|_{L^{q/k}}^{s/k} d\tau \right)^{k/s} \lesssim$$
$$t^{2\alpha - \frac{\alpha n(k-1)}{\sigma p}} \|f\|_{L^{s/k}([0,T]; L^{q/k}(\mathbb{R}^n))}$$

   thanks to

$$2\alpha - 1 - \frac{\alpha n}{\sigma}\left(\frac{k}{q} - \frac{1}{p}\right) + \frac{1}{\phi} = 2\alpha - \frac{\alpha n(k-1)}{\sigma p}$$

   because of (p, q, s) be an admissible triplet with respect to (α, σ) and $\frac{1}{\phi} = 1 - \frac{k}{s}$.
3. We take the supremum for all $t \in [0, T]$.

Then, we estimate the $L^s([0,T]; L^q(\mathbb{R}^n))$ norm:

1. For $k < q < kq$ from the previous estimate (20) we have

$$\left\| (-\Delta)^{\frac{\eta}{2}} M_{\sigma,n}(T-\tau,\cdot|\alpha,\mu) * f(\tau,\cdot) \right\|_{L^q} \lesssim (T-\tau)^{2\alpha-1-\frac{\alpha n(k-1)}{\sigma q}} \|f\|_{L^{q/k}} \tag{21}$$

   where $1 + \frac{1}{q} = \frac{1}{r} + \frac{k}{q}$.

2. By Young's inequality (convolution on τ) we obtain

$$\|U^{nl}(t,x)\|_{L^s([0,T];L^q(\mathbb{R}^n))} \lesssim \left(\int_0^T \tau^{\left(2\alpha-1-\frac{\alpha n(k-1)}{\sigma q}\right)\psi}\, d\tau\right)^{1/\psi} \left(\int_0^T \|f\|_{L^{q/k}}^{s/k}\, d\tau\right)^{k/s} \lesssim T^{2\alpha-\frac{\alpha n(k-1)}{\sigma p}} \|f\|_{L^{s/k}([0,T];L^{q/k}(\mathbb{R}^n))}$$

thanks to

$$2\alpha - 1 - \frac{\alpha n(k-1)}{\sigma q} + \frac{1}{\psi} = 2\alpha - \frac{\alpha n(k-1)}{\sigma p}$$

because of (p, q, s) be an admissible triplet with respect to (α, σ) and $\frac{1}{\psi} = 1 - \frac{(k-1)}{s}$.

□

Lastly, for $q \geq kp$ follows $p \leq kp \leq q$ with $k > 1$. Hence, we can apply the Lebesgue interpolation theorem with $\theta = \frac{q-kp}{(q-p)k}$ and Hölder's inequality to show the corresponding estimate of the $L^\infty([0,T]; L^p(\mathbb{R}^n))$ norm and $L^s([0,T]; L^q(\mathbb{R}^n))$ norm for $U^{nl}(t,x)$ given by (19). □

We now present the first theorem.

Theorem 1 *Let $\eta = \frac{(1-\alpha)\sigma}{\alpha}$ and $\sigma > \frac{\alpha}{2\alpha-1}$ for $\frac{1}{2} < \alpha \leq 1$, (p, q, s) is an admissible triplet with respect to (α, σ) and $p_0 = \frac{n(k-1)}{2\sigma}$.*
Assume that $f(u) = |u|^k$, for $k > 1$, $u_0 \in \dot{H}^{\eta,p}(\mathbb{R}^n)$ and $u_1 \in L^p(\mathbb{R}^n)$.

(i) For $p > p_0$ there exist $T > 0$ and a unique mild solution $(-\Delta)^{\frac{\eta}{2}} u(t,\cdot) \in C([0,T]; L^p(\mathbb{R}^n)) \cap L^s([0,T]; L^q(\mathbb{R}^n))$;
(ii) For $p = p_0$ there exists a global small data solution $(-\Delta)^{\frac{\eta}{2}} u(t,\cdot) \in C([0,\infty[; L^p(\mathbb{R}^n)) \cap L^s([0,\infty[; L^q(\mathbb{R}^n))$.

Proof For simplicity we only give a sketch of the proof. From (6) we write a Duhamel's type formula for the derivatives

$$(-\Delta)^{\frac{\eta}{2}} u = \left(N_{\sigma,n}(t,\cdot|\alpha,\mu) * (-\Delta)^{\frac{\eta}{2}} u_0\right)(x) + ((-\Delta)^{\frac{\eta}{2}} J_{\sigma,n}(t,\cdot|\alpha,\mu) * u_1)(x) + U^{nl} \tag{22}$$

where $U^{nl}(t,x)$ is given by (19). By using Lemma 3 and Proposition 1 we apply a fixed point argument (by Banach contraction mapping theorem) for a nonlinear operator

$$Fu(t,x) = \int_0^t \left[(-\Delta)^{\frac{\eta}{2}} M_{\sigma,n}(t-\tau,\cdot|\alpha,\mu) * |u(\tau,\cdot|^k \right](x)\, d\tau \tag{23}$$

defined in the following evolution space: $X([0,T]) := C([0,T]; L^p(\mathbb{R}^n)) \cap L^s([0,T]; L^q(\mathbb{R}^n))$ endowed with the norm

$$\left\|(-\Delta)^{\frac{\eta}{2}} u\right\|_{X([0,T])} := \sup_{t\in[0,T]} \left\|(-\Delta)^{\frac{\eta}{2}} u\right\|_{L^\infty([0,T];L^p)} + \sup_{(p,q,s)} \left\|(-\Delta)^{\frac{\eta}{2}} u\right\|_{L^s([0,T];L^q)} \tag{24}$$

in which (p, q, s) is an admissible triplet with respect to (α, σ). For$p > p_0$ we can choose T small enough such that

$$0 < T < \max\left(\|(-\Delta)^{\frac{\eta}{2}} u_0\|_{L^p}, \|u_1\|_{L^p}\right)^{n/(\sigma p)-2/(k-1)}.$$

For $p = p_0$ the existence of global solution is obvious because the contraction mapping does not depend of T and the initial data are sufficiently small. □

We have a similar result to Proposition 1 to use it later (Theorem 2).

Proposition 2 (Estimate of the Nonlinear Part) *Let (p, q, s) is an admissible triplet with respect to (α, σ) and $k > 1$. If $f \in C_{s/k}([0,T]; L^{q/k}(\mathbb{R}^n))$ with $1 < k < q$ then*

$$\left\|U^{nl}(t,x)\right\|_{L^\infty([0,T];L^p(\mathbb{R}^n))} + \|U^{nl}(t,x)\|_{C_s([0,T];L^q(\mathbb{R}^n))} \lesssim T^{2\alpha - \frac{\alpha n(k-1)}{p\sigma}} \|f\|_{C_{s/k}([0,T];L^{q/k}(\mathbb{R}^n))}$$

for $k < q < kp$ and

$$\left\|U^{nl}(t,x)\right\|_{L^\infty([0,T];L^p(\mathbb{R}^n))} + \|U^{nl}(t,x)\|_{C_s([0,T];L^q(\mathbb{R}^n))} \lesssim T^{2\alpha - \frac{\alpha n(k-1)}{p\sigma}} \||f|^{1/k}\|^{k\theta}_{L^\infty([0,T];L^p(\mathbb{R}^n))} \||f|^{1/k}\|^{k(1-\theta)}_{C_{s/k}([0,T];L^{q/k}(\mathbb{R}^n))}$$

for $q \geq kp$ with $\theta = \frac{q-kp}{(q-p)k}$.

Proof For $k < q < kp$ with $k > 1$:

1. By using the assumption $f \in C_{s/k}([0,T]; L^{q/k}(\mathbb{R}^n))$ we estimate

$$\int_0^t \left\|(-\Delta)^{\frac{\eta}{2}} M_{\sigma,n}(t-\tau, \cdot|\alpha,\mu) * f(\tau,\cdot)\right\|_{L^p} d\tau$$
$$\lesssim \int_0^t (t-\tau)^{2\alpha-1-\frac{\alpha n(k-1)}{\sigma p}} \tau^{-k/s}\tau^{k/s} \|f\|_{L^{q/k}} d\tau$$

$$= t^{2\alpha - \frac{\alpha n(k-1)}{\sigma p} - \frac{k}{s}} \|f\|_{C_{s/k}([0,T];L^{q/k}(\mathbb{R}^n))} \int_0^1 (1-y)^{2\alpha - 1 - \frac{\alpha n(k-1)}{\sigma p}} y^{-k/s} dy$$

$$= C t^{2\alpha - \frac{\alpha n(k-1)}{\sigma p}} \|f\|_{C_{s/k}([0,T];L^{q/k}(\mathbb{R}^n))}$$

since (p, q, s) is an admissible triplet with respect to (α, σ).

2. We take the supremum for all $t \in [0, T]$.

The remaining proof is similar to the that one in Proposition 1. □

We state a second theorem, whose proof is analogous to that one in Theorem 1.

Theorem 2 *Let* $\eta = \frac{(1-\alpha)\sigma}{\alpha}$ *and* $\sigma > \frac{\alpha}{2\alpha-1}$ *for* $\frac{1}{2} < \alpha \leq 1$, (p, q, s) *is an admissible triplet with respect to* (α, σ) *and* $p_0 = \frac{n(k-1)}{2\sigma}$.
Assume that $\|f(t,\cdot)\|_{L^{q/k}} \lesssim t^{-k/s}$ *for* $t > 0$ *with* $k > 1$, $u_0 \in \dot{H}^{\eta,p}(\mathbb{R}^n)$ *and* $u_1 \in L^p(\mathbb{R}^n)$.

(i) For $p > p_0$ *there exist* $T > 0$ *and a unique mild solution* $(-\Delta)^{\frac{\eta}{2}} u(t,\cdot) \in C([0,T]; L^p(\mathbb{R}^n)) \cap C_s([0,T]; L^q(\mathbb{R}^n))$;
(ii) For $p = p_0$ *there exists a global small data solution* $(-\Delta)^{\frac{\eta}{2}} u(t,\cdot) \in C([0,\infty[; L^p(\mathbb{R}^n)) \cap C_s([0,\infty[; L^q(\mathbb{R}^n))$.

5 Conclusions

- The Fox-H functions have an important role due to their natural connection to the fractional calculus. The general approach consists in the computation of L^r decay estimates for the fundamental solutions (the convolution kernels in our case), using detailed asymptotic analysis of these functions. Perhaps it is a very technique procedure and not so useful to derive decay estimates (see [17]). However, our approach is based on tools from the harmonic analysis.
- We use Strichartz estimates with boundness arguments imposed to $f(t,x)$, which are similar to those appearing in the so-called *Christ-Kiselev lemma* (see [24]).
- We can include a more general inhomogeneous term: $I_{0^+}^{\beta-2\alpha} f(t,x)$, with $2\alpha \leq \beta < 2\alpha + 1$ (see [15]) .
- The formulation of Cauchy problems correspond to a borderline case of the eq.:

$$\partial_t^\gamma u(t,x) + \mu(-\Delta)^\theta \partial_t^\alpha u(t,x) + (-\Delta)^\sigma u(t,x) = I_{0^+}^\delta f(t,x) \tag{25}$$

with $0 < \theta \leq \sigma$, $0 < \alpha \leq 1$, $\alpha < \gamma \leq 2$ and $0 \leq \delta < 1$.

Acknowledgments The second author is partially supported by "FCT—Fundação para a Ciência e a Tecnologia, I.P." Grant/Award Number: UIDB/05037/2020.

References

1. Ammari, K., Nicaise, S.: Stabilization of a transmission wave/plate equation. J. Differ. Equ. **249**(3), 707–727 (2010)
2. D'Abbicco, M.: A wave equation with structural damping and nonlinear memory. Nonlinear Differ. Equ. Appl. **21**, 751–773 (2014). https://doi.org/10.1007/s00030-014-0265-2
3. D'Abbicco, M., Ebert, M.R.: A new phenomenon in the critical exponent for structurally damped semi-linear evolution equations. Nonlinear Anal. Theory Methods Appl. **149**, 1–40 (2017)
4. D'Abbicco, M., Ebert, M.R.: L^p-L^q estimates for a parameter-dependent multiplier with oscillatory and diffusive components. J. Math. Anal. Appl. **504**(1), 125393 (2021)
5. D'Abbicco, M., Ebert, M.R.: The critical exponent for semilinear σ-evolution equations with a strong non-effective damping, Nonlinear Anal. **215**, 112637 (2022)
6. D'Abbicco, M., Girardi, G.: A structurally damped σ-evolution equation with nonlinear memory. Math. Methods Appl. Sci. 1–19 (2020). https://doi.org/10.1002/mma.6633
7. D'Abbicco, M., Girardi, G.: Asymptotic profile for a two-terms time fractional diffusion problem. Fract. Calc. Appl. Anal. **25**, 1199–1228 (2022). https://doi.org/10.1007/s13540-022-00041-3
8. D'Abbicco, M., Girardi, G.: Decay estimates for a perturbed two-terms space-time fractional diffusive problem. Evol. Equ. Control Theory **12**(4), 1056–1082 (2023). https://doi.org/10.3934/eect.2022060
9. D'Abbicco, M., Ebert, M.R., Picon, T.H.: The critical exponent(s) for the semilinear fractional diffusive equation. J. Fourier Anal. and Appl. **25**(3), 696–731 (2019)
10. De Carli, L.: On the L^p-L^q norm of the Hankel transform and related operators. J. Math. Anal. Appl. **348**(1), 366–382 (2008)
11. Carvalho, A. , Cholewa, J., Dlotko, T.: Strongly damped wave problems: bootstrapping and regularity of solutions. J. Differ. Equ. **244**(9), 2310–2333 (2008)
12. Denk, R., Schnaubelt, R.: A structurally damped plate equation with Dirichlet-Neumann boundary conditions. J. Differ. Equ. **259**(4), 1323–1353 (2015)
13. Dipierro, S., Valdinoci, E., Vespri, V.: Decay estimates for evolutionary equations with fractional time-diffusion. J. Evol. Equ. **19**, 435–462 (2019). https://doi.org/10.1007/s00028-019-00482-z
14. Du, M., Wang, Z., Hu, H.: Measuring memory with the order of fractional derivative. Sci. Rep. **3**(1), 1–3 (2013)
15. Faustino, N., Marques, J.: Structurally damped $\sigma-$evolution equations with power-law memory (2022). ArXiv preprint: 2212.10463. https://arxiv.org/abs/2212.10463
16. Karch, G.: Selfsimilar profiles in large time asymptotics of solutions to damped wave equations. Stud. Math. **143**(2), 175–197 (2000)
17. Kemppainen, J., Siljander, J., Zacher, R.: Representation of solutions and large-time behavior for fully nonlocal diffusion equations. J. Differ. Equ. **263**(1), 149–201 (2017)
18. Kilbas, A.A., Srivastava, H.M., Trujillo, J.J.: Theory and Applications of Fractional Differential Equations. Elsevier, New York (2006)
19. Martinez, B.B., Denk, R. , Monzón, J.H., Kammerlander, F., Nendel, M.: Regularity and asymptotic behavior for a damped plate membrane transmission problem. J. Math. Anal. Appl. **474**(2), 1082–1103 (2019)
20. Patrizi S., Valdinoci, E.: Long-time behavior for crystal dislocation dynamics. Math. Models Methods Appl. Sci. **27**(12), 2185–2228 (2017)
21. Pham, D.T., Mezadek, M.K., Reissig, M.: Global existence for semilinear structurally damped σ-evolution models. J. Math. Anal. Appl. **431**(1), 569–596 (2015)
22. Russell, D.L.: Mathematical models for the elastic beam and their control-theoretic implications. In: Semigroups, Theory and Applications, vol. II. Pitman Research Notes in Mathematics Series, vol. 152, pp. 177–216. Longman Essex, Harlow (1986)

23. Stinga, P.R.: Fractional derivatives: Fourier, elephants, memory effects, viscoelastic materials and anomalous diffusions (2022). ArXiv preprint: 2212.02279
24. Tao, T.: Nonlinear Dispersive Equations: Local and Global Analysis, vol. 106. American Mathematical Society, Providence (2006)
25. Wang, B.: Random attractors for non-autonomous stochastic wave equations with multiplicative noise. Discrete Continuous Dyn. Syst.**34**(1), 269–300 (2014)

Revisit on Global Existence of Solutions for Semilinear Damped Wave Equations in $\mathbb{R}^N$ with Noncompactly Supported Initial Data

Yuta Wakasugi

Abstract In this note, we study the Cauchy problem of the semilinear damped wave equation and our aim is the small data global existence for noncompactly supported initial data. For this problem, Ikehata and Tanizawa [5] introduced the energy method with the exponential-type weight function $e^{|x|^2/(1+t)}$, which is the so-called Ikehata–Todorova–Yordanov type weight. In this paper, we suggest another weight function of the form $(1+|x|^2/(1+t))^{\lambda}$, which allows us to treat polynomially decaying initial data and give a simpler proof than the previous studies treating such initial data.

1 Introduction

In this note, we study the Cauchy problem of the semilinear damped wave equation

$$\begin{cases} u_{tt} - \Delta u + u_t = |u|^p, & (t,x) \in (0,\infty) \times \mathbb{R}^N, \\ u(0,x) = u_0(x), \ u_t(0,x) = u_1(x), & x \in \mathbb{R}^N, \end{cases} \tag{1}$$

where $u = u(t,x)$ is a real-valued unknown, $x = (x_1, \ldots, x_N) \in \mathbb{R}^N$, $N \geq 1$, $u_t = \partial u/\partial t$, $u_{tt} = \partial^2 u/\partial t^2$, $\Delta u = \partial^2 u/\partial x_1^2 + \cdots + \partial^2 u/\partial x_N^2$, $p > 1$, and u_0, u_1 are given initial data.

The purpose is to show the small data global existence of the solution with noncompactly supported initial data. For this problem, Ikehata and Tanizawa [5] introduced the energy method with the exponential-type weight function $e^{|x|^2/(1+t)}$, which is the so-called Ikehata–Todorova–Yordanov type weight. They proved the small data global existence for $p > p_F(N) := 1 + 2/N$ and exponentially decaying

Y. Wakasugi (✉)
Laboratory of Mathematics, Graduate School of Advanced Science and Engineering, Hiroshima University, Higashihiroshima, Japan
e-mail: wakasugi@hiroshima-u.ac.jp

M. R. Ebert et al. (eds.), *New Tools in Mathematical Analysis and Applications*, Research Perspectives, https://doi.org/10.1007/978-3-031-77050-0_23

initial data. This result removed the compactness assumption on the support of initial data in the earlier result by Todorova and Yordanov [9].

In this note, we suggest another weight function of the form $(1+\frac{|x|^2}{1+t})^{\lambda}$ with sufficiently large λ. This allows us to weaken the assumption of [5] to polynomially decaying initial data. Actually, this type of weight was already used in Ikehata et al. [6] to study the case of absorbing nonlinearity. However, it seems to have been overlooked that this technique is also effective in the case of souring nonlinearity.

To state our results, we first define the mild solution. Let $\mathcal{D}(t)$ be the fundamental solution of the linear damped wave equation, that is, $\mathcal{D}(t)$ is defined by the Fourier multiplier

$$\mathcal{D}(t)=\mathcal{F}^{-1}e^{-\frac{t}{2}}\frac{\sinh(t\sqrt{1/4-|\xi|^2})}{\sqrt{1/4-|\xi|^2}}\mathcal{F},$$

where $\mathcal{F}, \mathcal{F}^{-1}$ denote the Fourier and inverse Fourier transforms, respectively.

Definition 1 Let $T\in(0,\infty]$ and $I=[0,T]$ if $T<\infty$ and $I=[0,\infty)$ if $T=\infty$. Let $(u_0,u_1)\in H^1(\mathbb{R}^N)\times L^2(\mathbb{R}^N)$. We say that $u\in C(I;H^1(\mathbb{R}^N))\cap C^1(I;L^2(\mathbb{R}^N))$ is a mild solution of the Cauchy problem (1) on I if u satisfies

$$u(t)=\mathcal{D}(t)(u_0+u_1)+\frac{\partial}{\partial t}\mathcal{D}(t)u_0+\int_0^t\mathcal{D}(t-s)|u(s)|^p\,ds$$

in $C(I;H^1(\mathbb{R}^N))\cap C^1(I;L^2(\mathbb{R}^N))$. In particular, if $T<\infty$, then u is called a local (in time) mild solution; if $T=\infty$, then u is called a global (in time) mild solution.

The main result of this note is the following.

Theorem 1 *Let $N\geq 1$ and let $p>1$ satisfy*

$$p_F(N)<p\leq\frac{N}{[N-2]_+},$$

where $p_F(N)=1+2/N$. Then, there exist constants $\lambda=\lambda(N,p)>0$ and $\varepsilon=\varepsilon(N,p,\lambda)>0$ such that if the initial data $(u_0,u_1)\in H^1(\mathbb{R}^N)\times L^2(\mathbb{R}^N)$ satisfy

$$\|(1+|x|)^{\lambda}u_1\|_{L^2}+\|(1+|x|)^{\lambda}\nabla_x u_0\|_{L^2}+\|(1+|x|)^{\lambda}u_0\|_{L^2}<\varepsilon, \tag{2}$$

then the Cauchy problem (1) *admits the unique global mild solution.*

Remark 1 The notation $p \le \frac{N}{[N-2]_+}$ means that $p < \infty$ for $N \le 2$ and $p \le \frac{N}{N-2}$ for $N \ge 3$. Also, we may take λ in the theorem so that $\lambda > \max\{1, \frac{N}{2}\}$ and

$$\lambda > \max\left\{\frac{2N-(N-2)p}{2p}, \frac{2N-8/N-(N-2)p}{4(p-p_F(N))}, \frac{q-1}{q}\frac{N+2-(N-2)p}{2(p-p_F(N))}\right\}$$

with $q = \max\{2, \frac{N}{2}(p-1)\}$ hold. Note that $\lambda > \frac{N}{2}$ ensures $u_0, \nabla_x u_0, u_1 \in L^1(\mathbb{R}^N)$.

Remark 2 From the proof of Theorem 1, in addition to $u \in C([0,\infty); H^1(\mathbb{R}^N)) \cap C^1([0,\infty); L^2(\mathbb{R}^N))$, we can see that the solution u has the bound $\|u\|_{X(\infty)} < \infty$, where the norm $\|u\|_{X(\infty)}$ is defined in Definition 3 below. This further implies that u satisfies the boundedness of weighted energy $\|(1+|x|)^\lambda u_t(t)\|_{L^2} + \|(1+|x|)^\lambda \nabla_x u(t)\|_{L^2} < \infty$ for every $t > 0$ and the decay estimate

$$\|\partial_t^k \partial_x^\alpha u(t)\|_{L^2} \lesssim (1+t)^{-\frac{N}{4}-k-\frac{|\alpha|}{2}} \quad (t > 0)$$

for $k \in \mathbb{Z}_{\ge 0}$ and $\alpha \in \mathbb{Z}_{\ge 0}^N$ satisfying $k + |\alpha| \le 1$.

Remark 3 The conclusion of Theorem 1 itself is weaker than those of Hayashi et al. [3] and Sobajima [8] in the sense that our order of the weight λ (see Remark 1 above) has to be larger than those of [3, 8]. However, the method of the proof of Theorem 1 is more elementary and it will also be applicable to the case of time-dependent damping such as

$$u_{tt} - \Delta u + \frac{\mu}{1+t} u_t = |u|^p.$$

It will slightly improve the assumption on the initial data in the result of D'Abbicco [2, Theorem 3]. Moreover, combining with the linear estimates for damped wave equations in general measure spaces proved in [4], we will be able to obtain global existence results for semilinear damped wave equations in more general domains or with some potentials, etc.

Finally, we introduce some notations used in this note. The set of nonnegative integers is expressed by $\mathbb{Z}_{\ge 0}$. The symbol $f \lesssim g$ stands for $f \le Cg$ with some constant $C > 0$. The gradient with respect to the spatial variables is written as ∇_x, that is, $\nabla_x u = (u_{x_1}, \ldots, u_{x_N})$. $L^q(\mathbb{R}^N)$ and $H^s(\mathbb{R}^N)$ denote the usual Lebesgue and Sobolev spaces, respectively.

2 Energy Estimate with a Polynomial-type Weight

The main idea of this note is to use the following weighted energy functional.

Definition 2 For $\lambda > 0$ and $A \geq 2\lambda$, we define the weight functions by

$$\psi(t,x) = A + \frac{|x|^2}{1+t}, \quad \Psi(t,x) = \psi(t,x)^{\lambda} \tag{3}$$

and the weighted energy of a function u by

$$E_{\Psi}(t) = \int_{\mathbb{R}^N} (|u_t(t,x)|^2 + |\nabla_x u(t,x)|^2)\Psi(t,x)\,dx.$$

Lemma 1 *The weight function Ψ satisfies $\Psi_t \leq 0$ and*

$$\frac{|\nabla_x \Psi|^2}{-\Psi_t} \leq 2\Psi.$$

Proof We calculate

$$\Psi_t = \lambda \psi^{\lambda-1}\psi_t = -\lambda \frac{|x|^2}{(1+t)^2}\psi^{\lambda-1} \leq 0.$$

Moreover, since $A \geq 2\lambda$, we have

$$\frac{|\nabla_x \Psi|^2}{-\Psi_t} = \frac{\lambda^2 \frac{4|x|^2}{(1+t)^2}\psi^{2(\lambda-1)}}{\lambda \frac{|x|^2}{(1+t)^2}\psi^{\lambda-1}} = 4\lambda\psi^{\lambda-1} = 4\frac{\lambda}{\psi}\Psi \leq 4\frac{\lambda}{A}\Psi \leq 2\Psi.$$

This completes the proof. □

Lemma 2 *Let u be a mild solution to* (1) *on* $[0, T]$. *Then, we have*

$$\begin{aligned} E_{\Psi}(t) \leq E_{\Psi}(0) &- \frac{2}{p+1}\int_{\mathbb{R}^N} |u_0(x)|^p u_0(x)\Psi(0,x)\,dx \\ &+ \frac{2}{p+1}\int_{\mathbb{R}^N} |u(t,x)|^p u(t,x)\Psi(t,x)\,dx \\ &- \frac{2}{p+1}\int_0^t \int_{\mathbb{R}^N} |u(s,x)|^p u(s,x)\Psi_t(s,x)\,dxds, \end{aligned}$$

provided that the right-hand side is finite.

Proof Differentiating $E_\Psi(t)$, applying integration by parts and using the Eq. (1), we have[1]

$$\begin{aligned}\frac{d}{dt}E_\Psi(t) &= 2\int_{\mathbb{R}^N}(u_t u_{tt} + \nabla_x u\cdot\nabla_x u_t)\Psi\,dx + \int_{\mathbb{R}^N}\left(|u_t|^2 + |\nabla_x u|^2\right)\Psi_t\,dx \\ &= -2\int_{\mathbb{R}^N}u_t^2\Psi\,dx + 2\int_{\mathbb{R}^N}|u|^p u_t\Psi\,dx \\ &\quad -2\int_{\mathbb{R}^N}u_t(\nabla_x u\cdot\nabla_x\Psi)\,dx + \int_{\mathbb{R}^N}\left(|u_t|^2 + |\nabla_x u|^2\right)\Psi_t\,dx.\end{aligned}$$

Here, by Lemma 1, we note that $\Psi_t \le 0$ and

$$-2u_t(\nabla_x u\cdot\nabla_x\Psi) + |\nabla_x u|^2\Psi_t = \frac{1}{\Psi_t}|\Psi_t\nabla_x u - u_t\nabla_x\Psi|^2 - \frac{|\nabla_x\Psi|^2}{\Psi_t}u_t^2 \le 2\Psi u_t^2$$

hold. Using them to the previous identity, we deduce

$$\begin{aligned}\frac{d}{dt}E_\Psi(t) &\le 2\int_{\mathbb{R}^N}|u|^p u_t\Psi\,dx \\ &= \frac{2}{p+1}\frac{d}{dt}\int_{\mathbb{R}^N}|u|^p u\Psi\,dx - \frac{2}{p+1}\int_{\mathbb{R}^N}|u|^p u\Psi_t\,dx.\end{aligned}$$

Integrating it over $[0, t]$ gives the conclusion. □

3 Proof of Theorem 1

3.1 Preliminaries

The following local existence result can be proved in completely the same way as Ikehata and Tanizawa [5, Proposition 2.1], and we omit the proof.

Lemma 3 *Let* $N \ge 1$, $1 < p \le \frac{N}{[N-2]_+}$. *Then for each* $(u_0, u_1) \in H^1(\mathbb{R}^N) \cap L^2(\mathbb{R}^N)$ *satisfying*

$$\|(1+|x|)^\lambda u_1\|_{L^2} + \|(1+|x|)^\lambda\nabla_x u_0\|_{L^2} + \|(1+|x|)^\lambda u_0\|_{L^2} < \infty,$$

there exists $T > 0$ *such that the Cauchy problem* (1) *admits a unique mild solution on* $[0, T]$ *satisfying*

$$\|\Psi^{1/2}u_t(t)\|_{L^2} + \|\Psi^{1/2}\nabla_x u(t)\|_{L^2} + \|\Psi^{1/2}u(t)\|_{L^2} < \infty$$

[1] By an approximation argument, we can justify the formal calculation here, see e.g., [10].

for $t \in [0, T]$. Moreover, if the maximal existence time T_m, the supremum of T such that there is a unique mild solution on $[0, T]$, is finite, then the solution must satisfy

$$\limsup_{t \uparrow T_m} \left(\|\Psi^{1/2} u_t(t)\|_{L^2} + \|\Psi^{1/2} \nabla_x u(t)\|_{L^2} + \|\Psi^{1/2} u(t)\|_{L^2} \right) = +\infty.$$

We define the following norm to construct the global solution.

Definition 3 We define the norm $\|\cdot\|_{X(T)}$ by

$$\|u\|_{X(T)} := \sup_{0<t<T} \Big[E_\Psi(t)^{\frac{1}{2}} + (1+t)^{\frac{N}{4}+1} \|u_t(t)\|_{L^2} + (1+t)^{\frac{N}{4}+\frac{1}{2}} \|\nabla_x u(t)\|_{L^2} + (1+t)^{\frac{N}{4}} \|u(t)\|_{L^2} \Big].$$

By the local existence result in Lemma 3 and noting

$$\|\Psi^{1/2} u(t)\|_{L^2} = \left\| \Psi^{1/2} \left(u_0 + \int_0^t u_t(s)\, ds \right) \right\|_{L^2} \lesssim \|(1+|x|)^\lambda u_0\|_{L^2} + T \|u\|_{X(T)},$$

it suffices to show the a priori estimate $\|u\|_{X(T)} \le C$ with some constant C independent of T under the smallness assumption (2).

In order to estimate the L^2-norms in $\|u\|_{X(T)}$, we apply the following Matsumura estimate.

Lemma 4 (Matsumura [7]) *For $k \in \mathbb{Z}_{\ge 0}$, $\alpha \in \mathbb{Z}_{\ge 0}^N$, $q \in [1, 2]$ and $g \in L^q(\mathbb{R}^N) \cap H^{k+|\alpha|-1}(\mathbb{R}^N)$, we have*

$$\|\partial_t^k \partial_x^\alpha \mathcal{D}(t) g\|_{L^2} \lesssim (1+t)^{-\frac{N}{2}\left(\frac{1}{q}-\frac{1}{2}\right)-k-\frac{|\alpha|}{2}} \left(\|g\|_{L^q} + e^{-\frac{t}{4}} \|g\|_{H^{k+|\alpha|-1}} \right).$$

Finally, to estimate the nonlinear term with polynomial weights, we employ the Caffarelli–Kohn–Nirenberg inequality.

Lemma 5 (Caffarelli et al. [1]) *Let $p, q, r, \alpha, \beta, \sigma, a \in \mathbb{R}$ satisfy $p, q \ge 1$, $r > 0$, $0 \le a \le 1$,*

$$\frac{1}{p} + \frac{\alpha}{N}, \frac{1}{q} + \frac{\beta}{N}, \frac{1}{r} + \frac{\gamma}{N} > 0,$$

where $\gamma = a\sigma + (1-a)\beta$. Then, there exists a positive constant C such that the following inequality holds for all $u \in C_0^\infty(\mathbb{R}^N)$

$$\||x|^\gamma u\|_{L^r} \le C \||x|^\alpha \nabla_x u\|_{L^p}^a \||x|^\beta u\|_{L^q}^{1-a}$$

if and only if the following relations hold:

$$\frac{1}{r} + \frac{\gamma}{N} = a\left(\frac{1}{p} + \frac{\alpha - 1}{N}\right) + (1-a)\left(\frac{1}{q} + \frac{\beta}{N}\right)$$

(this is the dimensional balance),

$$0 \le \alpha - \sigma \quad \text{if} \quad a > 0,$$

and

$$\alpha - \sigma \le 1 \quad \text{if} \quad a > 0 \quad \text{and} \quad \frac{1}{p} + \frac{\alpha - 1}{N} = \frac{1}{r} + \frac{\gamma}{N}.$$

In particular, the case $\alpha = \beta = \gamma = 0$ is called the Gagliardo–Nirenberg inequality.

3.2 *Proof of Theorem 1*

In what follows, we suppose that λ is taken to be suitably large so that the following calculation to be valid, while we will not describe the precise conditions on λ (see also Remark 1). We first prove that the semilinear terms in Lemma 2 is bounded by $\|u\|_{X(T)}$.

Lemma 6 *Let u be a mild solution to* (1) *on* $[0, T]$. *Then, we have*

$$\int_{\mathbb{R}^N} |u(t,x)|^{p+1} \Psi(t,x)\, dx + \int_0^t \int_{\mathbb{R}^N} |u(s,x)|^{p+1} |\Psi_t(s,x)|\, dxds \le C\|u\|_{X(T)}^{p+1},$$

with some constant $C > 0$, provided that the right-hand side is finite.

Proof We consider the first term of the left-hand side. Noting

$$\int_{\mathbb{R}^N} |u|^{p+1} \Psi\, dx \lesssim \|u\|_{L^{p+1}}^{p+1} + (1+t)^{-\lambda} \||x|^{\frac{2\lambda}{p+1}} u\|_{L^{p+1}}^{p+1} \tag{4}$$

and applying Lemma 5, we have

$$\|u\|_{L^{p+1}} \lesssim \|\nabla_x u\|_{L^2}^{\theta} \|u\|_{L^2}^{1-\theta}, \quad \theta = \frac{N(p-1)}{2(p+1)}, \tag{5}$$

$$\||x|^{\frac{2\lambda}{p+1}} u\|_{L^{p+1}} \lesssim \||x|^{\lambda} \nabla_x u\|_{L^2}^{\Theta} \|u\|_{L^q}^{1-\Theta}, \quad \Theta = \frac{\frac{1}{p+1} + \frac{2\lambda}{N(p+1)} - \frac{1}{q}}{\frac{1}{2} - \frac{1}{q} + \frac{\lambda - 1}{N}}, \tag{6}$$

where $q = \max\{2, \frac{N}{2}(p-1)\}$. Here, we note that the conditions $\lambda > 1$ and $\lambda > \frac{q-1}{q}\frac{N+2-(N-2)p}{2(p-p_F(N))}$ and an elementary computation imply $\Theta \in (0,1)$. Moreover, we remark that the condition $0 \le \alpha - \sigma$ if $a > 0$ in Lemma 5 requires $p \le 1 + \frac{2q}{N}$, which is valid from the above choice of q. Furthermore, $H^1(\mathbb{R}^N) \subset L^q(\mathbb{R}^N)$ holds, since $p \le 1 + \frac{4}{[N-2]_+}$. They are the reasons why we take $q = \max\{2, \frac{N}{2}(p-1)\}$. The Gagliardo–Nirenberg inequality further implies

$$\|u\|_{L^q} \lesssim \|\nabla_x u\|_{L^2}^{\mu} \|u\|_{L^2}^{1-\mu}, \quad \mu = N\left(\frac{1}{2} - \frac{1}{q}\right). \tag{7}$$

Therefore, we conclude

$$\begin{aligned}
&\int_{\mathbb{R}^N} |u|^{p+1} \Psi \, dx \\
&\lesssim \left(\|\nabla_x u\|_{L^2}^{\theta} \|u\|_{L^2}^{1-\theta} \right)^{p+1} \\
&\quad + (1+t)^{\lambda\left(\frac{p+1}{2}\Theta - 1\right)} \left(\left\| \frac{|x|^{\lambda}}{(1+t)^{\lambda/2}} \nabla_x u \right\|_{L^2}^{\Theta} (\|\nabla_x u\|_{L^2}^{\mu} \|u\|_{L^2}^{1-\mu})^{1-\Theta} \right)^{p+1} \\
&\lesssim \left[(1+t)^{-\left(\frac{N}{4}+\frac{\theta}{2}\right)(p+1)} + (1+t)^{\lambda\left(\frac{p+1}{2}\Theta - 1\right) - \left(\frac{N}{4}+\frac{\mu}{2}\right)(1-\Theta)(p+1)} \right] \|u\|_{X(T)}^{p+1} \\
&\lesssim \|u\|_{X(T)}^{p+1},
\end{aligned}$$

since a straightforward computation shows $\lambda\left(\frac{p+1}{2}\Theta - 1\right) - \left(\frac{N}{4} + \frac{\mu}{2}\right)(1-\Theta)(p+1) < 0$ if $\lambda > \dfrac{q-1}{q}\dfrac{N+2-(N-2)p}{2(p-p_F(N)}$. Moreover, noticing $|\Psi_t| \lesssim (1+t)^{-1}\Psi$, we can estimate the second term of the left-hand side of Lemma 6 in the same way. This completes the proof. □

Next, we estimate the term $(1+t)^{N/4}\|u\|_{L^2}$ in the norm $\|u\|_{X(T)}$.

Lemma 7 *Let u be a mild solution to* (1) *on* $[0,T]$. *Then, we have*

$$(1+t)^{\frac{N}{4}} \left\| \int_0^t \mathcal{D}(t-s)|u(s)|^p \, ds \right\|_{L^2} \lesssim \|u\|_{X(T)}^p.$$

Proof By Lemma 4, we estimate

$$(1+t)^{\frac{N}{4}} \left\| \int_0^t \mathcal{D}(t-s)|u(s)|^p \, ds \right\|_{L^2}$$
$$\lesssim (1+t)^{\frac{N}{4}} \int_0^{\frac{t}{2}} (1+t-s)^{-\frac{N}{4}} \left(\||u(s)|^p\|_{L^1} + e^{-\frac{t-s}{4}} \||u(s)|^p\|_{L^2} \right) ds$$
$$+ (1+t)^{\frac{N}{4}} \int_{\frac{t}{2}}^t \||u(s)|^p\|_{L^2} \, ds.$$

In order to estimate the term $\||u(s)|^p\|_{L^1} = \|u(s)\|_{L^p}^p$, we divide the case into $p < 2$ and $p \geq 2$. When $p < 2$, applying Lemma 5, we have

$$\|u\|_{L^p} \lesssim \||x|^\lambda \nabla_x u\|_{L^2}^\theta \|u\|_{L^2}^{1-\theta}, \quad \theta = \frac{N(2-p)}{2p(\lambda-1)}.$$

We remark that $\theta < 1$ holds, since $\lambda > \dfrac{2N-(N-2)p}{2p}$. Thus, we have

$$\|u\|_{L^p}^p \lesssim (1+t)^{\frac{\lambda p}{2}\theta - \frac{N}{4}p(1-\theta)} \left(\left\| \left(\frac{|x|}{\sqrt{1+t}} \right)^\lambda \nabla_x u \right\|_{L^2}^\theta \left((1+t)^{\frac{N}{4}} \|u\|_{L^2} \right)^{1-\theta} \right)^p$$
$$\lesssim (1+t)^{\frac{\lambda p}{2}\theta - \frac{N}{4}p(1-\theta)} \|u\|_{X(T)}^p$$

A direct computation shows that $\frac{\lambda p}{2}\theta - \frac{N}{4}p(1-\theta) < -1$, since $\lambda > \frac{2N-8/N-(N-2)p}{4(p-p_F(N))}$. When $p \geq 2$, the Gagliardo–Nirenberg inequality implies

$$\|u\|_{L^p} \lesssim \|\nabla_x u\|_{L^2}^\theta \|u\|_{L^2}^{1-\theta}, \quad \theta = \frac{N(p-2)}{2p}.$$

Then, we easily obtain

$$\|u\|_{L^p}^p \lesssim (1+t)^{-\left(\frac{N}{4}+\frac{\theta}{2}\right)} \|u\|_{X(T)}^p$$

with $-\left(\frac{N}{4}+\frac{\theta}{2}\right) < -1$, since $p > p_F(N)$. Therefore, in both cases, we conclude

$$\int_0^{\frac{t}{2}} \||u(s)|^p\|_{L^1} \, ds \lesssim \|u\|_{X(T)}^p.$$

Next, we estimate the term $\| |u(s)|^p \|_{L^2} = \|u(s)\|_{L^{2p}}^p$. Applying the Gagliardo–Nirenberg inequality, we have

$$\|u\|_{L^{2p}} \lesssim \|\nabla_x u\|_{L^2}^{\theta} \|u\|_{L^2}^{1-\theta}, \quad \theta = \frac{N(p-1)}{2p}.$$

We remark that $\theta \le 1$ holds, since $p \le \frac{N}{[N-2]_+}$. Using the above estimate and repeating a similar computation as above, we have

$$(1+t)^{\frac{N}{4}} \int_0^{\frac{t}{2}} e^{-\frac{t-s}{4}} \| |u(s)|^p \|_{L^2}\, ds + (1+t)^{\frac{N}{4}} \int_{\frac{t}{2}}^t \| |u(s)|^p \|_{L^2}\, ds \lesssim \|u\|_{X(T)}^p,$$

which completes the proof. □

In the same way, we can estimate the other terms in the norm $\|u\|_{X(T)}$.

Lemma 8 *Let u be a mild solution to* (1) *on* $[0,T]$. *For nonnegative integers k, j satisfying $k + j \le 1$, we have*

$$(1+t)^{\frac{N}{4}+k+\frac{j}{2}} \left\| \int_0^t \partial_t^k \nabla_x^j \mathcal{D}(t-s) |u(s)|^p\, ds \right\|_{L^2} \lesssim \|u\|_{X(T)}^p.$$

Proof By Lemma 4, we have

$$\begin{aligned}
&(1+t)^{\frac{N}{4}+k+\frac{j}{2}} \left\| \int_0^t \partial_t^k \nabla_x^j \mathcal{D}(t-s) |u(s)|^p\, ds \right\|_{L^2} \\
&\lesssim (1+t)^{\frac{N}{4}+k+\frac{j}{2}} \int_0^{\frac{t}{2}} (1+t-s)^{-\frac{N}{4}-k-\frac{j}{2}} \left(\| |u(s)|^p \|_{L^1} + e^{-\frac{t-s}{4}} \| |u(x)|^p \|_{L^2} \right) ds \\
&\quad + (1+t)^{\frac{N}{4}+k+\frac{j}{2}} \int_{\frac{t}{2}}^t (1+t-s)^{-k-\frac{j}{2}} \| |u(s)|^p \|_{L^2}\, ds.
\end{aligned}$$

The rest part is completely the same as Lemma 7 and we omit the detail. □

Now let us prove Theorem 1. By Lemmas 2 and 6, we have

$$E_\psi(t)^{1/2} \lesssim E_\psi(0)^{1/2} + \|(1+|x|)^{\frac{2\lambda}{p+1}} u_0\|_{L^{p+1}}^{\frac{p+1}{2}} + \|u\|_{X(T)}^{\frac{p+1}{2}}.$$

By applying the same inequalities as (4)–(7) to the second term, the right-hand side can be further bounded by $C(\varepsilon + \varepsilon^{\frac{p+1}{2}}) + C\|u\|_{X(T)}^{\frac{p+1}{2}}$. Next, by Lemma 4, we can easily prove that

$$\begin{aligned}
&(1+t)^{\frac{N}{4}+1} \|u_t^L(t)\|_{L^2} + (1+t)^{\frac{N}{4}+\frac{1}{2}} \|\nabla_x u^L(t)\|_{L^2} + (1+t)^{\frac{N}{4}} \|u^L(t)\|_{L^2} \\
&\lesssim \|u_0\|_{L^1} + \|u_0\|_{H^1} + \|u_1\|_{L^1} + \|u_1\|_{L^2} \lesssim \varepsilon
\end{aligned}$$

for the linear part of the solution $u^L := \mathcal{D}(t)(u_0 + u_1) + \frac{\partial}{\partial t}\mathcal{D}(t)u_0$. Moreover, by Lemmas 7 and 8, we have

$$\begin{aligned}
&(1+t)^{\frac{N}{4}+1}\left\| \int_0^t \partial_t \mathcal{D}(t-s)|u(s)|^p\, ds \right\|_{L^2} \\
&\quad + (1+t)^{\frac{N}{4}+\frac{1}{2}}\left\| \int_0^t \nabla_x \mathcal{D}(t-s)|u(s)|^p\, ds \right\|_{L^2} \\
&\quad + (1+t)^{\frac{N}{4}}\left\| \int_0^t \mathcal{D}(t-s)|u(s)|^p\, ds \right\|_{L^2} \\
&\lesssim \|u\|_{X(T)}^p.
\end{aligned}$$

Putting all the above estimates together, we conclude

$$\|u\|_{X(T)} \lesssim \varepsilon + \varepsilon^{\frac{p+1}{2}} + \|u\|_{X(T)}^{\frac{p+1}{2}} + \|u\|_{X(T)}^p,$$

where the implicit constant is independent of T. From this and a standard argument give the a priori estimate $\|u\|_{X(T)} \leq C$ with some constant $C > 0$ independent of T, provided that ε is sufficiently small, and the proof is complete.

Acknowledgments This work was supported by JSPS KAKENHI Grant Number JP20K14346. The author is deeply grateful to Professors Masahiro Ikeda, Koichi Taniguchi and Motohiro Sobajima for the fruitful discussion. Also, the author would like to thank the referee for useful comments.

References

1. Caffarelli, L., Kohn, R., Nirenberg, L.: First order interpolation inequalities with weights. Compositio Math. **53**, 259–275 (1984)
2. D'Abbicco, M.: The threshold of effective damping for semilinear wave equations. Math. Methods Appl. Sci. **38**, 1032–1045 (2015)
3. Hayashi, N., Kaikina, E.I., Naumkin, P.I.: Damped wave equation with super critical nonlinearities. Differ. Integr. Equ. **17**, 637–652 (2004)
4. Ikeda, M., Taniguchi, K., Wakasugi, Y.: Global existence and asymptotic behavior for semilinear damped wave equations on measure spaces. arXiv:2106.10322v3
5. Ikehata, R., Tanizawa, K.: Global existence of solutions for semilinear damped wave equations in $\mathbf{R}^N$ with noncompactly supported initial data. Nonlinear Anal. **61**, 1189–1208 (2005)
6. Ikehata, R., Nishihara, K., Zhao, H.: Global asymptotics of solutions to the Cauchy problem for the damped wave equation with absorption. J. Differ. Equ. **226**, 1–29 (2006)
7. Matsumura, A.: On the asymptotic behavior of solutions of semi-linear wave equations. Publ. RIMS Kyoto University **12**, 169–189 (1976)
8. Sobajima, M.: Global existence of solutions to semilinear damped wave equation with slowly decaying initial data in exterior domain. Differ. Integr. Equ. **32**, 615–638 (2019)

9. Todorova, G., Yordanov, B.: Critical exponent for a nonlinear wave equation with damping. J. Differ. Equ. **174**, 464–489 (2001)
10. Wakasugi, Y.: Decay property of solutions to the wave equation with space-dependent damping, absorbing nonlinearity, and polynomially decaying data. Math. Methods Appl. Sci. **46**, 7067–7107 (2023)

Part VIII
Stochastic Processes

Hedging in a Market with Jumps: A FBSDE Approach

Evelina Shamarova and Rui Sá Pereira

Abstract We propose a model for hedging in a market with jumps for a large investor. The dynamics of the stock prices and the value process is governed by forward-backward SDEs driven by Teugels martingales. Unlike known FBSDE market models, ours accounts for jumps in stock prices. Moreover, it allows to find an optimal hedging strategy.

1 Introduction

In this article, we propose a forward-backward-SDE (FBSDE) model for hedging in a market with a large investor and where the stock prices are allowed to jump. The attempts to price contingent claims have their origin in the work of [1], where the authors give a formula to price a European call option assuming that the stock price evolves as a geometric Brownian motion. One of the assumptions of the classical Black and Scholes model is that no individual investor action is able to influence market prices. The importance of accounting for the existence of large investors, however, has been increasing, given the prevalence of electronic trading and, in particular, high frequency trading which makes possible to issue thousands of orders over short periods of time. This problem was addressed in [2], where the authors present an FBSDE market model for a large investor, but in a Brownian market environment, where the stock prices, being modeled as geometric Brownian motions, are not allowed to have jumps. However, the documented evidence of jumps in the distribution of the returns (see, e.g., [3]) suggests that a geometric Brownian motion is not entirely suited to model the evolution of stock prices in real markets. In particular, in periods of heavy market turbulence, such as the "flash

E. Shamarova (✉)
Departamento de Matemática, Universidade Federal da Paraíba, João Pessoa, Brazil
e-mail: evelina.shamarova@academico.ufpb.br

R. S. Pereira
Departamento de Matemática, Universidade do Porto, Porto, Portugal
e-mail: manuelsapereira@gmail.com

M. R. Ebert et al. (eds.), *New Tools in Mathematical Analysis and Applications*,
Research Perspectives, https://doi.org/10.1007/978-3-031-77050-0_24

crash" in May 2010, when the main US indexes temporarily dropped by more than 9 per cent, hedging strategies resulting from Brownian models leave investors exposed to a significant downside risk.

It is known that markets, where the stock prices are modeled involving Lévy processes are, in general, incomplete, so contingent claims may not admit self-financing replicating strategies. The first attempt to define optimal strategies in the context of incomplete markets was made in [4], where the authors propose an optimal strategy as the one that minimizes, in a certain sense, the injection of capital needed.

In our FBSDE market model, the evolution of the d-dimensional stock price $S_t = \{S_t^i\}_{i=1}^d$ is governed by a SDE driven by m independent Brownian motions and $d-m$ martingales picked from the system of orthonormalized Teugels martingales $\{H_t^{(ik)}\}_{i=1,\,k\in\mathbb{N}}^l$ such that for each $i \in \{1, \dots, l\}$, the family $\{H_t^{(ik)}\}_{k=1}^\infty$ is associated to a Lévy process L_t^i. All processes L_t^i are assumed independent and purely discontinuous. We refer the reader to [5] (p. 763) for details on the martingales $H_t^{(ik)}$. Remark that different stock prices S_t^i can jump at different times. Further, the value process V_t and the portfolio process $\pi_t = \{\pi_t^i\}_{i=1}^d$ evolve according to a backward SDE with the final condition $h(S_T)$ which is the payoff at maturity T.

Our model involves the martingales $H_t^{(ik)}$ because they are independent, strongly orthonormal, purely discontinuous, but most importantly, the system $\{H_t^{(ik)}\}_{i=1,\,k\in\mathbb{N}}^l$, completed with the Brownian motions $\{B_t^i\}_{i=1}^m$, possesses the predictable representation property. The latter allows to decompose the discounted value process into a sum of the value of the hedging portfolio and a strongly orthogonal martingale. Therefore, our model allows to find a hedging strategy which is optimal in the sense of [8]. It is worth to mention that due to the presence of $H_t^{(ik)}$'s, the SDEs representing the evolution of stocks become, in fact, driven by power-jump martingales built on the basis of the underlying Lévy processes (see [5], p. 763). The presence of these "power-jump" terms may reflect "skewness", "kurtosis", and other volatile behavior or extremal movements of the market.

Thus, the main contribution of this work is introducing a model which accounts for jumps in stock prices and allows to find an optimal hedging strategy in the context of incomplete markets.

2 FBSDE Model for Hedging in a Market with Jumps

In what follows, we present our model. Let $(\Omega, \mathcal{F}, P)$ be a probability space, $\{B_t^i\}_{i=1}^m$ be independent real-valued Brownian motions, and $\{L_t^i\}_{i=1}^l$ be independent purely discontinuous real-valued Lévy processes with the Lévy measures ν^i satisfying

$$\int_{\mathbb{R}} (1 \wedge x^2)\nu^i(dx) < \infty \quad \text{and} \quad \int_{|x|\geqslant\varepsilon} e^{\lambda|x|}\nu^i(dx) < C \tag{1}$$

for some positive ε, λ, and C. Define the filtration $\mathcal{F}_t = \sigma\{B_s^i, 0 \leqslant s \leqslant t,\ 1 \leqslant i \leqslant m\} \vee \sigma\{L_s^i, 0 \leqslant s \leqslant t,\ 1 \leqslant i \leqslant l\} \vee \mathcal{N}$, where $\mathcal{N}$ is the collection of all P-null sets. We agree that all Lévy measures ν considered in this work, satisfy the condition $\nu(\{0\}) = 0$.

Let, for each $i \in \{1, \ldots, l\}$, $\{H_t^{(ik)}\}_{k=1}^{\infty}$ be the family of orthonormalized Teugels martingales associated to the Lévy process L_t^i. Lemma 1 below provides a useful representation for the orthonormalized Teugels martingales $H_t^{(i)}$ associated to an arbitrary one-dimensional Lévy process ℓ_t with the Lévy triple (b, a, λ), $a = (a_1, \ldots, a_M)$, and the Lévy measure λ satisfying (1) with ν_i substituted by λ. As in [5], we introduce the polynomials $q_{i-1}(x)$ obtained by the orthonormalization of the system $\{1, x, x^2, \ldots\}$ with respect to the measure $x^2\lambda(dx) + |a|^2\delta_0(dx)$, where δ_0 is the Dirac measure. Furthermore, we define $p_i(x) = xq_{i-1}(x)$.

Lemma 1 *Let ℓ be a one-dimensional Lévy process with the Lévy-Itô decomposition $\ell_t = bt + \sum_{i=1}^M a_i\beta_i(t) + \int_{|x|\leqslant 1} x\tilde{\mu}(t, dx) + \int_{|x|>1} x\mu(t, dx)$, where $\{\beta_i(t)\}_{i=1}^M$ are independent real-valued standard Brownian motions. Then, it holds that $H_t^{(i)} = q_{i-1}(0)\sum_{j=1}^M a_j\beta_j(t) + \int_{\mathbb{R}} p_i(x)\tilde{\mu}(t, dx)$. In particular, if ℓ_t is purely discontinuous, then $H_t^{(i)} = \int_{\mathbb{R}} p_i(x)\tilde{\mu}(t, dx)$.*

Proof Define $\tilde{p}_i(x) = p_i(x) - xq_{i-1}(0)$. We will use the following representation for $H_t^{(i)}$ obtained in [5] (p. 763):

$$H_t^{(i)} = q_{i-1}(0)\ell_t + \sum_{0<s\leqslant t}\tilde{p}_i(\Delta\ell_s) - t\,\mathbb{E}\Big[\sum_{0<s\leqslant 1}\tilde{p}_i(\Delta\ell_s)\Big] - tq_{i-1}(0)\mathbb{E}[\ell_1].$$

Since $\ell_t = \ell_t^c + \sum_{0\leqslant s\leqslant t}\Delta\ell_s$, where ℓ_t^c is the continuous part of ℓ_t, we obtain

$$\begin{aligned} H_t^{(i)} &= q_{i-1}(0)\ell_t^c + \sum_{0<s\leqslant t} p_i(\Delta\ell_s) - \mathbb{E}\big[\sum_{0<s\leqslant t}\tilde{p}_i(\Delta\ell_s)\big] - q_{i-1}(0)\mathbb{E}[\ell_t] \\ &= q_{i-1}(0)\big[\ell_t^c - \mathbb{E}[\ell_t^c]\big] + \sum_{0<s\leqslant t} p_i(\Delta\ell_s) - \mathbb{E}\big[\sum_{0<s\leqslant t} p_i(\Delta\ell_s)\big] \\ &= q_{i-1}(0)\sum_{j=1}^M a_j\beta_j(t) + \int_{\mathbb{R}} p_i(x)\tilde{\mu}(t, dx). \end{aligned}$$

□

Let us proceed with the description of the model. Fix a finite-time horizon $T > 0$ and consider a market consisting of d risky assets (stocks) and risk-free money on a deposit. We assume that the price process of the risk-free deposit evolves according to the equation

$$dD_t = r(t, S_t, V_t, \pi_t)D_t dt, \qquad D_0 = 1, \tag{2}$$

where $r : [0, T] \times \mathbb{R}^d \times \mathbb{R} \times \mathbb{R}^d \to \mathbb{R}$ is the interest rate, $S_t = \{S_t^i\}_{i=1}^d$ is the d-dimensional risky asset price process, V_t is the (real-valued) value process, and $\pi_t = \{\pi_t^i\}_{i=1}^d$ is the portfolio process with π_t^i being the number assets of the ith stock. The evolution of S_t^i is assumed to be governed by the SDE

$$dS_t^i = S_t^i\big\{\tilde{f}_i(t, S_t, V_t, \pi_t)dt + \sum_{j=1}^d \sigma_i^{\alpha_j}(t, S_{t-}, V_{t-})dM_t^{\alpha_j}\big\} \tag{3}$$

with the non-random initial condition $S_0^i > 0$. In (3), $\tilde{f}_i$ and $\sigma_i^{\alpha_j}$ are real-valued functions defined on spaces of appropriate dimensions. Further, for $j = 1, 2, \ldots, m$, $\alpha_j = j$ and $M_t^{\alpha_j} = B_t^j$, while for $j = m+1, \ldots, d$, α_j's are arbitrarily picked multiindexes from the set $\{(ik), i = 1, \ldots, l, k = 2, 4, \ldots\}$ and $M_t^{\alpha_j} = H_t^{\alpha_j}$. Remark that the index k takes only even values.

The value process V_t represents the wealth of a "large" investor who holds d stocks and money on a deposit. The investor is assumed large, so the coefficients in our model would depend on V_t, S_t, and π_t.

We define an *admissible hedging strategy* as a pair of predictable processes (π_t, π_t^0) such that $V_t = \sum_{i=1}^d \pi_t^i S_t^i + \pi_t^0 D_t$ and $V_T = h(S_T)$, where $h(S_T)$ is the payoff at maturity T. Note that the solution of (2) takes the form $D_t = \exp\{\int_0^t r_s ds\}$, where $r_s = r(s, S_s, V_s, \pi_s)$. Let $A_t = \exp\{-\int_0^t r_s ds\}$. Define $\hat{S}_t^i = A_t S_t^i$ and $\hat{V}_t = A_t V_t$ to be the *discounted stock price* and *discounted value process*, respectively. Furthermore, we define the *cumulative cost process* $C_t = \hat{V}_t - \sum_{i=1}^d \int_0^t \pi_s^i d\hat{S}_s^i$. We say that the strategy is *optimal*, if it is admissible and C_t is a square-integrable martingale strongly orthogonal to the martingale part of each $\hat{S}_t^i$.

Lemma 2 *The representations* $\hat{V}_t = \sum_{i=1}^d \int_0^t \pi_s^i d\hat{S}_s^i + C_t$ *and*

$$V_t = V_0 + \sum_{i=1}^{d} \int_0^t \pi_s^i dS_s^i + \int_0^t \pi_s^0 dD_s + \int_0^t D_s dC_s \tag{4}$$

are equivalent.

Proof Since $\langle V, A \rangle_t = \langle S^i, A \rangle_t = [V, A]_t = [S^i, A]_t = 0$, then by Itô's product formula, $d\hat{S}_t^i = A_t dS_t^i - A_t r_t S_t^i dt$ and $d\hat{V}_t = A_t dV_t - r_t A_t V_t dt$. Substituting these expressions into the equation for $\hat{V}_t$ we obtain $dV_t = \sum_{i=1}^d \pi_t^i dS_t^i + (V_t - \sum_{i=1}^d \pi_t^i S_t^i) r_t dt + D_t dC_t = \sum_{i=1}^d \pi_t^i dS_t^i + \pi_t^0 dD_t + D_t dC_t$. □

Now we derive a backward SDE (BSDE) for the process V_t with representation (4). First, we substitute dD_t and dS_t^i with the right-hand sides of Eqs. (2) and (3), respectively. Since $V_T = h(S_T)$, from (4) we obtain

$$V_t - \mathbb{E}[h(S_T) - \int_t^T g(s, S_s, V_s, Z_s^{(\alpha)}) ds - \sum_{j=1}^{d} \int_0^T Z_s^{\alpha_j} dM_s^{\alpha_j} - \int_0^T D_s dC_s | \mathcal{F}_t]$$

$$= \sum_{j=1}^{d} \int_0^t Z_s^{\alpha_j} dM_s^{\alpha_j} + \int_0^t D_s dC_s = \sum_{k=1}^{l} \sum_{j=1}^{\infty} \int_0^t \hat{Z}_s^{(kj)} dH_s^{(kj)}, \tag{5}$$

where $g(t, s, v, z) = \tilde{g}(t, s, v, \pi)$ with $\pi = \mathbf{s}^{-1} \sigma(t, s, v)^{-1} z$ and $\sigma(t, s, v)$ being the $d \times d$ matrix with the element $\sigma_i^{\alpha_j}$ in the jth line and the ith column, $\tilde{g}(t, s, v, \pi) = \sum_{i=1}^d s_i \pi_i \tilde{f}_i(t, s, v, \pi) + (v - \sum_{i=1}^d s_i \pi_i) r(t, s, v, \pi)$, $\pi = (\pi_i)_{i=1}^d$,

$s = (s_i)_{i=1}^d$, and $\mathbf{s} = \text{diag}\{s_1, \dots, s_n\}$. Further, (α) denotes the set of multiindexes $(\alpha) = \{\alpha_1, \dots, \alpha_d\}$ and $Z_t^{(\alpha)} = (Z_t^{\alpha_1}, \dots, Z_t^{\alpha_d})$. The last identity in (5) follows from the predictable representation property of the system $\{H_t^{(kj)}\}_{j=1}^\infty$ for a fixed k (with $\hat{Z}_t^{(kj)}$ being predictable processes) and from the independence of the Lévy processes L_t^k, $k = 1, \dots, l$. The BSDE for V_t follows from (5):

$$V_t = h(S_T) - \int_t^T g(s, S_s, V_s, Z_s^{(\alpha)})ds - \sum_{j=1}^d \int_t^T Z_s^{\alpha_j} dM_s^{\alpha_j} - \sum_{(kj)\notin(\alpha)} \int_t^T \hat{Z}_s^{(kj)} dH_s^{(kj)}. \tag{6}$$

Making the change of variable $\pi = \mathbf{s}^{-1}\sigma(t, s, v)^{-1}z$ and introducing the functions $f_i(t, s, v, z) = \tilde{f}_i(t, s, v, \mathbf{s}^{-1}\sigma(t, s, v)^{-1}z)$, we transform SDE (3) to

$$dS_t^i = S_t^i\{f_i(t, S_t, V_t, Z_t^{(\alpha)})dt + \sum_{j=1}^d \sigma_i^{\alpha_j}(t, S_t, V_t)dM_t^{\alpha_j}\}. \tag{7}$$

Lemma 3 *FBSDEs* (6)–(7) *are equivalent to*

$$\begin{cases} S_t^i = S_0^i + \int_0^t S_s^i f_i(s, S_s, V_s, \mathcal{Z}_s, \hat{\mathcal{Z}}_s(\cdot))ds + \int_0^t S_s^i \sigma_i(s, S_s, V_s)\, dB_s \\ \qquad + \int_0^t \int_{\mathbb{R}^l} S_s^i \psi_i(s, S_{s-}, V_{s-}, u)\tilde{N}(ds, du), \qquad i = 1, \dots, d, \\ V_t = h(S_T) - \int_t^T g(s, S_s, V_s, \mathcal{Z}_s, \hat{\mathcal{Z}}_s(\cdot))ds - \int_t^T \mathcal{Z}_s dB_s \\ \qquad - \int_t^T \int_{\mathbb{R}^l} \hat{\mathcal{Z}}_s(u)\tilde{N}(ds, du), \end{cases} \tag{8}$$

where for $u = (u_1, \dots, u_l)$, $\psi_i(t, s, v, u) = \sum_{q=m+1}^d \sigma_i^{\alpha_q}(t, s, v)p_{\alpha_q}(u_{\alpha_q})$ *with* $u_{\alpha_q} = u_k$ *if* $\alpha_q = (kj)$. *Further,* $\mathcal{Z}_t = (Z_t^{\alpha_1}, \dots, Z_t^{\alpha_m})$ *and for each* $k \in \{1, \dots, l\}$, $\hat{Z}_t^{(kj)}$ *are the components of the decomposition of* $\hat{\mathcal{Z}}_t(0, \dots, u_k, \dots, 0)$ *with respect to the basis of polynomials* $p_{(kj)}(u_k)$ *in the space* $L_2(\nu^k(du_k))$, *while* $(Z_t^{\alpha_{m+1}}, \dots, Z_t^{\alpha_d}) = \{\hat{Z}_t^{(kj)}\}_{(kj)\in(\alpha)}$. *Finally,* $\tilde{N}$ *is the compensated Poisson random measure for the Lévy process* $(L_t^1, \dots, L_t^l)$.

Remark With a slight abuse of notation, in the coefficients f and g, we write $\mathcal{Z}_t$ instead of $(Z_t^{\alpha_1}, \dots, Z_t^{\alpha_m})$ and $\hat{\mathcal{Z}}_t(\cdot)$ instead of $(Z_t^{\alpha_{m+1}}, \dots, Z_t^{\alpha_d})$. The dependence on $\hat{\mathcal{Z}}_t(\cdot)$ is understood as the dependence on its $d - m$ components $(Z_t^{\alpha_{m+1}}, \dots, Z_t^{\alpha_d})$.

Proof of Lemma 3 Note that for each k, the system $\{H_t^{(kj)}\}_{j=1}^{\infty}$ has the predictable representation property. Therefore,

$$\int_t^T \int_{\mathbb{R}^l} \hat{\mathcal{Z}}_s(u)\tilde{N}(ds, du) = \sum_{k=1}^{l} \int_t^T \int_{R_k} \hat{\mathcal{Z}}_s(0, \ldots, u_k, \ldots, 0)\tilde{N}^k(ds, du_k)$$

$$= \sum_{k=1}^{l} \sum_{j=1}^{\infty} \int_t^T \hat{Z}_s^{(kj)} dH_s^{(kj)},$$

where $R_k = \{te_k, t \in \mathbb{R}\}$ with $\{e_k\}_{k=1}^l$ being an orthonormal basis in $\mathbb{R}^l$, and $\tilde{N}^k(t, \cdot)$ is the compensated Poisson random measure for L_t^k, which, by the independence of L_t^k's, is the restriction of $\tilde{N}(t, \cdot)$ to R_k. Since, by Lemma 1, $H_t^{(kj)} = \int_{R_k} p_{(kj)}(u_k)\tilde{N}^k(t, du_k)$, we obtain that in $L_2(\nu^k)$, $\hat{\mathcal{Z}}_t(0, \ldots, u_k, \ldots, 0) = \sum_{j=1}^{\infty} \hat{Z}_t^{(kj)} p_{(kj)}(u_k)$ a.s. Moreover, for each k, the system of polynomials $\{p_{(kj)}\}_{j=1}^{\infty}$ is orthonormal in $L_2(\nu^k)$ by the orthonormality of $H_t^{(kj)}$'s. Finally, since for each (kj), $p_{(kj)}(0) = 0$, we obtain

$$\int_{\mathbb{R}^l} \sum_{q=m+1}^{d} \sigma_i^{\alpha_q}(t, x, y) p_{\alpha_q}(u_{\alpha_q})\tilde{N}(dt, du)$$

$$= \sum_{q=m+1}^{d} \sigma_i^{\alpha_q}(t, x, y) \int_{R_{\alpha_q}} p_{\alpha_q}(u_{\alpha_q})\tilde{N}^{\alpha_k}(dt, du_{\alpha_q}) = \sum_{k=m+1}^{d} \sigma_i^{\alpha_q}(t, x, y) dH_t^{\alpha_q}.$$

where $R_{\alpha_q} = R_k$, $\tilde{N}^{\alpha_q} = \tilde{N}^k$, and $u_{\alpha_q} = u_k$ for $\alpha_q = (kj)$. □

Remark Since $\{\alpha_j\}_{j=m+1}^d$ are multiindexes picked from the set $\{(ik), i = 1, \ldots, l, k = 2, 4, \ldots\}$, then each polynomial p_{α_j} is of even degree, and, therefore, achieves a finite global minimum, which we denote by A_j.

Assumption (A1) below guarantees the existence and uniqueness of solution to FBSDEs (6)–(7).

(A1) The coefficients of FBSDEs (8) satisfy the hypotheses of Theorem 3.1 in [9] (p. 436).

Lemma 4 *Assume (A1). Then, there exists a unique $\mathcal{F}_t$-adapted solution $(S_t, V_t, Z_t^{(\alpha)})$ to FBSDEs* (6)–(7) *such that (S_t, V_t) has càdlàg paths and $Z_t^{(\alpha)}$ is predictable.*

Proof Under (A1), Theorem 3.1 in [9] guarantees the existence of a unique $\mathcal{F}_t$-adapted solution $(S_t, V_t, \mathcal{Z}_t, \hat{\mathcal{Z}}_t(\cdot))$ to FBSDEs (8) such that (S_t, V_t) is càdlàg and $(\mathcal{Z}_t, \hat{\mathcal{Z}}_t(\cdot))$ is predictable. By Lemma 3, this is equivalent to the existence of a unique $\mathcal{F}_t$-adapted solution $(S_t, V_t, Z_t^{(\alpha)})$ to (6)–(7). □

Assumptions (A2)–(A4) below guarantee the positivity of the prices S_t^i, the non-negativity of V_t, and the existence of the optimal strategy.

(A2) $\det\{\sigma(t,s,v)\} \neq 0$ for all $(t,s,v) \in [0,T] \times \mathbb{R}^d \times \mathbb{R}$.

(A3) For all $(t,s,v) \in [0,T] \times \mathbb{R}^d \times \mathbb{R}$, $i \in \{1,\dots d\}$, and $j \in \{m+1,\dots,d\}$, $\sigma_i^{\alpha_j}(t,s,v) > 0$. Moreover, if $A_j < 0$, then $\sigma_i^{\alpha_j}(t,x,v)|A_j| < (d-m)^{-1}$.

(A4) If $(S_t, V_t, \mathcal{Z}_t, \hat{\mathcal{Z}}_t(\cdot))$ is the $\mathcal{F}_t$-adapted solution to FBSDEs (8), then the random function $(\omega,t,y,z,\hat{z}) \mapsto g(t,S_t,y,z,\hat{z})$ satisfies condition (A_γ) in [7] (p. 1362). Moreover, $h(S_T) \geqslant 0$ a.s.

The main result of our paper is the following.

Theorem 1 *Let (A1)–(A4) hold, and let $(S_t, V_t, Z_t^{(\alpha)})$ be the solution to FBSDEs (6)–(7). Then, $S_t^i > 0$, $i = 1,\dots,d$, and $V_t \geqslant 0$ a.s. Moreover, the pair of stochastic processes (π_t, π_t^0), where $\pi_t = \operatorname{diag}\{S_t^1,\dots,S_t^d\}^{-1}\sigma^{-1}(t,S_t,V_t)Z_t^{(\alpha)}$ and $\pi_t^0 = \hat{V}_t - \sum_{i=1}^d \pi_s^i \hat{S}_t^i$, is the optimal hedging strategy.*

Proof Note that the above representation for π_t holds by construction. Next, by the representation for the function ψ_i, obtained in Lemma 3, and by (A3), $\inf_{t>0}\psi_i(t,S_{t-},V_{t-},\Delta L_t) > -1$. Therefore, S_t^i can be represented by the Doléans-Dade exponential which is finite a.s.:

$$S_t^i = S_0^i\, e^{\int_0^t \left(\tilde{f}_i(s,S_s,V_s,\pi_s) - \frac{\|\sigma_i(s,S_s,V_s)\|^2}{2}\right)ds + \int_0^t \sigma_i(s,S_s,V_s)dB_s + \int_0^t \int_{\mathbb{R}^l} \psi_i(s,S_{s-},V_{s-},u)\tilde{N}(ds,du)}$$
$$\times \prod_{0\leqslant s\leqslant t} \left(1+\psi_i(s,S_{s-},V_{s-},\Delta L_s)\right) e^{-\psi_i(s,S_{s-},V_{s-},\Delta L_s)},$$

where $\sigma_i = (\sigma_i^{\alpha_j})_{j=1}^d$. Therefore, for all i, $S_t^i > 0$ a.s. Let us prove the non-negativity of V_t. To this end, we apply the comparison theorem from [7] (Theorem 2.5, p. 1362) to the BSDE in (8), considered with respect to $(V_t, \mathcal{Z}_t, \hat{\mathcal{Z}}_t(\cdot))$, whereas the process S_t is fixed and assumed known from Lemma 4. Note that, by the definition, $g(t,S_t,0,0,0) = 0$. Therefore, we compare the solution $(V_t, \mathcal{Z}_t, \hat{\mathcal{Z}}_t(\cdot))$ with the identically zero solution to the BSDE whose generator is the same as in (8) but the final condition is zero. Remark that (A4) implies the assumptions of the comparison theorem in [7]. Thus, by Theorem 2.5 in [7], $V_t \geqslant 0$ for all t a.s.

Note that, by (5), $C_t = V_0 + \sum_{(kj)\notin(\alpha)} \int_0^t A_s \hat{Z}_s^{(kj)} dH_s^{(kj)}$, and, therefore, it is a square integrable martingale. Moreover, C_t is (weakly) orthogonal to the stable subspace $\mathcal{S}$ generated by $\{M_t^{\alpha_j}\}_{j=1}^d$, which follows from Theorem 35 of [6] (p. 149) and from the strong orthogonality of the martingales $M_t^{\alpha_j}$. By Theorem 36 of [6] (p. 150), C_t is strongly orthogonal to $\mathcal{S}$. It remains to note that the martingale parts of $\{\hat{S}_t^i\}_{i=1}^d$ belong to $\mathcal{S}$. □

Corollary 1 *The Föllmer-Schweizer decomposition of the discounted contingent claim $A_T h(S_T)$ takes the form*

$$A_T h(S_T) = V_0 + \sum_{i=1}^{d} \int_0^T \pi_t^i \, d\hat{S}_t^i + \sum_{(kj)\notin(\alpha)} \int_0^T A_t \hat{Z}_t^{(kj)} dH_t^{(kj)}.$$

References

1. Black, F., Scholes, M.: The pricing of options and corporate liabilities. J. Pol. Econ. **81**, 637–659 (1973)
2. Cvitanić, J., Ma, J.: Hedging options for a large investor and Forward-Backward SDE's. Ann. Appl. Probab. **6**(2), 370–398 (1996)
3. Eberlein, E., Keler, U.: Hyperbolic distributions in finance.Bernoulli **1**(3), 281–299 (1995)
4. Föllmer, H., Schweizer, M.: Hedging of contingent claims under incomplete information. In: Davis, M.H.A., Elliott, R.J. (eds.) Applied Stochastic Analysis. Edition: Stochastics Monographs, vol. 5, pp. 389–414. Gordon and Breach, London (1991)
5. Nualart, D., Schoutens, W.: Backward stochastic differential equations and Feynman-Kac formula for Lévy processes with applications in finance. Bernoulli **7**(5), 761–776 (2001)
6. Protter, P.: Stochastic Integration and Differential Equations, a New Approach. Springer, Berlin (1992)
7. Royer, M.: Backward stochastic differential equations with jumps and related non-linear expectations. Stoch. Process Their Appl. **116**, 1358–1376 (2006)
8. Schweizer, M.: Local risk-minimization for multidimensional assets and payment streams. Advances in Mathematics of Finance. Polish Academy of Sciences, vol. 83. Banach Center Publications, Institute of Mathematics, Warsaw (2008)
9. Wu, Z.: Forward-Backward Stochastic Differential Equations with Brownian motion and Poisson process. Acta Math. Appl. Sin. **15**(4), 433–443 (1999)

Global Smooth Solutions for the Stochastic Navier-Stokes Equations with Super-linear Stratonovich Noise in the 3D Torus

Marco Bagnara

Abstract We consider the Navier-Stokes equations in the 3D torus perturbed by a super-linear noise in Stratonovich form. We show global-in-time existence and uniqueness of smooth solutions, ruling out the possibility of finite-time blow-ups for such equations. The result is an application of the abstract framework developed in (Bagnara A suitable nonlinear Stratonovich noise prevents blow-up in the Euler equations and other SPDEs. Preprint, 2023. arXiv:2312.10446), where it is shown that a suitable Stratonovich noise prevents the blow-up possibly induced by a drift with super-linear growth in a class of SPDEs.

1 Introduction

We consider the incompressible Navier-Stokes equations on the three-dimensional torus $\mathbb{T}^3$. The Navier-Stokes system describes the motion of an incompressible viscous fluid. It reads as

$$\begin{cases} \partial_t u + (u \cdot \nabla)u - \nu \Delta u + \nabla p = 0 & (t, x) \in [0, T] \times \mathbb{T}^3, \\ \operatorname{div} u = 0 & (t, x) \in [0, T] \times \mathbb{T}^3, \\ u(0, x) = u_0(x) & x \in \mathbb{T}^3, \\ \fint_{\mathbb{T}^3} u(t, x)\, \mathrm{d}x = 0 & (t, x) \in [0, T] \times \mathbb{T}^3, \end{cases} \tag{1}$$

where $u : [0, T] \times \mathbb{T}^3 \to \mathbb{R}^3$ is the velocity field, $p : [0, T] \times \mathbb{T}^3 \to \mathbb{R}$ is the pressure field and $\nu > 0$ is the so-called viscosity parameter. The problem should be understood with periodic boundary conditions.

The study of the Navier-Stokes system has begun with the seminal work of Leray [10], where he showed existence of global weak solutions in the whole three-

M. Bagnara (✉)
Scuola Normale Superiore, Pisa, Italy
e-mail: marco.bagnara@sns.it

M. R. Ebert et al. (eds.), *New Tools in Mathematical Analysis and Applications*, Research Perspectives, https://doi.org/10.1007/978-3-031-77050-0_25

dimensional space and has been later extended to bounded domains by Hopf in [8]. If on the one hand local well–posedness for smooth initial data has been established, see [9], on the other global well-posedness remains a well-known open problem (Millenium Prize).

In recent years, remarkably new advancements have appeared in the literature, suggesting an answer for the global well-posedness in the negative. With no claim of completeness, we mention, for instance, [1], where the authors exhibit two distinct Leray solutions with zero initial velocity and identical body force, and [4] where it is shown, through convex integration techniques, that solutions are not unique in the class of weak solutions with finite kinetic energy. Similar negative results have also be proven in the stochastic setting, see, for instance, [7], where the authors, again exploiting convex integration tecniques, provide global-in-time existence and non-uniqueness of probabilistically strong solutions to the three-dimensional Navier–Stokes system driven by space-time white noise.

The lack of a global well-posedness result in the deterministic 3D Navier-Stokes equations has led to the question of whether noise can provide well-posedness. This concept, known as regularization by noise, has largely spread in the fluid dynamics literature. We refer to the lecture notes [5] and references therein for an overview of the topic. Here, in particular, we study whether a suitable noise can prevent solutions from blowing up in high-order Sobolev norms. Recently, this direction has been explored in many manuscripts both under Ito and Stratonovich perturbations. The underlying idea is that a suitable strong enough noise is able to prevent the blow-up in many SPDEs. Concerning the Ito setting, in [11] a general criterion is applied to both stochastic Euler and Navier-Stokes equations on the torus, while in [3] the case of the Euler equations on a bounded domain is presented. On the other hand, for the Stratonovich case, global solutions for the Euler equations in a bounded domain have been exhibit in [2]. In the present work, standing on the abstract framework developed in the latter, we address the Navier-Stokes equations in the periodic domain.

2 Setting and Main Result

In this section we introduce the stochastic Navier-Stokes equations and we state the main result of this manuscript. Consider the stochastic Navier-Stokes equation on the three-dimensional torus $\mathbb{T}^3$ perturbed by a stochastic forcing in Stratonovich form:

$$\begin{cases} du + (u \cdot \nabla)u\, dt - \nu \Delta u\, dt + \nabla p\, dt = \sigma(u) \circ dW_t & (t, x) \in [0, T] \times \mathbb{T}^3, \\ \operatorname{div} u = 0 & (t, x) \in [0, T] \times \mathbb{T}^3, \\ u(0, x) = u_0(x) & x \in \mathbb{T}^3, \\ \int_{\mathbb{T}^3} u(t, x)\, \mathrm{d}x = 0 & (t, x) \in [0, T] \times \mathbb{T}^3. \end{cases} \tag{2}$$

Here, $u : [0, T] \times \mathbb{T}^3 \times \Omega \to \mathbb{R}^3$ is the velocity field, $p : [0, T] \times \mathbb{T}^3 \times \Omega \to \mathbb{R}$ is the pressure field, $\nu > 0$ is the viscosity parameter and W is a Q-Wiener process on a filtered probability space $(\Omega, \mathcal{A}, (\mathcal{F}_t)_t, \mathbb{P})$ satisfying the standard assumption. Moreover, u_0 is the deterministic divergence free initial datum, with zero spatial average. The problem is understood with periodic boundary conditions. Both u and p are unknown functions, whereas the noise coefficient σ will be specified later.

Let $s \in \mathbb{N}$, we define the following (standard) Sobolev function spaces

$$\mathcal{H}^s(\mathbb{T}^3) = \left\{ u \in H^s(\mathbb{T}^3) \mid \operatorname{div} u = 0, \int_{\mathbb{T}^3} u \, \mathrm{d}x = 0 \right\},$$

where $H^s(\mathbb{T}^3)$ refers to the set of functions whose weak derivatives up to order-s belong to $L^2(\mathbb{T}^3)$. We use the convention that $H^0(\mathbb{T}^3) = L^2(\mathbb{T}^3)$. We endow $\mathcal{H}^s(\mathbb{T}^3)$ with the $H^s(\mathbb{T}^3)$-scalar product. Note that $\mathcal{H}^s(\mathbb{T}^3)$ is a separable Hilbert space for every $s \in \mathbb{N}$, as it is a closed subspace of the separable Hilbert space $H^s(\mathbb{T}^3)$. With this notation at hand, we now better specify W and σ.

Assumption 1 *W is a Q-Wiener process with values in $\mathcal{H}^s(\mathbb{T}^3)$ for a trace-class operator $Q : \mathcal{H}^s(\mathbb{T}^3) \to \mathcal{H}^s(\mathbb{T}^3)$. Let $W = \sum_{k\geq 1} \sqrt{\lambda_k} w_k W^k$, where $\{w_k\}_{k\in\mathbb{N}}$ is an orthonormal basis of $\mathcal{H}^s(\mathbb{T}^3)$ made of eigenvectors of Q, $\{\lambda_k\}_{k\in\mathbb{N}}$ is the associated sequence of non-negative eigenvalues and $\{W^k\}_{k\in\mathbb{N}}$ are independent one dimensional Brownian motions defined on the filtered probability space $(\Omega, \mathcal{A}, (\mathcal{F}_t)_t, \mathbb{P})$. We require*

$$\sum_{k=1}^{\infty} \lambda_k \|w_k\|^2_{H^{s+2}(\mathbb{T}^3)} < \infty.$$

Assumption 2 *The diffusion coefficient $\sigma : \mathcal{H}^s(\mathbb{T}^3) \to L(\mathcal{H}^s(\mathbb{T}^3), \mathcal{H}^s(\mathbb{T}^3))$ has the form*

$$\sigma(u) = c_N \psi(\|u\|_{H^s}) \|u\|^{\eta}_{H^s} \left(I - \alpha \frac{u \otimes u}{\|u\|^2_{H^s}} \right) \tag{3}$$

for $\eta > 3/2$, $c_N > 0$ (or alternatively $\eta = 3/2$, c_N large enough) and α satisfies

$$1 < \alpha < 1 + \frac{1}{\eta - 1} \frac{\operatorname{Tr} Q - \|Q\|}{\|Q\|}. \tag{4}$$

In (3), $\psi : \mathbb{R} \to [0, 1]$ is a smooth even function such that, for some $R > 0$, $\psi(s) \equiv 0$ for $|s| \leq R/2$ and $\psi(s) \equiv 1$ for $|s| \geq R$, while, in (4), $\|Q\|$ denotes the operational norm of Q as a map from $\mathcal{H}^s(\mathbb{T}^3)$ into itself.

Recall the Leray projector $\mathcal{P}$, which is the orthogonal projection of $L^2(\mathbb{T}^3)$ onto its closed subspace $\mathcal{H}^0(\mathbb{T}^3)$. Equivalently, for every $v \in L^2(\mathbb{T}^3)$, we have $\mathcal{P}v =$

$v - \mathcal{Q}v$, where $\mathcal{Q}v = -\nabla w$, and w is uniquely defined, up to an additive constant, by the elliptic Neumann problem

$$-\Delta w = \operatorname{div} v \qquad x \in D.$$

Remark 1 Crucially for this work, due to periodic boundary conditions on the torus, if v is with zero mean and such that $\operatorname{div} v = 0$, then $\mathcal{P}v = v$.

Applying $\mathcal{P}$ to the first equation in (2), we can eliminate the pressure gradient obtaining

$$\begin{cases} du = \mathcal{P}\left((-u \cdot \nabla)u + \nu \Delta u\right) dt + \sigma(u) \circ dW_t & (t, x) \in [0, T] \times \mathbb{T}^3, \\ u(0, x) = u_0(x) & x \in \mathbb{T}^3. \end{cases} \tag{5}$$

Note that applying $\mathcal{P}$ is not restrictive, since one can retrieve the pressure p a posteriori, starting from a solution u to (5), by solving an elliptic problem.

The main result is the following.

Theorem 1 *Let $s \in \mathbb{N}$ and $s \geq 4$. Under the Assumptions 1 and 2, for every deterministic initial condition $u_0 \in \mathcal{H}^{s+2}(\mathbb{T}^3)$, the stochastic Navier-Stokes equation* (5) *has a unique, global-in-time, strong solution in $L^\infty([0, T]; \mathcal{H}^{s+2}(\mathbb{T}^3)) \cap C([0, T]; \mathcal{H}^s(\mathbb{T}^3))$, $\mathbb{P}$-a.s., for all $T > 0$.*

Remark 2 As can be better seen in Sect. 3, the proof of Theorem 1 makes heavy use of the periodic domain to avoid boundary terms appearing when integrating by parts the laplacian. However, in a general domain D, we expect that a more refined argument, which exploits the regularity coming from the laplacian term, should be able to provide existence and uniqueness of solutions in $L^2([0, T]; \mathcal{H}^2(D)) \cap C([0, T]; \mathcal{H}^1(D))$, $\mathbb{P}$-a.s., for all $T > 0$, given an initial datum $u_0 \in \mathcal{H}^2(D)$.

3 Proof of the Main Result

The proof of Theorem 1 largely exploits the framework developed in [2]. For the reader's convenience, we mention here their setting and main result. They consider three separable Hilbert space E_1, E_0 and E_{-1}, such that E_1 is compactly embedded in E_0 and E_0 is continuously embedded in E_{-1}. Then, they address the abstract stochastic differential equation,

$$dX_t = b(X_t)dt + \sigma(X_t) \circ dW_t, \tag{6}$$

where $b : E_0 \to E_{-1}$ is a Borel function, $\sigma : E_0 \to L(E_0, E_0)$ is a C^1 function and W is a Q-Wiener process on $(\Omega, \mathcal{A}, (\mathcal{F}_t)_t, \mathbb{P})$ with values in E_0, for a given trace-class operator $Q : E_0 \to E_0$. We now present their set of assumption.

Assumption 3 (On the Wiener Process) *This assumption is the analogous of Assumption 1. It sufficient to replace* $\mathcal{H}^s(\mathbb{T}^3)$ *with* E_0 *and* $\mathcal{H}^{s+2}(\mathbb{T}^3)$ *with* E_1.

Assumption 4 (On the Spaces) *For the separable Hilbert spaces* E_{-1}, E_0, E_1 *introduced before, the compact embedding* $E_1 \subset\subset E_0$ *and the continuous embedding* $E_0 \hookrightarrow E_{-1}$ *are both dense. Moreover, there exist* $\theta \in (0,1)$ *and* $M > 0$ *such that* $\|v\|_{E_0} \leq M\|v\|_{E_1}^{1-\theta}\|v\|_{E_{-1}}^{\theta}$ *for all* $v \in E_1$.

Assumption 5 (On the Galerkin Basis) *There exists a sequence of elements* $\{e_n\}_{n\in\mathbb{N}_+}$ *in* E_1 *whose span is dense in* E_{-1} *such that, if* $\Pi_n : E_{-1} \to E^n$ *denotes the orthogonal projector from* E_{-1} *to its subspace* $E^n := span\{e_1, \dots e_n\}$, *then for all* $x \in E_1$

$$\|\Pi_n x\|_{E_1} \leq C\|x\|_{E_1},$$

for a constant $C > 0$ *independent of* $n \in \mathbb{N}_+$ *and* x.

Assumption 6 (On the Drift Coefficient) *The drift coefficient* b *satisfies the following:*

(a) $b : E_0 \to E_{-1}$ *is continuous and bounded on balls and* $b(E_1) \subseteq E_0$.
(b) There exist $R \geq 1$, $m > 2$ *and a non-decreasing* $g : \mathbb{R} \to \mathbb{R}$, *such that*

$$\begin{cases} \langle \Pi_n b(x), x\rangle_{E_1} \leq C + g(\|x\|_{E_0})\|x\|_{E_1}^2 & \forall x \in E^n,\ \forall n \in \mathbb{N}, \\ \langle b(x), x\rangle_{E_0} \leq g(\|x\|_{E_0})\|x\|_{E_0}^2 & \forall x \in B_{R;E_0}^c \cap E_1, \\ g(\|x\|_{E_0}) \leq c_D\|x\|_{E_0}^{m-2} & \forall x \in B_{R;E_0}^c \cap E_1. \end{cases}$$

(c) The projected drift $\Pi_n b : E^n \to E^n$ *is locally Lipschitz for every* $n \in \mathbb{N}$.
(d) There exists an increasing function $k : \mathbb{R} \to \mathbb{R}_+$ *such that*

$$\langle b(x) - b(y), x - y\rangle_{E_0} \leqslant k(\|x\|_{E_1} \vee \|y\|_{E_1})\|x - y\|_{E_0}^2$$

for all $x, y \in E_0 \cap E_1$.

Assumption 7 (On the Diffusion Coefficient) *This assumption is the analogous of Assumption 2. It is sufficient to replace* $\mathcal{H}^s(\mathbb{T}^3)$ *with* E_0.

The following result can be found in [2, Theorem 2.12].

Theorem 2 *Under Assumptions 3, 4, 5, 6 and 7, let in* (3) $\eta > m/2$, $c_N > 0$ *(or alternatively* $\eta = m/2$, c_N *large enough) and* α *such that*

$$1 < \alpha < 1 + \frac{1}{\eta - 1}\frac{\operatorname{Tr} Q - \|Q\|}{\|Q\|}. \tag{7}$$

Then, for every deterministic initial condition $x_0 \in E_1$*, there exists a unique, global-in-time, strong solution of* (6) *in* $L^\infty([0,T]; E_1) \cap C([0,T]; E_0)$, $\mathbb{P}$*-a.s., for all* $T > 0$.

Theorem 2 shows that there exists a sufficiently strong noise able to prevent the blow-up in a fairly large class of SPDEs. Its proof is based on the Lyapunov function method, in particular by choosing as Lyapunov function a smooth approximation of $\log \|X_t\|_{E_0}$. The feature of such Lyapunov function is its negative concavity along the radial direction, resulting in a negative sign in front of the second order term appearing in Itô's formula. In such a way, a fast-growing noise is able to compete against a super-linear drift, avoiding the blow-up possibly induced by such drift.

With this result in mind, we are left to show that the stochastic Navier-Stokes equation (5) can be seen as an instance of this framework. In such a way, Theorem 1 will be a corollary of the abstract Theorem 2.

We start by choosing $E_1 = \mathcal{H}^{s+2}(\mathbb{T}^3)$, $E_0 = \mathcal{H}^s(\mathbb{T}^3)$ and $E_{-1} = \mathcal{H}^{s-2}(\mathbb{T}^3)$ for $s \geq 4$. The following (classical) lemma shows that Assumption 4 is straightforwardly satisfied for this choice of the spaces E_1, E_0 and E_{-1}.

Lemma 1 *Let* $s \in \mathbb{N}$ *and* $s \geq 2$*. Then,* $\mathcal{H}^{s+2}(\mathbb{T}^3)$ *is compactly embedded in* $\mathcal{H}^s(\mathbb{T}^3)$*, and* $\mathcal{H}^s(\mathbb{T}^3)$ *is continuously embedded in* $\mathcal{H}^{s-2}(\mathbb{T}^3)$*. Both embeddings are dense. Moreover, we have*

$$\|v\|_{H^s(\mathbb{T}^3)} \leq \|v\|^{\frac{1}{2}}_{H^{s+2}(\mathbb{T}^3)} \|v\|^{\frac{1}{2}}_{H^{s-2}(\mathbb{T}^3)}, \qquad \forall v \in \mathcal{H}^{s+2}(\mathbb{T}^3).$$

Proof The proof is completely analogous to [3, Lemma 5.4]. □

The next step is to provide a suitable Galerkin basis. Let $\mathbb{Z}^3_0 := \mathbb{Z}^3 \setminus \{0\}$ be the non-zero lattice points and let $\{b_k\}_{k \in \mathbb{Z}^3_0}$ be the usual trigonometric basis of $L^2(\mathbb{T}^3, \mathbb{R})$. Let A^1_k and A^2_k be unitary vectors such that $\{\frac{k}{|k|}, A^1_k, A^2_k\}$ is an orthonormal basis of $\mathbb{R}^3$ and define

$$e^i_k(x) := A^i_k b_k(x), \qquad x \in \mathbb{T}^3,\ k \in \mathbb{Z}^3_0.$$

Then $B = \{e^i_k : k \in \mathbb{Z}^3_0,\ i \in \{1,2\}\}$ is an orthonormal basis of $\mathcal{H}^0(\mathbb{T}^3)$, which is made of vectors orthogonal in $\mathcal{H}^m(\mathbb{T}^3)$ for every $m \geq 0$. We define

$$E^n := \operatorname{span}\left\{e^i_k : k \in \mathbb{Z}^3_0,\ |k| \leq n,\ i \in \{1,2\}\right\}$$

to be the finite-dimensional vector space suitable for Galerkin approximations. Moreover, we also define Π_n to be the orthogonal projection from $\mathcal{H}^{s-2}(\mathbb{T}^3)$ into E^n. Thanks to the orthogonality property of the elements of B with respect to all scalar products in $\mathcal{H}^m(\mathbb{T}^3)$ for every $m \geq 0$, it follows that Π_n is an orthogonal projection with respect to all Sobolev scalar products, hence

$$\|\Pi_n x\|_{\mathcal{H}^m} \leq \|x\|_{\mathcal{H}^m}.$$

In particular, Assumption 4 is satisfied. We are only left to show the crucial assumption, i.e. the one concerning the properties of the drift coefficient. For the stochastic Navier-Stokes equation under investigation the drift coefficient takes the form

$$\begin{aligned} b(u) &= -\mathcal{P}(u \cdot \nabla u) + \nu \mathcal{P} \Delta u \\ > &= -\mathcal{P}(u \cdot \nabla u) + \nu \Delta u, \end{aligned}$$

where we exploited Remark 1 to get rid of the Leray projector in front of the laplacian. We recall a few classical estimates (see, for instance, [6]). Let $m \in \mathbb{N}$ be such that $m > d/2$, where d is the dimension of the domain ($d = 3$ in our case). Then,

- (Moser estimate) $\|uv\|_{H^m} \leq C(\|u\|_{L^\infty}\|v\|_{H^m} + \|u\|_{H^m}\|v\|_{L^\infty})$,
- (Sobolev embedding) $\|u\|_{L^\infty} \leqslant C\|u\|_{H^m}$ and $\|u\|_{W^{1,\infty}} \leqslant C\|u\|_{H^{m+1}}$.

Hence, for all integers m such that $m > d/2$, we infer

$$\|(u \cdot \nabla)v\|_{H^m} \leq C(\|u\|_{L^\infty}\|v\|_{H^{m+1}} + \|u\|_{H^m}\|v\|_{W^{1,\infty}}) \leq C\|u\|_{H^{m+1}}\|v\|_{H^{m+1}} \tag{8}$$

whenever $u, v \in \mathcal{H}^{m+1}(\mathbb{T}^3)$. Moreover, if $m > d/2+1$, then by [6, Lemma 2.1 (c)],

$$\langle \mathcal{P}[(u \cdot \nabla)v], v\rangle_{H^m} \leq C(\|u\|_{W^{1,\infty}}\|v\|_{H^m} + \|u\|_{H^m}\|v\|_{W^{1,\infty}})\|v\|_{H^m}, \tag{9}$$

when $u \in \mathcal{H}^m(\mathbb{T}^3)$ and $v \in \mathcal{H}^{m+1}(\mathbb{T}^3)$. We proceed checking Assumption 6 item by item. Assumption 6-(a) follows from the continuity of the Leray projector, estimate (8) and a simple triangle inequality, provided that $s \geq 4$. We move to Assumption 6-(b). First, notice that thanks to the orthogonality property of Π_n with respect to all Sobolev scalar products

$$\langle \Pi_n b(u), u\rangle_{H^{s+2}} = \langle b(u), u\rangle_{H^{s+2}}.$$

Recalling estimate (9)

$$\langle \mathcal{P}(u \cdot \nabla u), u\rangle_{H^{s+2}} \leq C\|u\|_{W^{1,\infty}}\|u\|^2_{H^{s+2}}, \tag{10}$$

and, exploiting integration by parts,

$$\begin{aligned}\langle \Delta u, u\rangle_{H^{s+2}} &= \sum_{|\alpha|\leq s+2}\sum_{i=1}^{3}\int_{\mathbb{T}^3} \partial^\alpha \partial_i \partial_i u \partial^\alpha u\, dx \\ > &= -\sum_{|\alpha|\leq s+2}\sum_{i=1}^{3}\int_{\mathbb{T}^3} \partial^\alpha \partial_i u \partial^\alpha \partial_i u\, dx = -\|\nabla u\|_{H^{s+2}}.\end{aligned} \tag{11}$$

Notice that in (11) we crucially exploited the periodic boundary conditions in the torus to neglect the boundary term in integration by parts. In particular, combining (10) and (11), we obtain, for $u \in E^n$,

$$\langle \Pi_n b(u), u\rangle_{H^{s+2}} \leq C\|u\|_{W^{1,\infty}}\|u\|^2_{H^{s+2}}$$

and similarly, for $u \in \mathcal{H}^{s+2}$,

$$\langle b(u), u\rangle_{H^s} \leq C\|u\|_{W^{1,\infty}}\|u\|^2_{H^s}.$$

Hence, exploiting the Sobolev embedding of $H^s(\mathbb{T}^3)$ into $W^{1,\infty}(\mathbb{T}^3)$, Assumption 6-(b) is met with

$$g(\|u\|_{H^s}) = c_D\|u\|_{H^s}, \tag{12}$$

for some $c_D > 0$, so that we can choose $m = 3$.

Assumption 6-(c) holds since $\Pi_n b$ is the sum of a linear and a quadratic function in a finite dimensional space.

Finally, utilizing (8), (9), the properties of $\mathcal{P}$ and the Sobolev embeddings, we have

$$\begin{aligned}&\langle \mathcal{P}[v\cdot\nabla v - w\cdot\nabla w], v-w\rangle_{H^s} \\ &\quad= \langle \mathcal{P}[v\cdot\nabla(v-w)], v-w\rangle_{H^s} + \langle \mathcal{P}[(v-w)\cdot\nabla w], v-w\rangle_{H^s} \\ &\quad\leqslant C(\|v\|_{W^{1,\infty}}\|v-w\|_{H^s} + \|v\|_{H^s}\|v-w\|_{W^{1,\infty}})\|v-w\|_{H^s} \\ &\qquad+ C(\|v-w\|_{L^\infty}\|w\|_{H^{s+1}} + \|v-w\|_{H^s}\|w\|_{W^{1,\infty}})\|v-w\|_{H^s} \\ &\quad\leqslant C(\|v\|_{H^s} + \|w\|_{H^{s+1}})\|v-w\|^2_{H^s},\end{aligned}$$

and

$$\langle \nu\Delta(u-v), u-v\rangle_{H^s} = -\nu\|\nabla(u-v)\|_{H^s} \leq 0,$$

which fulfils Assumption 6-(d). This conclude the proof.

References

1. Albritton, D., Brué, E., Colombo, M.: Non-uniqueness of Leray solutions of the forced Navier-Stokes equations. Ann. Math. (2) **196**(1), 415–455 (2022). https://doi.org/10.4007/annals.2022.196.1.3
2. Bagnara, M.: A suitable nonlinear Stratonovich noise prevents blow-up in the Euler equations and other SPDEs. Preprint (2023). arXiv:2312.10446
3. Bagnara, M., Maurelli, M., Xu, F.: No blow-up by nonlinear Itô noise for the Euler equations. Preprint (2023). arXiv:2305.09852
4. Buckmaster, T., Vicol, V.: Nonuniqueness of weak solutions to the Navier-Stokes equation. Ann. Math. (2) **189**(1), 101–144 (2019). https://doi.org/10.4007/annals.2019.189.1.3
5. Flandoli, F.: Random perturbation of PDEs and fluid dynamic models. Lecture Notes in Mathematics, vol. 2015. Springer, Heidelberg (2011). https://doi.org/10.1007/978-3-642-18231-0. Lectures from the 40th Probability Summer School held in Saint-Flour, 2010, École d'Été de Probabilités de Saint-Flour. [Saint-Flour Probability Summer School]
6. Glatt-Holtz, N.E., Vicol, V.C.: Local and global existence of smooth solutions for the stochastic Euler equations with multiplicative noise. Ann. Probab. **42**(1), 80–145 (2014). https://doi.org/10.1214/12-AOP773
7. Hofmanová, M., Zhu, R., Zhu, X.: Global existence and non-uniqueness for 3D Navier-Stokes equations with space-time white noise. Arch. Ration. Mech. Anal. **247**(3), Paper No. 46, 70 (2023). https://doi.org/10.1007/s00205-023-01872-x
8. Hopf, E.: Über die Anfangswertaufgabe für die hydrodynamischen Grundgleichungen. Math. Nachr. **4**, 213–231 (1951). https://doi.org/10.1002/mana.3210040121
9. Kato, T., Ponce, G.: Commutator estimates and the Euler and Navier-Stokes equations. Commun. Pure Appl. Math. **41**(7), 891–907 (1988). https://doi.org/10.1002/cpa.3160410704
10. Leray, J.: Sur le mouvement d'un liquide visqueux emplissant l'espace. Acta Math. **63**(1), 193–248 (1934). https://doi.org/10.1007/BF02547354
11. Tang, H., Wang, F.Y.: A General Framework for Solving Singular SPDEs with Applications to Fluid Models Driven by Pseudo-differential Noise. Preprint (2022). arXiv:2208.08312

Part IX
Wavelet Theory and Its Related Topics

On Characterization of the Gabor Wavelet Transform of Analytic Functionals

Keiko Fujita

Abstract The wavelet transformation is usually considered for the square integrable functions. Since the Gabor function is of exponential type, we can consider the Gabor wavelet transformation to the analytic functionals. As the Gabor function is Gaussian type, the Gabor wavelet transformation is similar to the windowed Fourier transformation. That is, the Gabor wavelet transformation is closely related to the Fourier transformation. Therefore our previous results on Fourier transformation are useful when we consider the Gabor transformation. In this paper, we will review our previous results and will consider the relationship between the Fourier transform and the Gabor wavelet transform of analytic functional on the sphere based on our previous results.

1 Introduction

For $z = (z_1, z_2, \cdots, z_{n+1}) \in \mathbf{C}^{n+1}$ and $w = (w_1, w_2, \cdots, w_{n+1}) \in \mathbf{C}^{n+1}$, we set

$$z \cdot w = z_1 w_1 + \cdots + z_{n+1} w_{n+1}, \quad z^2 = z \cdot z, \ \|z\|^2 = z \cdot \overline{z}.$$

The Fourier transform of an integrable function f on $\mathbf{R}^{n+1}$ is defined by

$$(\mathcal{F}f)_{\mathbf{R}^{n+1}}(\omega) = \int_{R^{n+1}} e^{-ix\cdot\omega}\overline{f(x)}dx,$$

and the windowed Fourier transform with respect to the window function w is defined by

$$(\mathcal{W}_w f)(\tau, \omega) = \int_{R^{n+1}} w(x-\tau)e^{-ix\cdot\omega}\overline{f(x)}dx.$$

K. Fujita (✉)
University of Toyama, Toyama, Japan
e-mail: keiko@sci.u-toyama.ac.jp

M. R. Ebert et al. (eds.), *New Tools in Mathematical Analysis and Applications*, Research Perspectives, https://doi.org/10.1007/978-3-031-77050-0_26

Let $\omega_0 \in \mathbf{C}^{n+1} \setminus \{0\}$ be fixed. Put

$$G_{\omega_0}(x) = e^{-x^2/2} e^{-ix\cdot\omega_0}.$$

For $a \in \mathbf{R}_+ = \{x : x > 0\}$, the Gabor wavelet transform of f is defined by

$$(\mathcal{G}_{w_0} f)_{\mathbf{R}^{n+1}}(\tau, a) = \frac{1}{a} \int_{\mathbf{R}^{n+1}} G_{\omega_0}\left(\frac{x-\tau}{a}\right) \overline{f(x)} dx, \quad \tau \in \mathbf{C}^{n+1}.$$

Similarly, for an integrable function f on $S^n = \{x \in \mathbf{R}^{n+1}; x^2 = 1\}$, the Fourier transform of f is defined by

$$(\mathcal{F} f)(\omega) = \int_{S^n} e^{-ix\cdot\omega} \overline{f(x)} d\Omega(x),$$

where $d\Omega$ is the normalized invariant measure on S^n.

For the square integrable functions f and g on S^n, we define a sesquilinear form $(f, g)_{S^n}$ by

$$(f, g)_{S^n} \equiv \int_{S^n} f(x)\overline{g(x)} d\Omega.$$

Then $(\cdot, \cdot)_{S^n}$ gives an inner product. We denote by $L^2(S^n)$ the space of square integrable functions on S^n with the inner product $(\cdot, \cdot)_{S^n}$.

We denote by $\mathcal{A}(S^n)$ the space of real analytic functions on S^n, and by $\mathcal{A}'(S^n)$ the space of analytic functionals (or hyperfunctions) on S^n. Let $\langle \cdot, \cdot \rangle$ be the canonical bilinear form of duality on $\mathcal{A}'(S^n) \times \mathcal{A}(S^n)$. Since $\mathcal{A}(S^n) \subset L^2(S^n) \subset \mathcal{A}'(S^n)$, for $g \in L^2(S^n)$ we can define $T_g \in \mathcal{A}'(S^n)$ by

$$\langle T_g, f \rangle = (f, g)_{S^n} = \int_{S^n} f(x)\overline{g(x)} d\Omega, \quad f \in \mathcal{A}(S^n).$$

Note that the mapping $g \mapsto T_g$ is a continuous antilinear injection. Since the exponential function $\exp(-ix \cdot \omega)$ is an analytic function in x or in ω, for $T \in \mathcal{A}'(S^n)$, we can define the Fourier transform of T by

$$(\mathcal{F}T)(\omega) = \langle T_x, \exp(-ix \cdot \omega) \rangle, \quad \omega \in C^{n+1}.$$

We call the mapping

$$\mathcal{F} : T \mapsto (\mathcal{F}T)(w) = \langle T_x, \exp(-ix \cdot \omega) \rangle, \tag{1}$$

the Fourier transformation.

In Sect. 2, we will consider the relationship between the Fourier transform and the Gabor wavelet transform. In Sect. 3, we will consider the expansion formula of the Gabor wavelet transform. At the last section, we remark on the inverse Fourier transformation.

2 Gabor Wavelet Transformation on the Sphere

In this section, first we will consider the windowed Fourier transformation on the sphere and consider the relationship between the windowed Fourier transform and the Fourier transform. Then we will consider the Gabor wavelet transformation on the sphere and consider the relationship between the Gabor wavelet transform and the Fourier transform.

2.1 Windowed Fourier Transformation on the Sphere

Let $b \in \mathbf{R} \setminus \{0\}$ and put

$$w_b(x) = e^{-x^2/(2b^2)} \equiv \exp(-x^2/(2b^2)).$$

For $f \in \mathcal{A}'(S^n)$, we define the windowed Fourier transformation $\mathcal{W}_b\mathcal{F}$ with the window function $w_b(x)$ by

$$\mathcal{W}_b\mathcal{F} : f \mapsto (\mathcal{W}_b\mathcal{F}f)(\tau, \omega) = \langle f, w_b(x-\tau)e^{-ix\cdot\omega}\rangle \quad \omega, \tau \in \mathbf{C}^{n+1}. \tag{2}$$

For $x \in S^n$ we have

$$e^{-ix\cdot\omega} w_b(x-\tau) = e^{\frac{-1-\tau^2}{2b^2}} e^{-ix\cdot\frac{\omega+i\tau}{b^2}} = e^{\frac{-1-\tau^2}{2b^2}} e^{-ix\cdot\omega} e^{x\cdot\tau/b^2}.$$

Thus we have

$$(\mathcal{W}_b\mathcal{F}f)(\tau, \omega) = e^{-\frac{1+\tau^2}{2b^2}} (\mathcal{F}f)(\omega + i\tau/b^2) = e^{-\frac{1+\tau^2}{2b^2}} (\mathcal{F}g)(\omega), \tag{3}$$

where we put $g(x) = \overline{e^{x\cdot\tau/b^2}} f(x)$. Therefore, the windowed Fourier transform of $f \in \mathcal{A}'(S^n)$ can be expressed by means of the Fourier transform of f or g. It means that the windowed Fourier transformation is decomposed into the mapping $S_1 : f \mapsto \overline{e^{x\cdot\tau/b^2}} f(x)$, $\mathcal{F}$ and the mapping $S_2 : (\mathcal{F}(S_1 f)) \mapsto e^{-\frac{1+\tau^2}{2b^2}} (\mathcal{F}(S_1 f))$; that is, $\mathcal{W}_b\mathcal{F} = S_2 \circ \mathcal{F} \circ S_1$.

Since S_1 and S_2 are the multiplication operators and we know the inverse mapping of the Fourier transformation, we can construct the inverse mapping of the windowed Fourier transformation.

2.2 Gabor Wavelet Transformation on the Sphere

For $f \in \mathcal{A}'(S^n)$ and $a \in \mathbf{R}_+ = \{x : x > 0\}$, similar to the case of $\mathbf{R}^{n+1}$, we define the Gabor wavelet transformation $\mathcal{G}_{\omega_0}$ on S^n by

$$\mathcal{G}_{\omega_0} : f \mapsto (\mathcal{G}_{\omega_0} f)(\tau, a) = \frac{1}{a}\left\langle f_x, G_{\omega_0}\left(\frac{x-\tau}{a}\right)\right\rangle, \quad \tau \in \mathbf{C}^{n+1}. \tag{4}$$

For $x \in S^n$ we have

$$G_{\omega_0}\left(\frac{x-\tau}{a}\right) = e^{-\frac{1+\tau^2}{2a^2}} e^{i\tau\cdot\frac{\omega_0}{a}} e^{-ix\cdot(\frac{\omega_0}{a}+i\frac{\tau}{a^2})} = e^{-\frac{1+\tau^2}{2a^2}} e^{i\tau\cdot\frac{\omega_0}{a}} e^{-ix\cdot\frac{\omega_0}{a}} e^{\frac{x\cdot\tau}{a^2}}.$$

Thus we have

$$(\mathcal{G}_{\omega_0} f)(\tau, a) = \frac{1}{a} e^{-\frac{1+\tau^2}{2a^2}} e^{i\tau\cdot\frac{\omega_0}{a}} (\mathcal{F} f)\left(\frac{\omega_0}{a} + i\frac{\tau}{a^2}\right) \tag{5}$$

$$= \frac{1}{a} e^{-\frac{1+\tau^2}{2a^2}} e^{i\tau\cdot\frac{\omega_0}{a}} (\mathcal{F} g)\left(\frac{\omega_0}{a}\right), \tag{6}$$

where we put $g(x) = \overline{\exp(x\cdot\tau/a^2)} f(x)$.

Since we constructed the inverse mapping of the Fourier transformation, by (5) or (6), we can construct the inverse mapping of the Gabor wavelet transformation.

3 Expansion Formula of the Gabor Wavelet Transform

In this section, first we will recall some notations to consider the Gabor wavelet transform of the analytic functionals. Next, we will review the expansion formula.

3.1 Spherical Harmonics Expansion

We will represent the images of (1), (2) and (4) by means of the infinite series of the Legendre polynomials and the Bessel polynomials. We recall some notations.

Let $P_{k,n}(t)$ be the Legendre polynomial of degree k and of dimension $n+1$:

$$P_{k,n}(t) = \left(\frac{-1}{2}\right)^2 \frac{\Gamma(\frac{n}{2})}{\Gamma(k+\frac{n}{2})}(1-t^2)^{\frac{2-n}{2}} \frac{d^k}{dt^k}(1-t^2)^{k+\frac{n-2}{2}}.$$

We define the extended Legendre polynomial by

$$P_{k,n}(z, w) = (\sqrt{z^2})^k (\sqrt{w^2})^k P_{k,n}\left(\frac{z}{\sqrt{z^2}} \cdot \frac{w}{\sqrt{w^2}}\right), \quad z, w \in \mathbf{C}^{n+1}. \tag{7}$$

Then $P_{k,n}(z, w)$ is a homogeneous harmonic polynomial of degree k in z and in w. That is, $\Delta_z P_{k,n}(z, w) \equiv (\frac{\partial^2}{\partial z_1^2} + \cdots + \frac{\partial^2}{\partial z_{n+1}^2}) P_{k,n}(z, w) = 0$. Note that $P_{k,n}(z, w) = P_{k,n}(w, z)$. We denote by $N(k, n)$ the dimension of the space of homogeneous harmonic polynomials of degree k. We know

$$N(k, n) = \frac{(2k + n - 1)(k + n - 2)!}{k!(n - 1)!} .$$

By the orthogonality of the Legendre polynomials, we have

$$N(k, n) \int_{S^n} P_{k,n}(x, z) P_{l,n}(x, w) d\Omega(x) = \delta_{kl} P_{k,n}(z, w) . \tag{8}$$

For $\nu \neq -1, -2, \cdots$, let

$$J_\nu(t) = \left(\frac{t}{2}\right)^\nu \sum_{l=0}^{\infty} \frac{1}{l!\Gamma(\nu + l + 1)} \left(\frac{it}{2}\right)^{2l}$$

be the Bessel function of order ν. We put

$$\begin{aligned} \tilde{j}_k(t) &= \Gamma\left(k + \frac{n+1}{2}\right) \left(\frac{t}{2}\right)^{-(k + \frac{n-1}{2})} J_{k + \frac{n-1}{2}}(t) \\ &= \sum_{l=0}^{\infty} \frac{\Gamma(k + (n + 1)/2)}{l!\Gamma(k + l + (n + 1)/2)} \left(\frac{it}{2}\right)^{2l} . \end{aligned}$$

Note that $\tilde{j}_k(0) = 1$ and $\tilde{j}_k(-t) = \tilde{j}_k(t)$. (See Lemma 5.13 in [5]).

By using the extended Legendre polynomials and the modified Bessel functions, the exponential function is represented as follows;

$$\exp(z \cdot w) = \sum_{k=0}^{\infty} \frac{\Gamma(\frac{n+1}{2}) N(k, n)}{2^k \Gamma(k + \frac{n+1}{2})} \tilde{j}_k(i\sqrt{z^2}\sqrt{w^2}) P_{k,n}(z, w). \tag{9}$$

Since $\tilde{j}_k(-t) = \tilde{j}_k(t)$, by (7) we have

$$\exp(-iz \cdot w) = \sum_{k=0}^{\infty} \frac{\Gamma(\frac{n+1}{2}) N(k, n)(-i)^k}{2^k \Gamma(k + \frac{n+1}{2})} \tilde{j}_k(\sqrt{z^2}\sqrt{w^2}) P_{k,n}(z, w).$$

3.2 Expansion Formula

For $f \in L^2(S^n)$, define

$$f_k(z) = N(k,n) \int_{S^n} f(y) P_{k,n}(z,y) d\Omega(y), \quad z \in \mathbf{C}^{n+1}.$$

Then for $x \in S^n$, we have $f(x) = \sum_{k=0}^{\infty} f_k(x)$ in the sense of $L^2(S^n)$. Similarly, for $f \in \mathcal{A}'(S^n)$, define

$$f_k(z) = N(k,n) \langle f_y, P_{k,n}(z,y) \rangle, \quad z \in \mathbf{C}^{n+1}.$$

Then for $x \in S^n$, $f(x) = \sum_{k=0}^{\infty} f_k(x)$ in the sense of $\mathcal{A}'(S^n)$.

We consider the images of $f_z(x) = P_{k,n}(x,\overline{z})$, $z \in \mathbf{C}^{n+1}$ under the Fourier transformation, the windowed Fourier transformation and the Gabor wavelet transformation. Note that $\overline{P_{k,n}(z,w)} = P_{k,n}(\overline{z},\overline{w})$. We recall $P_{k,n}(z,w) = P_{k,n}(w,z)$. By (8) and (9), for $x \in S^n$ we have

$$\begin{aligned}(\mathcal{F} f_z)(\omega) &= \int_{S^n} \sum_{l=0}^{\infty} \frac{\Gamma((n+1)/2) N(l,n)(-i)^l}{2^l \Gamma(l+\frac{n+1}{2})} \\ &\quad \times \tilde{j}_l(\sqrt{\omega^2}) P_{l,n}(x,\omega), \overline{P_{k,n}(x,z)} d\Omega(x) \\ &= C_k \tilde{j}_k(\sqrt{\omega^2}) P_{k,n}(\omega,\overline{z}) = C_k \tilde{j}_k(\sqrt{\omega^2}) f_z(\omega).\end{aligned}$$

where we put $C_k = (-i)^k \Gamma((n+1)/2)/(2^k \Gamma(k+(n+1)/2))$. Thus for $f \in \mathcal{A}'(S^n)$, we have

$$(\mathcal{F} f)(\omega) = \sum_{k=0}^{\infty} C_k \tilde{j}_k(\sqrt{\omega^2}) f_k(\omega)$$

in the sense of $\mathcal{A}'(S^n)$. Therefore for $f \in \mathcal{A}'(S^n)$, the windowed Fourier transform is represented by the following infinite series;

$$\begin{aligned}(\mathcal{W}_b \mathcal{F} f)(\tau,\omega) &= e^{-\frac{1+\tau^2}{2b^2}} \sum_{k=0}^{\infty} C_k \tilde{j}_k \left(\sqrt{\left(\omega + i\frac{\tau}{b^2}\right)^2} \right) f_k\left(\omega + i\frac{\tau}{b^2}\right), \\ &= e^{-\frac{1+\tau^2}{2b^2}} \sum_{k=0}^{\infty} C_k \tilde{j}_k \left(\sqrt{\omega^2}\right) g_k(\omega), \quad \tau, \omega \in \mathbf{C}^{n+1},\end{aligned}$$

where $g_k(x) = N(k,n)\langle g_y, P_{k,n}(x,y)\rangle$ for $g(x) = \overline{e^{x\cdot\tau/b^2}} f(x)$, and the Gabor wavelet transform is represented by the following infinite series;

$$(\mathcal{G}_{\omega_0} f)(\tau, a) = \frac{1}{a} e^{-\frac{1+\tau^2}{2a^2}} e^{i\tau\cdot\frac{\omega_0}{a}} \sum_{k=0}^{\infty} C_k \tilde{j}_k \left(\sqrt{\left(\frac{\omega_0}{a} + i\frac{\tau}{a^2}\right)^2} \right) f_k \left(\frac{\omega_0}{a} + i\frac{\tau}{a^2} \right),$$

$$= \frac{1}{a} e^{-\frac{1+\tau^2}{2a^2}} e^{i\tau\cdot\frac{\omega_0}{a}} \sum_{k=0}^{\infty} C_k \tilde{j}_k \left(\sqrt{\left(\frac{\omega_0}{a}\right)^2} \right) h_k \left(\frac{\omega_0}{a} \right),$$

where $h_k(x) = N(k,n)\langle h_y, P_{k,n}(x,y)\rangle$ for $h(x) = \overline{\exp(x\cdot\tau/a^2)} f(x)$.

Appendix

In this section, to construct the inverse Gabor transformation on the sphere, we will recall the inverse Fourier transformation.

In [2], [3] and [4], we treated the inverse Gabor wavelet transformation on S^2. For $z, w \in \mathbf{C}^{n+1}$, put

$$E(z, w) = \sum_{k=0}^{\infty} \frac{1}{2^k k! (-i)^k \tilde{j}_k(\sqrt{z^2}\sqrt{w^2})} P_{k,n}(z, w).$$

For $0 < s < \infty$, let

$$K_\nu(s) = K_{-\nu}(s) = \int_0^\infty \exp(-s\cosh t)\cosh \nu t dt,$$

be the modified Bessel function. Put

$$\rho_r(s) = \begin{cases} \displaystyle\sum_{l=0}^{(n-1)/2} a_l r^{l+n+1} s^{l+1} K_l(rs), & n \text{ is odd}, \\ \displaystyle\sum_{l=0}^{n/2} a_l r^{l+n+1/2} s^{l+1} K_{l-1/2}(rs), & n \text{ is even}, \end{cases}$$

where the constants a_l are defined by

$$\int_0^\infty s^{2l+n-1} \rho_n(s) ds = \frac{N(l,n) l! \Gamma(l + (n+1)/2) 2^{2l}}{\Gamma((n+1)/2)}, \quad l = 0, 1, 2, \cdots.$$

In [1] we define a measure $d\mu$ on $\mathbf{R}^{n+1}$ by

$$\int_{\mathbf{R}^{n+1}} f(x)d\mu(x) = \int_0^\infty \int_{S^n} f(s\omega)d\Omega(\omega)s^{n-1}\rho(s)ds.$$

For $z \in S^n$, the mapping

$$F \mapsto \int_{\mathbf{R}^{n+1}} F(x)\overline{E(z,x)}d\mu(x) = \int_{\mathbf{R}^{n+1}} F(x)E(z,x)d\mu(x) \tag{10}$$

gives the inverse mapping of the Fourier transformation defined by (1).

By (3), (5) and (6), the inverse mapping of the Windowed Fourier transformation (2) and the inverse mapping of the Gabor wavelet transformation (4) can be constructed by using (10).

References

1. Fujita, K.: Hilbert spaces related to harmonic functions. Tôhoku Math. J. **48**, 149–163 (1996). http://doi.org/10.2748/tmj/1178225416
2. Fujita, K.: On an inverse transformation of the Gabor transformation on the sphere. Proceedings of the Seventh International Conference on Information, Information, pp. 45–48 (2015). ISBN 4-901329-08-1
3. Fujita, K.: Gabor transformation on the Sphere and its inverse transformation. New Trends in Analysis and Interdisciplinary Applications. Trends in Mathematics, Proceedings of the 10th ISAAC Congress, pp. 573–580. Springer International Publishing (2017). http://doi.org/10.1007/978-3-319-48812-7_72
4. Fujita, K.: Some topics on the gabor wavelet transformation. Current Trends in Analysis, Its Applications and Computation. Proceedings of the 12th ISAAC Congress, pp. 671–679. Aveiro, Portugal, 2019. Springer Nature Switzerland AG. Part of Springer Nature (2022)
5. Morimoto, M.: Analytic Functionals on the Sphere. Translation of Mathematical Monographs, vol. 178. American Mathematical Society, Providence (1998)

The p-adic Fourier Transform of the Locally Constant Functions

Munehiro Kobayashi and Toshio Suzuki

Abstract p-adic numbers have applications in several areas of mathematics and physics. In this study, we consider the p-adic Fourier transform of the locally constant functions and the test functions. We can obtain the matrix representation of the p-adic Fourier transform and this allows the Fourier transform to be computed efficiently.

1 Introduction

The p-adic numbers were first introduced by Kurt Hensel in 1897. These numbers play a crucial role in various branches of modern mathematics, including algebraic theory and representation theory. In the context of a prime number p, the p-adic numbers serve as an extension of the rational numbers. Notably, they exhibit significant differences from the real numbers while sharing certain properties with them.

A p-adic number is characterized by its representation in digits based on the prime number p, in contrast to the decimal system. Unlike decimal numbers, p-adic numbers extend infinitely in the direction of increasing integer digits rather than toward a decimal point. The p-adic analysis is anticipated to find applications not only in number theory but also in areas such as quantum mechanics and mathematical physics (see [2]).

Let p be a prime number. Every non-zero $x \in \mathbf{Q}$ can be represented by

$$x = p^{\mathrm{val}(x)} \frac{m}{n}, \tag{1}$$

M. Kobayashi
Schilf Institute Co., Ltd., Chiyoda-ku, Japan
e-mail: kobayashi@schilf-institute.co.jp

T. Suzuki (✉)
Kanagawa University, Rokkakubashi, Kanagawa-ku, Yokohama-shi, Kanagawa, Japan
e-mail: tosuzuki@kanagawa-u.ac.jp

M. R. Ebert et al. (eds.), *New Tools in Mathematical Analysis and Applications*,
Research Perspectives, https://doi.org/10.1007/978-3-031-77050-0_27

where $m, n \in \mathbf{Z}$ are not divisible by p and $\mathrm{val}(x)$ is an integer determined uniquely. Then the p-adic norm of $x \in \mathbf{Q}$ in (1) is given by

$$|x|_p = \begin{cases} 0 & (x = 0), \\ p^{-\mathrm{val}(x)} & (x \neq 0). \end{cases}$$

This p-adic norm satisfies the properties of a non-Archimedean norm. The field of real numbers $\mathbf{R}$ is defined by the completion of the field of rational numbers $\mathbf{Q}$ with respect to the Euclidean metric determined by the absolute value $|\cdot|$. On the other hand, the p-adic field $\mathbf{Q}_p$ is obtained by completing the field of rational numbers with respect to the p-adic norm.

Every non-zero p-adic number $x \in \mathbf{Q}_p$ such that $|x|_p = p^{-\mathrm{val}(x)}$ can be uniquely represented in the canonical form

$$x = x_{\mathrm{val}(x)} p^{\mathrm{val}(x)} + x_{\mathrm{val}(x)+1} p^{\mathrm{val}(x)+1} + x_{\mathrm{val}(x)+2} p^{\mathrm{val}(x)+2} + \cdots \tag{2}$$

$$= \begin{cases} [\cdots x_{\mathrm{val}(x)+1} x_{\mathrm{val}(x)} \underbrace{0 \cdots 0}_{\mathrm{val}(x)}]_{\mathbf{Q}_p} & \text{if } \mathrm{val}(x) \geqslant 0, \\ [\cdots x_1 x_0 \,.\, x_{-1} \cdots x_{\mathrm{val}(x)}]_{\mathbf{Q}_p} & \text{otherwise.} \end{cases} \tag{3}$$

where $0 \leqslant x_k \leqslant p-1$, $x_{\mathrm{val}(x)} \neq 0$. We call (2) p-adic expansion of x, and denote the expansion by (3).

For $\gamma \in \mathbf{Z}$, the sphere S_γ and the ball B_γ with radius p^γ and centered on $a \in \mathbf{Q}_p$ are defined as

$$S_\gamma(a) = \{x \in \mathbf{Q}_p \mid |x-a|_p = p^\gamma\}, \quad B_\gamma(a) = \{x \in \mathbf{Q}_p \mid |x-a|_p \leqslant p^\gamma\},$$

respectively. If $a = 0$, then we write $S_\gamma(0) = S_\gamma$, $B_\gamma(0) = B_\gamma$. For $x \in \mathbf{Q}_p$, $\{x\}_p$ denotes the fractional part of $x \in \mathbf{Q}_p$, that is,

$$\{x\}_p = \begin{cases} 0 & \text{if } x \in B_0, \\ p^{\mathrm{val}(x)} \displaystyle\sum_{j=0}^{|\mathrm{val}(x)|-1} x_j p^j & \text{if } x \notin B_0. \end{cases}$$

In this paper, we study the Fourier transform of the complex valued function $f : \mathbf{Q}_p \to \mathbf{C}$. Although the definition of the p-adic Fourier transform is not much different from the ordinary Fourier transform, the properties of the p-adic Fourier theory for such functions seem to be very different from those for functions on $\mathbf{R}$.

2 The p-adic Fourier Transform

Since $\mathbf{Q}_p$ is a locally compact commutative group with respect to addition, there exists the Haar measure dx. This measure dx is normalized by the unit ball B_0, that is,

$$\int_{B_0} dx = 1$$

(see [2]). To consider the p-adic Fourier transform, we introduce the locally constant function and the test function. A complex valued function on $\mathbf{Q}_p$ is called locally constant if for any $x \in \mathbf{Q}_p$ there exists $l \in \mathbf{Z}$ such that,

$$f(x + x') = f(x) \quad \text{for} \quad |x'|_p \leqslant p^l.$$

We call the largest number $l = l(f) \in \mathbf{Z}$ the parameter of constancy of a function f. A locally constant function with compact support is called a test function, and we denote the set of test functions by $\mathcal{D}$. Especially, we denote by $\mathcal{D}_N^l$ the set of locally constant functions with the parameter of constancy l and with support in B_N.

For $f \in \mathcal{D}$, we define the Fourier transform as

$$\hat{f}(\xi) = \int_{\mathbf{Q}_p} f(x)\chi_p(x\xi)\, dx \tag{4}$$

where χ_p is an additive character on $\mathbf{Q}_p$, that is, $\chi_p(x) = \exp(2\pi i\{x\}_p)$. If $f \in \mathcal{D}_N^l$ is a function depending only the p-adic norm $|x|_p$, that is, $f(x) = f(p^{\mathrm{val}(x)})$, then its Fourier transform can be calculated as

$$\hat{f}(\xi) = \int_{\mathbf{Q}_p} f(x)\chi_p(x\xi)\, dx = \sum_{\gamma \in \mathbf{Z}} f(p^\gamma)\lambda(\xi, \gamma)$$

where $\lambda(\xi, \gamma) = \int_{S_{-\gamma}} \chi_p(x\xi)\, dx$ and we can obtain the Fourier transform $\hat{f}$ as the summation of f and λ. From this formula, we have that the Fourier transform of the function which depends only the norm becomes the function which depends only the norm. In particular, various transforms used in time-frequency analysis, including the p-adic wavelet transform, the p-adic window Fourier transform, the p-adic Stockwell transform, etc., are rewritten in the same way as above, using summation (see [1]). In the following, we further compute the p-adic Fourier transform of the test function to see its properties.

3 The Fourier Transform of a Test Function

Let Ω be the characteristic function of B_0 and $\Delta_l = \Omega(p^l x)$, that is, Δ_l represents the characteristic function of B_l. Also, we put $\mathcal{P}_N^l \subset \{0, \cdots, p-1\}^l$ be the set of all sequences of length l with the first N elements being zero. Every test function $f \in \mathcal{D}$ with the parameter of constancy $(-l) \in \mathbf{Z}$ can be represented as

$$f(x) = \sum_{P \in \mathcal{P}_N^l} c_P \Delta_{-l}(x - a_P),$$

where N and l are natural numbers such that $0 \leqslant N < l$ (see [2]). Here, we used coefficients $c_P \in \mathbf{C}$ and the rational numbers $a_P \in \mathbf{Q}$ defined by

$$a_P = [d_{l-1} \cdots d_N \underbrace{0 \cdots 0}_{N}]_{\mathbf{Q}_p}, \quad \text{where } P = (\underbrace{0, \cdots 0}_{N}, d_N, \cdots, d_{l-1}) \in \mathcal{P}_N^l.$$

Then, the support of the function f is the ball B_{-N} and the parameter of constancy is $-l$, that is, $f \in \mathcal{D}_{-N}^{-l}$.

With these settings, we calculate the Fourier transform of f.

$$\begin{aligned}
\hat{f}(\xi) &= \int_{\mathbf{Q}_p} f(x) \chi_p(x\xi) dx \\
&= \int_{\mathbf{Q}_p} \left(\sum_{P \in \mathcal{P}_N^l} c_P \Delta_{-l}(x - a_P) \right) \chi_p(x\xi) dx \\
&= \sum_{P \in \mathcal{P}_N^l} c_P \int_{\mathbf{Q}_p} \Delta_{-l}(x - a_P) \chi_p(x\xi) dx \\
&= \sum_{P \in \mathcal{P}_N^l} c_P \int_{\mathbf{Q}_p} \Delta_{-l}(x) \chi_p(x\xi) \chi_p(a_P \xi) dx \\
&= \sum_{P \in \mathcal{P}_N^l} c_P \chi_p(a_P \xi) \int_{B_{-l}} \chi_p(x\xi) dx \\
&= \sum_{P \in \mathcal{P}_N^l} c_P \chi_p(a_P \xi) p^{-l} \Delta_l(\xi) \\
&= p^{-l} \sum_{P \in \mathcal{P}_N^l} c_P \exp(2\pi i \{a_P \xi\}_p) \Delta_l(\xi)
\end{aligned}$$

We show that the value of the term $\exp(2\pi i\{a_P\xi\}_p)\Delta_l(\xi)$ on the rightmost side depends only on the $-l$-th, $\cdots$, $-(N+1)$-th digits of the p-adic expansion of ξ. First, recall the expansion $a_P = [d_{l-1}\cdots d_N 0\cdots 0]_{\mathbf{Q}_p}$ and the sequence P

$$P = (\underbrace{0,\cdots 0}_{N}, d_N, \cdots, d_{l-1}) \in \mathcal{P}_N^l.$$

Also, if $\Delta_l(\xi) = 1$ holds, we have that the p-adic valuation $\mathrm{val}(\xi)$ is equal to or larger than $-l$, and so the expansion of ξ can be written as $\xi = [\cdots\xi_k\cdots\xi_0 . \xi_{-1}\cdots\xi_{-l}]_{\mathbf{Q}_p}$. Note that we remain symbolically using the same expansion starting from ξ_{-l} even in the case $\mathrm{val}(\xi) > -l$ with $\xi_{-l} = \cdots = \xi_{\mathrm{val}(\xi)-1} = 0$.

Then, observe $a_P = p^N \sum_{i=N}^{l-1} d_i p^{i-N}$ and $\sum_{i=N}^{l-1} d_i p^{i-N} \in B_0$. Further, the representation $\xi = p^{-N}\xi_{\geqslant -N} + p^{-N-1}\sum_{j=-l}^{-N-1} \xi_j p^{j+N+1}$ holds for some $\xi_{\geqslant -N} \in B_0$. Hence, we have

$$\begin{aligned} a_P\xi &= \left(p^N \sum_{i=N}^{l-1} d_i p^{i-N}\right)\left(p^{-N}\xi_{\geqslant -N} + p^{-N-1}\sum_{j=-l}^{-N-1} \xi_j p^{j+N+1}\right) \\ &= \left(\sum_{i=N}^{l-1} d_i p^{i-N}\right)\xi_{\geqslant -N} + p^{-1}\left(\sum_{i=N}^{l-1} d_i p^{i-N}\right)\left(\sum_{j=-l}^{-N-1} \xi_j p^{j+N+1}\right). \end{aligned}$$

Notice that $\left(\sum_{i=N}^{l-1} d_i p^{i-N}\right)\xi_{\geqslant -N} \in B_0$, and so it holds that

$$\{a_P\xi\}_p = \left\{p^{-1}\left(\sum_{i=N}^{l-1} d_i p^{i-N}\right)\left(\sum_{j=-l}^{-N-1} \xi_j p^{j+N+1}\right)\right\}_p.$$

The above shows the value of the term $\{a_P\xi\}_p$ only depends on $\xi_{-l}, \cdots, \xi_{-N-1}$ from the expansion of ξ as intended.

Finally, we obtain a representation to calculate the p-adic discrete Fourier transform of f. For $Q \in \mathcal{P}_N^l$ we put $\xi_Q \in \mathbf{Q}_p$ as

$$\xi_Q = [\cdots 0 . \underbrace{0\cdots 0}_{N} d'_{-N-1}\cdots d'_{-l}]_{\mathbf{Q}_p}, \text{ where } Q = (\underbrace{0,\cdots,0}_{N}, d'_{-N-1}, \cdots, d'_{-l}).$$

For any $\xi \in \mathbf{Q}_p$, the value $\hat{f}(\xi)$ equals $\hat{f}(\xi_Q)$ for some $Q \in \mathcal{P}_N^l$. Let $\hat{c}_Q$ be $\hat{f}(\xi_Q)$, and we have the representation

$$\begin{aligned} \hat{c}_Q &= p^{-l}\sum_{P\in\mathcal{P}_N^l} c_P \exp(2\pi i\{a_P\xi_Q\}_p)\Delta_l(\xi_Q) \\ &= p^{-l}\sum_{P\in\mathcal{P}_N^l} c_P \exp(2\pi i a_P\xi_Q). \end{aligned}$$

Recall that c_P is a complex number and both a_P and ξ_Q are rational. Furthermore, the representation shows $\hat{f}$ has the support B_l and the parameter of constancy is N, that is, $\hat{f} \in \mathcal{D}_l^N$.

From the point of view of computation algorithms, the obtained representation can be calculated similarly by the discrete Fourier transform on a sequence of complex numbers of length p^{l-N}. Especially, in the case of the base $p = 2$, the above representation can be calculated using the fast Fourier transform with the computational complexity of order $O\left((l-N)2^{l-N}\right)$. In the case $p \neq 2$, the Cooley-Tukey algorithm can be used to calculate with the complexity of order $O\left((l-N)2^{l-N}\right)$ as well.

4 The Fourier Transform of a p-adic Radial Locally Constant Function

For the characteristic function Ω, we put $\Pi(x) = \Omega(x) - \Omega(px)$. That is, Π is the characteristic function of S_0. For integers $0 \leqslant N < l$, we define the function $f \in \mathcal{D}_{-N}^{-l-1}$ depending only the norm as

$$f(x) = \sum_{N \leqslant \gamma \leqslant l} a_\gamma \Pi(p^{-\gamma} x).$$

It is known that

$$\lambda(\xi, \gamma) = \int_{S_{-\gamma}} \chi_p(x\xi)dx = \begin{cases} p^{\gamma-1}(p-1) & (\gamma \leqslant \mathrm{val}(\xi)), \\ -p^{\gamma-1} & (\gamma = \mathrm{val}(\xi)+1), \\ 0 & (\gamma \geqslant \mathrm{val}(\xi)+2) \end{cases}$$
$$= p^\gamma \Omega(p^{-\gamma}\xi) - p^{\gamma-1}\Omega(p^{-\gamma+1}\xi) \tag{5}$$

(see [1, 2]). Then, using (5), its Fourier transform can be calculated as

$$\begin{aligned} \hat{f}(\xi) &= \sum_{\gamma \in \mathbf{Z}} \int_{S_{-\gamma}} f(x)\chi_p(x\xi)\,dx \\ &= \sum_{\gamma \in \mathbf{Z}} f(p^{-\gamma}) \int_{S_{-\gamma}} \chi_p(x\xi)dx \\ &= \sum_{N \leqslant \gamma \leqslant l} a_\gamma \left(p^{-\gamma}\Omega(p^\gamma \xi) - p^{-\gamma-1}\Omega(p^{\gamma+1}\xi)\right). \end{aligned}$$

In summary, we obtain the following result.

$$\hat{f}(\xi) = \begin{cases} 0 & (\mathrm{val}(\xi) < -l-1), \\ -a_l p^{-l-1} & (\mathrm{val}(\xi) = -l-1), \\ -a_{-\mathrm{val}(\xi)-1} p^{\mathrm{val}(\xi)} + \displaystyle\sum_{-\mathrm{val}(\xi) \leqslant \gamma \leqslant l} a_\gamma (p-1) p^{-\gamma-1} & (-l-1 < \mathrm{val}(\xi) < -N-1), \\ -a_N p^{-N-1} + \displaystyle\sum_{N+1 \leqslant \gamma \leqslant l} a_\gamma (p-1) p^{-\gamma-1} & (-N-1 \leqslant \mathrm{val}(\xi)). \end{cases}$$

This $\hat{f}(\xi)$ is also the function which depends only on the norm. The support of $\hat{f}(\xi)$ is B_{l+1}, and the parameter of constancy of $\hat{f}(\xi)$ is $N+1$. That is, $\hat{f} \in \mathcal{D}_{l+1}^{N+1} \subset \mathcal{D}_{l+1}^{N}$. For k satisfying $-l-1 \leqslant k \leqslant -N-1$, by putting $\hat{a}_k = \hat{f}(p^k)$, we can express the Fourier transform of $f \in \mathcal{D}_N^l$ as the following matrix calculation:

$$\begin{pmatrix} \hat{a}_{-N-1} \\ \hat{a}_{-N-2} \\ \hat{a}_{-N-3} \\ \vdots \\ \hat{a}_{-l+1} \\ \hat{a}_{-l} \\ \hat{a}_{-l-1} \end{pmatrix} =$$

$$\begin{pmatrix} -p^{-N-1} & (p-1)p^{-N-2} & (p-1)p^{-N-3} & \cdots & (p-1)p^{-l+1} & (p-1)p^{-l} & (p-1)p^{-l-1} \\ 0 & -p^{-N-2} & (p-1)p^{-N-3} & \cdots & (p-1)p^{-l+1} & (p-1)p^{-l} & (p-1)p^{-l-1} \\ 0 & 0 & -p^{-N-3} & \cdots & (p-1)p^{-l+1} & (p-1)p^{-l} & (p-1)p^{-l-1} \\ \vdots & \vdots & \vdots & \ddots & \vdots & \vdots & \vdots \\ 0 & 0 & 0 & \cdots & -p^{-l+1} & (p-1)p^{-l} & (p-1)p^{-l-1} \\ 0 & 0 & 0 & \cdots & 0 & -p^{-l} & (p-1)p^{-l-1} \\ 0 & 0 & 0 & \cdots & 0 & 0 & -p^{-l-1} \end{pmatrix} \begin{pmatrix} a_N \\ a_{N+1} \\ a_{N+2} \\ \vdots \\ a_{l-2} \\ a_{l-1} \\ a_l \end{pmatrix}.$$

5 Conclusion

In this paper, we focused on the p-adic Fourier transform. We have shown that, under some assumptions, the p-adic Fourier transform can be calculated with the same way as the fast Fourier transform with the computational complexity of order $O\left((l-N)2^{l-N}\right)$. Especially, with more assumptions, we showed that the p-adic Fourier transform can be represented as the upper triangular matrix computation.

References

1. Suzuki, T.: The Construction of p-Adic Stockwell Transform and its Spectra. p-Adic Numbers Ultrametric Anal. Appl. **13**(2), 166–173 (2021)
2. Vladimirov, V.S., Volovich, I.V., Zelenov, E.I.: p-Adic Analysis and Mathematical Physics. World Scientific, Singapore (1994)

Fractional Neural Network Interpolation Operator for Irregular Grid Points

Manju Sharma and Uaday Singh

Abstract The interpolation and approximation of a function by the neural network operator for an irregular grid of data points are frequently used in the applications of neural networks such as image and signal processing. In this paper, we prove interpolation and approximation results for a continuous function defined on a box-domain of $\mathbb{R}^s$ by neural network operators for multivariate data of irregular grid points. We measure the rate of approximation in terms of the modulus of continuity of the functions being approximated. We define this neural network interpolation operator by using the linear combinations of sigmoidal functions from a new class of sigmoidal functions and by using the fractional mean values of the function instead of its sample values.

1 Introduction

The neural networks are a class of mathematical models and a kind of information processing system which have been widely studied and applied in the numerous areas such as automatic control, signal processing, pattern recognition, quadratic optimization and many others. A neural network is a network of neurons arranged in layers. Each neuron stores a certain weight, bias and evaluates a certain activation function. The first layer is the input layer and the last layer is the output layer. The layers between the first and the last layer are the hidden layers. A neural network with a single hidden layer containing n neurons, s dimensional input and 1 dimensional output is defined as

$$M^{n,\eta}(\mathbf{x}) = \sum_{m=1}^{n} b_m \eta(\mathbf{u}_m.\mathbf{x} + v_m),$$

M. Sharma · U. Singh (✉)
Department of Mathematics, Indian institute of Technology Roorkee, Roorkee, India
e-mail: manju_s@ma.iitr.ac.in; uaday.singh@ma.iitr.ac.in

M. R. Ebert et al. (eds.), *New Tools in Mathematical Analysis and Applications*, Research Perspectives, https://doi.org/10.1007/978-3-031-77050-0_28

where $\mathbf{x} = (x_1, x_2, \ldots, x_s) \in \mathbb{R}^s$ is the input vector; $\{b_m \in \mathbb{R} : 1 \leqslant m \leqslant n, m \in \mathbb{N}\}$ are the coefficients; $\{\mathbf{u}_m = (u_{m1}, u_{m2}, \ldots, u_{ms}) \in \mathbb{R}^s : 1 \leqslant m \leqslant n, m \in \mathbb{N}\}$ are the weight vectors and $\{v_m \in \mathbb{R} : 1 \leqslant m \leqslant n, m \in \mathbb{N}\}$ are the thresholds of neural network with the activation function η. It is well known that single hidden layer neural networks are universal approximators [11].

The theory of neural network operators(NNOs) has been introduced to study the constructive approximation process by neural networks(NNs). The main advantage of using NNOs is that the coefficients, the weights, and the thresholds needed to approximate the objective function are known. Firstly, the NNOs were introduced by Anastassiou [1] to put the study of Cardaliaguet and Euvrard [5] in the framework of NNOs. After this many authors studied NNOs for different choices of activation functions and normed spaces of the objective functions [2–4, 8, 9].

The problem of data interpolation is strictly related to the theory of training neural networks. The concept of interpolation by using NNOs is introduced by Costarelli by taking the ramp function as the activation function [6]. Costarelli [7] further introduced a neural network interpolation operator(NNIO) activated by a sigmoidal function generated by the one-dimensional central B-spline of order $s \in \mathbb{N}^+$. Qian and Yu [13] introduced an NNIO activated by a newly defined class of sigmoidal functions. Recently, some NNIOs have been introduced by using Lagrange's interpolating polynomial and fractional mean values for a set of uniformly spaced data observations [12, 14]. But in applications of NNs, the values of the training set are not always uniformly spaced. In this direction, Costarelli [7] also defined an NNIO activated by sigmoidal functions generated by using the central B-spline of order s and extended the results for a type of irregular grid of data observations.

The main purpose of this paper is to define an NNIO activated by the sigmoidal function from a newly defined class to interpolate and approximate the multivariate functions for irregular grid points by using the fractional mean value of the function.

2 Preliminaries

Let us start with some notations for multivariate calculus. We consider the vectors $\bar{u} \in \mathbb{R}^s$ with Euclidean norm $\|\bar{u}\|_2$ defined as

$$\|\bar{u}\|_2 = \left(u_1^2 + \cdots + u_s^2\right)^{1/2}.$$

The operations are defined in componentwise manner, i.e., for $\bar{u}, \bar{v} \in \mathbb{R}^s$ and $c \in \mathbb{R}$, we define the scaler multiplication by

$$c\bar{u} := (cu_1, cu_2, \ldots, cu_s),$$

sum by

$$\bar{u} + \bar{v} := (u_1 + v_1, \ldots, u_s + v_s),$$

product by

$$\bar{u}\bar{v} := (u_1 v_1, \ldots, u_s v_s),$$

and division by

$$\bar{u}/\bar{v} := (u_1/v_1, \ldots, u_s/v_s), v_i \neq 0, i = 1, \ldots, s.$$

Further some standard multivariate notations are also used, i.e., for $\bar{u} \in \mathbb{R}^s$, we write

$$\langle \bar{u} \rangle = \prod_{i=1}^{s} u_i,$$

$\bar{u} \leqslant \bar{v}$ we mean $u_i \leqslant v_i$ for $i = 1, \ldots, s$,

$$\bar{u}^{\bar{v}} = \prod_{i=1}^{s} u_i^{v_i}.$$

Now let us recall briefly the properties of the fractional integral and fractional mean value which we use in this paper.

For fixed $\alpha \in \mathbb{R}_+$ and $g \in L^1([c, d], \mathbb{R})$. The left side Riemann-Liouville fractional integral of order α denoted by J_{c+}^{α} is defined as

$$J_{c+}^{\alpha} g(x) = \frac{1}{\Gamma(\alpha)} \int_c^x (x - t)^{\alpha - 1} g(t) dt, \quad x > c.$$

Fractional integral J_{c+}^{α} satisfies the following properties.

1. $J_{c+}^{\alpha} \left(J_{c+}^{\beta} g(x) \right) = J_{c+}^{\alpha+\beta} g(x)$
2. $J_{c+}^{\alpha} (b_1 g_1(x) + b_2 g_2(x)) = b_1 J_{c+}^{\alpha} g_1(x) + b_2 J_{c+}^{\alpha} g_2(x)$, for $b_1, b_2 > 0, \alpha, \beta > 0$ and $g_1, g_2 \in L^1([c, d], \mathbb{R})$.

The mean value theorem of fractional integral J_{c+}^{α} for multivariate case can be formulated as:

Let $\mathcal{R} := \prod_{i=1}^{s} [c_i, d_i], c_i, d_i \in \mathbb{R}, i = 1, \ldots, s$ and $g : \mathcal{R} \to \mathbb{R}$ be a continuous function. Then, there exists some $a_i \in [c_i, d_i]$ such that

$$g(a_1, \ldots, a_s) = \prod_{i=1}^{s} \frac{\alpha^s}{(d_i - c_i)^{\alpha}} \int_{c_1}^{d_1} \cdots \int_{c_s}^{d_s} (d_i - x_i)^{\alpha - 1} g(x_1, \ldots, x_s) dx_i.$$

For more details, one can refer to [10].

3 Construction of Fractional Neural Network Interpolation Operator

We start this section with some important definitions.

Definition 1 Let $\mathcal{C}(\mathcal{R})$ denote the set of multivariate continuous functions defined on $\mathcal{R} = [a_1, b_1] \times \ldots \times [a_s, b_s]$. For $\delta > 0$, the modulus of continuity of $g \in \mathcal{C}(\mathcal{R})$ is defined as:

$$\omega(g, \delta) := \sup_{\bar{x}, \bar{y} \in \mathcal{R}, \|\bar{x}-\bar{y}\|_2 \leqslant \delta} |g(\bar{x}) - g(\bar{y})|,$$

where $\|\cdot\|_2$ denotes the usual Euclidean norm on $\mathbb{R}^s$.

Definition 2 Let $\delta > 0$. Then the first order L_p modulus of continuity of $g \in L^p(\mathcal{R})$, $1 \leqslant p < +\infty$ is defined as

$$\omega_p(g, \delta) := \sup_{0 \leqslant \|\bar{h}\|_2 \leqslant \delta} \left\{ \|g(\cdot + \bar{h}) - g(\cdot)\|_p \right\},$$

where $\|g\|_p = \left(\int_{\mathcal{R}} |g(\bar{x})|^p d\bar{x}\right)^{\frac{1}{p}}$, $1 \leqslant p < +\infty$.

Definition 3 A measurable function $\eta : \mathbb{R} \to \mathbb{R}$ is said to be a sigmoidal function if:

$$\lim_{x \to -\infty} \eta(x) = 0 \text{ and } \lim_{x \to \infty} \eta(x) = 1.$$

Definition 4 We say that the sigmoidal function $\eta \in \mathcal{C}(m)$, $m \in \mathbb{R}^+$, if it satisfies the following:

1. $\eta(x)$ is nondecreasing;
2. $\eta(x) = 1$ for $x \geqslant m$, and $\eta(x) = 0$ for $x \leqslant -m$;
3. $\eta(x) \neq 1$ for $-m < x < m$;
4. $\eta(x) + \eta(-x) = 1$ for $x \in (-m, m)$.

We define the function

$$\sigma(x) := \eta(x + m) - \eta(x - m), \; x \in \mathbb{R}$$

as the translation of $\eta \in \mathcal{C}(m)$.

For $\eta \in \mathcal{C}(m)$, σ satisfies certain properties given below in the form of a lemma:

Lemma 1 *Assume that* $\eta \in \mathcal{C}(m)$. *Then,*

D1: $\sigma(x)$ *is nonnegative and even function.*
D2: $\sigma(x)$ *is non-decreasing for* $x < 0$ *and non-increasing for* $x \geqslant 0$.
D3: $supp(\sigma) \subseteq [-2m, 2m]$.

D4: $\sigma(x) + \sigma(x - 2m) = 1$ *for* $x \in [0, 2m]$.
D5: $\sigma(x) > 0$ *for* $x \in (0, 2m)$.

For the proof of properties D1–D4 of Lemma 1 one can refer [13] and D5 can be easily verified by using the condition (3) of Definition 4. For $\eta \in \mathcal{C}(m)$, we define the multivariate density function ϕ as following product

$$\phi(\bar{x}) := \sigma(x_1).\sigma(x_2)\ldots\sigma(x_s),\ \bar{x} := (x_1, \ldots, x_s) \in \mathcal{R}.$$

By using the definition, we can easily observe that $\sup_{\bar{x}\in\mathbb{R}^s} |\phi(\bar{x})| =: \|\phi\|_\infty \leqslant 1$ and $\sup_{\bar{x}\in\mathcal{R}} |\phi(\bar{x})| =: \|\phi\|_\infty \leqslant 1$ [14].

We introduce multivariate neural network interpolation operators by using the normalized density function ϕ. Throughout this paper, we write $I_s = \{1, 2, \ldots, s\} \subset \mathbb{N}$ and $Z_n = \{0, 1, \ldots, n\} \subset \mathbb{N} \cup \{0\}$. For any $i = 1, \ldots, s$, we consider the $n + 1$ nodes $y_0^i = a_i < y_1^i < \cdots < y_{n-1}^i < y_n^i = b_i$ for every fixed $n \in \mathbb{N}^+$. Suppose that corresponding to n, there exist $0 < d_1 \leqslant d_2$ with $d_2 < 2d_1$ such that for every $k_i \in I_{n-1}$:

$$d_1 \leqslant y_{k_i+1}^i - y_{k_i}^i \leqslant d_2 \text{ for every } i \in I_s. \tag{1}$$

Observe that the nodes $y_{k_i}^i$ are uniformly spaced when $d_1 = d_2$. We denote by $\bar{y}_{\bar{k}} := \left(y_{k_1}^1, \ldots, y_{k_s}^s\right)$, where $\bar{k} := (k_1, \ldots, k_s)$. We specify $\bar{y}_{\bar{n}+k} = \bar{y}_{\bar{n}+k-2}$ for $k = 1, 2, \ldots, n$.

Definition 5 Let $\{\bar{y}_{\bar{k}} : \bar{k} = (k_1, k_2, \ldots, k_s)$ and $k_i = 0, 1, \ldots, n\} \subset \mathcal{R}$ be the set of irregular grid points satisfying the above condition (1). Let $f : \mathcal{R} \subset \mathbb{R}^s \to \mathbb{R}$ be a bounded and measurable function. The fractional NNIO activated by the density function ϕ, defined above, is defined by

$$K_n^\alpha(f, \bar{x}) := \frac{\sum\limits_{\bar{k}=0}^{\bar{n}} \phi\left(\frac{2m(\bar{x}-\bar{y}_{\bar{k}})}{d_1}\right)\left[\frac{\alpha^s \int\limits_{\bar{y}_{\bar{k}}}^{\bar{y}_{\bar{k}+1}} (\bar{y}_{\bar{k}+1}-\bar{t})^{\alpha-1} f(\bar{t})d\bar{t}}{(\bar{y}_{\bar{k}+1}-\bar{y}_{\bar{k}})^\alpha}\right]}{\sum\limits_{\bar{k}=0}^{\bar{n}} \phi\left(\frac{2m(\bar{x}-\bar{y}_{\bar{k}})}{d_1}\right)},$$

where by $\sum\limits_{\bar{k}=0}^{n}$ we mean $\sum\limits_{k_1=0}^{n} \cdots \sum\limits_{k_s=0}^{n}$.

Now, we will prove the well-definedness of the operator K_n^α for any bounded function $f : \mathcal{R} \to \mathbb{R}$. Using the definition of ϕ for every $\bar{x} \in \mathcal{R}$, we have:

$$\begin{aligned}
&\sum_{k_1=0}^{n} \cdots \sum_{k_s=0}^{n} \phi\left(\frac{2m}{d_1}(\bar{x} - \bar{y}_{\bar{k}})\right) \\
&= \sum_{k_1=0}^{n} \cdots \sum_{k_s=0}^{n} \phi\left(\frac{2m}{d_1}\left(x_1 - y_{k_1}^1\right), \ldots, \frac{2m}{d_1}\left(x_s - y_{k_s}^s\right)\right) \\
&\geqslant \phi\left(\frac{2m}{d_1}\left(x_1 - y_{k_1}^1\right), \ldots, \frac{2m}{d_1}\left(x_s - y_{k_s}^s\right)\right),
\end{aligned}$$

for any $\bar{y}_{\bar{k}}$. We can choose a $\bar{y}_{\bar{k}}$ such that $2m\left|x_i - y_{k_i}^i\right| \leqslant md_2$ for every $i \in I_s$.

$$\begin{aligned}
\phi\left(\frac{2m}{d_1}\left(x_1 - y_{k_1}^1\right), \ldots, \frac{2m}{d_1}\left(x_s - y_{k_s}^s\right)\right) &\geqslant \phi\left(\frac{md_2}{d_1}, \ldots, \frac{md_2}{d_1}\right) \\
&= \sigma\left(\frac{md_2}{d_1}\right) \cdot \sigma\left(\frac{md_2}{d_1}\right) \cdots \sigma\left(\frac{md_2}{d_1}\right) \\
&= \left[\sigma\left(\frac{md_2}{d_1}\right)\right]^s.
\end{aligned}$$

Thus, we have

$$\sum_{k_1=0}^{n} \cdots \sum_{k_s=0}^{n} \phi\left(\frac{2m}{d_1}\left(x_1 - y_{k_1}^1\right), \ldots, \frac{2m}{d_1}\left(x_s - y_{k_s}^s\right)\right) \geqslant \left[\sigma\left(\frac{md_2}{d_1}\right)\right]^s > 0, \tag{2}$$

since $0 < \frac{md_2}{d_1} < 2m$.

Moreover, for every bounded and measurable function $f : \mathcal{R} \to \mathbb{R}$ we have:

$$\left|K_n^\alpha(f, \bar{x})\right| \leqslant \|f\|_\infty \frac{\sum_{k_1=0}^{n} \cdots \sum_{k_s=0}^{n} \phi\left(\frac{2m}{d_1}(\bar{x} - \bar{y}_{\bar{k}})\right)}{\sum_{k_1=0}^{n} \cdots \sum_{k_s=0}^{n} \phi\left(\frac{2m}{d_1}(\bar{x} - \bar{y}_{\bar{k}})\right)} = \|f\|_\infty < +\infty.$$

Hence, K_n^α is a well-defined operator.

Lemma 2 *Let $\alpha > 0$ be fixed. Then operator K_n^α satisfies the normalization property*

$$K_n^\alpha(\mathbf{1}, \bar{x}) = 1, \bar{x} \in \mathcal{R}, \ \mathcal{R} := \prod_{i=1}^{s}[a_i, b_i]$$

where $\mathbf{1}(.)$ denotes the multivariate constant function on $\mathbb{R}^s$.

Proof Let $\alpha \in \mathbb{R}_+$ be fixed. Now consider

$$K_n^\alpha(\mathbf{1}, \bar{x}) = \frac{\sum_{\bar{k}=0}^{\bar{n}} \phi\left(\frac{2m(\bar{x}-\bar{y}_{\bar{k}})}{d_1}\right) \left[\frac{\alpha^s \int_{\bar{y}_{\bar{k}}}^{\bar{y}_{\bar{k}+1}} (\bar{y}_{\bar{k}+1}-\bar{t})^{\alpha-1} d\bar{t}}{(\bar{y}_{\bar{k}+1}-\bar{y}_{\bar{k}})^\alpha}\right]}{\sum_{\bar{k}=0}^{n} \phi\left(\frac{2m(\bar{x}-\bar{y}_{\bar{k}})}{d_1}\right)}$$

$$= \frac{\sum_{\bar{k}=0}^{\bar{n}} \phi\left(\frac{2m(\bar{x}-\bar{y}_{\bar{k}})}{d_1}\right)}{\sum_{\bar{k}=0}^{\bar{n}} \phi\left(\frac{2m(\bar{x}-\bar{y}_{\bar{k}})}{d_1}\right)} = 1.$$

□

Next, we are going to prove the interpolation of function by the operator K_n^α over the set of irregular grid points.

Theorem 1 *Let $f : \mathcal{R} \to \mathbb{R}$ be a continuous function. Then,*

$$K_n^\alpha(f, \bar{y}_{\bar{j}}) = f(\bar{y}_{\bar{j}}), \alpha \in \mathbb{R}_+$$

for every $\bar{j} := (j_1, j_2, \ldots, j_s)$, where $j_i = 0, 1, \ldots, n, i = 1, \ldots, s$.

Proof Let $\bar{j} := (j_1, j_2, \ldots, j_s)$ with $j_i = 0, 1, \ldots, n, i = 1, \ldots, s$ be fixed and $\alpha \in \mathbb{R}_+$. Since for $\bar{k} \neq \bar{j}$, for every $1 \leqslant i \leqslant s$ such that

$$\frac{2m|y^i_{j_i} - y^i_{k_i}|}{d_1} \geqslant \frac{2md_1}{d_1} \geqslant 2m,$$

i.e., for $\bar{k} \neq \bar{j}$ we have,

$$\phi\left(\frac{2m(\bar{y}_{\bar{j}} - \bar{y}_{\bar{k}})}{d_1}\right) = 0. \tag{3}$$

Now, using (3), we can easily observe that

$$K_n^\alpha(f,\bar{y}_{\bar{j}}) = \frac{\sum\limits_{\bar{k}=0}^{\bar{n}} \phi\left(\frac{2m(\bar{y}_{\bar{j}}-\bar{y}_{\bar{k}})}{d_1}\right)\left[\frac{\alpha^s \int\limits_{\bar{y}_{\bar{k}}}^{\bar{y}_{\bar{k}+1}} (\bar{y}_{\bar{k}+1}-\bar{t})^{\alpha-1} f(\bar{t})d\bar{t}}{(\bar{y}_{\bar{k}+1}-\bar{y}_{\bar{k}})^\alpha}\right]}{\sum\limits_{\bar{k}=0}^{\bar{n}} \phi\left(\frac{2m(\bar{y}_{\bar{j}}-\bar{y}_{\bar{k}})}{d_1}\right)}$$

$$= \frac{\phi\left(\frac{2m(\bar{y}_{\bar{j}}-\bar{y}_{\bar{j}})}{d_1}\right)\left(\frac{\alpha^s \int\limits_{\bar{y}_{\bar{j}}}^{\bar{y}_{\bar{j}+1}} (\bar{y}_{\bar{j}+1}-\bar{t})^{\alpha-1} f(\bar{t})d\bar{t}}{(\bar{y}_{\bar{j}+1}-\bar{y}_{\bar{j}})^\alpha}\right)}{\phi\left(\frac{2m(\bar{y}_{\bar{j}}-\bar{y}_{\bar{j}})}{d_1}\right)}$$

$$= f(\bar{y}_{\bar{j}})$$

for every $\bar{j} := (j_1, j_2, \ldots, j_s)$, where $j_i = 0, 1, \ldots, n, i = 1, \ldots, s$. □

Now, we prove the direct approximation theorem and estimate the rate of approximation in terms of the modulus of continuity defined above.

Theorem 2 *Let $f : \mathcal{R} \to \mathbb{R}$ be a continuous function. Then for fixed $\alpha \in \mathbb{R}_+$,*

$$\|K_n^\alpha(f,\cdot) - f\|_\infty \leqslant 2\sigma\left(\frac{md_2}{d_1}\right)^{-s} \omega(f, d_2).$$

Proof Let $\alpha \in \mathbb{R}_+$ and $\bar{x} \in \mathcal{R}$ such that $\bar{y}_{\bar{j}} \leqslant \bar{x} \leqslant \bar{y}_{\bar{j}+1}$, $\bar{j} = (j_1, j_2, \ldots, j_s)$. In view of (D4), $\phi\left(\frac{2m(\bar{x}-\bar{y}_{\bar{k}})}{d_1}\right) = 0$ for $\bar{k} \neq \bar{j}, \bar{j}+1$. By using this fact and Lemma 2 we have,

$$|K_n^\alpha(f,\bar{x}) - f(\bar{x})|$$

$$\leqslant \frac{\sum\limits_{\bar{k}=0}^{\bar{n}} \phi\left(\frac{2m(\bar{x}-\bar{y}_{\bar{k}})}{d_1}\right)\left[\frac{\int\limits_{\bar{y}_{\bar{k}}}^{\bar{y}_{\bar{k}+1}} \alpha^s (\bar{y}_{\bar{k}+1}-\bar{t})^{\alpha-1} |f(\bar{t})-f(\bar{x})|d\bar{t}}{(\bar{y}_{\bar{k}+1}-\bar{y}_{\bar{k}})^\alpha}\right]}{\sum\limits_{\bar{k}=0}^{n} \phi\left(\frac{2m(\bar{x}-\bar{y}_{\bar{k}})}{d_1}\right)}$$

$$\leqslant \frac{1}{\sigma(\frac{md_2}{d_1})^s}\left[\sum_{\bar{k}=0}^{\bar{n}}\phi\left(\frac{2m(\bar{x}-\bar{y}_{\bar{k}})}{d_1}\right)\left(\frac{\int\limits_{\bar{y}_{\bar{k}}}^{\bar{y}_{\bar{k}+1}}\alpha^s(\bar{y}_{\bar{k}+1}-\bar{t})^{\alpha-1}|f(\bar{t})-f(\bar{x})|d\bar{t}}{(\bar{y}_{\bar{k}+1}-\bar{y}_{\bar{k}})^{\alpha}}\right)\right]$$

$$= \frac{1}{\sigma(\frac{md_2}{d_1})^s}\phi\left(\frac{2m(\bar{x}-\bar{y}_{\bar{j}})}{d_1}\right)\left(\frac{\int\limits_{\bar{y}_{\bar{j}}}^{\bar{y}_{\bar{j}+1}}\alpha^s(\bar{y}_{\bar{j}+1}-\bar{t})^{\alpha-1}|f(\bar{t})-f(\bar{x})|d\bar{t}}{(\bar{y}_{\bar{j}+1}-\bar{y}_{\bar{j}})^{\alpha}}\right)$$

$$+\frac{1}{\sigma(\frac{md_2}{d_1})^s}\phi\left(\frac{2m(\bar{x}-\bar{y}_{\bar{j}+1})}{d_1}\right)\left(\frac{\int\limits_{\bar{y}_{\bar{j}+1}}^{\bar{y}_{\bar{j}+2}}\alpha^s(\bar{y}_{\bar{j}+2}-\bar{t})^{\alpha-1}|f(\bar{t})-f(\bar{x})|d\bar{t}}{(\bar{y}_{\bar{j}+2}-\bar{y}_{\bar{j}+1})^{\alpha}}\right)$$

$$\leqslant \frac{\omega(f,d_2)}{\sigma(\frac{md_2}{d_1})^s}\left(\phi\left(\frac{2m(\bar{x}-\bar{y}_{\bar{j}+1})}{d_1}\right)+\phi\left(\frac{2m(\bar{x}-\bar{y}_{\bar{j}+2})}{d_1}\right)\right)$$

$$\leqslant 2\omega(f,d_2)\sigma\left(\frac{md_2}{d_1}\right)^{-s}.$$

□

Theorem 3 *Let $f \in L^p(\mathcal{R})$, $1 \leqslant p < +\infty$, and let $\alpha > 0$ be fixed. Then,*

$$\|K_n^{\alpha}(f,\cdot)-f\|_p \leqslant 2^{\frac{1-s}{p}+1}\alpha^s\sigma\left(\frac{md_2}{d_1}\right)^{-s}\omega_p(f,d_2)(m)^{-s/p}\|\phi\|_1{}^{1/p},$$

for $\phi \in L^1(\mathbb{R})$.

Proof Firstly, we note that,

$$f\left(\bar{t}+\bar{x}-\bar{y}_{\bar{k}}\right) \leqslant \left|f\left(\bar{t}+\bar{x}-\bar{y}_{\bar{k}}\right)-f(\bar{t})\right|+f(\bar{t}),$$

where $\bar{t}, \bar{x} \in \mathbb{R}^s$. We set

$$W_n^{\alpha}(f,\bar{x}) := \frac{\sum\limits_{\bar{k}=0}^{\bar{n}}\phi\left(\frac{2m(\bar{x}-\bar{y}_{\bar{k}})}{d_1}\right)\left[\frac{\int\limits_{\bar{y}_{\bar{k}}}^{\bar{y}_{\bar{k}+1}}\alpha^s(\bar{y}_{\bar{k}+1}-\bar{t})^{\alpha-1}f(\bar{t}+\bar{x}-\bar{y}_{\bar{k}})d\bar{t}}{(\bar{y}_{\bar{k}+1}-\bar{y}_{\bar{k}})^{\alpha}}\right]}{\sum\limits_{\bar{k}=0}^{n}\phi\left(\frac{2m(\bar{x}-\bar{y}_{\bar{k}})}{d_1}\right)}.$$

Now, for any $f \in L^p(\mathcal{R})$,

$$\|K_n^\alpha(f,\cdot)-f\|_p \leqslant \|K_n^\alpha(f,\cdot)-W_n^\alpha(f,\cdot)\|_p+\|W_n^\alpha(f,\cdot)-f\|_p =: U_1+U_2. \quad (4)$$

Now, we estimate U_1.

$$U_1^p = \int_{\mathcal{R}} \left| \frac{\sum\limits_{\bar{k}=0}^{\bar{n}} \phi\left(\frac{2m(\bar{x}-\bar{y}_{\bar{k}})}{d_1}\right) \left[\frac{\int\limits_{\bar{y}_{\bar{k}}}^{\bar{y}_{\bar{k}+1}} \alpha^s(\bar{y}_{\bar{k}+1}-\bar{t})^{\alpha-1}|f(\bar{t})-f(\bar{t}+\bar{x}-\bar{y}_{\bar{k}})|d\bar{t}}{(\bar{y}_{\bar{k}+1}-\bar{y}_{\bar{k}})^\alpha} \right]}{\sum\limits_{\bar{k}=0}^{\bar{n}} \phi\left(\frac{2m(\bar{x}-\bar{y}_{\bar{k}})}{d_1}\right)} \right|^p d\bar{x}.$$

Using Jensen's inequality, we have

$$U_1^p \leqslant \frac{1}{\sigma\left(\frac{md_2}{d_1}\right)^{sp}} \int_{\mathcal{R}} \sum_{\bar{k}=0}^{\bar{n}} \left|\phi\left(\frac{2m(\bar{x}-\bar{y}_{\bar{k}})}{d_1}\right) \times \left[\frac{\int\limits_{\bar{y}_{\bar{k}}}^{\bar{y}_{\bar{k}+1}} \alpha^s(\bar{y}_{\bar{k}+1}-\bar{t})^{\alpha-1}|f(\bar{t})-f(\bar{t}+\bar{x}-\bar{y}_{\bar{k}})|d\bar{t}}{(\bar{y}_{\bar{k}+1}-\bar{y}_{\bar{k}})^\alpha} \right]^p \right| d\bar{x}.$$

Now let $\bar{y}_{\bar{j}} \leqslant \bar{x} \leqslant \bar{y}_{\bar{j}+1}$ then for $\bar{k} \neq \bar{j}, \bar{j}+1$, we have

$$\phi\left(\frac{2m(\bar{x}-\bar{y}_{\bar{k}})}{d_1}\right) = 0.$$

Now using this fact, we have

$$U_1^p \leqslant \frac{1}{\sigma\left(\frac{md_2}{d_1}\right)^{sp}} \int_{\mathcal{R}} \left|\phi\left(\frac{2m(\bar{x}-\bar{y}_{\bar{j}})}{d_1}\right) \times \left[\frac{\int\limits_{\bar{y}_{\bar{j}}}^{\bar{y}_{\bar{j}+1}} \alpha^s(\bar{y}_{\bar{j}+1}-\bar{t})^{\alpha-1}|f(\bar{t})-f(\bar{t}+\bar{x}-\bar{y}_{\bar{j}})|d\bar{t}}{(\bar{y}_{\bar{j}+1}-\bar{y}_{\bar{j}})^\alpha} \right]^p \right| d\bar{x}$$

$$+\frac{1}{\sigma\left(\frac{md_2}{d_1}\right)^{sp}}\int_{\mathcal{R}}\left|\phi\left(\frac{2m(\bar{x}-\bar{y}_{\bar{j}+1})}{d_1}\right)\right.$$

$$\left.\times\left[\frac{\int_{\bar{y}_{\bar{j}+1}}^{\bar{y}_{\bar{j}+2}}\alpha^s(\bar{y}_{\bar{j}+2}-\bar{t})^{\alpha-1}|f(\bar{t})-f(\bar{t}+\bar{x}-\bar{y}_{\bar{j}+1})|d\bar{t}}{(\bar{y}_{\bar{j}+2}-\bar{y}_{\bar{j}+1})^{\alpha}}\right]^p\right|d\bar{x}$$

$$\leqslant\frac{1}{\sigma\left(\frac{md_2}{d_1}\right)^{sp}}\int_{\mathcal{R}}\left[\phi\left(\frac{2m(\bar{x}-\bar{y}_{\bar{j}})}{d_1}\right)\right.$$

$$\left.\times\left[\frac{\int_{\bar{y}_{\bar{j}}}^{\bar{y}_{\bar{j}+1}}\alpha^s(\bar{y}_{\bar{j}+1}-\bar{t})^{\alpha-1}|f(\bar{t})-f(\bar{t}+\bar{x}-\bar{y}_{\bar{j}})|d\bar{t}}{(\bar{y}_{\bar{j}+1}-\bar{y}_{\bar{j}})^{\alpha}}\right]^p d\bar{x}\right]$$

$$+\frac{1}{\sigma\left(\frac{md_2}{d_1}\right)^{sp}}\int_{\mathcal{R}}\left[\phi\left(\frac{2m(\bar{x}-\bar{y}_{\bar{j}+1})}{d_1}\right)\right.$$

$$\left.\times\left[\frac{\int_{\bar{y}_{\bar{j}+1}}^{\bar{y}_{\bar{j}+2}}\alpha^s(\bar{y}_{\bar{j}+2}-\bar{t})^{\alpha-1}|f(\bar{t})-f(\bar{t}+\bar{x}-\bar{y}_{\bar{j}+1})|d\bar{t}}{(\bar{y}_{\bar{j}+2}-\bar{y}_{\bar{j}+1})^{\alpha}}\right]^p d\bar{x}\right]$$

$$\leqslant\frac{\alpha^{sp}\omega_p(f,d_2)^p}{\sigma\left(\frac{md_2}{d_1}\right)^{sp}}\int_{\mathcal{R}}\left(\frac{\phi\left(\frac{2m(\bar{x}-\bar{y}_{\bar{j}})}{d_1}\right)}{(\bar{y}_{\bar{j}}-\bar{y}_{\bar{j}})}+\frac{\phi\left(\frac{2m(\bar{x}-\bar{y}_{\bar{j}+1})}{d_1}\right)}{(\bar{y}_{\bar{j}+2}-\bar{y}_{\bar{j}+1})}\right)d\bar{x}.$$

By changing the variables $\frac{2m(\bar{x}-\bar{y}_{\bar{j}})}{d_1}=\bar{z_1}$ and $\frac{2m(\bar{x}-\bar{y}_{\bar{j}+1})}{d_1}=\bar{z_2}$. We have

$$U_1^p\leqslant\frac{2\|\phi\|_1\alpha^{sp}\omega_p(f,d_2)^p(d_1)^s}{\sigma\left(\frac{md_2}{d_1}\right)^{sp}(2md_1)^s}$$

$$\leqslant\frac{2\alpha^{sp}\|\phi\|_1\omega_p(f,d_2)^p}{\sigma\left(\frac{md_2}{d_1}\right)^{sp}(2m)^s}.$$

So, we have

$$U_1 \leqslant 2^{\frac{1-s}{p}} \alpha^s \sigma \left(\frac{md_2}{d_1}\right)^{-s} \omega_p(f, d_2)(m)^{-s/p} \|\phi\|_1{}^{1/p}.$$

By following a similar manner, we get the estimate,

$$U_2 \leqslant 2^{\frac{1-s}{p}} \alpha^s \sigma \left(\frac{md_2}{d_1}\right)^{-s} \omega_p(f, d_2)(m)^{-s/p} \|\phi\|_1{}^{1/p}.$$

Substituting the above estimates in (4), we have the result. □

4 Conclusion

In this study, we have used K_n^{α} to interpolate and approximate a multivariate function on the grid of irregular data points. The approximation results also hold for the case $d_1 = d_2$, i.e., for the set of uniformly spaced grid points. In this paper, we considered a type of irregular grid points. In our future work, we will work on the more general set of irregular grid points. The fractional neural network interpolation operators developed in this study over irregular data points offer a potent tool for enhancing image super-resolution (SR) reconstruction. Indeed, within the realm of digital image and signal processing encompassing tasks such as image enhancement, mosaic creation, medical image analysis, face recognition, signal denoising, and pattern recognition interpolation neural networks have garnered substantial usage.

References

1. Anastassiou, G.A.: Rate of convergence of some neural network operators to the unit-univariate case. J. Math. Anal. Appl. **212**, 237–262 (1997)
2. Anastassiou, G.A.: Intelligent Systems: Approximation by Artificial Neural Networks, vol. 19. Springer, Berlin (2011)
3. Anastassiou, G.A.: Multivariate hyperbolic tangent neural network approximation. Comput. Math. Appl. **61**, 809–821 (2011)
4. Anastassiou, G.A.: Univariate hyperbolic tangent neural network approximation. Math. Comput. Model. **53**, 1111–1132 (2011)
5. Cardaliaguet, P., Euvrard, G.: Approximation of a function and its derivative with a neural network. Neural Netw. **5**, 207–220 (1992)
6. Costarelli, D.: Interpolation by neural network operators activated by ramp functions. J. Math. Anal. Appl. **419**, 574–582 (2014)
7. Costarelli, D.: Neural network operators: constructive interpolation of multivariate functions. Neural Netw. **67**, 28–36 (2015)

8. Costarelli, D., Spigler, R.: Solving volterra integral equations of the second kind by sigmoidal functions approximation. J. Integral Equ. Appl. **25**, 193–222 (2013)
9. Costarelli, D., Spigler, R., et al.: Constructive approximation by superposition of sigmoidal functions. Anal. Theory Appl. **29**, 169–196 (2013)
10. Diethelm, K.: The mean value theorems and a Nagumo-type uniqueness theorem for Caputo's fractional calculus. Fract. Calc. Appl. **15**, 304–313 (2012)
11. Ismailov, V.E.: A three layer neural network can represent any multivariate function. J. Math. Anal. Appl. **523**, 127096 (2023)
12. Kadak, U.: Multivariate neural network interpolation operators. J. Comput. Appl. Math. **414**, 114426 (2022)
13. Qian, Y., Yu, D.: Rates of approximation by neural network interpolation operators. Appl. Math. Comput. **418**, 126781 (2022)
14. Wang, G., Yu, D., Zhou, P.: Neural network interpolation operators optimized by Lagrange polynomial. Neural Netw. **153**, 179–191 (2022)

The manufacturer's authorised representative in the EU is Springer Nature Customer Service Centre GmbH, Europaplatz 3, 69115 Heidelberg, Germany. If you have any concerns regarding our products, please contact ProductSafety@springernature.com

Printed and bound by CPI Group (UK) Ltd, Croydon, CR0 4YY
23/09/2026
02221537-0002